Evolutionary Algorithms in Engineering and Computer Science

Evolutionary Algorithms in Engineering and Computer Science

Recent Advances in Genetic Algorithms, Evolution Strategies, Evolutionary Programming, Genetic Programming and Industrial Applications

Edited by

K. Miettinen
University of Jyväskylä, Finland

P. Neittaanmäki
University of Jyväskylä, Finland

M. M. Mäkelä
University of Jyväskylä, Finland

J. Périaux
Dassault Aviation, France

JOHN WILEY & SONS, LTD
Chichester • Weinheim • New York • Brisbane • Singapore • Toronto

Baffins Lane, Chichester,
West Sussex PO19 1UD, England

National 01243 779777
International (+44) 1243 779777

e-mail (for orders and customer service enquiries): cs-books@wiley.co.uk

Visit our Home Page on http://www.wiley.co.uk
or
http://www.wiley.com

Other Wiley Editorial Offices

John Wiley & Sons, Inc., 605 Third Avenue,
New York, NY 10158-0012, USA

Wiley-VCH Verlag GmbH, Pappelallee 3,
D-69469 Weinheim, Germany

Jacaranda Wiley Ltd, 33 Park Road, Milton,
Queensland 4064, Australia.

John Wiley & Sons (Asia) Pte Ltd, 2 Clementi Loop #02-01,
Jin Xing Distripark, Singapore 0512

John Wiley & Sons (Canada) Ltd, 22 Worcester Road,
Rexdale, Ontario M9W 1L1, Canada

British Library Cataloguing in Publication Data

A catalogue record for this book is available from the British Library

ISBN 0 471 99902 4

Produced from camera-ready copy supplied by the author
Printed and bound in Great Britain by Bookcraft (Bath) Ltd
This book is printed on acid-free paper responsibly manufactured from sustainable forestry, in which at least two trees are planted for each one used for paper production.

Contents

Preface

This book collects the papers of the invited lecturers of EUROGEN99, the Short Course on Evolutionary Algorithms in Engineering and Computer Science, held at the University of Jyväskylä, Finland, between May 30 and June 3, 1999. In addition, this book contains several industrial presentations given during the Short Course by contributors belonging to the European Thematic Network INGENET or other European projects.

EUROGEN99 is the third in this series of EUROGEN Short Courses. The first course took place in Las Palmas in 1995 and the second in Trieste in 1997. EUROGEN95 concentrated on the theory of genetic algorithms whereas EUROGEN97 was more focused on applications. The theme of EUROGEN99 was chosen to be Recent Advances in Genetic Algorithms, Evolution Strategies, Evolutionary Programming, Genetic Programming and Industrial Applications. Special emphasis was given to practical industrial and engineering applications.

Evolution algorithms are computer science-oriented techniques which imitate nature according to Darwin's principle of "survival of the fittest." A population of solutions is treated with genetic operations, namely selection, crossover and mutation in order to produce improved generations. This evolution approach can be applied in multidisciplinary areas. Among them are fluid dynamics, electrical engineering, engineering design, electromagnetic, scheduling, pattern recognition and signal processing. All these applications are considered in this volume together with theoretical and numerical aspects of evolutionary computation for global optimization problems. The papers aim at creating a bridge between artificial intelligence, scientific computing and engineering so that both the performance and the robustness of solution methods for solving industrial problems can be improved.

This book contains three parts. Part I is a collection of introductory and methodological papers whereas Part II deals mainly with numerical aspects of evolutionary algorithms and contains some applications. Finally, Part III is entirely devoted to industrial problems.

The first two papers in Part I by Juha Haataja and David P. Fogel give an introduction to genetic algorithms and evolutionary computation, respectively. They are followed by two collections of recent developments and open issues in these topics by Kenneth De Jong and David P. Fogel.

Introductory papers are followed by methodological developments by Zbigniew Michalewicz et al., Colin Reeves et al. and Marco Tomassini. The first-mentioned paper discusses constraint-handling techniques and introduces a test case generator whereas the second paper embeds path tracing and neighbourhood search techniques in genetic algorithms. Parallel and distributed computing aspects are considered in the third paper.

Kalyanmoy Deb discusses evolutionary algorithms for multi-criterion optimization. In addition, results with test examples and a design problem of a welded beam are reported. Next, Marco Dorigo et al. introduce algorithms for the traveling salesman problem based on ant colony optimization.

Genetic programming and its applications in analog electrical circuits are studied by John Koza et al. and Forrest H. Bennett et al., respectively.

Part II is devoted to application-oriented approaches. In George S. Dulikravich et al., a hybrid constrained optimizer is applied in several multidisciplinary aero-thermal and aerodynamical problems.

In the next two papers, genetic and evolutionary algorithms are applied to engineering and optimal shape design problems. Marek Rudnicki et al. concentrate on electrical engineering problems whereas Miguel Cerrolaza et al. deal with finite and boundary element applications in optimal shape design.

More application-oriented approaches follow when Evelyne Lutton introduces genetic algorithms combined with fractals in the framework of signal analysis. Finally, Part II is concluded by the paper of Henri Luchian focusing on evolutionary approaches for pattern recognition and clustering problems.

Part III describes problems encountered in the following industrial areas: aeronautics, energy, electronics, paper machinery and foundry processes. To start with, evolution strategies applied in test cases related to airfoil optimization are introduced by DASA and Informatik Centrum Dortmund (ICD) from Germany. Next, the optimization problem of an active noise control system inside an aircraft is treated by the University of Patras, Greece and Dornier Gmbh, Germany. Further, generator scheduling in power systems is considered by UNELCO and Centre for Numeric Applications in Engineering (CEANI) from Spain. Next, partitioning methods for 3-D CFD-oriented unstructured grids are developed by National Technical University of Athens (NTUA), Greece and Dassault Aviation, France.

Furthermore, Valmet and the University of Jyväskylä from Finland describe the design problem of the headbox of a paper machine followed by a parallel multiobjective approach to computational fluid dynamics by INRIA and Dassault Aviation from France. Next, the application of multi objective genetic algorithm to foundry processes is described by the University of Trieste and Engin Soft Trading Srl from Italy. Finally, Centre for Applied Microelectronics (CMA) and CEANI from Spain concentrate on circuit partitioning in complex electronic systems.

This volume focused on evolutionary computing offers an accessible introduction to one of the most important fields of information and biotechnologies. With the continuous increase of speed of distributed parallel computers the applications of evolutionary algorithms will become routine in the coming years.

We hope that the contents of this book will influence the new generation of scientists and engineers in solving a new class of complex technological, economical and societal problems at the dawn of the 21st century.

THE EDITORS

Acknowledgements

We are grateful to the invited speakers of the EUROGEN99 Short Course for contributing their papers to this volume. Similarly, we thank the speakers of the industrial INGENET sessions for their papers. In addition, we appreciate the co-operative efforts of the members of the European Scientific Committee.

EUROGEN99 is one event among the on-going activities of the thematic network INGENET and we are indebted to the European Union (DGXII) for the financial support and, in particular, Dietrich Knoezer and Jyrki Suominen for their continuous interest and encouragement in this emergent information technology. We also acknowledge the support of the Department of Mathematical Information Technology at the University of Jyväskylä, city of Jyväskylä, Valmet Corporation, Dassault Aviation and ECCOMAS.

Special thanks should be extended to the organizers of the previous EUROGEN Short Courses, and professor Carlo Poloni, in particular, for useful information and practical advice.

We wish to express our gratitude to Elina Laiho-Logren and Marja-Leena Rantalainen for secretarial services and the latter, in particular, for her remarkable editorial work, Jari Toivanen for all his efforts and Markku Könkkölä, Tapani Tarvainen, Tuomas Kautto, Mika Videnoja and Janne Kanner for technical services and support.

Finally, two of the editors wish to acknowledge the personal support of the Academy of Finland (grants 22346, 8583).

Part I

Methodological aspects

1 Using Genetic Algorithms for Optimization: Technology Transfer in Action

J. HAATAJA

Center for Scientific Computing (CSC), Finland

For every living creature that succeeds in getting a footing in life there are thousands or millions that perish.

—J. W. N. Sullivan

Men have become the tools of their tools.

—Thoreau

1.1 WHY AND HOW TO OPTIMIZE?

Optimization aims for efficient allocation of scarce resources. You find examples of optimization in any scientific or engineering discipline [Haa95b, Haa98a, HK99]. For example, the minimum principle of physics produces many kinds of optimization problems [LSF70]. Agriculturists want to improve their cattle so that the cows are more healthy and produce better milk. Engineers are interested in making better paper machines, or in optimizing the fuel consumption of engines. As another type of example, consider designing optimal schedules for school classes or aircraft crews, or minimizing the lengths of wires in an electrical device.

The bottleneck of optimization is finding an algorithm which solves a given class of problems. There are no methods which can solve all optimization problems. Consider the following function $f(x) : [x^l, x^u] \to [0, 1]$:

$$f(x) = \begin{cases} 1, & \text{if } \|x - a\| < \epsilon, \ \ \epsilon > 0, \\ 0, & \text{elsewhere.} \end{cases} \tag{1.1}$$

We can make the maximisation of f harder by decreasing ϵ or by making the interval $[x^l, x^u]$ larger. Thus, difficult test problems can be generated relatively easily. The success or failure of a method is measured, of course, in solving real-world problems.

Because a good solution now is better than a perfect solution in the future, any method which can be adapted to a variety of problems is of great interest and of great practical use. Thus an algorithm may be useful even if it doesn't guarantee optimal solutions.

The time needed for finding a solution is important — consider publications and grants in scientific research, or products and orders in industry. Nowadays computers are cheap, humans are expensive. Therefore, we want to solve optimization problems by combining human creativity and the raw processing power of the computer.

An example of using different optimization methods is the numerical solution of the Ginzburg-Landau equations. The first successful approach [DGR90] used simulated annealing, which is a probabilistic method [AK89]. Next, the steepest descent method was used [WH91]. Larger systems were solved using more efficient gradient-based methods [GSB+92]. After this, researchers were able to solve (limited) three-dimensional models with a truncated Newton's method [JP93].

In the Ginzburg-Landau problem the optimization approach was demonstrated first with a general-purpose heuristic. After this, more efficient problem-specific methods were used to solve larger cases. However, this required the development of new algorithms and approaches for this problem, which took several years.

1.2 FROM PROBLEMS TO SOLUTIONS

Usually we do not have a ready-made optimization problem waiting for a solution. Rather, we have a phenomenon we are interested in. The phenomenon may be a physical process, a financial situation or an abstract mathematical construct. Moving from the phenomenon to a mathematical optimization problem is a challenge in itself.

The first step is to explore the phenomenon using mathematical tools. After we have a (perhaps simplified) model, we may proceed to simulate the phenomenon. We often must simplify the model to make simulations. Furthermore, we may have to design new computational methods to solve the problem in a reasonable time. When we have found a solution, we have to compare the results with the actual phenomenon to verify our model and solution.

Many kinds of systems can be described by optimization models. Even if our model does not include optimization, we may be later interested in optimizing the system. For example, after solving a fluid dynamics problem, we may want to optimize the geometry to get maximal flow.

When we have an optimization problem, we have to find a solution method. Here we have a wide variety of choices to make. First, we have to find out what kind of problem we have: Are the functions smooth, non-smooth or even non-continuous? Do we have discrete or continuous variables? Is the problem convex or not? Do we have many local maxima or it there just one maximum? Can we find a good initial guess for the solution?

We may have a standard optimization problem, which are discussed in many excellent textbooks [BSS93, Ber96, SN96]. Then we can use standard software packages [Haa95b, Haa98b, MW93] to solve the problem. Of course, before announcing that the problem is solved, we have to check for the scaling of the problem, formulate the convergence criteria, and find ways of verifying the solution.

However, in many real-world cases it is not possible to formulate a standard problem. Also, there may be so many local maxima that local optimization methods are useless. Optimization is hard, but finding the global maximum is much harder.

In these cases it is worthwhile to try genetic algorithms. However, tailoring a GA to the problem may be far from nontrivial. It may be reasonable to look at other methods also: simulated annealing, tabu search and other heuristics, or local optimization with several different initial guesses.

1.3 EVOLUTION AND OPTIMIZATION

Evolution is a collective phenomenon of adapting to the environment and passing genes to next generations [Rid96]. If a combination of genes gives a reasonable chance of survival and of finding a mate, this set of genes may survive to the next generation, even if the combination is non-optimal.

Evolution, unfortunately, is not optimization. Consider the human eye, where the light sensors are "backwards". Or consider the appendix, or the way we are eating and breathing through our mouths. These nonoptimal solutions are due to the genes which our distant ancestors happened to have.

Genetic algorithms (GA) are based on principles from genetics and evolution. GAs can be used to solve optimization problems, although the greatest promise may be in machine learning and adaptive systems [Gol89, Mic98].

Why to use ideas taken from evolution to solve optimization problems?

The power of mathematics is in technology transfer: we have models and methods which can describe many different phenomena and solve a wide variety of problems. GAs are one example of mathematical technology transfer: by simulating evolution we are actually able to solve optimization problems from a variety of sources. Of course, we may have to modify our understanding of evolution to solve our problems, but the basic principles do apply.

Instead of a single sample from the solution space, GAs maintain a population of vectors. New solution vectors are generated using selection, recombination and mutation. The new vectors are evaluated for their fitness, and the old population is replaced with a new one. This is repeated until the algorithm converges or we run out of time or patience.

Often a good enough solution will do. If the search space of all possible solution is big enough, we may be satisfied with a solution which is better than a random guess.

Genetic algorithms are adaptable to different situations. If you can find a good coding for the problem, GAs can be quite efficient and reliable. However, there are cases where other methods are more appropriate. For example, if the problem is smooth and derivatives are available, combining a gradient-based local optimization method with a global strategy will be hard to beat.

1.4 A SIMPLE GA

Genetic algorithms are problems solving strategies, which can be used as optimization methods. Here is an outline for a simple genetic algorithm:

> Initialize the population $\{\mathbf{x}^i\}$, $i = 1, \ldots, l$.
> Calculate the values of the object function: $f(\mathbf{x}^i)$, $i = 1, \ldots, l$.
> repeat
>   for $i = 1, 2, \ldots, l$
>   Recombine two selected individuals $\mathbf{x}^a$ and $\mathbf{x}^b$ to generate $\mathbf{x}'$.
>   Generate mutations: $\mathbf{x}' \to \mathbf{x}''$.
>   Evaluate the object function: $f_i = f(\mathbf{x}'')$.
>   Store the new individual in the new population.
>   end
> until a good enough solution if found

GAs work by combining the recombination, selection and mutation operations. The selection pressure drives the population towards better solutions. However, the algorithm needs also to escape from local optima, and here mutation comes into play.

Genetic algorithms mimic evolution, and they often behave like evolution in nature. Because of this, we may have to make GAs less true to life to solve optimization problems.

Suppose we are interested in finding the maximum of a continuous function $f(\mathbf{x}) : \mathbf{R}^n \to \mathbf{R}$. In this article we consider box-constrained problems, where $\mathbf{x}^l \leq \mathbf{x} \leq \mathbf{x}^u$.

As an example, consider the function

$$\begin{aligned} f(\mathbf{x}) &= \left((x_1 - 100)^2 + (x_2 - 100)^2\right)/4000 - \\ &\quad - \cos(x_1 - 100)\cos\left((x_2 - 100)/\sqrt{2}\right) + 1. \end{aligned}$$

This is the so-called Griewank function, which is illustrated in Figure 1.1. Note the small-scale details on the right. This function has thousands of local maxima. The global maximum is at the point $\mathbf{x}^* = (100, 100)^T$.

The Griewank function is a nice test for global optimization methods because it has several local maxima. We'll use this problem in Section 1.7 to test the Matlab implementation of the simple GA.

One of the open problems of GAs is the optimal coding of a particular optimization task. There are many choices to make: are we using binary or real coding, what are the crossover, mutation and selection operators, and how the values of the object function are mapped into function values.

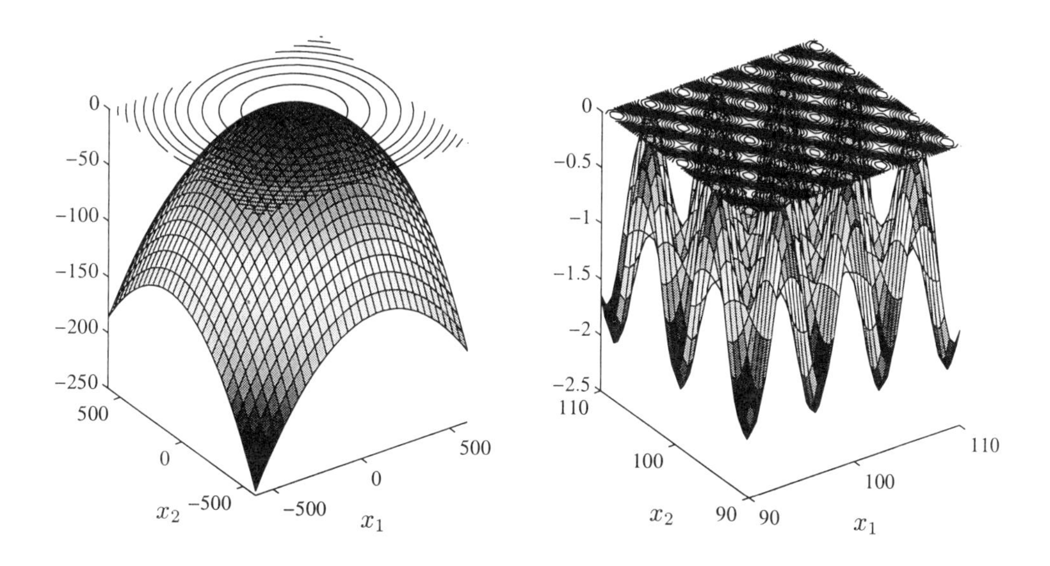

Figure 1.1 The Griewank function. The figure on the right show details not visible on the large-scale picture on the left

The traditional genetic algorithms don't process real numbers. Instead, a binary representation is used. In practice, using binary coding for real variables is wasteful. Often the performance of binary coded genetic algorithms is not satisfactory either. Thus other approaches have been tried [Mic98, Sch95, Dav91].

The coding of the optimization problem may be the most important factor in the success or failure of GAs. Finding a good coding makes it easy to tune the algorithm. If we are using a bad coding, GAs may perform worse than other probabilistic search methods.

In this article we'll study a real-coded GA. I have made available a Matlab implementation on the Web [Haaa]. Matlab is a good environment for getting familiar with GAs. The matrix syntax and the visualization possibilities are especially useful. However, Matlab is slow compared with compiled languages, so I recommend Matlab only as a learning and prototyping tool.

The real-coded GA processes a population of vectors $\{\mathbf{x}^i\}$, $\mathbf{x}^i \in \mathbf{R}^n$, $i = 1, 2, \ldots, l$. In the Matlab implementation two individuals are combined using the box crossover:

$$x'_i = (1 - c)x^a_i + cx^b_i.$$

Here $\mathbf{x}^a$ and $\mathbf{x}^b$ are the selected parents and $\mathbf{x}'$ is the new individual. The box crossover is done for selected component indexes i using the probability P_{cross}. Otherwise, the value from parent $\mathbf{x}^a$ is used. The coefficient c is a uniform random variable between $(-0.5, 1.5)$. The range of c is not set at $(0, 1)$ to avoid averaging behaviour. Thus the successive populations will not, by default, converge towards the middle of the initial population. (The averaging behaviour may explain the success of some optimization algorithms in problems where the maximum lies in the center of the feasible region.)

The parents are selected using tournament selection. For each of the parents, pick randomly two individuals from the population and select the better one with probability P_{tour}. Because we only need to compare the fitness values to find out which individual is better, there is no need to scale the values of the object function to some specific range.

The Matlab routine implements mutation using normal distribution for some components of the vector $\mathbf{x}'$:

$$x_i'' = x_i' + r(x_i^u - x_i^l).$$

Here the random number r is normally distributed with zero mean and standard deviation $|x_i^u - x_i^l|/10$. Vectors $\mathbf{x}^l$ and $\mathbf{x}^u$ are the box constraints of the optimization problem. Each component of $\mathbf{x}'$ is mutated with probability P_{mut}.

The algorithm also includes a version of elitism: take the best individual in the old population and compare it with a random individual of the new population. If the new individual is worse, replace it with the best individual from the old population.

1.5 OTHER METHODS FOR GLOBAL OPTIMIZATION

Before giving examples of using the simple GA discussed in the previous section, let us review a few other global optimization methods.

Evolutionary strategies (ES) have been extensively studied in Europe for more than twenty years [Sch95]. These methods were developed for the optimization of continuous functions. Differential evolution is a recent example of applying the ES concepts to solving optimization problems [PS97]. A Fortran 95 implementation of DE is found in the CSC textbook on Fortran 90/95 [HRR98].

Simulated annealing was developed based on physical cooling phenomena [AK89, vLA88, OvG89, Dav87]. There are quite a few implementations of the method available [PTVF92, CMMR87, Ing93, IR92]. Genetic annealing is an attempt to combine genetic algorithms and simulated annealing [Pri94, Haa95a].

Search concentration methods are another kind of heuristic, which can be tailored to solve spesific kinds of problems. The Smin1 software is an implementation of this method [PPH+86]. Smin1 includes a polytope method (also called "simplex method") as a local optimizer. The efficiency of Smin1 derives from "meta-optimization": the parameters of the method have been fine-tuned using a large set of example problems from various sources. Also, the software package makes three independent runs to verify the solution. Unfortunately, this software package is not freely available, and is somewhat hard to use.

Sigma is a freely available package, published as algorithm 667 in ACM Transactions on Mathematical Software [APPZ88]. This package is available at Netlib (www.netlib.org). Sigma uses stochastic integration techniques to solve optimization problems. I have compared Smin1 and Sigma with the genetic annealing method in a previous article [Haa95a].

In this paper we'll compare a genetic algorithm with two simple global optimization strategies based on the following outline:

1. Select an initial guess $\mathbf{x}^1 \in \mathcal{F} \subset \mathbf{R}^n$ and set $k = 1$.
2. Form the sampling distribution g_k and select a point $\mathbf{y}$ using the distribution g_k.
3. Set $\mathbf{x}^{k+1} = S(\mathbf{x}^k, \mathbf{y})$.
4. Compute $f(\mathbf{x}^{k+1})$. If the result is good enough, stop.
5. Set $k = k + 1$ and go to step 2.

Here we'll only consider two implementations of this algorithm: random search and local optimization with random initial guess. These two random search methods offer a baseline comparison with other optimization methods.

The random search method works as follows: sample the feasible region with uniform distribution, and always select the best point found as the new solution candidate. This algorithm finds the global optimum with probability 1 in the limit $k \to \infty$. Of course, in practice this method is inefficient.

We can analyze the probability of finding the optimum with this method. Consider the function from Eq. (1.1) on page 3. The probability of finding the global maximum with this method using M independent trials is

$$1 - \left(\frac{||x^u - x^l|| - 2\epsilon}{||x^u - x^l||} \right)^M .$$

Here we suppose that the interval $||x - a|| < \epsilon$ is inside the region $x^l \leq x \leq x^u$. The same calculation can be made in the n-dimensional case.

In the local optimization with random initial guess we'll use the fminu routine from the Matlab Optimization Toolbox. This routine is based on the BFGS secant method (see Section 1.6). Depending on the initial guess, this method will find a local optimum. We'll make several runs, of which we'll select the best result. The limit of total function evaluations (including line search and estimating the derivatives) is the same as with the other methods.

1.6 GRADIENT-BASED LOCAL OPTIMIZATION METHODS

Genetic algorithms are direct search methods: they use only the function values and do not require derivatives. When your object function is smooth and you need efficient local optimization, you should consider using standard gradient or Hessian based methods [Haa95b, BSS93, Ber96, SN96]. However, the performance and reliability of the different gradient methods varies considerably [Haa94].

Of course, gradient methods are not the only local optimization methods available. Recently there has been renewed interest in direct search methods also for local optimization [Pow98, Wri96].

Here we'll only consider gradient-based methods for local optimization. We assume a smooth objective function (i.e., continuous first and second derivatives). The object function is denoted by $f(\mathbf{x}) : \mathbf{R}^n \to \mathbf{R}$. The gradient vector $\nabla f(\mathbf{x})$ contains the first derivatives:

$$\nabla f(\mathbf{x}) = \begin{pmatrix} \frac{\partial f(\mathbf{x})}{\partial x_1} \\ \vdots \\ \frac{\partial f(\mathbf{x})}{\partial x_n} \end{pmatrix}.$$

For example, if $f(\mathbf{x}) = -100(x_2 - x_1^2)^2 - (1 - x_1)^2$, we get

$$\nabla f(\mathbf{x}) = \left(400x_1(x_2 - x_1^2) - 2(x_1 - 1),\ -200(x_2 - x_1^2)\right)^T.$$

The Hessian matrix $H(\mathbf{x}) = \nabla^T \nabla f(\mathbf{x})$ contains the second derivatives of the object function:

$$H(\mathbf{x}) = \nabla \nabla^T f(\mathbf{x}) = \begin{pmatrix} \frac{\partial^2 f(\mathbf{x})}{\partial^2 x_1} & \cdots & \frac{\partial^2 f(\mathbf{x})}{\partial x_1 \partial x_n} \\ \vdots & \ddots & \vdots \\ \frac{\partial^2 f(\mathbf{x})}{\partial x_1 \partial x_n} & \cdots & \frac{\partial^2 f(\mathbf{x})}{\partial^2 x_n} \end{pmatrix}.$$

For the previous example, the Hessian is

$$H(\mathbf{x}) = \begin{pmatrix} -1200x_1^2 + 400x_2 - 2 & 400x_1 \\ 400x_1 & -200 \end{pmatrix}.$$

Some methods need only the gradient vector, but in the Newton's method we need the Hessian matrix. Both the gradient and the Hessian can be approximated using finite differences, but this makes the methods less efficient and reliable.

Here is a general framework for the gradient methods (note that these methods search for the minimum, not maximum):

> Select an initial guess $\mathbf{x}^1$ and set $k = 1$.
> repeat
>   Solve the search direction $\mathbf{p}^k$ from the equation (1.2) or (1.3) below.
>   Find the next iteration point using Eq. (1.4) below: $\mathbf{x}^{k+1} = \mathbf{x}^k + \lambda_k \mathbf{p}^k$.
>   Set $k = k + 1$.
> until $\|\mathbf{x}^k - \mathbf{x}^{k-1}\| < \epsilon$

Depending on the details of the algorithm, we get several different methods. In conjugate gradient methods the search direction $\mathbf{p}^k$ is found as follows:

$$\mathbf{p}^k = -\nabla f(\mathbf{x}^k) + \beta_k \mathbf{p}^{k-1}. \tag{1.2}$$

In secant methods we use the equation

$$B_k\, \mathbf{p}^k = -\nabla f(\mathbf{x}^k), \tag{1.3}$$

where the matrix B_k estimates the Hessian. The matrix B_k is updated in each iteration. If you define B_k to be the identity matrix, you get the steepest descent method. If the matrix B_k is the Hessian $H(\mathbf{x}^k)$, you get the Newton's method.

The length λ_k of the search step is often computed using a line search:

$$\lambda_k = \arg \min_{\lambda > 0} f(\mathbf{x}^k + \lambda \mathbf{p}^k). \tag{1.4}$$

This is a one-dimensional optimization problem. Alternatively, you can use methods based on a trust region [Haa95b].

In general, you should never use the steepest descent method due to it's poor performance. Instead, a conjugate gradient method should be used. If the second derivatives are easy to compute, a (modified) Newton's method may be the most efficient. Also secant methods can be faster than conjugate gradient methods, but there is a memory tradeoff [Noc92, Haa95b].

The gradient-based methods described above are local optimizers. Consider the function

$$\begin{aligned} f(\mathbf{x}) &= 3\,(1-x_1)^2 \exp(-x_1^2 - (x_2+1)^2) - 10\,(\frac{x_1}{5} - x_1^3 - x_2^5) \\ &\quad \times \exp(-x_1^2 - x_2^2) - \frac{1}{3}\exp(-(x_1+1)^2 - x_2^2) + 10. \end{aligned}$$

This so-called "peaks" function has three local maxima. Of course, an optimization routine only knows the values of the function at a limited number of points. This is illustrated in Figure 1.2. The global maximum is at the point $\mathbf{x}^* \approx (-0.0106, 1.5803)^T$.

Because this function is smooth, we can use a gradient-based method to solve the problem. Figure 1.3 shows how the BFGS secant method converges to different local minima depending on the initial guess. This is not necessarily a bad tactic for solving the problem. Local optimization methods can be combined with other methods to get a good tradeoff between performance and reliability.

1.7 A COMPARISON OF METHODS

Here we'll compare the following three methods:

- a simple GA
- random search with uniform sampling (RS)
- local optimization (Matlab routine fminu) with random initial guesses (BFGS+RS).

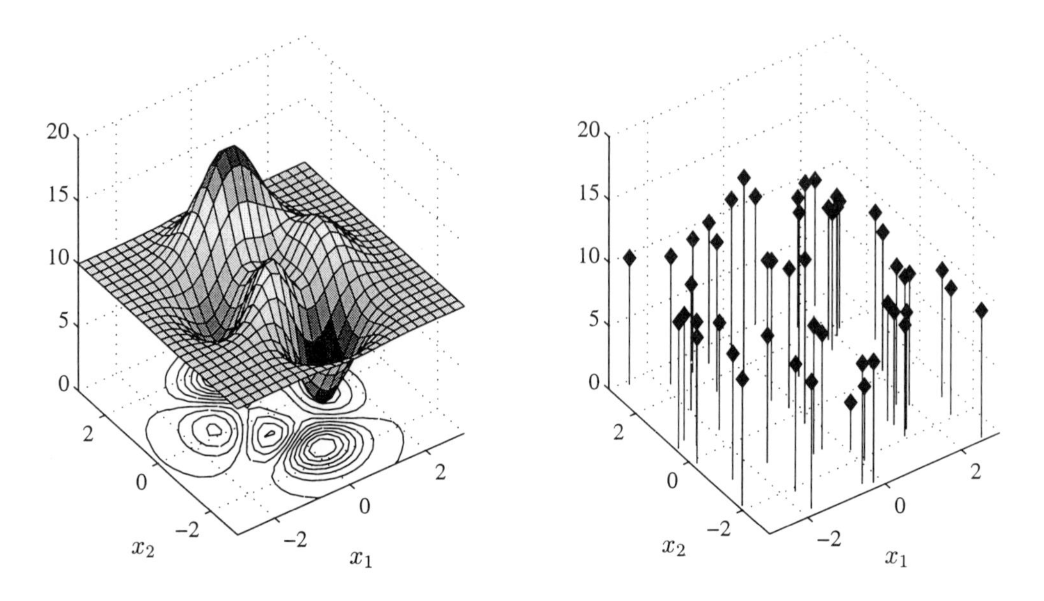

Figure 1.2 A function with three local maxima (surface plot on the left, function values at random points on the right).

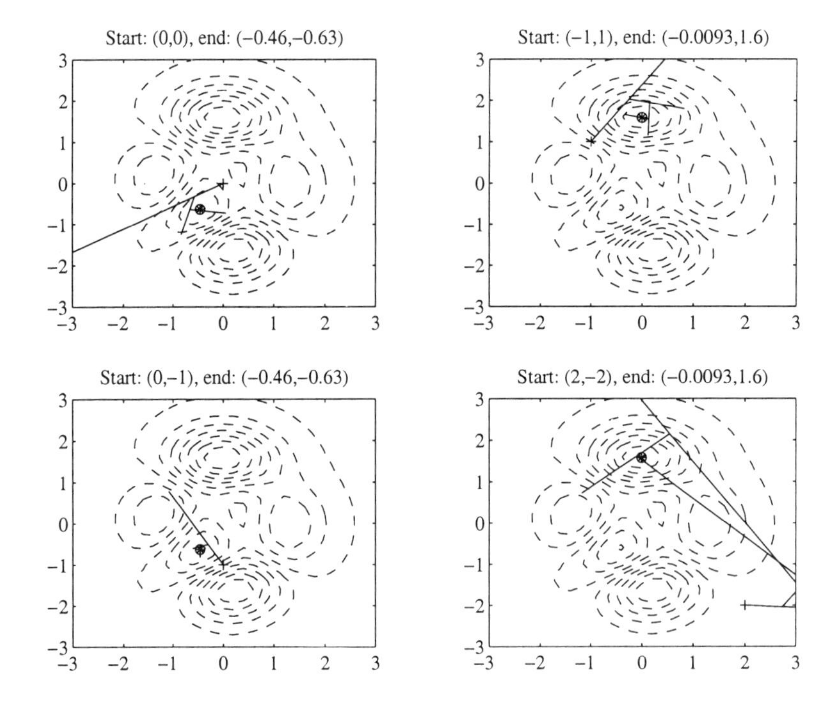

Figure 1.3 Optimization with a the routine fminu the Matlab Optimization Toolbox. The routine is based on the BFGS secant method.

We'll use the following functions for testing the methods:

$$\begin{aligned}
f_1(\mathbf{x}) &= -100(x_2 - x_1^2)^2 - (1 - x_1)^2, \\
f_2(\mathbf{x}) &= 3\,(1 - x_1)^2 \exp(-x_1^2 - (x_2 + 1)^2) - 10\,(\frac{x_1}{5} - x_1^3 - x_2^5) \\
&\qquad \times \exp(-x_1^2 - x_2^2) - \frac{1}{3} \exp(-(x_1 + 1)^2 - x_2^2) + 10, \\
f_3(\mathbf{x}) &= \left((x_1 - 100)^2 + (x_2 - 100)^2\right) /4000 - \\
&\qquad \cos(x_1 - 100) \cos\left((x_2 - 100)/\sqrt{2}\right) + 1.
\end{aligned}$$

All the functions are two-dimensional, which makes it easy to visualize them and to see the algorithms in action. Therefore, these function are well suited for teaching and demonstration purposes.

The function f_1 is the well-known Rosenbrock function, which is often used to test local optimization methods (Figure 1.4). Here we are using the inverted version of the function to get a maximization problem. The function f_2 is the "peaks" function from Matlab (Figure 1.2). The function f_3 is called the Griewank function (see Figure 1.1).

In the following I have done 50 independent runs of the three algorithms for each test problem. I have defined the run to be successful when the reported solution $\mathbf{x}^M$ is close to the global optimum $\mathbf{x}^*$:

$$||\mathbf{x}^M - \mathbf{x}^*|| < 0.2.$$

Also, the value of the object function has to be larger than -0.1, 18.0 or -0.01 for the functions f_1, f_2 and f_3, respectively. For the genetic algorithm one should check this condition for all vectors in the final population to see if we have found an acceptable solution.

The parameters of the genetic algorithm were $P_{\text{cross}} = 0.8$, $P_{\text{mut}} = 0.02$ and $P_{\text{tour}} = 0.7$ in all cases. The size of the population was set to $l = 30$ individuals, and the number of generations m varied between 20 and 300. For the random search methods, the number of function evaluations was limited to $l \times m$.

1.7.1 The Rosenbrock function

The Rosenbrock function has a single optimum at the point $\mathbf{x}^* = (1, 1)^T$. Thus, there is in fact no need for a global optimization method. However, this function is difficult to optimize because of the great difference between the function values at the bottom and top of the "ridge" (see Figure 1.4).

The number of GA generations was $m = 200$. In the random search and the local optimization, the number of function evaluations was limited to $30 \times 200 = 6000$. The results are summarized in Table 1.1. The column "Best" reports the best value found in the 50 trials. Also the worst and median values are reported. The success rate tells how many of the trials succeeded in finding the global maximum.

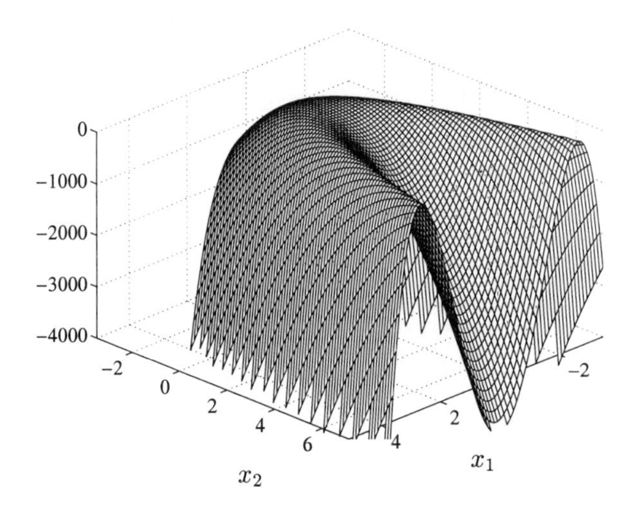

Figure 1.4 The two-dimensional inverted Rosenbrock function $f(\mathbf{x}) = -100(x_2 - x_1^2)^2 - (1 - x_1)^2$. The global maximum is at the point $\mathbf{x}^* = (1,1)^T$.

Table 1.1 Comparison of methods for solving the Rosenbrock problem.

Method	Best	Worst	Median	Succ. rate
GA	-4.2e-9	-0.036	-0.0013	96%
RS	-4.9e-4	-0.17	-0.047	48%
BFGS+RS	-7.8e-7	-0.17	-0.0046	100%

Figure 1.5 shows the behaviour of the GA in the best and worst of the 50 trials. The success rate of the GA was 96%, so only two trials failed to find the solution. However, even in the worst case the reported result was near the global solution, as seen in Fig. 1.5.

This problem is not easy for gradient-based methods either, due to the "banana" shape of the ridge where the function values are greatest. However, the BFGS method was successful in all trials, perhaps because the Rosenbrock function is used as a standard test in local optimization.

The random search method found the local optimum in about half of the trials. The high success rate is due to the small search area.

1.7.2 The "peaks" problem

The number of generations was $m = 20$. The number of function evaluations was limited to $30 \times 20 = 600$. The results are summarized in Table 1.2.

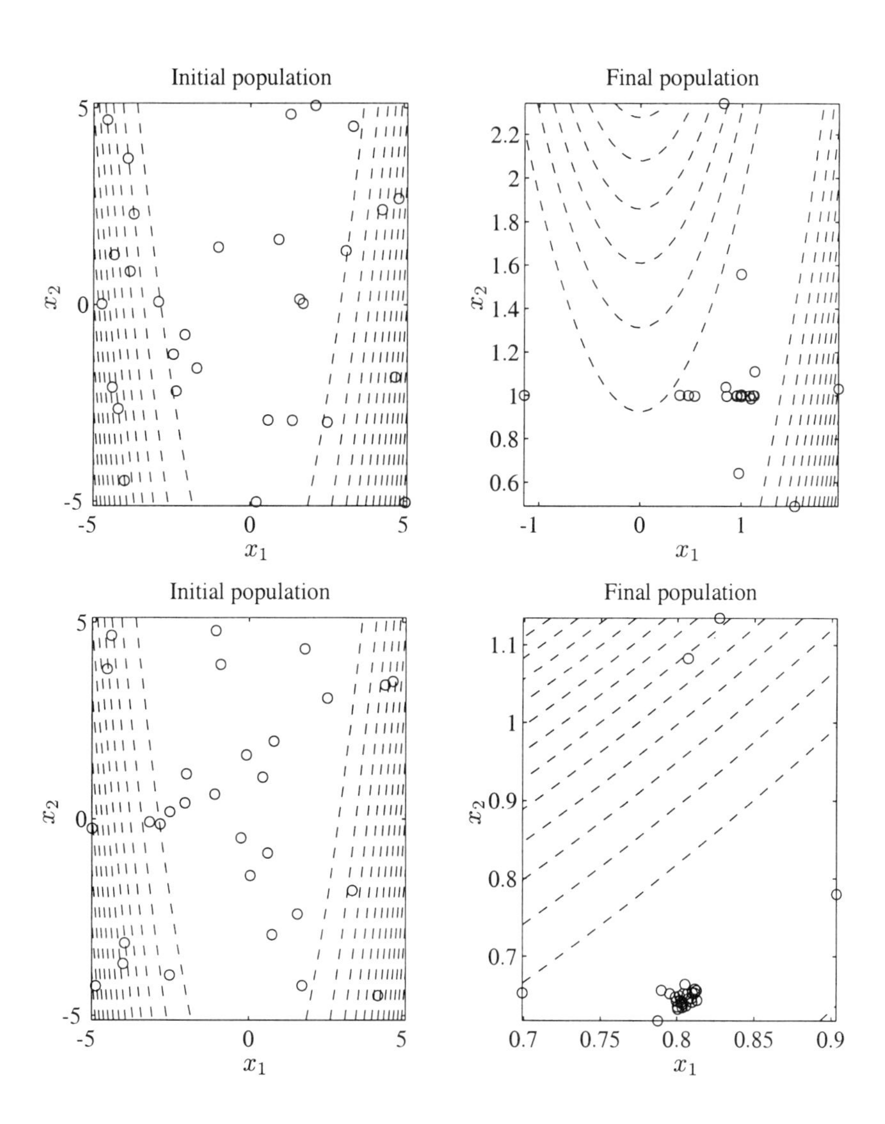

Figure 1.5 Behavior of the GA in the Rosenbrock problem. In the upper right plot the algorithm found the global maximum at the point $(1,1)^T$. In the lower right plot the maximum value was reported at the point $(0.81, 0.66)^T$.

Table 1.2 Comparison of methods for solving the Peaks problem (function f_2).

Method	Best	Worst	Mean	Succ. rate
GA	18.106	18.062	18.103	100%
RS	18.105	17.483	17.885	40%
BFGS+RS	18.106	18.106	18.106	98%

In this problem the GA was successful in all tests. The BFGS routine was unsuccessful in one of the 50 trials. The number of local optimization runs varied between 1 and 22. In the unsuccessful trial the algorithm got stuck and never found a maximum.

The random search found the optimum in 40% of the trials. Thus, we can estimate that the probability of finding the maximum with 600 random guesses is about 40%.

This problem is quite easy for all the methods, although the function has several local maxima. The most efficient method is a gradient-based local search with a good initial guess.

1.7.3 The Griewank problem

The number of generations was $m = 300$. In the random search and the local optimization, the number of function evaluations was limited to $30 \times 300 = 9000$. The results are summarized in Table 1.3

Table 1.3 Comparison of methods for solving the Griewank problem (function f_3).

Method	Best	Worst	Mean	Succ. rate
GA	-2.1e-7	-0.0099	-0.0040	66%
RS	-0.060	-0.69	-0.23	0%
BFGS+RS	-5.9e-14	-0.36	-0.068	2%

In the Griewank problem about 2/3 of the GA runs reported a maximum sufficiently close to the global maximum of the problem. In the other runs, the final populations contained individuals which were in the local maxima nearest to the global maximum.

To get better reliability, we can increase the number of generations, or we can use local optimization started from the values represented by the final population. Making several independent runs of the GA is another way to verify the solution.

In fact, we could use the GA as a preprocessor for a local optimizer. About 50–100 generations should be enough to pinpoint promising areas for local optimization. Of course, we can include local optimization in the GA itself, but this might be inefficient, at least in the initial stages when we are far from the true solution.

The random search didn't find the global maximum. In fact, we can compute that the probability of finding the global maximum with 9000 random guesses is less than 0.1%. Similarly, the probability of finding the solution in 50 trials is less than 4%.

The local search with random initial guess was successful in only one of the 50 trials. However, if we had used the GA for generating the initial guesses, the local search would have found the global optimum with only a few trials.

1.8 PARALLEL COMPUTING AND GAS

Even though GAs are not the fastest optimization methods, the inherent parallelism makes GAs attractive [Gol89, Mic98]. Basically, only the crossover and selection operations are not parallel. The overhead of these operations can be made small by splitting the global population into subpopulations which evolve relatively independently ("the island model").

Parallelization of genetic algorithms can be done using MPI or some other message-passing library [Haab, For95, HM97, HS98]. In a master-slave implementation the master process distributes the work among the processors. However, the master process may become a bottleneck, because of synchronization and bookkeeping. It is perhaps easier to change the basic genetic algorithm by exchanging individuals at the end of each iteration (or every n iterations). The subpopulations can be arranged in any topology. For example, in a two-dimensional cartesian grid each subpopulation has four neighbors. MPI offers ready-made routines for setting up this topology and for exchanging individuals [For95, HM97].

Parallel GA are still a subject to be explored. There are several interesting questions to be answered: How often should individuals be exchanged between populations? What is the convergence rate of the island model compared to a GA with one big population? Also, if the object function takes different amounts of time to compute at different points, one has to solve the question of load balancing between the subpopulations.

1.9 FURTHER INFORMATION

The recent review of GAs be Melanie Mitchell [Mic98] is very readable and concise, although the emphasis is not on optimization. Of course, anyone interested in genetic algorithms should have access to the books by Goldberg, Davis, Michalewicz, and Schwefel [Gol89, Dav91, Mic96, Sch95].

CSC has published several textbooks and guides on numerical methods and optimization [HKR93, Haa95b]. We have also published guidebooks on parallel computing [HM97, Saa95], programming with Fortran 90/95 [HRR98], and mathematical software [Haa98b].

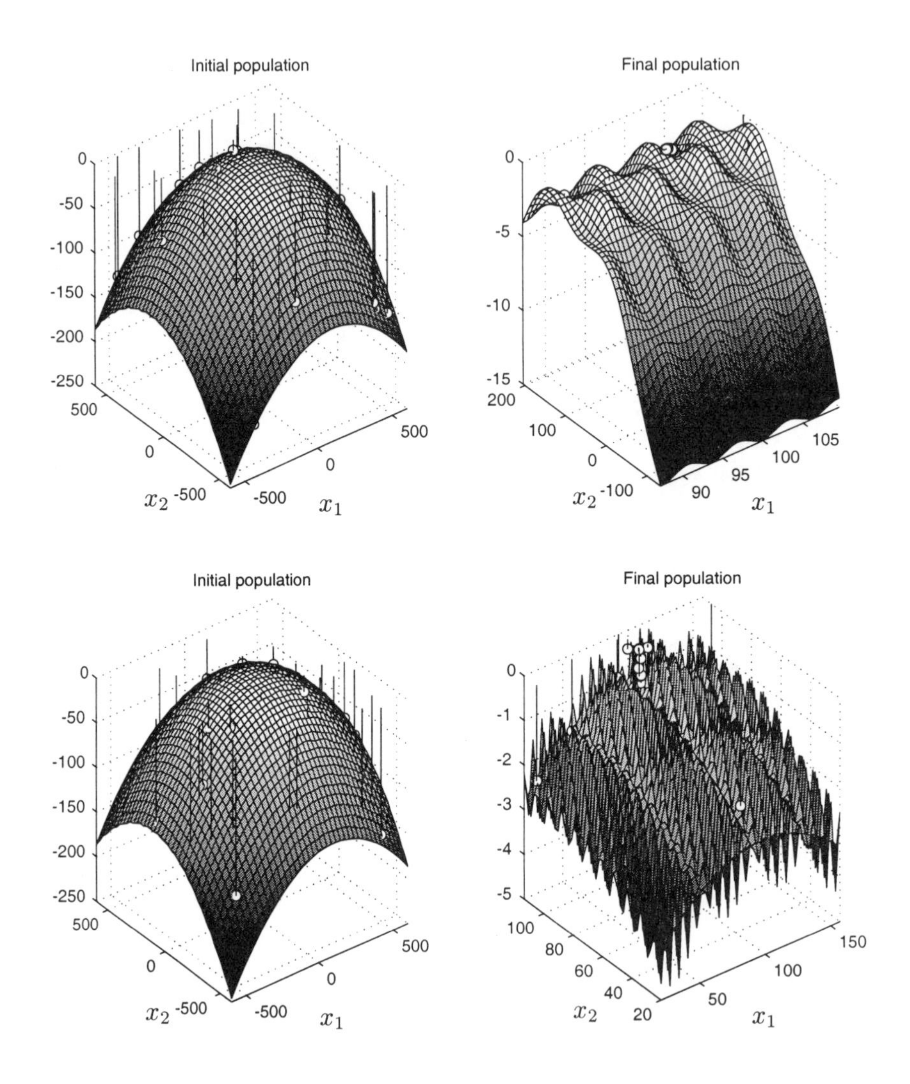

Figure 1.6 Behavior of the GA in the Griewank problem. In the upper right plot the algorithm found the global maximum at the point $(100, 100)^T$. In the lower right plot the maximum value was reported at the point $(93.7, 100.0)^T$.

I have included a Matlab implementation of a binary-coded GA in the guidebook on solving optimization problems [Haa95b]. Our textbook on Fortran 90/95 contains an implementation of a hybdid optimization method called differential evolution [HRR98, PS97]. Examples of using genetic algorithms are available on the Web [Haaa].

There are several Web sites dedicated to evolutionary computing. A good starting point is The Hitch-Hiker's Guide to Evolutionary Computation:

http://research.de.uu.net:8080/encore/www/

Miscellaneous information on genetic algorithms is available at The Genetic Algorithms Archive:

http://www.aic.nrl.navy.mil:80/galist/

REFERENCES

[AK89] Emile Aarts and Jan Korst. Simulated Annealing and Boltzmann Machines. Wiley, 1989.

[APPZ88] F. Aluffi-Pentini, V. Parisi, and F. Zirilli. SIGMA — A Stochastic-Integration Global Minimization Algorithm. ACM Transactions on Mathematical Software, 14(4):366–380, December 1988.

[Ber96] Dimitri P. Bertsekas. Nonlinear Programming. Athena Scientific, 1996.

[BSS93] Mokhtar S. Bazaraa, Hanif D. Sherali, and C. M. Shetty. Nonlinear Programming: Theory and Algorithms. Wiley, 2nd edition, 1993.

[CMMR87] A. Corana, M. Marchesi, C. Martini, and S. Ridella. Minimizing Multimodal Functions of Continuous Variables with the "Simulated Annealing" Algorithm. ACM Transactions on Mathematical Software, 13(3):262–280, September 1987.

[Dav87] Lawrence Davis. Genetic algorithms and simulated annealing. Pitman, 1987.

[Dav91] Lawrence Davis, editor. Handbook of Genetic Algorithms. Van Hostrand Reinhold, 1991.

[DGR90] M. M. Doria, J. E. Gubernatis, and D. Rainer. Solving the Ginzburg-Landau equations by simulated annealing. Physical Review B, 41(10):6335, 1990.

[For95] Message-Passing Interface Forum. MPI: A Message-Passing Interface Standard. University of Tennessee, 1995.

[Gol89] David E. Goldberg. Genetic algorithms in search, optimization, and machine learning. Addison-Wesley, 1989.

[GSB+92] J. Garner, M. Spanbauer, R. Benedek, K. J. Strandburg, S. Wright, and P. Plassmann. Critical fields of Josephson-coupled superconducting multilayers. Physical Review B, 45(14):7973, 1992.

[Haaa] Juha Haataja. Examples of using genetic algorithms. URL http://www.csc.fi/Movies/GA.html.

[Haab] Juha Haataja. Examples of using MPI (Message Passing Interface). URL http://www.csc.fi/programming/examples/mpi/.

[Haa94] Juha Haataja. Suurten optimointitehtävien ratkaiseminen. Licentiate thesis, Helsinki University of Technology, 1994. Title in English: Solving large-scale optimization problems.

[Haa95a] Juha Haataja. A comparison of software for global optimization. In Jarmo T. Alander, editor, Proceedings of the First Nordic Workshop on Genetic Algorithms and Their Applications. University of Vaasa, 1995.

[Haa95b] Juha Haataja. Optimointitehtävien ratkaiseminen. Center for Scientific Computing, 1995. Written in Finnish. URL http://www.csc.fi/math_topics/opt/.

[Haa98a] Juha Haataja, editor. Alkuräjähdyksestä kännykkään — näkökulmia laskennalliseen tieteeseen. Center for Scientific Computing, 1998. Written in Finnish.

[Haa98b] Juha Haataja, editor. Matemaattiset ohjelmistot. Center for Scientific Computing, 3rd edition, 1998. Written in Finnish. URL http://www.csc.fi/oppaat/mat.ohj/.

[HK99] Juha Haataja and Aila Kinnunen, editors. CSC Report on Scientific Computing 1997–1998. Center for Scientific Computing, 1999. To appear.

[HKR93] Juha Haataja, Juhani Käpyaho, and Jussi Rahola. Numeeriset menetelmät. Center for Scientific Computing, 1993. Written in Finnish.

[HM97] Juha Haataja and Kaj Mustikkamäki. Rinnakkaisohjelmointi MPI:llä. Center for Scientific Computing, 1997. Written in Finnish. URL http://www.csc.fi/oppaat/mpi/.

[HRR98] Juha Haataja, Jussi Rahola, and Juha Ruokolainen. Fortran 90/95. Center for Scientific Computing, 2nd edition, 1998. Written in Finnish. URL http://www.csc.fi/oppaat/f95/.

[HS98] Juha Haataja and Ville Savolainen, editors. Cray T3E User's Guide. Center for Scientific Computing, 2nd edition, 1998. URL http://www.csc.fi/oppaat/t3e/.

[Ing93] Lester Ingber. Simulated annealing: Practice versus theory. Mathematical and Computer Modelling, 18(11):29–57, 1993.

[IR92] Lester Ingber and Bruce Rosen. Genetic Algorithms and Very Fast Simulated Reannealing: A Comparison. Mathematical and Computer Modelling, 16(11):87–100, 1992.

[JP93] Mark T. Jones and Paul E. Plassmann. Computation of Equilibrium Vortex Structures for Type-II Superconductors. The International Journal of Supercomputer Applications, 7(2):129–143, 1993.

[LSF70] Robert B. Leighton, Matthew Sands, and Richard Feynman. Feynman lectures on physics. 1970. Chapter 19: The Principle of Least Action.

[Mic96] Zbigniew Michalewicz. Genetic Algorithms + Data Structures = Evolution Programs. Springer, 1996.

[Mic98] Melanie Michell. An Introduction to Genetic Algorithms. MIT Press, 1998.

[MW93] Jorge J. Moré and Stephen J. Wright. Optimization Software Guide. SIAM, 1993. Frontiers in applied mathematics 14.

[Noc92] Jorge Nocedal. Theory of algorithms for unconstrained optimization. In Acta Numerica, pages 199–242. 1992.

[OvG89] R. H. J. M. Otten and L. P. P. P. van Ginneken. The Annealing Algorithm. Kluwer Academic Publishers, 1989.

[Pow98] M. J. D. Powell. Direct search algorithms for optimization calculations. In Acta Numerica 1998. Cambridge University Press, 1998. To be published.

[PPH+86] S. Palosaari, S. Parviainen, J. Hiironen, J. Reunanen, and P. Neittaanmäki. A Random Search Algorithm for Constrained Global Optimization. Acta Polytechnica Scandinavica, 172, 1986.

[Pri94] Kenneth V. Price. Genetic Annealing. Dr. Dobb's Journal, pages 127–132, October 1994.

[PS97] Kenneth Price and Rainer Storm. Differential Evolution. Dr. Dobb's Journal, pages 18–24, April 1997.

[PTVF92] William H. Press, Saul A. Teukolsky, William T. Vetterling, and Brian P. Flannery. Numerical Recipes in C/Fortran — The Art of Scientific Computing. Cambridge University Press, 2nd edition, 1992.

[Rid96] Mark Ridley. Evolution. Blackwell Science, 2nd edition, 1996.

[Saa95] Sami Saarinen. Rinnakkaislaskennan perusteet PVM-ympäristössä. Center for Scientific Computing, 1995.

[Sch95] Hans-Paul Schwefel. Evolution and Optimum Seeking. Wiley, 1995.

[SN96] Ariela Sofer and Stephen G. Nash. Linear and Nonlinear Programming. McGraw Hill, 1996.

[vLA88] P. J. M. van Laarhoven and E. H. L. Aarts. Simulated Annealing: Theory and Applications. D. Reidel Publishing Company, 1988.

[WH91] Z. D. Wang and C.-R. Hu. Numerical relaxation approach for solving the general Ginzburg-Landau equations for type-II superconductors. Physical Review B, 44(21):11918, 1991.

[Wri96] Margaret H. Wright. Direct search methods: once scorned, now respectable. In D. F. Griffiths and G. A. Watson, editors, Numerical Analysis 1995 (Proceedings of the 1995 Dundee Biennial Conference in Numerical Analysis), pages 191–208. Addison-Wesley, 1996.

2 An Introduction to Evolutionary Computation and Some Applications

D. B. FOGEL

Natural Selection, Inc.
3333 N. Torrey Pines Ct., Suite 200
La Jolla, CA 92037
E-mail: dfogel@natural-selection.com

Abstract. Evolutionary computation is becoming common in the solution of difficult, real-world problems in industry, medicine, and defense. This paper introduces evolutionary computation and reviews some of the practical advantages to using evolutionary algorithms as compared with classic methods of optimization or artificial intelligence. Specific advantages include the flexibility of the procedures, as well as the ability to self-adapt the search for optimum solutions on the fly. In addition, several recent applications of evolutionary computation are reviewed. As desktop computers increase in speed, the application of evolutionary algorithms will become routine.

2.1 INTRODUCTION

Darwinian evolution is intrinsically a robust search and optimization mechanism. Evolved biota demonstrate optimized complex behavior at every level: the cell, the organ, the individual, and the population. The problems that biological species have solved are typified by chaos, chance, temporality, and nonlinear interactivities. These are also characteristics of problems that have proved to be especially intractable to classic methods of optimization. The evolutionary process can be applied to problems where heuristic solutions are not available or generally lead to unsatisfactory results. As a result, evolutionary algorithms have recently received increased interest, particularly with regard to the manner in which they may be applied for practical problem solving.

Evolutionary computation, the term now used to describe the field of investigation that concerns all evolutionary algorithms, offers practical advantages to the researcher facing difficult optimization problems. These advantages are multifold, including the simplicity of the approach, its robust response to changing circumstance, its flexibility, and many other facets. This paper summarizes some of these advantages and offers suggestions in designing evolutionary algorithms for real-world problem solving. It is assumed that the reader is familiar with the basic concepts of evolutionary algorithms, and is referred to Fogel (1995), Bäck (1996), and Michalewicz (1996) for introductions.

2.2 ADVANTAGES OF EVOLUTIONARY COMPUTATION

2.2.1 Conceptual simplicity

A primary advantage of evolutionary computation is that it is conceptually simple. The main flow chart that describes every evolutionary algorithm applied for function optimization is depicted in Figure 2.1. The algorithm consists of initialization, which may be a purely random sampling of possible solutions, followed by iterative variation and selection in light of a performance index. This figure of merit must assign a numeric value to any possible solution such that two competing solutions can be rank ordered. Finer granularity is not required. Thus the criterion need not be specified with the precision that is required of some other methods. In particular, no gradient information needs to be presented to the algorithm. Over iterations of random variation and selection, the population can be made to converge to optimal solutions (Fogel, 1994; Rudolph, 1994; and others).

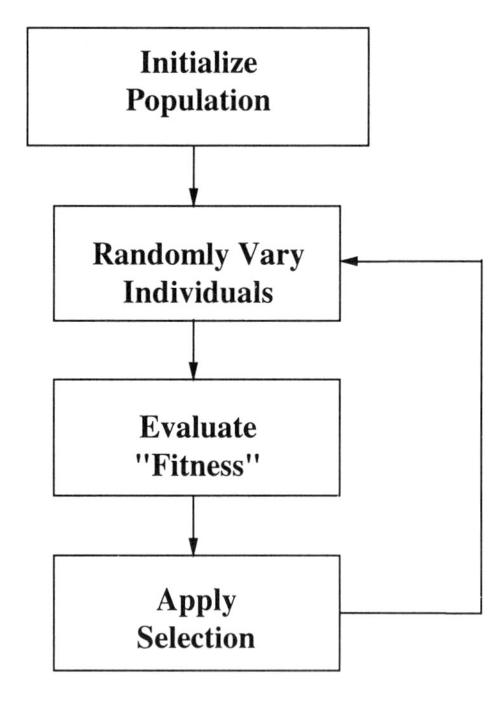

Figure 2.1 The main flowchart of the vast majority of evolutionary algorithms. A population of candidate solutions to a problem at hand is initialized. This often is accomplished by randomly sampling from the space of possible solutions. New solutions are created by randomly varying existing solutions. This random variation may include mutation and/or recombination. Competing solutions are evaluated in light of a performance index describing their "fitness" (or equivalently, their error). Selection is then applied to determine which solutions will be maintained into the next generation, and with what frequency. These new "parents" are then subjected to random variation, and the process iterates.

The evolutionary search is similar to the view offered by Wright (1932) involving "adaptive landscapes." A response surface describes the fitness assigned to alternative genotypes as they interact in an environment (Figure 2.2). Each peak corresponds to an optimized collection of behaviors (phenotypes), and thus one or more sets of optimized genotypes. Evolution probabilistically proceeds up the slopes of the topography toward peaks as selection culls inappropriate phenotypic variants. Others (Atmar, 1979; Raven and Johnson, 1986, pp. 400-401) have suggested that it is more appropriate to view the adaptive landscape from an inverted position. The peaks become troughs, "minimized prediction error entropy wells" (Atmar, 1979). Such a viewpoint is intuitively appealing. Searching for peaks depicts evolution as a slowly advancing, tedious, uncertain process. Moreover, there appears to be a certain fragility to an evolving phyletic line; an optimized population might be expected to quickly fall off the peak under slight perturbations. The inverted topography leaves an altogether different impression. Populations advance rapidly, falling down the walls of the error troughs until their cohesive set of interrelated behaviors is optimized. The topography is generally in flux, as a function of environmental erosion and variation, as well as other evolving organisms, and stagnation may never set in. Regardless of which perspective is taken, maximizing or minimizing, the basic evolutionary algorithm is the same: a search for the extrema of a functional describing the objective worth of alternative candidate solutions to the problem at hand.

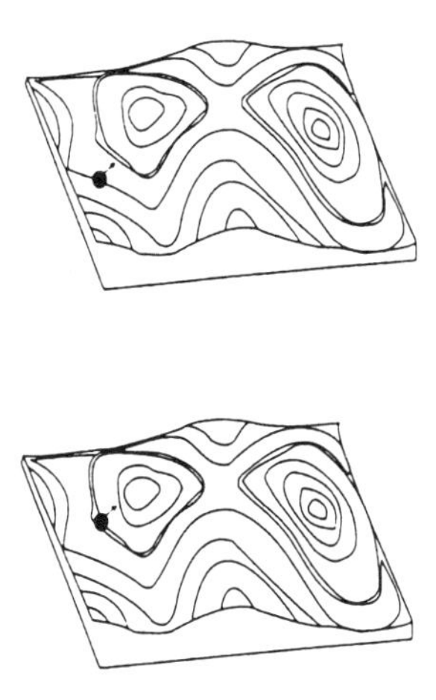

Figure 2.2 Evolution on an inverted adaptive topography. A landscape is abstracted to represent the fitness of alternative phenotypes and, as a consequence, alternative genotypes. Rather than viewing the individuals or populations as maximizing fitness and thereby climbing peaks on the landscape, a more intuitive perspective may be obtained by inverting the topography. Populations proceed down the slopes of the topography toward valleys of minimal predictive error.

The procedure may be written as the difference equation:

$$\mathbf{x}[t+1] = s(v(\mathbf{x}[t])) \quad (1)$$

where $\mathbf{x}[t]$ is the population at time t under a representation $\mathbf{x}$, v is a random variation operator, and s is the selection operator (Fogel and Ghozeil, 1996). There are a variety of possible representations, variation operators, and selection methods (Bäck et al., 1997). Not more than 10 years ago, there was a general recommendation that the best representation was a binary coding, as this provided the greatest "implicit parallelism" (see Goldberg, 1989 for a discussion; more detail is beyond the scope of this paper). But this representation was often cumbersome to implement (consider encoding a traveling salesman problem as a string of symbols from {0,1}), and empirical results did not support any necessity, or even benefit, to binary representations (e.g., Davis, 1991; Michalewicz, 1992; Koza, 1992). Moreover, the suggestions that recombining alternative solutions through crossover operators, and amplifying solutions based on the relative fitness also did not obtain empirical support (e.g., Fogel and Atmar, 1990; Bäck and Schwefel, 1993; Fogel and Stayton, 1994; and many others). Recent mathematical results have proved that there can be no best choice for these facets of an evolutionary algorithm that would hold across all problems (Wolpert and Macready, 1997), and even that there is no best choice of representation for any individual problem (Fogel and Ghozeil, 1997). The effectiveness of an evolutionary algorithm depends on the interplay between the operators s and v as applied to a chosen representation $\mathbf{x}$ and initialization $\mathbf{x}[0]$. This dependence provides freedom to the human operator to tailor the evolutionary approach for their particular problem of interest.

2.2.2 Broad applicability

Evolutionary algorithms can be applied to virtually any problem that can be formulated as a function optimization task. It requires a data structure to represent solutions, a performance index to evaluate solutions, and variation operators to generate new solutions from old solutions (selection is also required but is less dependent on human preferences). The state space of possible solutions can be disjoint and can encompass infeasible regions, and the performance index can be time varying, or even a function of competing solutions extant in the population. The human designer can choose a representation that follows their intuition. In this sense, the procedure is representation independent, in contrast with other numerical techniques which might be applicable for only continuous values or other constrained sets. Representation should allow for variation operators that maintain a behavioral link between parent and offspring. Small changes in the structure of a parent should lead to small changes in the resulting offspring, and likewise large changes should engender gross alterations. A continuum of possible changes should be allowed such that the effective "step size" of the algorithm can be tuned, perhaps online in a self-adaptive manner (discussed later). This flexibility allows for applying essentially the same procedure to discrete combinatorial problems, continuous-valued parameter optimization problems, mixed-integer problems, and so forth.

2.2.3 Outperform classic methods on real problems

Real-world function optimization problems often (1) impose nonlinear constraints, (2) require payoff functions that are not concerned with least squared error, (3) involve nonstationary conditions, (4) incorporate noisy observations or random processing, or include other vagaries that do not conform well to the prerequisites of classic optimization techniques. The response surfaces posed in real-world problems are often multi-modal, and gradient-based methods rapidly converge to local optima (or perhaps saddle points) which may yield insufficient performance. For simpler problems, where the response surface is, say, strongly convex, evolutionary algorithms do not perform as well as traditional optimization methods (Bäck, 1996). But this is to be expected as these techniques were designed to take advantage of the convex property of such surfaces. Schwefel (1995) has shown in a series of empirical comparisons that in the obverse condition of applying classical methods to multi-modal functions, evolutionary algorithms offer a significant advantage. In addition, in the often encountered case of applying linear programming to problems with nonlinear constraints, this offers an almost certainly incorrect result because the assumptions required for the technique are violated. In contrast, evolutionary computation can directly incorporate arbitrary constraints (Michalewicz, 1996).

Moreover, the problem of defining the payoff function for optimization lies at the heart of success or failure: Inappropriate descriptions of the performance index lead to generating the right answer for the wrong problem. Within classic statistical methods, concern is often devoted to minimizing the squared error between forecast and actual data. But in practice, equally correct predictions are not of equal worth, and errors of identical magnitude are not equally costly. Consider the case of correctly predicting that a particular customer will ask to purchase 10 units of a particular product (e.g., an aircraft engine). This is typically worth less than correctly predicting that the customer will seek to purchase 100 units of that product, yet both predictions engender zero error and are weighted equally in classic statistics. Further, the error of predicting the customer will demand 10 units and having them actually demand 100 units is not of equal cost to the manufacturer as predicting the customer will demand 100 units and having them demand 10. One error leaves a missed opportunity cost while the other leaves 90 units in a warehouse. Yet again, under a squared error criterion, these two situations are treated identically. In contrast, within evolutionary algorithms, any definable payoff function can be used to judge the appropriateness of alternative behaviors. There is no restriction that the criteria be differentiable, smooth, or continuous.

2.2.4 Potential to use knowledge and hybridize with other methods

It is always reasonable to incorporate domain-specific knowledge into an algorithm when addressing particular real-world problems. Specialized algorithms can outperform unspecialized algorithms on a restricted domain of interest (Wolpert and Macready, 1997). Evolutionary algorithms offer a framework such that it is comparably easy to incorporate such knowledge. For example, specific variation operators may be known to be useful when applied to particular representations (e.g., 2-OPT on the traveling salesman problem). These can be directly applied as mutation or recombination operations. Knowledge can also be implemented into the performance index, in the form of known physical or chemical properties (e.g., van der Waals interactions, Gehlhaar et al., 1995). Incorporating such information focuses the evolutionary search, yielding a more efficient exploration of the state space of possible solutions.

Evolutionary algorithms can also be combined with more traditional optimization techniques. This may be as simple as the use of a conjugate-gradient minimization used after primary search with an evolutionary algorithm (e.g., Gehlhaar et al., 1995), or it may involve simultaneous application of algorithms (e.g., the use of evolutionary search for the structure of a model coupled with gradient search for parameter values, Harp et al., 1989). There may also be a benefit to seeding an initial population with solutions derived from other procedures (e.g., a greedy algorithm, Fogel and Fogel, 1996). Further, evolutionary computation can be used to optimize the performance of neural networks (Angeline et al., 1994), fuzzy systems (Haffner and Sebald, 1993), production systems (Wilson, 1995), and other program structures (Koza, 1992; Angeline and Fogel, 1997). In many cases, the limitations of conventional approaches (e.g., the requirement for differentiable hidden nodes when using back propagation to train a neural network) can be avoided.

2.2.5 Parallelism

Evolution is a highly parallel process. As distributed processing computers become more readily available, there will be a corresponding increased potential for applying evolutionary algorithms to more complex problems. It is often the case that individual solutions can be evaluated independently of the evaluations assigned to competing solutions. The evaluation of each solution can be handled in parallel and only selection (which requires at least pairwise competition) requires some serial processing. In effect, the running time required for an application may be inversely proportional to the number of processors. Regardless of these future advantages, current desktop computing machines provide sufficient computational speed to generate solutions to difficult problems in reasonable time (e.g., the evolution of a neural network for classifying features of breast carcinoma involving over 5 million separate function evaluations requires only about three hours on a 200 MHz 604e PowerPC, Fogel et al., 1997).

2.2.6 Robust to dynamic changes

Traditional methods of optimization are not robust to dynamic changes in the environment and often require a complete restart in order to provide a solution (e.g., dynamic programming). In contrast, evolutionary algorithms can be used to adapt solutions to changing circumstance. The available population of evolved solutions provides a basis for further improvement and in most cases it is not necessary, nor desirable, to reinitialize the population at random. Indeed, this procedure of adapting in the face of a dynamic environment can be used to advantage. For example, Wieland (1990) used a genetic algorithm to evolve recurrent neural networks to control a cart-pole system comprising two poles (Figure 2.3). The degree of difficulty depended on the relative pole lengths (i.e., the closer the poles were to each other in length, the more difficult the control problem). Wieland (1990) developed controllers for a case of one pole of length 1.0 m and the other of 0.9 m by successively controlling systems where the shorter pole was started at 0.1 m and incremented sequentially to 0.9 m. At each increment, the evolved population of networks served as the basis for a new set of controllers. A similar procedure was offered in Saravanan and Fogel (1994), and Fogel (1996).

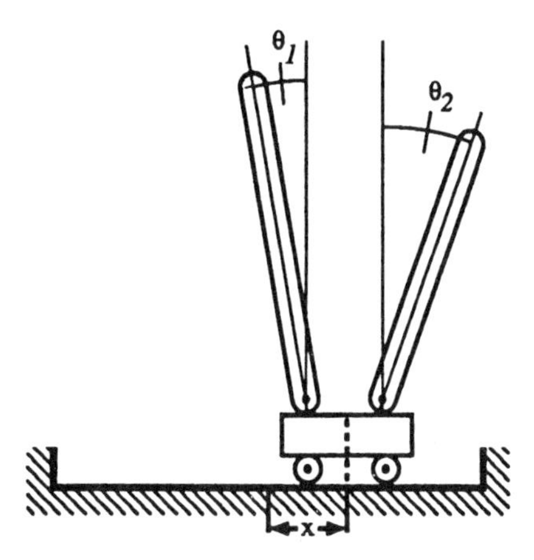

Figure 2.3 A cart with two poles. The objective is to maintain the cart between the limits of the track while not allowing either pole to exceed a specified maximum angle of deflection. The only control available is a force with which to push or pull on the cart. The difficulty of the problem is dependent on the similarity in pole lengths. Wieland (1990) and Saravanan and Fogel (1994) used evolutionary algorithms to optimize neural networks to control this plant for pole lengths of 1.0 m and 0.9 m. The evolutionary procedure required starting with poles of 1.0 m and 0.1 m, and iteratively incrementing the length of the shorter pole in a series of dynamic environments. In each case, the most recent evolved population served as the basis for new trials, even when the pole length was altered.

The ability to adapt on the fly to changing circumstance is of critical importance to practical problem solving. For example, suppose that a particular simulation provides perfect fidelity to an industrial production setting. All workstations and processes are modeled exactly, and an algorithm is used to find a "perfect" schedule to maximize production. This perfect schedule will, however, never be implemented in practice because by the time it is brought forward for consideration, the plant will have changed: machines may have broken down, personnel may not have reported to work or failed to keep adequate records of prior work in progress, other obligations may require redirecting the utilization of equipment, and so forth. The "perfect" plan is obsolete before it is ever implemented. Rather than spend considerable computational effort to find such perfect plans, a better prescription is to spend less computational effort to discover suitable plans that are robust to expected anomalies and can be evolved on the fly when unexpected events occur.

2.2.7 Capability for self-optimization

Most classic optimization techniques require appropriate settings of exogenous variables. This is true of evolutionary algorithms as well. However, there is a long history of using the evolutionary process itself to optimize these parameters as part of the search for optimal solutions (Reed et al., 1967; Rosenberg, 1967; and others). For example, suppose a search problem requires finding the real-valued vector that minimizes a particular functional $F(\mathbf{x})$, where $\mathbf{x}$ is a vector in n dimensions. A typical evolutionary algorithm (Fogel, 1995) would use Gaussian random variation on current parent solutions to generate offspring:

$$x'_i = x_i + \sigma_i N(0,1) \qquad (2)$$

where the subscript indicates the ith dimension, and σ_i is the standard deviation of a Gaussian random variable. Setting the "step size" of the search in each dimension is critical to the success of the procedure (Figure 2.4). This can be accomplished in the following two-step fashion:

$$\sigma'_i = \sigma_i exp(\tau N(0,1)+\tau' N_i(0,1)) \tag{3}$$

$$x'_i = x_i + \sigma'_i N_i(0,1) \tag{4}$$

where $\tau \propto (2n)^{0.5}$ and $\tau' \propto (2n^{0.5})^{0.5}$ (Bäck and Schwefel, 1993). In this manner, the standard deviations are subject to variation and selection at a second level (i.e., the level of how well they guide the search for optima of the functional $F(\mathbf{x})$). This general procedure has also been found effective in addressing discrete optimization problems (Angeline et al., 1996; Chellapilla and Fogel, 1997; and others). Essentially, the effect is much like a temperature schedule in simulated annealing, however, the schedule is set by the algorithm as it explores the state space, rather than *a priori* by a human operator.

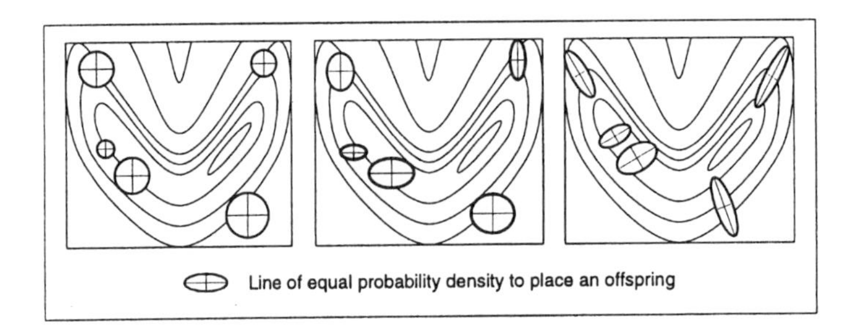

Figure 2.4 When using Gaussian mutations in all dimensions (as in evolution strategies or evolutionary programming), the contours of equal probability density for placing offspring are depicted above (Bäck, 1996). In the left panel, all standard deviations in each dimension are equal, resulting in circular contours. In the middle panel, the standard deviations in each dimension may vary, but the perturbation in each dimension is independent of the others (zero covariance), resulting in elliptical contours. In the right panel, arbitrary covariance is applied, resulting in contours that are rotatable ellipses. The method of self-adaptation described in text can be extended to adapt arbitrary covariances, thereby allowing the evolutionary algorithm to adapt to the changes in the response surface during the search for the optimum position on the surface. Similar procedures have been offered for self-adaptation when solving discrete combinatorial problems.

2.2.8 Able to solve problems that have no known solutions

Perhaps the greatest advantage of evolutionary algorithms comes from the ability to address problems for which there are no human experts. Although human expertise should be used when it is available, it often proves less than adequate for automating problem-solving routines. Troubles with such expert systems are well known: the experts may not agree, may not be self-consistent, may not be qualified, or may simply be in error. Research in artificial intelligence has fragmented into a collection of methods and tricks for solving particular problems in restricted domains of interest. Certainly, these methods have been successfully applied to specific problems (e.g., the chess program Deep Blue). But most of these applications require human expertise. They may be impressively applied to difficult problems requiring great computational speed, but they generally do not advance our understanding of intelligence. "They solve problems, but they do not solve the problem of how to solve problems," (Fogel, 1995, p. 259). In contrast, evolution provides a method for solving the problem of how to solve problems. It is a recapitulation of the scientific method (Fogel et al., 1966) that can be used to learn fundamental aspects of any measurable environment.

2.3 CURRENT DEVELOPMENTS AND APPLICATIONS

The available space permits only a brief discussion of three current developments in evolutionary algorithms. Readers are urged to review Bäck et al. (1997) for a more complete cataloguing of case studies in real-world applications. Here, three examples are presented in medicine, electrical engineering, and robots. For the sake of novice readers, greater detail is provided for the first example than for the others, in the hope that this will facilitate a more comprehensive understanding of the general approach.

2.3.1 Breast cancer detection

Carcinoma of the breast is second only to lung cancer as a tumor-related cause of death in women. There are now more than 180,000 new cases and 45,000 deaths annually in the United States alone (Boring et al., 1993). It begins as a focal curable disease, but it is usually not identifiable by palpation at this stage, and mammography remains the mainstay in effective screening. It has been estimated that the mortality from breast carcinoma could be decreased by as much as 25 percent if all women in the appropriate age groups were regularly screened (Strax, 1989).

Intra- and inter-observer disagreement and inconsistencies in mammographic interpretation (Elmore et al., 1994; Ciccone et al., 1992) have led to an interest in using computerized pattern recognition algorithms, such as artificial neural networks (ANNs) (Haykin, 1994; and others), to assist the radiologist in the assessment of mammograms. ANNs hold promise for improving the accuracy of determining those patients where further assessment and possible biopsy is indicated.

ANNs are computational models based on the neuronal structure of natural organisms. They are stimulus-response transfer functions that map an input space to a specified output space. They are typically used to generalize such an input-output mapping over a set of specific examples. For example, as will be described here, the input can be radiographic features from mammograms, with the output being an indication of the likelihood of a malignancy. Given a network architecture (i.e., type of network, the number of nodes in each layer, the weighted connections between the nodes, and so forth), and a training set of input patterns, the collection of variable weights determines the output of the network to each presented pattern. The error between the actual output of the network and the desired target output defines a potentially multimodal response surface over a multidimensional hyperspace (the dimension is equal to the number of weights).

A commonly employed method for finding weight sets in such applications is error back propagation, which is essentially a gradient method. As such, it is subject to entrapment in locally optimal solutions, and the resulting weight sets are often unsuitable for practical applications. Numerical optimization techniques that do not suffer from such entrapment can be used to advantage in these cases. Evolutionary algorithms offer one such technique.

To provide a specific example, Fogel et al. (1997) used data that were collected by assessing film screen mammograms in light of a set of 12 radiographic features (Table 2.1). These features were assessed in 112 cases of suspicious breast masses, all of which were subsequently examined by open surgical biopsy with the associated pathology indicating whether or not a malignant condition had been found. In all, 63 cases were associated with a biopsy-proven malignancy, while 49 cases were indicated to be negative by biopsy.

Table 2.1 The features and rating system used for assessing mammograms in Fogel et al. (1997). Assessment was made by Eugene Wasson, M.D.

Feature	Rating
Mass size	either zero or in mm
Mass margin (each subparameter rated as none (0), low (1), medium (2), or high (3))	(a) Well circumscribed, (b) Micro-lobulated, (c) Obscured, (d) Indistinct, (e) Spiculated
Architectural distortion	none or distortion
Calcification number	none (0), < 5 (1), 5-10 (2), or > 10 (3)
Calcification morphology	none (0), not suspicious (1), moderately suspicious (2), or highly suspicious (3)
Calcification density	none (0), dense (1), mixed (2), faint (3)
Calcification distribution	none (0), scattered (1), intermediate (2), clustered (3)
Asymmetric density	either zero or in mm

These data were processed using a simple feedforward ANN restricted to two hidden sigmoid nodes (following the maxim of parsimony, this being the simplest architecture that can take advantage of the nonlinear properties of the nodes), with a single linear output node, resulting in 31 adjustable weights. The networks were trained using an evolutionary algorithm in a leave-one-out cross validation procedure. For each complete cross validation where each sample pattern was held out for testing and then replaced in a series of 112 separate training procedures, a population of 250 networks of the chosen architecture were initialized at random by sampling weight values from a uniform random variable distributed over [-0.5,0.5]. Each weight set (i.e. candidate solution) also incorporated an associated self-adaptive mutational vector used to determine the random variation imposed during the generation of offspring networks (described below). Each of these self-adaptive parameters was initialized to a value of 0.01. Each weight set was evaluated based on how well the ANN classified the 111 available training patterns, where a diagnosis of malignancy was assigned a target value of 1.0 and a benign condition was assigned a target of 0.0. The performance of each network was determined as the sum of the squared error between the output and the target value taken over the 111 available patterns.

After evaluating all existing (parent) networks, the 250 weight sets were used to create 250 offspring sets (one offspring per parent). This was accomplished in a two-step procedure. For each parent, the self-adaptive parameters were updated as:

$$\sigma'_i = \sigma_i \exp\left(\tau N(0,1) + \tau' N_i(0,1)\right) \quad (5)$$

where $\tau = \frac{1}{\sqrt{2n}}$, $\tau = \frac{1}{\sqrt{2\sqrt{n}}}$, $N(0,1)$ is a standard normal random variable sampled once for all 31 parameters of the vector σ, and $N_i(0,1)$ is a standard normal random variable sampled anew for each parameter. The settings for τ and τ' have been demonstrated to be fairly robust (Bäck and Schwefel, 1993). These updated self-adaptive parameters were then used to generate new weight values for the offspring according to the rule:

$$x'_i = x_i + \sigma'_i C \qquad (6)$$

where C is a standard Cauchy random variable (determined as the ratio of two independent standard Gaussian random variables). The Cauchy mutation allows for a significant chance of generating saltations but still provides a reasonable probability that offspring networks will reside in proximity to their parents. All of the offspring weight sets were evaluated in the same manner as their parents.

Selection was applied to eliminate half of the total parent and offspring weight sets based on their observed error performance. A pairwise tournament was conducted where each candidate weight set was compared against a random sample from the population. The sample size was chosen to be 10 (a greater sample size indicates more stringent selection pressure). For each of the 10 comparisons, if the weight set had an associated classification error score that was lower than the randomly sampled opponent it received a "win." After all weight sets had participated in this tournament, those that received the greatest number of wins were retained as parents of the next generation. This process affords a probabilistic selection, allowing for the possibility of climbing up and out of hills and valleys on the error response surface.

This process was iterated for 200 generations, whereupon the best available network as measured by the training performance was used to classify the held-out input feature vector. The result of this classification was recorded (i.e., the output value of the network and the associated target value) and the process was restarted by replacing the held-out vector and removing the next vector in succession until all 111 patterns had been classified. Note that each final classification was made using a network that was not trained on the pattern in question.

Each complete cross validation was repeated 16 times, with different randomly selected populations of initial weights, to determine the reliability of the overall procedure. A typical trade-off of the probability of detection, P(D), and false positive, P(FP), is shown in Figure 2.5. This compares favorably with other efforts to train neural networks for breast cancer detection (e.g.,, Wu et al., 1993; Baker et al., 1995) in that roughly equivalent power to correctly reject negative film screens is attained at the P(D) = 0.95 level using about an order-of-magnitude fewer weights. The use of evolutionary algorithms allows for the discovery of simpler models, and this may be particularly important in medical applications because simpler models may be easier to interpret and explain.

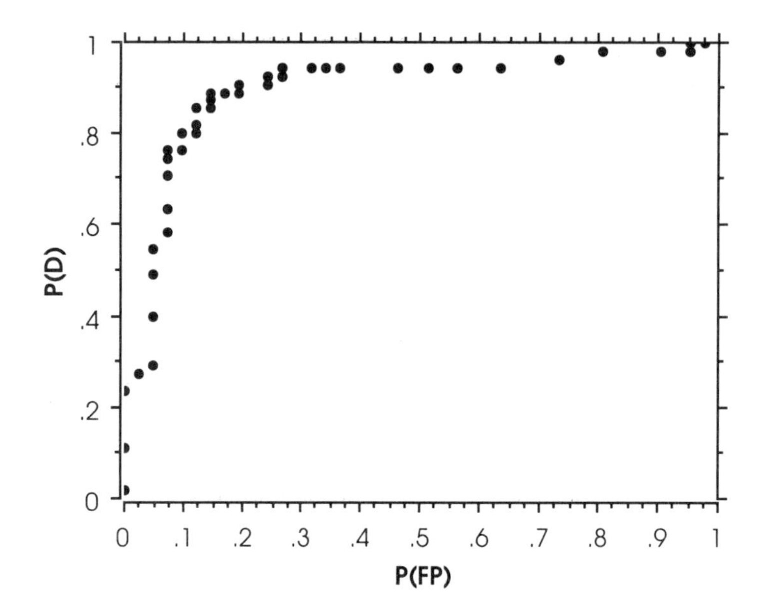

Figure 2.5 A typical ROC curve (raw data) generated in one complete cross validation where each of 112 patterns was classified in turn, based on training over the remaining 111 patterns. Each point represents the probability of detection, P(D), and probability of false positive, P(FP), that is attained as the threshold for classifying a result as malignant is increased systematically over [0,1].

2.3.2 Neuronal control of autonomous agents

Recent attention has been devoted to designing algorithms that allow robots to adapt to changing circumstances on-the-fly (e.g., Xiao et al., 1997; and others). Salomon (1997) has studied the use of evolutionary optimization to adapt the behavior of Khepera robots so as to facilitate navigation around obstacles. The Khepera robots is relatively small (55 mm in diameter and 32 mm in height) and has eight infrared (IR) and ambient light sensors and two motors which can be controlled independently (Figure 2.6). In the studied task, the robot must learn to complete a circuit in an enclosed arena while avoiding bumping walls. The robot is controlled using a "receptive field" neural architecture, in which computational nodes perform Gaussian filters (rather than sigmoid filters as is common in perceptrons). Fitness of a robot trajectory is measured by the smoothness of speed and rotation (i.e., stopping or spinning too rapidly leads to a very small fitness contribution, as does crashing into a wall), as well as the rate of speed to traverse the course (i.e., the faster the better). Salomon (1997) indicated success in evolving the robot's trajectory using very small populations (three parents, six offspring) in an evolution strategy, with a learning time that is more than an order of magnitude improved over previous efforts to use genetic algorithms (Floreano and Mondada, 1996) to accomplish essentially the same task.

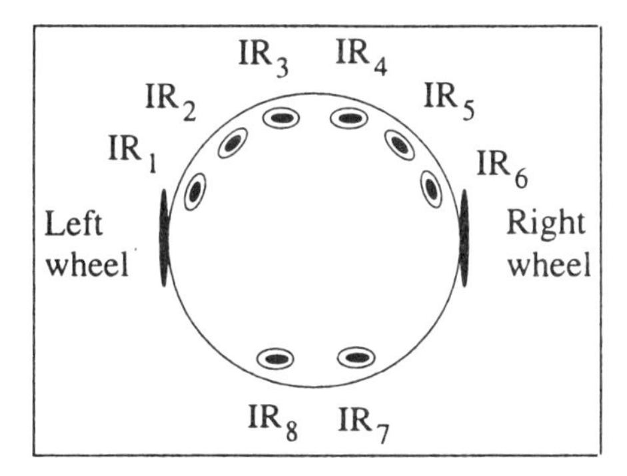

Figure 2.6 The sensor/motor system for the Khepera robot. There are eight IR and ambient light sensors, and two independent motors which can drive the left and right wheels.

2.3.3 Freeway ramp control

Ramp control has long been considered an effective way to reduce freeway congestion (Newman et al., 1969) and its implementation via ramp metering (i.e., regulating the rate at which cars enter a freeway) has become commonplace on many of America's roads. Robinson and Doctor (1989) cited several benefits to ramp metering in a nationwide case study: (1) a decrease in the number of accidents, (2) an increase in the average freeway speed, and (3) reduced travel times in light of suburban growth. Much of the current research in this area has focused on elaborate ramp control strategies that can use information from a centralized traffic coordination control facility. What is needed is a robust, real-time ramp metering strategy that is simple to implement and able to alleviate both recurring and nonrecurring congestion based on the available freeway sensor information.

McDonnell et al. (1995) offered an evolutionary approach to ramp control. The most commonly used sensor-based ramp metering controller is based on the demand-capacity algorithm, which adjusts the ramp metering rates based on the difference between downstream capacity and the sum of mainline and ramp demands. This difference is used in determining the number of vehicles per hour (vph) that should be allowed onto the freeway. For example, if mainline demand is 4800 vph and downstream capacity is 6000 vph, then a metering period of $T = 3$ seconds, or 1200 vph, is the maximum discharge rate that ensure capacity is not exceeded.

In contrast to a system-wide demand-capacity type of metering McDonnell et al. (1995) proposed a rule-based system that operated on both feedback information from the downstream freeway sensors and feedforward information from the upstream freeway sensors. If the upstream freeway sensors indicate an approaching high mainline volume, the ramp meter can reduce the discharge rate so that the downstream capacity is not exceeded. Similarly, high occupancy rates from the downstream freeway sensors may also warrant a reduction in the ramp discharge rate. A simple example of this is given in Figure 2.7, which shows upstream, adjacent, and downstream sensor station. Figure 2.8 provides a list of rules which correspond to the sensors and middle onramp meter of Figure 2.7.

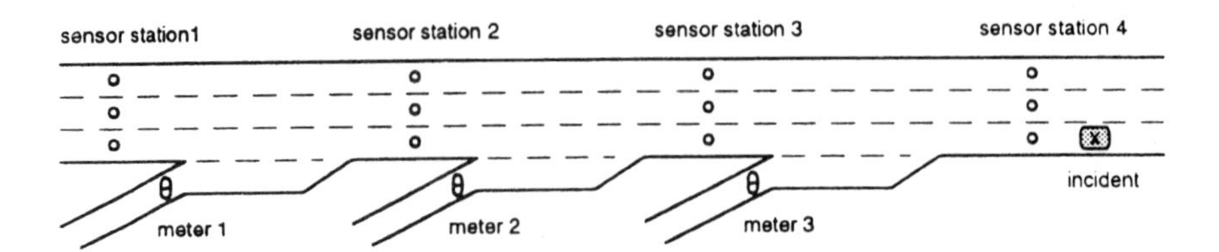

Figure 2.7 The uniform freeway segment used in McDonnell et al. (1995). Traffic moves from left to right. There are loop detectors (sensors) in each lane just prior to each onramp. These provide a measure of the flow rate over the lane. An incident is preprogrammed to occur in the first lane (from the right) after the 4th sensor station.

	Sensor Occupancy Percentages			Headway
Rule	*sensor i-1*	*sensor i*	*sensor i+1*	*meter i*
1	L	L	L	L
2	L	L	M	L
3	L	L	H	M
4	L	M	L	ML
5	L	M	M	M
6	L	M	H	MH
7	L	H	L	H
8	L	H	M	H
9	L	H	H	H
10	M	L	L	L
11	M	L	M	ML
12	M	L	H	M
13	M	M	L	ML
14	M	M	M	M
15	M	M	H	MH
16	M	H	L	H
17	M	H	M	H
18	M	H	H	H
19	H	L	L	ML
20	H	L	M	M
21	H	L	H	M
22	H	M	L	M
23	H	M	M	MH
24	H	M	H	H
25	H	H	L	H
26	H	H	M	H
27	H	H	H	H
28	x	L	L	NM

Figure 2.8 The rules for mapping sensor occupancy rates to metering rates. The occupancy categories are defined as low (L), medium (M), and high (H). The headway levels are defined as low (L), medium-low (ML), medium (M), medium-high (MH), and high (H). These headway levels are assigned reasonably, but arbitrarily. "x" is a wild care or "don't care" condition. Each of the sensor stations (*i*-1, *i*, *i*+1) are assigned as shown in Figure 2.7. Note that rule 28 supersedes rules 1, 10, and 19. The rationale behind using rule 28 is that, should an upstream incident cause low occupancies on the downstream links, then traffic should be let on at a higher rate. Discrimination between a wave of high occupancy traffic and an incident yields a more appropriate implementation of this diversionary tactic. NM refers to no metering.

The average occupancies of each sensor station are converted to linguistic variables according to the linear mapping shown in Figure 2.9. The linguistic variable conditions assigned to the average occupancy rates of the suite of sensor stations are converted to a meter rate using rules such as IF (conditions $C(i\text{-}1)$ and $C(i)$ and $C(i+1)$) THEN (headway = $H(i)$). The metering rates in Figure 2.8 were arbitrarily quantified according the rules:

High Rate = 15 sec

Medium High Rate = 12 sec

Medium Rate = 9 sec

Medium Low Rate = 6 sec

Low Rate = 3 sec

The evolutionary computation problem was to adaptive determine appropriate settings for the thresholds indicated in Figure 2.9.

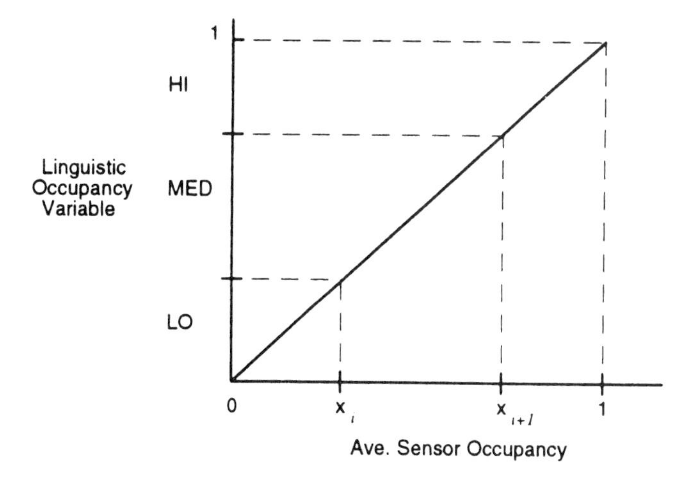

Figure 2.9 The sensor occupancy to linguistic occupancy variable mapping. The threshold values, x_i and x_{i+1}, are determined using evolutionary algorithms.

The measure of effectiveness (MOE) was taken to be the total number of vehicles exiting the system. The freeway used is shown in Figure 2.7 and simulated using INTRAS, a highly detailed traffic simulation program. The location of an incident, when it was simulated, is indicated downstream from sensor station 4. One-mile spacings were used between each ramp with the sensor locations half-way in between. The incident was located roughly 100 meters downstream of sensor station 4. Figure 2.10 shows the number of vehicles that were able to flow through the freeway system after an incident using a variety of rule for metering. As observed, the case of evolved rules allowed the greatest number of vehicles to pass with the first 30 minutes after the incident. Current efforts are aimed at engaging evolutionary algorithms for the real-time control of actual freeway onramps.

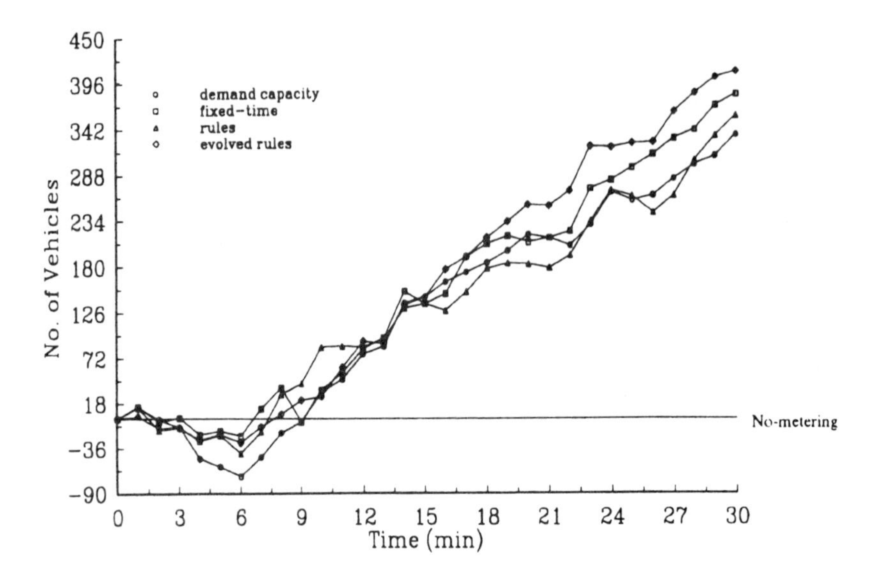

Figure 2.10 Cumulative throughput increase (in the number vehicles) over the 30-minute INTRAS simulation for each ramp control strategy relative to no metering at all when a high demand is made on the freeway. After nine minutes, all of the ramp control strategies yielded a higher vehicle exit count than the no metering strategy, however, the long term best performing strategy was that offered by the evolutionary algorithm. "Demand capacity" refers to the method outlined in the text. "Fixed-time" sets the metering to T = 12 seconds (which yielded the highest performance of any fixed-time procedure). "Rules" sets the thresholds to 0.05 and 0.2, which were subsequently evolved and improved.

2.4 CONCLUSIONS

Although the history of evolutionary computation dates back to the 1950s and 1960s (Fogel, 1998), only within the last decade have evolutionary algorithms become practicable for solving real-world problems on desktop computers (Bäck et al., 1997). As computers continue to deliver accelerated performance, these applications will only become more routine. The flexibility of evolutionary algorithms to address general optimization problems using virtually any reasonable representation and performance index, with variation operators that can be tailored for the problem at hand, and selection mechanisms tuned for the appropriate level of stringency, gives these techniques an advantage over classic numerical optimization procedures. Moreover, the two-step procedure to self-adapt parameters that control the evolutionary search frees the human operator from having to handcraft solutions, which would often be time consuming or simply infeasible. Evolutionary algorithms offer a set of procedures that may be usefully applied to problems that have resisted solution by common techniques, and can be hybridized with such techniques when such combinations appear beneficial.

Acknowledgments

The author would like to thank the conference organizers for inviting this presentation and T. Bäck for permission to reprint his figure from Bäck (1996).

REFERENCES

Angeline, P.J., Fogel, D.B.: An evolutionary program for the identification of dynamical systems. *SPIE Aerosence 97, Symp. on Neural Networks*, S.K. Rogers and D. Ruck (eds.), Vol. 3077 (1997) 409-417.

Angeline, P.J., Fogel, D.B., Fogel, L.J.: A comparison of self-adaptation methods for finite state machines in a dynamic environment. *Evolutionary Programming V*, L.J. Fogel, P.J. Angeline, T. Bäck (eds.), MIT Press, Cambridge, MA (1996) 441-449.

Angeline, P.J., Saunders, G.M., Pollack, J.B.: An evolutionary algorithm that constructs recurrent neural networks. *IEEE Trans. Neural Networks*, **5** (1994) 54-65.

Atmar, W.: The inevitability of evolutionary invention. (1979) unpublished manuscript.

Bäck, T.: *Evolutionary Algorithms in Theory and Practice*, Oxford, NY (1996).

Bäck, T., Fogel, D.B., Michalewicz, Z. (eds.): *Handbook of Evolutionary Computation*, Oxford, NY (1997).

Bäck, T., Schwefel, H.-P.: An overview of evolutionary algorithms for parameter optimization. *Evol. Comp.* **1** (1993) 1-24.

Baker, J.A., Kornguth, P.J., Lo, J.Y., Williford, M.E., and Floyd, C.E.: Breast cancer: prediction with artificial neural networks based on BI-RADS standardized lexicon. *Radiology*, **196** (1995) 817-822.

Boring, C.C., Squires, T.S., and Tong, T.: Cancer statistics. *CA: Cancer Journal for Clinicians*, **43** (1993) 7-26.

Chellapilla, K., Fogel, D.B.: Exploring self-adaptive methods to improve the efficiency of generating approximate solutions to traveling salesman problems using evolutionary programming. *Evolutionary Programming VI,* P.J. Angeline, R.G. Reynolds, J.R. McDonnell, and R. Eberhart (eds.), Springer, Berlin (1997) 361-371.

Ciccone, G., Vineis, P., Frigerio, A., and Segman, N.: Inter-observer and intra-observer variability of mammogram interpretation: a field study. *Eur J. Cancer*, **28A** (1992) 1054-1058.

Davis, L. (ed.): *Handbook of Genetic Algorithms*, Van Nostrand Reinhold, NY (1991).

Elmore, J.G., Wells, C.K., Lee, C.H., Howard, D.H., and Feinstein, A.R.: Variability in radiologists' interpretations of mammograms. *N. England J. Med.*, **331** (1994) 1493-1499.

Floreano, D., and Mondada, F.: Evolution of homing navigation in a real mobile robot. *IEEE Trans. Systems, Man and Cybenetics-Part B*, **26** (1996) 396-407.

Fogel, D.B.: Asymptotic convergence properties of genetic algorithms and evolutionary programming: analysis and experiments. *Cybern. & Syst.*, **25** (1994) 389-407.

Fogel, D.B.:*Evolutionary Computation: Toward a New Philosophy of Machine Intelligence*, IEEE Press, NY (1995).

Fogel, D.B.: A 'correction' to some cart-pole experiments. *Evolutionary Programming VI*, L.J. Fogel, P.J. Angeline, and T. Bäck (eds.), MIT Press, Cambridge, MA (1996) 67-71.

Fogel, D.B. (ed.): *Evolutionary Computation: The Fossil Record*, IEEE Press, Piscataway, NJ (1998)`.

Fogel, D.B., Atmar, J.W.: Comparing genetic operators with Gaussian mutations in simulated evolutionary processes using linear systems. *Biol. Cybern.*, **63** (1990) 111-114.

Fogel, D.B., Fogel, L.J.: Using evolutionary programming to schedule tasks on a suite of heterogeneous computers. *Comp. & Oper. Res.*, **23** (1996) 527-534.

Fogel, D.B., Ghozeil, A.: Using fitness distributions to design more efficient evolutionary computations. *Proc. of 1996 IEEE Conf. on Evol. Comp.*, Keynote Lecture, IEEE Press, NY (1996) 11-19.

Fogel, D.B., Ghozeil, A.: A note on representations and variation operators," *IEEE Trans. Evol. Comp.*, **1** (1997) 159-161.

Fogel, D.B., Stayton, L.C.: On the effectiveness of crossover in simulated evolutionary optimization. *BioSystems*, **32** (1994) 171-182.

Fogel, D.B., Wasson, E.C., Boughton, E.M., and Porto, V.W.: A step toward computer-assisted mammography using evolutionary programming and neural networks. *Cancer Lett.*, **119** (1997) 93-97.

Fogel, L.J., Owens, A.J., Walsh, M.J.: *Artificial Intelligence through Simulated Evolution*, John Wiley, NY (1996).

Gehlhaar, D.K., Verkhivker, G.M., Rejto, P.A., Sherman, C.J., Fogel, D.B., Fogel, L.J., Freer, S.T.: Molecular recognition of the inhibitor AG-1343 by HIV-1 protease: conformationally flexible docking by evolutionary programming. *Chem. & Biol.*, **2** (1995) 317-324.

Goldberg, D.E.: *Genetic Algorithms in Search, Optimization and Machine Learning*, Addison-Wesley, Reading, MA (1989).

Haffner, S.B., Sebald, A.V.: Computer-aided design of fuzzy HVAC controllers using evolutionary programming. *Proc. of the 2nd Ann. Conf. on Evolutionary Programming*, D.B. Fogel and W. Atmar (eds.), Evolutionary Programming Society, La Jolla, CA (1993) 98-107.

Harp, S.A., Samad, T., Guha, A.: Towards the genetic synthesis of neural networks. *Proc. of the 3rd Intern. Conf. on Genetic Algorithms*, J.D. Schaffer (ed.), Morgan Kaufmann, San Mateo, CA (1989) 360-369.

Haykin, S.: *Neural Networks*, Macmillan, NY (1994).

Koza, J.R.:*Genetic Programming*, MIT Press, Cambridge, MA (1992).

McDonnell, J.R., Fogel, D.B., Rindt, C., Recker, W.W., and Fogel, L.J.: Using evolutionary programming to control metering rates on freeway ramps. *Evolutionary Algorithms in*

Management Applications, J. Biethahn and V. Nissen (eds.), Springer, Berlin (1995) 305-327.

Michalewicz, Z.:*Genetic Algorithms + Data Structures = Evolution Programs*, Springer, Berlin (1992).

Michalewicz, Z.:*Genetic Algorithms + Data Structures = Evolution Programs*, 3rd ed., Springer, Berlin (1996).

Newman, L., Dunnet, A., and Meis, G.J.: Freeway ramp control — What it can and cannot do. *Traffic Eng.*, June (1969) 14-21.

Raven, P.H., Johnson, G.B.: *Biology*, Times Mirror, St. Louis, MO (1986).

Reed, J., Toombs, R., Barricelli, N.A.: Simulation of biological evolution and machine learning. *J. Theor. Biol.*, 17 (1967) 319-342.

Robinson, J., and Doctor, M.: Ramp metering: does it really work? *ITE 1989 Compendium of Technical Papers.* (1989) 503-508.

Rosenberg, R.: Simulation of genetic populations with biochemical properties," Ph.D. Dissertation, Univ. of Michigan, Ann Arbor (1967).

Rudolph, G.: Convergence analysis of canonical genetic algorithms. *IEEE Trans. Neural Networks*, **5** (1994) 96-101.

Salomon, R.: Scaling behavior of the evolution strategy when evolving neuronal control architectures for autonomous agents. *Evolutionary Programming VI*, P.J. Angeline, R.G. Reynolds, J.R. McDonnell, and R. Eberhart (Eds.), Springer, Berlin (1997) 47-57.

Saravanan, N., Fogel, D.B.: Evolving neurocontrollers using evolutionary programming. *IEEE Conf. on Evol. Comp.*, Vol. 1, IEEE Press, Piscataway, NJ (1994) 217-222.

Schwefel, H.-P.: *Evolution and Optimum Seeking*, John Wiley, NY (1995).

Strax, P.: *Make Sure that You do not have Breast Cancer.* St. Martin's, NY (1989).

Wieland, A.P.: Evolving controls for unstable systems. *Connectionist Models: Proceedings of the 1990 Summer School*, D.S. Touretzky, J.L. Elman, T.J. Sejnowski, and G.E. Hinton (eds.), Morgan Kaufmann, San Mateo, CA (1990) 91-102.

Wilson, S.W.: Classifier fitness based on accuracy. *Evol. Comp.*, **3** (1995) 149-175.

Wolpert, D.H., Macready, W.G.: No free lunch theorems for optimization. *IEEE Trans. Evol. Comp.*, **1** (1997) 67-82.

Wright, S.: The roles of mutation, inbreeding, crossbreeding, and selection in evolution. *Proc. 6th Int. Cong. Genetics*, Vol. 1, Ithaca (1932) 356-366.

Wu, Y., Giger, M.L., Doi, K., Vyborny, C.J., Schmidt, R.A., and Metz, C.E.: Application of neural networks in mammography: applications in decision making in the diagnosis of breast cancer. *Radiology*, **187** (1993) 81-87.

Xiao, J., Michalewicz, Z., Zhang, L., and Torjanowski, K.: Adaptive evolutionary planner/navigator for mobile robots. *IEEE Trans. Evolutionary Computation*, **1** (1997) 18-28.

3 Evolutionary Computation: Recent Developments and Open Issues

K. DE JONG

Computer Science Department
George Mason University
Fairfax, VA 22030 USA
E-mail: kdejong@gmu.edu

3.1 INTRODUCTION

The field of evolutionary computation (EC) is in a stage of tremendous growth as witnessed by the increasing number of conferences, workshops, and papers in the area as well as the emergence of several journals dedicated to the field. It is becoming increasingly difficult to keep track of and understand the wide variety of new algorithms and new applications.

I believe there is, however, there is a coherent structure to the EC field that can help us understand where we are and where we're headed. The purpose of this paper is to present that view, use it to assess the field, and then discuss important new directions for research and applications.

3.2 THE HISTORICAL ROOTS OF EC

One useful way of understanding where we are is to understand how we got here, i.e., our historical roots. In the case of evolutionary computation, there have been three historical paradigms that have served as the basis for much of the activity in the field: genetic algorithms (GAs), evolution strategies (ESs), and evolutionary programming (EP).

GAs owe their name to an early emphasis on representing and manipulating individuals in terms of their genetic makeup rather than using a phenotypic representation. Much of the early work used a universal internal representation involving fixed-length binary strings with "genetic" operators like mutation and crossover defined to operate in a domain-independent fashion at this level without any knowledge of the phenotypic interpretation of the strings [28], [11]. This universality was also reflected in a strong emphasis on the design of robust adaptive systems with a broad range of applications. Equally important was the early emphasis on theoretical analysis resulting in "the schema theorem" and characterizations of the role and importance of crossover.

By contrast, ESs were developed with a strong focus on building systems capable of solving difficult real-valued parameter optimization problems [38], [43]. The "natural" representation was a vector of real-valued "genes" which were manipulated primarily by mutation operators designed to perturb the real-valued parameters in useful ways. Analysis played a strong role here as well with initial theorems on convergence to global optima, rates of convergence, and other ES properties such as the "1/5" rule.

Universality was also a central theme of the early work on EP. The direction this took was the idea of representing individuals phenotypically as finite state machines capable of responding to environmental stimuli, and developing operators (primarily mutation) for effecting structural and behavioral change over time [19]. These ideas were then applied to a broad range of problems including prediction problems, optimization, and machine learning.

These early characterizations, however, are no longer all that useful in describing the enormous variety of current activities on the field. GA practitioners are seldom constrained to universal fixed-length binary implementations. ES practitioners have incorporated recombination operators into their systems. EP is used for much more than just the evolution of finite state machines. Entire new subareas such as genetic programming [32] have developed. The literature is filled with provocative new terms and ideas such as "messy GAs" [22].

As a consequence, labels such as GA, EP, and ES are not all that helpful anymore in understanding where the field is today and where it's headed. In fact, in some cases the labels can actually be a hindrance, such as engaging in a debate as to whether a GA that uses a real-valued representation is still a GA.

My opinion is that we can make much better sense of things if we focus on the basic elements common to all evolutionary algorithms (EAs), and use that to understand particular differences in approaches and to help expose interesting new directions. I'll try to do so in the remainder of this paper.

3.3 BASIC EA COMPONENTS

If I am asked what are the basic components of an evolutionary algorithm such that, if I discarded any one of them, I would no longer consider it "evolutionary", my answer is:

- A population of individuals.
- A notion of fitness.
- A notion of population dynamics (births, deaths) biased by fitness.
- A notion of inheritance of properties from parent to child.

This, of course, could just as well describe any evolutionary system. In the case of *evolutionary computation*, our goal is to use such algorithms to solve problems, and so there are additional aspects involving the problem domain and how one chooses to use an EA to solve such problems. We explore these basic components in more detail in the following subsections, noting differences in approaches, and identifying some important unanswered questions.

3.3.1 Modeling the dynamics of population evolution

At a high level of abstraction we think of evolutionary processes in terms of the ability of more fit individuals to have a stronger influence on the future makeup of the population by surviving longer and by producing more offspring which continue to assert influence after the parents have disappeared. How these notions are turned into computational models varies quite dramatically within the EC community. This variance hinges on several important design decisions discussed briefly in the following subsections.

3.3.1.1 Choosing population sizes Most current EAs assume a constant population size N which is specified as a user-controlled input parameter. So called "steady state" EAs rigidly enforce this limit in the sense that each time an offspring is produced resulting in $N + 1$ individuals, a selection process is invoked to reduce the population size back to N. By contrast, "generational" EAs permit more elasticity in the population size by allowing $K \gg 1$ offspring to be produced before a selection process is invoked to delete K individuals.

Although we understand that the size of an EA's population can affect its ability to solve problems, we have only the beginnings of a theory strong enough to provide *a priori* guidance in choosing an appropriate fixed size (e.g., [21]), not much theory regarding appropriate levels of elasticity (K), and even less understanding as to the merits of dynamically adjusting the population size.

3.3.1.2 Deletion strategies The processes used to delete individuals varies significantly from one EA to another and includes strategies such as uniform random deletion, deletion of the K worst, and inverse fitness-proportional deletion. It is clear that "elitist" deletion strategies which are too strongly biased towards removing the worst can lead to premature loss of diversity and suboptimal solutions. It is equally clear that too little fitness bias results in unfocused and meandering search. Finding a proper balance is important but difficult to determine *a priori* with current theory.

3.3.1.3 Parental selection Similar issues arise with respect to choosing which parents will produce offspring. Biasing the selection too strongly towards the best individuals results in too narrow a search focus, while too little bias produces a lack of needed focus. Current methods include uniform random selection, rank-proportional selection, and fitness-proportional selection.

We understand these selection strategies in isolation quite well [2], [5]. However, it is clear that parental selection and individual deletion strategies must complement each other in terms of the overall effect they have on the exploration/exploitation balance. We have some theory here for particular cases such as Holland's "optimal allocation of trials" characterization of traditional GAs [28], and the "1/5" rule for ESs [38], but much stronger results are needed.

3.3.1.4 Reproduction and inheritance In addition to these selection processes, the mechanisms used for reproduction also affect the balance between exploration and exploitation. At one extreme one can imagine a system in which offspring are exact replicas of parents (asexual reproduction with no mutation) resulting in rapid growth in the proportions of the best individuals in the population, but with no exploration beyond the initial population members. At the other extreme, one can imagine a system in which the offspring have little resemblance to their parents, maximizing exploration at the expense of inheriting useful parental characteristics.

The EC community has focused primarily on two reproductive mechanisms which fall in between these two extremes: 1-parent reproduction with mutation and 2-parent reproduction with recombination and mutation. Historically, the EP and ES communities have emphasized the former while the GA community has emphasized the latter.

However, these traditional views are breaking down rapidly. The ES community has found recombination to be useful, particularly in evolving adaptive mutation rates [3]. Various members of the GA community have reported improved results by not using recombination [10], by not using mutation [32], or by adding new and more powerful mutation operators [14]. More recently the virtues of N-parent recombination ($N > 2$) have been explored [13].

As before, we have the tantalizing beginnings of a theory to help understand and guide the use and further development of reproductive mechanisms. Beginning with Holland's initial work, the GA community has analyzed in considerable detail the role of crossover and mutation (see, for example, [24], [50], [7], [12]. The ES community has developed theoretical models for optimal mutation rates with respect to convergence and convergence rates in the context of function optimization [44].

However, the rapid growth of the field is pressing these theories hard with "anomalous results" [20] and new directions not covered by current theory. One of the important issues not well understood is the benefit of adaptive reproductive operators. There are now a variety of empirical studies which show the effectiveness of adaptive mutation rates (e.g., [15], [3], or [18]) as well as adaptive recombination mechanisms (e.g., [41] or [8]).

3.3.2 Choice of representation

One of the most critical decisions made in applying evolutionary techniques to a particular class of problems is the specification of the space to be explored by an EA. This is accomplished by defining a mapping between points in the problem space and points in an internal representation space.

The EC community differs widely on opinions and strategies for selecting appropriate representations, ranging from universal binary encodings to problem-specific encodings for TSP problems and real-valued parameter optimization problems. The tradeoffs are fairly obvious in that universal encodings have a much broader range of applicability, but are frequently outperformed by problem-specific representations which require extra effort to implement and exploit additional knowledge about a particular problem class (see, for example, [34]).

Although there are strong historical associations between GAs and binary string representations, between ESs and vectors of real numbers, and between EP and finite state machines, it is now quite common to use representations other than the traditional ones in order to effectively evolve more complex objects such as symbolic rules, Lisp code, or neural networks. Claiming one EA approach is better than another on a particular class of problems is not meaningful any more without motivating and specifying (among other things) the representations chosen.

What is needed, but has been difficult to obtain, are theoretical results on representation theory. Holland's schema analysis [28] and Radcliffe's generalization to formae [37] are examples of how theory can help guide representation choices. Similarly "fitness correlation" [33] and operator-oriented views of internal fitness landscapes [31] emphasize the tightly coupled interaction between choosing a representation for the fitness landscape and the operators used to explore it. Clearly, much more work is required if effective representations are to be easily selectable.

3.3.3 Characteristics of fitness landscapes

The majority of the EC work to date has been with problem domains in which the fitness landscape is time-invariant and the fitness of individuals can be computed independently from other members of the current population. This is a direct result of the pervasiveness of optimization problems and the usefulness of evolutionary algorithms (EAs) in solving them. This has led to considerable insight into the behavior of EAs on such surfaces including such notions as "GA-easy", "GA-hard", and "deception".

Much of this work has involved optimization problems that are unconstrained or lightly constrained (e.g., upper and lower bounds on the variables). The situation becomes more difficult as the complexity of the constraints increases. The ability to exploit constraint knowledge is frequently the key to successful applications, and that in turn can imply creative, non-standard representations and operators [35]. How to do this effectively is still an interesting and open research issue.

Things become even more interesting and open ended if we attack problem classes in which the fitness landscape varies over time. There are at least three important problem classes of this type for which research results are badly needed: autonomously changing landscapes, the evolution of cooperative behavior, and ecological problems.

Problems involving autonomously changing landscapes frequently arise when fitness is defined in terms of one or more autonomous entities in the environment whose behavior can change independently of any of the search activity of an EA. Typical examples are mechanical devices which age, breakdown, etc, or changes in weather patterns which dramatically change the "fitness" of a particular ship on the open sea. If we apply typical optimization-oriented EAs to such problems, the strong pressures to converge generally result in a loss of the population diversity needed to respond to such changes. We currently have very little insight regarding how to design EAs for such problems.

Rule learning systems [29], [25], iterated prisoner's dilemma problems [1], [17], and immune system models [16] are examples of problems in which fitness is a function of how well an individual complements other individuals in the population. Rather than searching for a single optimal individual, the goal is to evolve groups of individuals (generalists, specialists, etc.) which collectively solve a particular problem.

If we apply typical optimization-oriented EAs to such problems, the strong pressures towards homogeneity in the population make it difficult to maintain different but cooperative individuals. Additional mechanisms for rewarding groups of individuals seem to be required (e.g., bucket brigades, profit sharing), but we have little in the way of theory to guide us.

Ecology-oriented problems present a third and perhaps most difficult class of landscapes in which the shape of the fitness landscape is directly affected by the evolutionary process itself. Perhaps a better way to think of this is in co-evolutionary terms in which multiple interacting evolutionary processes are at work modeling the availability of resources [30], prey-predator relationships, host-parasite interactions [27], and so on. Very few of our insights from the optimization world appear to carry over here.

The interest in using EAs to solve problems like these which violate traditional assumptions continues to grow. We already have examples of EAs which are are powerful function optimizers, but which are completely ineffective for evolving cooperative behavior or tracking a changing landscape. Modified EAs are now being developed for these new problem classes, but are also much less useful as traditional optimizers. These developments have created both the need and the opportunity to gain a deeper understanding of the behavior of EAs.

3.4 NEW AND IMPORTANT DIRECTIONS FOR EC RESEARCH

In the previous section, we summarized the current state of the art with respect to fundamental EC issues and indicated where additional research on these issues is required. In this section, we discuss some more speculative areas which are likely to play an important role in the near future.

3.4.1 Representation and morphogenesis

In the earlier section on representation issues we discussed the tradeoffs between problem-independent and problem-specific representations. Closely related to this is the biological distinction between the more universal genotypic descriptions of individuals in the form of plans for generating them and the phenotypic descriptions of the actual generated structures.

Historically, much of the EA work has involved the evolution of fairly simple structures could be represented in phenotypic form or be easily mapped onto simple genotypic representations. However, as we attempt to evolve increasingly more complex structures such as Lisp code [32] or neural networks [10], it becomes increasingly difficult to define forms of mutation and recombination which are capable of producing structurally sound and interesting new individuals. If we look to nature for inspiration, we don't see many evolutionary operators at the phenotype level (e.g., swapping arms and legs!). Rather, changes occur at the genotype level and the effects of those changes instantiated via growth and maturation. If we hope to evolve such complexity, we may need to adopt more universal encodings coupled with a process of morphogenesis (e.g., [9] or [26]).

3.4.2 Inclusion of Lamarckian properties

Although EAs may be inspired by biological systems, many interesting properties arise when we include features not available to those systems. One common example is the inclusion of Lamarckian operators, which allow the inheritance of characteristics acquired during the lifetime of an individual.

In the EC world this is beginning to show up in the form of hybrid systems in which individuals themselves go through a learning and/or adaptation phase as part of their fitness evaluation, and the results of that adaptation are passed on to their offspring (e.g., see Grefenstette (1991)).

Although initial empirical results are encouraging, we presently have no good way of analyzing such systems at a more abstract level.

3.4.3 Non-random mating and speciation

Currently, most EAs incorporate a random mating scheme in which the species or sex of an individual is not relevant. One problem with this, as with real biological systems, is that the offspring of parents from two species are often not viable. As we move to more complex systems which attempt to evolve cooperating behavior and which may have more than one evolutionary process active simultaneously, the roles of non-random mating and speciation will become an important issue.

Some solutions to these problems have been suggested, such as crowding [11], sharing [23], and tagging [6]. Unfortunately, these solutions tend to make fairly strong assumptions, such as the number of species and/or the distribution of niches in the environment. For some problems these assumptions are reasonable. However, in many cases such properties are not known *a priori* and must evolve as well [47].

3.4.4 Decentralized, highly parallel models

Because of the natural parallelism within an EA, much recent work has concentrated on the implementation of EAs on both fine and coarse grained parallel machines. Clearly, such implementations hold promise of significant decreases in the execution time of EAs.

More interestingly, though, for the topic of this paper, are the evolutionary effects that can be naturally implemented with parallel machines, namely, speciation, nicheing, and punctuated equilibria. For example, non-random mating may be easily implemented by enforcing parents to be neighbors with respect to the topology of the parallel architecture. Species emerge as local neighborhoods within that topology. Subpopulations in equilibrium are "punctuated" by easily implemented migration patterns from neighboring subpopulations.

However, each such change to an EA significantly changes its semantics and the resulting behavior. Our admittedly weak theory about traditional EAs needs to be strengthened and extended to help us in better understanding and designing these parallel implementations. In the case of finely grained, neighborhood models some significant progress is being made along these lines [40].

3.4.5 Self-adapting systems

Another theme that has been arising with increasing frequency is the inclusion of self-adapting mechanisms with EAs to control parameters involving the internal representation, mutation, recombination, and population size. This trend is due in part to the absence of strong predictive theories which specify such things *a priori*. It is also a reflection of the fact that EAs are being applied to more complex and time-varying fitness landscapes.

Some important issues that need to be solved involve the self-adaptation mechanism itself. For example, do we use an EA or some other mechanism? If we use an EA, how do we use fitness as a performance feedback for self-adaptation?

On a positive note, the EC community has already empirically illustrated the viability of self-adaptation of mutation and recombination as noted earlier, as well as adaptive representations like Argot [45], messy GAs [22], dynamic parameter encoding schemes [42], and Delta coding [51]. Recent work of Turner [49] suggests that simple performance-based mechanisms can be effectively used to dynamically tune parent selection and operator usage.

3.4.6 Coevolutionary systems

Hillis' work [27] on the improvements achievable by co-evolving parasites along with the actual individuals of interest gives an exciting glimpse of the behavioral complexity and power of such techniques. Holland's Echo system [30] reflects an even more complex ecological setting with renewable resources and predators. More recently, Rosin [39] and Potter [36] have shown the benefits of both "competitive" and "cooperative" co-evolutionary models.

Each of these systems suggests an important future role for co-evolution in EAs, but they raise more questions than they answer concerning a principled method for designing such systems as well as the kinds of problems for which this additional level of complexity is both necessary and effective.

3.4.7 Theory

One of the most frustrating aspects of evolutionary computation is the difficulty in producing strong theoretical results. EAs are, in general, stochastic non-linear systems for which we have only a limited set of mathematical tools. Progress continues to be made, but slowly, on several fronts [4] including the use of Markov models [48] and tools from statistical mechanics [46]. However, there is still a big gap between the simplifying assumptions needed for the analyses and the complexity of the EAs in actual use.

3.5 SUMMARY AND CONCLUSIONS

This is an exciting time for the EC field. The increased level of EC activity has resulted in an infusion of new ideas and applications which are challenging old tenets and moving the field well beyond its historical roots. As a result of this rapidly changing EC landscape, a new characterization of the field is required based on core issues and important new directions to be explored.

We have attempted to articulate this new view by summarizing the current state of the field, and also pointing out important open issues which need further research. We believe that a view of this sort is an important and necessary part of the continued growth of the field.

REFERENCES

1. R. Axelrod. The evolution of strategies in the iterated prisoner's dilemma. In L. Davis, editor, *Genetic Algorithms and Simulated Annealing*, pages 32–41. Morgan Kaufmann, 1987.

2. T. Bäck. Generalized convergence models for tournament and (μ, λ) selection. In L. Eshelman, editor, *Proceedings of the Sixth International Conference on Genetic Algorithms*, pages 2–9. Morgan Kaufmann, 1995.

3. T. Bäck and H.-P. Schwefel. An overview of evolutionary algorithms for parameter optimization. *Evolutionary Computation*, 1(1):1–23, 1993.

4. R. Belew and M. Vose, editors. *Foundations of Genetic Algorithms 4*. Morgan Kaufmann, 1996.

5. T. Blickle and L. Thiele. A mathematical analysis of tournament selection. In L. Eshelman, editor, *Proceedings of the Sixth International Conference on Genetic Algorithms*, pages 9–16. Morgan Kaufmann, 1995.

6. L. Booker. *Intelligent Behavior as an adaptation to the task environment*. PhD thesis, University of Michigan, Ann Arbor, 1982.

7. L. Booker. Recombination distributions for genetic algorithms. In D. Whitley, editor, *Foundations of Genetic Algorithms 2*, pages 29–44. Morgan Kaufmann, 1992.

8. L. Davis. Adapting operator probabilities in genetic algorithms. In J. D. Schaffer, editor, *Proceedings of the Third International Conference on Genetic Algorithms*, pages 60–69. Morgan Kaufmann, 1989.

9. R. Dawkins. *The Blind Watchmaker*. W. W. Norton, 1987.

10. H. de Garis. Genetic programming: modular evolution for darwin machines. In *Proceedings of the International Joint Conference on Neural Networks*, pages 194–197. Lawrence Erlbaum, 1990.

11. K. De Jong. *Analysis of the behavior of a class of genetic adaptive systems*. PhD thesis, University of Michigan, Ann Arbor, 1975.

12. K. De Jong and W. Spears. A formal analysis of the role of multi-point crossover in genetic algorithms. *Annals of Mathematics and Artificial Intelligence*, 5(1):1–26, 1992.

13. G. Eiben. Multi-parent's niche: N-ary crossovers on nk-landscapes. In *Proceedings of the Fourth International Conference on Parallel Problem Solving from Nature*, pages 319–335. Springer Verlag, 1996.

14. L. Eshelman and D. Schaffer. Preventing premature convergence in genetic algorithms by preventing incest. In R. K. Belew and L. B. Booker, editors, *Proceedings of the Fourth International Conference on Genetic Algorithms*, pages 115–122. Morgan Kaufmann, 1991.

15. T. Fogarty. Varying the probability of mutation in the genetic algorithm. In J. D. Schaffer, editor, *Proceedings of the Third International Conference on Genetic Algorithms*, pages 104–109. Morgan Kaufmann, 1989.

16. D. Fogel. Using genetic algorithms to explore pattern recognition in the immune system. *Evolutionary Computation*, 1(3):191–212, 1993.

17. D. Fogel. On the relationship between the duration of an encounter and the evolution of cooperation in the iterated prisoner's dilemma. *Evolutionary Computation*, 3(3):349–363, 1995.

18. David B. Fogel. *Evolutionary Computation*. IEEE Press, 1995.

19. L. Fogel, A. Owens, and M. Walsh. *Artificial intelligence through simulated evolution*. John Wiley, 1966.

20. S. Forrest and M. Mitchell. Relative building block fitness and the building block hypothesis. In D. Whitley, editor, *Foundations of Genetic Algorithms 2*, pages 109–126. Morgan Kaufmann, 1992.

21. D. Goldberg, K. Deb, and J. Clark. Accounting for noise in sizing of populations. In D. Whitley, editor, *Foundations of Genetic Algorithms 2*, pages 127–140. Morgan Kaufmann, 1992.

22. D. Goldberg, K. Deb, and B. Korb. Don't worry, be messy. In R. K. Belew and L. B. Booker, editors, *Proceedings of the Fourth International Conference on Genetic Algorithms*, pages 24–30. Morgan Kaufmann, 1991.

23. D. Goldberg and J. Richardson. Genetic algorithms with sharing for multimodal function optimization. In J. Grefenstette, editor, *Proceedings of the Second International Conference on Genetic Algorithms*, pages 41–49. Lawrence Erlbaum, 1987.

24. David E. Goldberg. *Genetic Algorithms in Search, Optimization, and Machine Learning*. Addison-Wesley, 1989.

25. J. Grefenstette, C. Ramsey, and A. Schultz. Learning sequential decision rules using simulation models and competition. *Machine Learning*, 5(4):355–381, 1990.

26. S. Harp, T. Samad, and A. Guha. Towards the genetic synthesis of neural networks. In J. D. Schaffer, editor, *Proceedings of the Third International Conference on Genetic Algorithms*, pages 360–369. Morgan Kaufmann, 1989.

27. D. Hillis. Co-evolving parasites improve simulated evolution as an optimization procedure. *Physica D*, 42:228–234, 1990.

28. J. Holland. *Adaptation in natural and artificial systems*. University of Michigan Press, 1975.

29. J. Holland. Escaping brittleness: The possibilities of general-purpose learning algorithms applied to parallel rule-based systems. In R. Michalski, J. Carbonell, and T. Mitchell, editors, *Machine Learning*, volume 2, pages 593–624. Morgan Kaufmann, 1986.

30. J. Holland. *Adaptation in natural and artificial systems, second edition*. MIT Press, 1992.

31. T. Jones. *Evolutionary algorithms, fitness landscapes, and search*. PhD thesis, University of New Mexico, 1995.

32. J. Koza. *Genetic Programming: On the programming of computers by means of natural selection*. Bradford Books, Cambridge, 1992.

33. B. Manderick, M. de Weger, and P. Spiessens. The genetic algorithm and the structure of the fitness landscape. In R. K. Belew and L. B. Booker, editors, *Proceedings of the Fourth International Conference on Genetic Algorithms*, pages 143–150. Morgan Kaufmann, 1991.

34. Z. Michalewicz. *Genetic Algorithms + Data Structures = Evolution Programs*. Springer-Verlag, 1994.

35. Z. Michalewicz and M. Schoenauer. Evolutionary algorithms for constrained optimization problems. *Evolutionary Computation*, 4(1):1–32, 1996.

36. M. Potter. *The Design and Analysis of a Computational Model of Cooperative Coevolution*. PhD thesis, George Mason University, 1997.

37. N. Radcliffe. Forma analysis and random respectful recombination. In R. K. Belew and L. B. Booker, editors, *Proceedings of the Fourth International Conference on Genetic Algorithms*, pages 222–229. Morgan Kaufmann, 1991.

38. I. Rechenberg. *Evolutionsstrategie: optimierung technischer systeme nach prinzipien der biologischen evolution*. Frommann-Holzboog, Stuttgart, 1973.

39. C. Rosin and R. Belew. Methods for competitive co-evolution: Finding opponents worth beating. In L. Eshelman, editor, *Proceedings of the Sixth International Conference on Genetic Algorithms*, pages 373–380. Morgan Kaufmann, 1995.

40. J. Sarma. *An Analysis of Decentralized and Spatially Distributed Genetic Algorithms*. PhD thesis, George Mason University, 1998.

41. D. Schaffer and A. Morishima. An adaptive crossover mechanism for genetic algorithms. In J. Grefenstette, editor, *Proceedings of the Second International Conference on Genetic Algorithms*, pages 36–40. Morgan Kaufmann, 1987.

42. N. Schraudolph and R. Belew. Dynamic parameter encoding for genetic algorithms. *Machine Learning*, 9(1):9–22, 1992.

43. H.-P. Schwefel. *Numerical optimization of computer models*. John Wiley and Sons, 1981.

44. H.-P. Schwefel. *Evolution and optimum seeking*. Wiley, 1995.

45. C. Shaefer. The argot strategy: adaptive representation genetic optimizer technique. In J. Grefenstette, editor, *Proceedings of the Second International Conference on Genetic Algorithms*, pages 50–58. Lawrence Erlbaum, 1987.

46. J. Shapiro and A. Pruegel-Bennet. Genetic algorithm dynamics in a two-well potential. In M. Vose and D. Whitley, editors, *Foundations of Genetic Algorithms 4*, pages 101–116. Morgan Kaufmann, 1996.

47. W. Spears. Simple subpopulation schemes. In *Proceedings of the Evolutionary Programming Conference*, pages 296–307. World Scientific, 1994.

48. W. Spears. *The Role of Mutation and Recombination in Evolutionary Algorithms*. PhD thesis, George Mason University, 1998.

49. M. Turner. *Performance-based Self-adaptive Evolutionary Behavior*. PhD thesis, George Washington University, 1998.

50. M. Vose and G. Liepins. Schema disruption. In R. Belew and L. Booker, editors, *Proceedings of the Fourth International Conference on Genetic Algorithms*, pages 237–242. Morgan Kaufmann, 1991.

51. D. Whitley, K. Mathias, and P. Fitzhorn. Delta coding: an iterative search strategy for genetic algorithms. In R. K. Belew and L. B. Booker, editors, *Proceeding of the Fourth International Conference on Genetic Algorithms*, pages 77–84. Morgan Kaufmann, 1991.

4 Some Recent Important Foundational Results in Evolutionary Computation

D. B. FOGEL

Natural Selection, Inc.
3333 N. Torrey Pines Ct., Suite 200
La Jolla, CA 92037, USA
E-mail: dfogel@natural-selection.com

Abstract. A review of four foundational results in evolutionary computation is offered. In each case, recent discoveries overturn conventional wisdom. The importance of these new findings is discussed in the context of future progress in the field of evolutionary computation.

4.1 INTRODUCTION

Several important results in evolutionary computation have been developed during the past 5–10 years. This paper addresses four such results that concern foundations of evolutionary computation. In particular, attention is focused on (1) the belief that it is possible to generate superior general problem solvers, (2) the notion that maximizing implicit parallelism is useful, (3) the fundamental nature of the schema theorem, and (4) the correctness of analysis of a two-armed bandit problem that underlies much work in one particular subset of evolutionary computation. In each case, the conventional wisdom from just a decade ago is seen to be either incomplete or incorrect. As a result, we must reconsider our foundations and take an integrated approach to the study of evolutionary computation as a whole.

4.2 NO FREE LUNCH THEOREM

It is natural to ask if there is a best evolutionary algorithm that would always give superior results across the possible range of problems. Is there some choice of variation operators and selection mechanisms that will always outperform all other choices regardless of the problem? Unfortunately, the answer is no: There is no best evolutionary algorithm. In mathematical terms, let an algorithm a be represented as a mapping from previously unvisited sets of points to a single new (i.e., previously unvisited) point in the state space of all possible points (solutions). Let $P\left(d_m^y | f, m, a\right)$ be the conditional probability of obtaining a particular sample d_m when algorithm a is iterated m times on cost function f. Given these preliminaries, Wolpert and Macready (1997) proved the so-called "no free lunch" theorem:

Theorem (No Free Lunch) For any pair of algorithms a_1 and a_2

$$\sum_f P\left(d_m^y | f, m, a_1\right) = \sum_f P\left(d_m^y | f, m, a_2\right)$$

(see Appendix A of Wolpert and Macready, 1997 for the proof; English (1996) showed that a similar no free lunch result holds whenever the values assigned to points are independent and identically distributed random variables). That is, the sum of the conditional probabilities of visiting each point d_m is the same over all cost functions f regardless of the algorithm chosen. The immediate corollary of this theorem is that for any performance measure $\Phi\left(d_m^y\right)$, the average over all f of $P\left(\Phi\left(d_m^y\right) | f, m, a\right)$ is independent of a. In other words, there is no best algorithm, whether or not that algorithm is "evolutionary," and moreover whatever an algorithm gains in performance on one class of problems is necessarily offset by that algorithm's performance on the remaining problems.

This simple theorem has engendered a great deal of controversy in the field of evolutionary computation, and some associated misunderstanding. There has been considerable effort expended in finding the "best" set of parameters and operators for evolutionary algorithms since at least the mid-1970s. These efforts have involved the type of recombination, the probabilities for crossover and mutation, the representation, the population size, and so forth. Most of this research has involved empirical trials on benchmark functions. But the no free lunch theorem essentially dictates that the conclusions made on the basis of such sampling are in the strict mathematical sense limited to only those functions studied. Efforts to find the best crossover rate, the best mutation operator, and so forth, in the absence of restricting attention to a particular class of problems are pointless.

For an algorithm to perform better than even random search (which is simply another algorithm) it must reflect something about the structure of the problem it faces. By consequence, it mismatches the structure of some other problem. Note too that it is not enough to simply identify that a problem has some specific structure associated with it: that structure must be appropriate to the algorithm at hand. Moreover, the structure must be specific. It is not enough to say, as is often heard, "I am concerned only with real-world problems, not all possible problems, and therefore the no free lunch theorem does not apply." What is the structure of "real-world" problems? Indeed, what is a real-world problem? The obvious vague quality of this description is immediately problematic. What constitutes a real-world problem now might not have been a problem at all, say, 100 years ago (e.g., what to watch on television on a Thursday night). Regardless, simply narrowing the domain of possible problems without identifying the correspondence between the set of problems considered and the algorithm at hand does not suffice to claim any advantage for a particular method of problem solving.

One apt example of how the match between an algorithm and the problem can be exploited was offered in De Jong et al. (1995). For a very simple problem of finding the two-bit vector $\mathbf{x}$ that maximizes the function:

$$f(\mathbf{x}) = \text{integer}(\mathbf{x}) + 1$$

where integer($\mathbf{x}$) returns 0 for [00], 1 for [01], 2 for [10], and 3 for [11], De Jong et al. (1995) employed an evolutionary algorithm with (1) one-point crossover at either a probability of 1.0 or 0.0, (2) a constant mutation rate of 0.1, and (3) a population of size five. In this trivial example, it was possible to calculate the exact probability that the global best vector would be contained in the population as a function of the number of generations. In this case, the use of crossover definitely increased the likelihood of discovering the best solution, and mutation alone was better than random search.

In this example the function *f* assigns values {1, 2, 3, 4} to the vectors {[00], [01], [10], [11]}, respectively, but this is not the only way to assign these fitness values to the possible strings. In fact, there are 4! = 24 different permutations that could be used. De Jong et al. (1995) showed that, in this case, the performance obtained with the 24 different permutations fall into three equivalence classes, each containing eight permutations that produce identical probability of success curves. In the second and third equivalence classes the use of crossover was seen to be detrimental to the likelihood of success, and in fact random search outperformed evolutionary search for the first 10–20 generations in the third equivalence class. In the first equivalence class, crossover could combine the second- and third-best vectors to generate the best vector. In the second and third equivalence classes, it could not usefully combine these vectors: the structure of the problem does not match the structure of the crossover operator. Any specific search operator can be rendered superior or inferior simply by changing the structure of the problem.

Intrinsic to every evolutionary algorithm is a representation for manipulating candidate solutions to the problem at hand. The no free lunch theorem establishes that there is no best evolutionary algorithm across all problems. The fundamental result is two-fold: 1) claims that evolutionary algorithms must rely on specific operators to be successful (e.g., the often heard claim that crossover is a necessary component of a successful evolutionary algorithm, as found in say Goldberg [1989]) is provably false, and 2) efforts to make generally superior evolutionary algorithms are misguided.

4.3 COMPUTATIONAL EQUIVALENCE OF REPRESENTATIONS

Holland (1975, p. 71) suggested that alternative representations in an evolutionary algorithm could be compared by calculating the number of schemata processed by the algorithm. In order to maximize intrinsic parallelism, it was recommended that representations should be chosen with the fewest "detectors with a range of many attributes." In other words, alphabets with low cardinality are to be favored because they generate more schemata. For example, six detectors with a range of 10 values can generate one million distinct representations, which is about the same as 20 detectors with a range of 2 values (2^{20} = 1,048,576). But the number of schemata processed is 11^6 (= 1,771,561) vs. 3^{20} (= $3.49 \infty 10^9$).[1] This increased number of schemata was suggested to give a "larger information flow" to reproductive plans such as genetic algorithms. Emphasis was therefore placed on binary representations, this offering the lowest possible cardinality and the greatest number of schemata.

To review, a schema is a template with fixed and variable symbols. Consider a string of symbols from an alphabet **A**. Suppose that some of the components of the string are held fixed while others are free to vary. Following Holland (1975), define a wild card symbol, # $\notin$ **A**, that matches any symbol from **A**. A string with fixed and/or variable symbols defines a schema, which is a set denoted by a string over the union of {#} and the alphabet **A** = {0,1}.

---

[1] The example is from Holland (1975, p. 71). The value for 3^{20} is corrected here.

Consider the schema [01##], which includes [0100], [0101], [0110], and [0111]. Holland (1975, pp. 64–74) offered that every evaluated string actually offers partial information about the expected fitness of all possible schemata in which that string resides. That is, if string [0000] is evaluated to have some fitness, then partial information is also received about the worth of sampling from variations in [####], [0###], [#0##], [#00#], [#0#0], and so forth. This characteristic is termed *intrinsic parallelism* (or *implicit parallelism*), in that through a single sample, information is gained with respect to many schemata.

Antonisse (1989) offered a different interpretation of the wild card symbol # that led to an alternative recommendation regarding the cardinality of a chosen representation. Rather than view the # symbol in a string as a "don't care" character, it can be viewed as indicating all possible subsets of symbols at the particular position. If $\mathbf{A} = \{0, 1, 2\}$, then the schema [000#] would indicate the sets {[0000] [0001]}, {[0000] [0002]}, {[0001] [0002]}, and {[0000] [0001] [0002]} as the # symbol indicates the possibilities of (a) 0 or 1, (b) 0 or 2, (c) 1 or 2, and (d) 0, 1, or 2. When schemata are viewed in this manner, the greater implicit parallelism comes from the use of more, not fewer, symbols.

Fogel and Ghozeil (1997a) showed that, in contrast to the arguments offered in both Holland (1975) and Antonisse (1989), there can be no intrinsic advantage to any choice of cardinality: Given some weak assumptions about the structure of the problem space[2] and representation, equivalent evolutionary algorithms can be generated for any choice of cardinality. Moreover, their theorems indicate that there can be no intrinsic advantage to using any particular two-parent search operator (e.g., a crossover) as there always exists an alternative two-parent operator that performs the equivalent function, regardless of the chosen representation.

The necessary notation and definitions are as follows: Let $T\colon V \rightarrow W$ denote an operator mapping elements from the set V to the set W. If $v \in V$ and $w \in W$ then the operator can be described as $T(v) = w$. If $S\colon W \rightarrow U$, then $S \circ T$ denotes the composition $S(T(v))$. $I_V\colon V \rightarrow V$ is the identity operator. A mapping T that is both one-one and onto is said to be a *bijection* and has a unique inverse (Devlin, 1993, p. 15). Moreover, if T and S are bijections, then the composition $S \circ T$ is also a bijection.

To provide the framework for the required analysis, let $\mathbf{A}$ be the set of all possible structures a_i, $i = 1,\ldots, N$ ($|\mathbf{A}| = N$). Let $x(a_i) \in \mathbf{X}$ be the x-representation of a_i, (i.e., $x\colon \mathbf{A} \rightarrow \mathbf{X}$) and similarly let $y(a_i) \in \mathbf{Y}$ be the y-representation of a_i ($|\mathbf{X}| = |\mathbf{Y}| = N$). Let x and y be bijections. Let an operator h_x be applied to each $x(a_k)$ such that the probability of a new structure from $\mathbf{X}$ is selected according to the random variable X with probability mass function (pmf) $P_{x(a_k)}(X = x(a_i))$, $i = 1, \ldots, N$.[3] Similarly, let an operator h_y be applied to each $y(a_k)$ such that the probability of a new structure from $\mathbf{Y}$ is selected according to the random variable Y with pmf $P_{y(a_k)}(Y = y(a_i))$, $i = 1, \ldots, N$. Thus $h_x\colon \mathbf{X} \rightarrow \mathbf{P}_\mathbf{X}$, $h_y\colon \mathbf{Y} \rightarrow \mathbf{P}_\mathbf{Y}$, where $\mathbf{P}_\mathbf{X}$

[2] Note that the results will hold for each individual instance of a problem space, and are not dependent on an average across a variety of problems (cf. Wolpert and Macready, 1997).

[3] Subscripts are assigned to pmfs throughout the development simply to aid in identifying each function. Each function is best viewed simply as an element of the space of pmfs, and equivalent pmfs may be duplicated with different names.

and $\mathbf{P_Y}$ are the spaces of pmfs specified over the possible x- and y-representations of $\mathbf{A}$. Assume h_x and h_y are bijections.

It is now possible to prove:

Theorem Given $\mathbf{A}$, $\mathbf{X}$, $\mathbf{Y}$, x, y, h_x, h_y, $\mathbf{P_X}$, and $\mathbf{P_Y}$ as provided above, then, for any specified h_x and space $\mathbf{P_X}$, it is possible to specify h_y and $\mathbf{P_Y}$ such that for all $k = 1, ..., N$,

$$P_{y(a_k)}(Y = y(a_i)) = P_{x(a_k)}(X = x(a_i)),\ i = 1, ..., N.$$

Proof: The operators x and h_x are bijections, therefore their inverses x^{-1} and h_x^{-1} exist and are unique. The function $(h_y \circ y \circ x^{-1} \circ h_x^{-1}){:}\mathbf{P_X} \rightarrow \mathbf{P_Y}$ is a bijection. By inspection, for $k = 1, ..., N$, set $P_{y(a_k)}(Y = y(a_i))$ equal to $P_{x(a_k)}(X = x(a_i))$, $i = 1, ..., N$. ◊

This theorem shows that for any evolutionary algorithm operating on solutions under a finite representation and using a search operator that is applied to a single parent to create a single offspring, an alternative search operator operating on the same single parent can be specified for any other finite representation such that the probability mass function describing the likelihood of generating each new solution is exactly equivalent. The theorem is easily extended to algorithms that operate on a population of individuals under the same restrictions for search operators because each individual generates a new offspring independently from every other individual in the extant population. Thus, for these algorithms under these conditions, there can be no intrinsic advantage to any particular choice of representation.

A form of this result can be extended to evolutionary algorithms that incorporate variation operators that use two individuals to create a single offspring (e.g., the crossover operator suggested in Holland [1975, p. 98]). Before proceeding, note the following:

Lemma If operators h_1 and h_2 are bijections mapping $\mathbf{X}$ into $\mathbf{Y}$ then the matrix operator:

$$\mathbf{h} = \begin{pmatrix} h_1 & 0 \\ 0 & h_2 \end{pmatrix}$$

acting on $[x_1\ x_2]^{\mathrm{T}}$ to produce

$$\begin{pmatrix} h_1(x_1) \\ h_2(x_2) \end{pmatrix} = \begin{bmatrix} y_1\ y_2 \end{bmatrix}^{\mathrm{T}}$$

is a bijection.

Proof: Since h_1 and h_2 are bijections, h_1^{-1} and h_2^{-1} exist and are unique. Thus:

$$\mathbf{h}^{-1} = \begin{pmatrix} h_1^{-1} & 0 \\ 0 & h_2^{-1} \end{pmatrix}$$

is the unique inverse of **h**, as $\mathbf{h} \circ \mathbf{h}^{-1} = I_X$. Since **h** is invertible it is a bijection. ◊

Given **A**, **X**, **Y**, x, and y, let $\mathbf{A}^n$ be the set of all possible collections of n structures $a_i \in \mathbf{A}$, with elements denoted by the sets a^n_k, $k = 1, ..., N^n$, allowing for repeated structures ($|\mathbf{A}^n| = N^n$). Let $x^n(a^n_k) \in \mathbf{X}^n$ be the x-representation of a^n_k, and similarly let $y^n(a^n_k) \in \mathbf{Y}^n$ be the y-representation of a^n_k. Let an operator h_x be applied to two elements of $x^n(a^n_k)$, $x(a^n_{k,j_1})$ and $x(a^n_{k,j_2})$, $j_1, j_2 \in \{1, ..., n\}$, for some k, such that a new structure from **X** is selected according to the random variable X with probability mass function (pmf): $P_{x(a^n_{k,j_1}),x(a^n_{k,j_2})}(X = x(a_i)), i = 1, ..., N$. Let an operator h_y be applied to an element of $y^n(a^n_k)$, $y(a^n_{k,j_1})$, and a natural number, $k_2 \in \{1, ..., N\}$, such that a new structure from **Y** is selected according to the random variable Y with pmf $P_{y(a^n_{k,j_1}),k_2}(Y = y(a_i)), i = 1, ..., N$. Thus h_x: $\mathbf{X}^2 \rightarrow \mathbf{P_X}$, h_y: $(\mathbf{Y},\mathbf{N}) \rightarrow \mathbf{P_Y}$, where $\mathbf{P_X}$ and $\mathbf{P_Y}$ are the spaces of pmfs specified over the possible x- and y-representations of **A**, and **N** is the natural numbers $[1,..., N]$. Assume h_x and h_y are bijections. Let an operator θ_x be applied to an element $x^n(a^n_k)$ such that a structure from **X** is selected according to the random variable X_θ with pmf $P_{x^n(a^n_k)}(X_\theta = x(a_i)), i = 1, ..., N$. Similarly, let an operator θ_y be applied to an element $y^n(a^n_k)$ such that a structure from **Y** is selected according to the random variable Y_θ with pmf $P_{y^n(a^n_k)}(Y_\theta = y(a_i)), i = 1, ..., N$. Also let an operator θ_N be applied to an element $y^n(a^n_k)$ such that a number from **N** is selected according to the random variable N_θ with pmf $P_{N;y^n(a^n_k)}(N_\theta = i), i = 1, ..., N$. Thus θ_x: $\mathbf{X}^n \rightarrow \mathbf{P_X}$, θ_y: $\mathbf{Y}^n \rightarrow \mathbf{P_Y}$, and θ_N: $\mathbf{Y}^n \rightarrow \mathbf{P_N}$, where $\mathbf{P_N}$ is the space of pmfs specified over the N natural numbers.

This leads to:

Theorem Given **A**, **X**, **Y**, **N**, x, y, $\mathbf{A}^n$, $\mathbf{X}^n$, $\mathbf{Y}^n$, h_x, h_y, θ_x, θ_y, θ_N, $\mathbf{P_X}$, $\mathbf{P_Y}$, and $\mathbf{P_N}$ as specified above, then:

(1) given any specified θ_x it is possible to specify some θ_y such that for all $k = 1, ..., N^n$,

$$P_{y^n(a^n_k)}(Y_\theta = y(a_i)) = P_{x^n(a^n_k)}(X_\theta = x(a_i)), \; i = 1, ..., N;$$

(2) given any specified θ_x it is possible to specify some θ_N such that for all $k = 1, ..., N^n$,

$$P_{N;y^n(a^n_k)}(N_\theta = i) = P_{x^n(a^n_k)}(X_\theta = x(a_i)), \; i = 1, ..., N;$$

(3) given any specified h_x it is possible to specify h_y such that for all $k = 1, ..., N^n$, and $k_2 = 1, ..., N$,

$$P_{y(a^n_{k,j_1}),k_2}(Y = y(a_i)) = P_{x(a^n_{k,j_1}),x(a^n_{k,j_2})}(X = x(a_i)),\ i = 1, \ldots, N.$$

Proof: Items (1) and (2) follow directly from Theorem 1. With regard to (3), x is a bijection, therefore

$$\mathbf{V_x} = \begin{pmatrix} x(\cdot) & 0 \\ 0 & x(\cdot) \end{pmatrix}$$

is an bijection (Lemma). The operator h_x is a bijection acting on $[x_1\ x_2]^T$, where x_1 and x_2 are elements of $\mathbf{X}$. Thus h_x^{-1} and $\mathbf{V_X}^{-1}$ exist and are unique. Let an operator c be applied to an element of $\mathbf{A}$, a_{k_2}, such that $c(a_{k_2}) = k_2$. Thus c is a bijection. Since y and c are bijections

$$\mathbf{V_y} = \begin{pmatrix} y(\cdot) & 0 \\ 0 & c(\cdot) \end{pmatrix}$$

is a bijection (Lemma). Since h_y is a bijection, $h_y \circ \mathbf{V_Y} \circ \mathbf{V_X}^{-1} \circ h_x^{-1}$: $\mathbf{P_X} \rightarrow \mathbf{P_Y}$ is a bijection. By inspection, for $k = 1, \ldots, N^n$, $k_2 = 1, \ldots, N$, and $j_1, j_2 = 1, \ldots, n$, set

$$P_{y(a^n_{k,j_1}),k_2}(Y = y(a_i)) \text{ equal to } P_{x(a^n_{k,j_1}),x(a^n_{k,j_2})}(X = x(a_i)),\ i = 1, \ldots, N. \quad \Diamond$$

This theorem shows that for evolutionary algorithms operating on candidate solutions under a finite representation and using a search operator that is applied to two parents to create a single offspring, an alternative search operator operating on a single parent and an index number can be specified for any other (or the same) finite representation such that the probability mass function describing the likelihood of generating new solutions is exactly equivalent. Consequently, there is no intrinsic advantage to choosing a binary representation to allow crossover to recombine the greatest number of schemata. Completely equivalent algorithms can be created regardless of the number of schemata afforded by a particular representation. In addition, the *mechanism* of crossover (e.g., as a method for combining schemata) cannot be afforded fundamental importance. For any two-parent crossover operator, there exists another two-parent operator having exactly the same functionality that operates essentially as a one-parent operator indexed by the second parent.

These theorems from Fogel and Ghozeil (1997a) provide an extension of the result offered in Battle and Vose (1993) where it was shown that isomorphisms exist between alternative instances of genetic algorithms for binary representations (i.e., the operations of crossover and mutation on a population under a particular binary representation in light of a fitness function can be mapped equivalently to alternative similar operators for any other binary representation). They also extend the results of Vose and Liepins (1991) and Radcliffe (1992), where it was shown that there can be no general advantage for any particular binary representation of $\mathbf{A}$. Although particular representations and operators may be more computationally tractable or efficient than others in certain cases, or may appeal to the designer's intuition, under the conditions studied here, no choice of representation, or one-point or two-point variation operator, can offer a capability not found in another choice of representation or analogous operator.

The theorems proved hold for finite cardinality of $\mathbf{A}$, however, it is straightforward to extend the scope to include spaces $\mathbf{A}$ with an infinite number of solutions by having a given representation partition **A** into a finite number of sets. It should also be straightforward to extend the application of the theorems to cases involving operators acting on more than two parents. The fundamental result is that there is no intrinsic benefit to maximizing implicit parallelism in an evolutionary algorithm.

4.4 SCHEMA THEOREM IN THE PRESENCE OF RANDOM VARIATION

The traditional method of selection in genetic algorithms (from say Holland [1975]) requires solutions to be reproduced in proportion to their fitness relative to the other solutions in the population (sometimes termed *roulette wheel selection* or *reproduction with emphasis*). That is, if the fitness of the *i*th string in the population at time t, $\mathbf{x}_i^t$, is denoted by $f(\mathbf{x}_i^t)$, then the probability of selecting the *i*th string for reproduction for each available slot in the population at time t+1 is given by:

$$P(\mathbf{x}_i^{t+1}) = \frac{f(\mathbf{x}_i^t)}{\sum_{i=1}^{n} f(\mathbf{x}_i^t)} \tag{1}$$

where there are n members of the population at time t. This procedure requires strictly positive fitness values.

The implementation of proportional selection leads to the well-known variant of the schema theorem (Holland, 1975):

$$E\ P(H,t+1) = P(H,t)\frac{f(H,t)}{\bar{f}_t} \tag{2}$$

where H is a particular schema (the notation of H is used to denote the schema as a hyperplane), $P(H,t)$ is the proportion of strings in the population at time t that are an instance of schema H, $f(H,t)$ is the mean fitness of strings in the population at time t that are an instance of H, $\bar{f}_t$ is the mean fitness of all strings in the population at time t, and $E\ P(H,t+1)$ denotes the expected proportion of strings in the population at time t+1 that will be instances of the schema H. Eq. (2) does not include any effects of recombination or mutation. It only describes the effects of proportional selection on a population in terms of the manner in which schemata are expected to increase or decrease over time.

The fitness associated with a schema depends on which instances of that schema are evaluated. Moreover, in real-world practice, the evaluation of a string will often include some random effects (e.g., observation error, time dependency, uncontrollable exogenous variables). That is, the observed fitness of a schema H (or any element of that schema) may not be described by a constant value, but rather by a random variable with an associated probability density function. Selection operates not on the mean of all possible samples of the schema H, but only on the fitness associated with each observed instance of the schema H in the population. It is therefore of interest to assess the expected allocation of trials to schemata when their observed fitness takes the form of a random variable. Fogel and Ghozeil (1997b) showed that this can result in the introduction of a bias such that the expected sampling from alternative schemata will not be in proportion to their mean fitness.

To simplify analysis, but without loss of generality, consider the task of choosing to replicate a string from one of two disjoint schemata, denoted by H and H'. For the immediate discussion, let the fitness of schema H be described by the random variable S having probability density function (pdf) $f_S(s)$ and likewise let the fitness of H' be given by the random variable S' having pdf $f_{S'}(s')$. The process of determining which schema to choose from, H or H', is conducted as follows:

(i) Samples are drawn from H and H' yielding values s and s' in accordance with S and S', respectively, and the probability of choosing to replicate a string in schema H is given under proportional selection by:

$$P(H) = \frac{s}{s + s'}. \tag{3}$$

Thus the probability of selecting schema H is a random variable:

$$Z = \frac{S}{S + S'} \tag{4}$$

which has a pdf that is positive over the range (0,1) for strictly positive random variables S and S', and zero otherwise.

(ii) The realized value z of the random variable Z as given in Eq. (4) is used to construct a Bernoulli random variable X with probability of choosing H set equal to z. The random variable X takes on a value of 1 if the decision is to sample from schema H, and 0 otherwise.

For n independent decisions of which schema to replicate, the two-step process is repeated n times, with different possible realizations of h and h', and therefore different realizations of z, (z_i, $i = 1,\ldots, n$) and a set of n different Bernoulli random variables X_i, $i = 1,\ldots, n$. Note that each of the n samples from the random variable Z are independent and identically distributed. Through straightforward manipulation, the expected proportion of the n decisions which are sampled from schema H (analogous to the expected proportion of schema H at time t+1 in Eq. (2)) is $E(Z)$. That is, to determine the expected proportion of schema H in n decisions, it suffices to determine the expectation of Z.

To do this requires determining the probability density function for Z, because it is well known that in general, for any two random variables X and Y with well-defined expectations:

$$E\left(\frac{X}{Y}\right) \neq \frac{E(X)}{E(Y)}$$

and therefore, in general:

$$E(Z) \neq \frac{E(S)}{E(S) + E(S')} \tag{5}$$

although it is possible that equality may occur in specific circumstances. The probability density function for Z is dependent on the probability density functions f_S and $f_{S'}$, and can be found by noting that:

$$Z = \frac{S}{S + S'}$$
$$= \frac{1}{1 + \dfrac{S'}{S}}. \tag{6}$$

Therefore, for independent random variables S and S', the probability density function of Z can be calculated using a series of three transforms: 1) determine the density of the random variable $W = \frac{S'}{S}$, 2) determine the density of the random variable $V = 1 + W$, and 3) determine the density of $Z = V^{-1}$. For some cases of discrete random variables S and S', (4) can be determined directly.

For notational convenience, let the random variable $X = S$, let $Y = S'$, and let X and Y be independent. It is of interest to identify any difference between:

$$E(Z) = E\left(\frac{X}{X + Y}\right) \tag{7}$$

and

$$\frac{E(X)}{E(X) + E(Y)} \tag{8}$$

for such a difference indicates that the expected sampling from alternative schemata H and H' will not be in proportion to their average schema fitness (cf. Rana et al., 1996). We have calculated this value for a number of cases. For the sake of space, only three will be presented in some detail here, while the others will be simply be listed for completeness.

Case 1: For X taking the value $\mu_X + \Delta$ with probability p and the value $\mu_X - \Delta$ with probability $1-p$, and Y being a delta function at μ_Y,

$$E(Z) = \frac{\mu_X + \dfrac{\Delta\left(\mu_Y(2p-1) - \Delta\right)}{\left(\mu_X + \mu_Y\right)}}{\left(\mu_X + \mu_Y\right) - \dfrac{\Delta^2}{\left(\mu_X + \mu_Y\right)}}.$$

The expectation of X is

$$E(X) = p\left(\mu_X + \Delta\right) + \left(1 - p\right)\left(\mu_X - \Delta\right)$$
$$= \mu_X + \Delta\left(2p - 1\right)$$

and $E(Y) = \mu_Y$.

For the situation where $\Delta = 0$:

$$E(Z) = \frac{\mu_X}{\mu_X + \mu_Y},$$

however, for $\Delta \neq 0$, Eq. (8) will not in general describe $E(Z)$. For example, for the case where $\mu_Y = 2$, $\mu_X = 1$, and $\Delta = 0.5$ (recall that Δ cannot be greater than μ_X because X is constrained to take on strictly positive values), Fig. 4.1 shows $E(Z)$ as a function of p. Fig. 4.1 also shows the expected proportion of samples from X that would be obtained by the ratio in Eq. (8), indicating that a bias is introduced in the expected sampling for all values of $p \neq 0$ or 1.

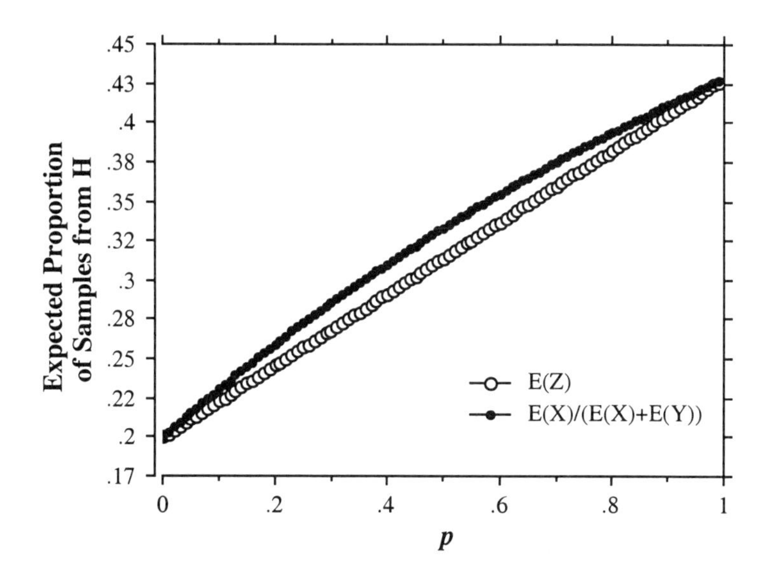

Figure 4.1 The expectation of the random variable $Z = X/(X + Y)$ and the ratio $E(X)/(E(X) + E(Y))$, where X and Y are the random variables describing the probability density function for sampling from schemata H and H', respectively, as a function of p for Case 1 with parameters $\mu_Y = 2$, $\mu_X = 1$, and $\Delta = 0.5$. All values of $p \neq 0$ or 1 introduce a bias in the expected sampling from H.

Case 2: For $X \sim U(0,2\mu_X)$ and $Y \sim U(0,2\mu_Y)$:

$$E(Z) = \frac{1}{2}\left[\frac{\mu_Y}{\mu_X}\ln\left(\frac{\mu_Y}{\mu_X + \mu_Y}\right) - \frac{\mu_X}{\mu_Y}\ln\left(\frac{\mu_X}{\mu_X + \mu_Y}\right) + 1\right]. \tag{9}$$

Figure 4.2 shows $E(Z)$ as a function of μ_Y (denoted by $E(Y)$) for $\mu_X = 1$. The graph indicates that $E(Z)$ is described by Eq. (8) only when $\mu_X = \mu_Y$, as can be confirmed by inspection of Eq. (9). This is a rather uninteresting condition, however, because when $\mu_X = \mu_Y$, the random variables describing the fitness of schema H and schema H' are identical, thus the decision of which schema to allocate more samples to is irrelevant.

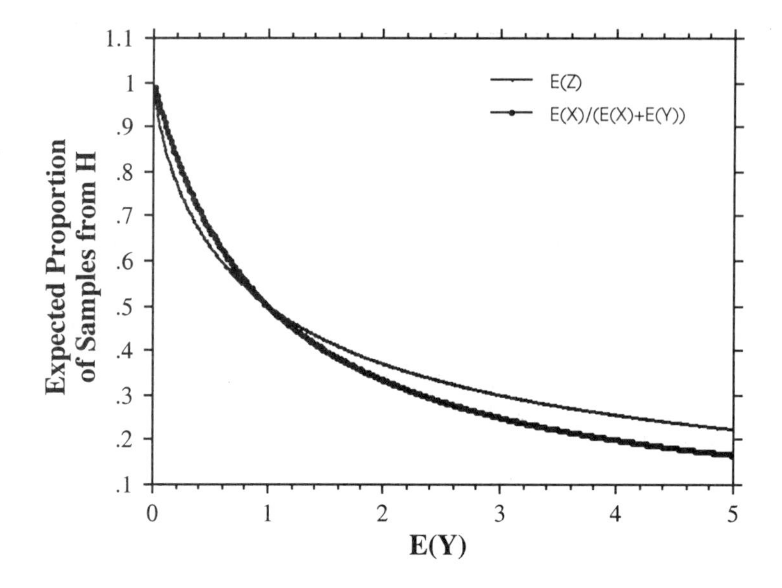

Figure 4.2 The expectation of Z as a function of μ_Y for $\mu_X = 1$ for Case 2 (sampling from uniform random fitness values). When $\mu_Y = \mu_X$ (i.e., $\mu_Y = 1$ here), no difference between $\mu_X/(\mu_X + \mu_Y)$ and $E(Z)$ is introduced, however, a bias is introduced for all other values of μ_Y.

Similar analysis has been conducted for the following cases:

(i) X as a delta function at μ_X, and Y taking the value $\mu_Y + \Delta$ with probability p and $\mu_Y - \Delta$ with probability $1-p$;

(ii) X taking value $\mu_X + \Delta_X$ with probability p_X and the value $\mu_X - \Delta_X$ with probability $1-p_X$, and Y taking value $\mu_Y + \Delta_Y$ with probability p_Y and the value $\mu_Y - \Delta_Y$ with probability $1-p_Y$;

(iii) $X \sim \exp(\lambda_X)$ and $Y \sim \exp(\lambda_Y)$, where exp refers to an exponential random variable;

(iv) $X \sim \gamma(1,c_X)$ and $Y \sim \gamma(1,c_Y)$, where γ refers to a gamma random variable;

(v) $X \sim N(0,1)$ and $Y \sim N(0,1)$, where N refers to a normal (Gaussian) random variable.

For cases (i–iii), as with the two analyzed above, a bias is generally introduced into the expected sampling from H. To provide a guide as to how the results can be derived, the development for case (iii) is provided in the appendix of Fogel and Ghozeil (1997b) and Figure 4.3 shows the relationship between Eqs. (7) and (8). Case (iv) results in a special circumstance where the expectation $E(Z)$ in fact is equal to $E(X)/(E(X) + E(Y))$ and therefore no bias is introduced. Finally, case (v) is admittedly unworkable under proportional selection because it may generate negative values for X or Y, but even if this were ignored (cf. Rana et al., 1996), the resulting density function of Z is Cauchy, and therefore the expectation of sampling from schema H does not exist.

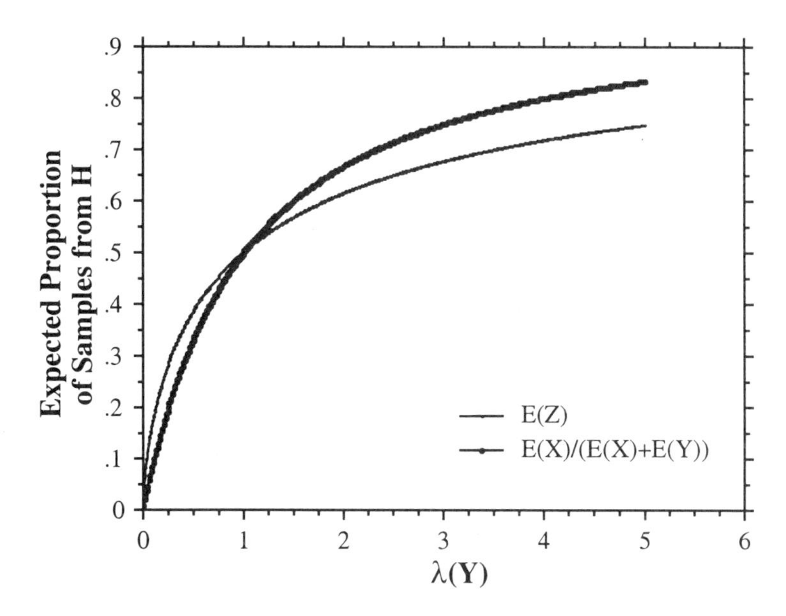

Figure 4.3 The expectation of Z as a function of $\lambda_Y = \mu_Y^{-1}$ for $\mu_X = 1$ for case (iii) (sampling from exponentially distributed random fitness values). When $\lambda_Y = \lambda_X$ (i.e., $\lambda_Y = 1$ here), no difference between $\mu_X/(\mu_X + \mu_Y)$ and $E(Z)$ is introduced, however, a bias is introduced for all other values of λ_Y.

The schema theorem of Holland (1975) refers only to the specific realized values of competing schemata in a population; it is not intended to handle the case when the fitness of these schemata are described by random variables. Under this circumstance, the analysis in Fogel and Ghozeil (1997b) extends a result from population genetics that describes the expected frequency of an allele under random conditions (Gillespie, 1977). It indicates that the expected proportion of a particular schema H in the population at the next time step is not generally governed by the ratio of the mean of that schema H to the sum of the means of schema H and schema H'. For the cases examined, the expression in Eq. (8) described $E(Z)$ only for specific Gamma distributed fitness scores, although it is possible that there are other unusual cases where this may occur. In general, there is no a priori reason to expect the schema theorem to adequately describe the mean sampling of alternative schemata when the fitness evaluation of those schemata is governed by random variation. This may occur in the form of observation noise on any particular string, random selection of individual strings from a particular schema, or a number of other mechanisms. Under such conditions, reliably estimating mean hyperplane performance is not sufficient to predict the expected proportion of samples that will be allocated to a particular hyperplane under proportional selection. The fundamental result is that the schema theorem cannot, in general, be used to reliably predict the average representation of schemata in future populations when schema fitness depends on random factors.

4.5 TWO-ARMED BANDITS AND THE OPTIMAL ALLOCATION OF TRIALS

Recalling the idea of sampling from alternative schemata, in order to optimally allocate trials to schemata, a loss function describing the worth of each trial must be formulated. Holland (1975, pp. 75–83) considered and analyzed the problem of minimizing expected losses while sampling from alternative schemata. The essence of this problem can be modeled as a two-armed bandit (slot machine): Each arm of the bandit is associated with a random payoff described by a mean and variance (much like the result of sampling from a particular schema where the payoff depends on which instance of that schema is sampled). The goal is to best allocate trials to the alternative arms conditioned on the payoffs that have already been observed in previous trials. Holland (1975, pp. 85–87) extended the case of two competing schemata to any number of competing schemata. The results of this analysis were used to guide the formulation of one early version of evolutionary algorithms so it is important to review this analysis and its consequences. This is particularly true because the formulation has recently been shown mathematically to be flawed (Rudolph, 1997; Macready and Wolpert, 1998).

Holland (1975, pp. 75–83) examined the two-armed bandit problem where there are two random variables, RV_1 and RV_2 (representing two slot machines) from which samples may be taken. The two random variables are assumed to be independent and possess some unknown means (μ_1, μ_2) and unknown variances (σ_1^2, σ_2^2). A trial is conducted by sampling from a chosen random variable; the result of the trial is the payoff. Suppose some number of trials has been devoted to each random variable (n_1 and n_2, respectively) and that the average payoff (the sum of the individual payoffs divided by the number of trials) from one random variable is greater than for the other. Let RV_{high} be the random variable for which the greater average payoff has been observed (not necessarily the random variable with the greater mean) and let RV_{low} be the other random variable. The objective in this problem is to allocate trials to each random variable so as to minimize the expected loss function:

$$L(n_1,n_2) = [q(n_1,n_2)n_1 + (1 - q(n_1,n_2))n_2]\cdot|\mu_1 - \mu_2| \tag{10}$$

where $L(n_1,n_2)$ describes the total expected loss from allocating n_1 samples to RV_1 and n_2 samples to RV_2, and

$$q(n_1,n_2) = \begin{array}{l} \Pr(\bar{x}_1 > \bar{x}_2),\ \text{if } \mu_1 < \mu_2 \\ \Pr(\bar{x}_1 < \bar{x}_2),\ \text{if } \mu_1 > \mu_2 \end{array}$$

where $\bar{x}_1$ and $\bar{x}_2$ designate the mean payoffs (sample means) for having allocated n_1 and n_2 samples to RV_1 and RV_2, respectively. Note that specific measurements and the number of samples yield explicit values for the sample means.

As summarized in Goldberg (1989, p. 37), Holland (1975, pp. 77–78) proved the following (notation consistent with Holland, 1975):

Theorem Given N trials to be allocated to two random variables with means $\mu_1 > \mu_2$ and variances σ_1^2 and σ_2^2, respectively, and the expected loss function described in Eq. (10), the minimum expected loss results when the number of trials allocated to the random variable with the lower observed average payoff is

$$n^* \cong b^2 \ln\left[\frac{N^2}{8\pi b^4 \ln N^2}\right] \tag{11}$$

where $b = \sigma_1/(\mu_1 - \mu_2)$. The number of trials to be allocated to the random variable with the higher observed average payoff is $N - n^*$.

If the assumptions associated with the proof (Holland, 1975, pp. 75–83) hold true, and if the mathematical framework were correct, the above analysis would apply equally well to a k-armed bandit as to a two-armed bandit.

Unfortunately, as offered in Macready and Wolpert (1998), the error in Holland (1975, pp. 77–85) stems from considering the unconditioned expected loss for allocating $N - 2n$ trials after allocating n to each bandit, rather than the expected loss conditioned on the information available after the $2n$ pulls. Macready and Wolpert (1998) showed that a simple greedy strategy which pulls the arm that maximizes the payoff for the next pull based on a Bayesian update from prior pulls outperforms the strategy offered in Holland (1975, p. 77). The tradeoff between exploitation (pulling the best Bayesian bandit) and exploration (pulling the other bandit in the hopes that the current Bayesian information is misleading) offered in Holland (1975) is not optimal for the problem studied. Macready and Wolpert (1998) remarked that the algorithms proposed in Holland (1975) "are based on the premise that one should engage in exploration, yet for the very problem invoked to justify genetic exploration, the strategy of not exploring at all performs better than n^* exploring algorithms..."

The ramifications of this condition are important: One of the keystones that has differentiated genetic algorithms from other forms of evolutionary computation has been the notion of optimal schema processing. Until recently, this was believed to have a firm theoretical footing. It now appears, however, that there is no basis for claiming that the n^* sampling offered in Holland (1975) is optimal for minimizing expected losses to competing schemata. In turn, neither is the use of proportional selection optimal for achieving a proper allocation of trials (recall that proportional selection in limit was believed to achieve an exponentially increasing allocation of trials to the observed best schemata, this in line with the mathematical analysis of the two-armed bandit). And as yet another consequence then, the schema theorem, which describes the expected allocation of trials to schemata in a single generation cannot be of any fundamental importance (cf. Goldberg, 1989). It is simply a formulation for describing one method of performing selection, but not a generally *optimal* method.

4.6 CONCLUSIONS

A great deal has been learned about evolutionary algorithms over just the past decade, or even the last five years. Prior convention had presumed:

(1) It would be possible to design generally superior search algorithms

(2) It would be beneficial to maximize implicit parallel processing of contending schemata

(3) The schema theorem can usefully predict the propagation of schemata even in noisy conditions

(4) The two-armed bandit analysis of Holland (1975) provided justification for the optimality of proportional selection

All of these presumptions have now been shown to be misguided or simply in error.

The current state of knowledge suggests that rather than focus on and emphasize differences between alternative methods of evolutionary computation (e.g., "genetic algorithms" vs. "evolutionary programming" vs. "evolution strategies"), a more prudent and productive formulation is to consider the field of evolutionary algorithms as a whole. In each method, a population of contending solutions undergoes variation and selection. The representation for solutions should follow from the task at hand (rather than say, use a binary string to maximize schema processing). The variation operators should reflect the structure of the problem (rather than the default use of any particular operation such as, say, one-point crossover). The form of selection should reflect the desired rate of convergence (rather than rely solely on proportional selection, which has no theoretical basis as being optimum). Although this circumstance leaves the practitioner of evolutionary computation without much specific direction, a lack of guidance is certainly to be favored over the dogmatic acceptance of incorrect guidance. The creativity to design effective evolutionary algorithms rests primarily with the practitioner, who must learn about the problem they face and tailor their algorithm to that problem.

Acknowledgments

The author thanks Z. Michalewicz, M.D. Vose, P.J. Angeline, L.J. Fogel, T. Bäck, H.-G. Beyer, T.M. English, W.G. Macready, V.W. Porto, A. Ghozeil, K. Chellapilla for comments on various parts of this review, and the conference organizers for their kind invitation to present it.

REFERENCES

J. Antonisse, "A new interpretation of schema notation that overturns the binary encoding constraint," *Proceedings of the Third Intern. Conf. on Genetic Algorithms*, J. D. Schaffer (ed.), Morgan Kaufmann, San Mateo, CA, pp. 86–91, 1989.

D. L. Battle and M. D. Vose, "Isomorphisms of genetic algorithms," *Artificial Intelligence*, vol. 60, pp. 155–165, 1993.

K. A. De Jong, W. M. Spears, and D. F. Gordon, "Using Markov chains to analyze GAFOs," In *Foundations of Genetic Algorithms 3*, L. D. Whitley and M. D. Vose (eds.), San Mateo, CA: Morgan Kaufman, 1995, pp. 115–137.

K. Devlin, *The Joy of Sets: Fundamentals of Contemporary Set Theory*, 2nd Ed., Springer-Verlag, Berlin, 1993.

T. M. English, "Evaluation of evolutionary and genetic optimizers: no free lunch," In *Evolutionary Programming V: Proc. of the 5th Annual Conference on Evolutionary*

Programming, L. J. Fogel, P. J. Angeline, and T. Bäck (eds.), Cambridge, MA: MIT Press, 1996, pp. 163–169.

D. B. Fogel and A. Ghozeil, "A note on representations and variation operators," *IEEE Transactions on Evolutionary Computation*, vol. 1:2, 1997a, pp. 159–161.

D. B. Fogel and A. Ghozeil, "Schema processing under proportional selection in the presence of random effects," *IEEE Transactions on Evolutionary Computation*, vol. 1:4, 1997b, pp. 290–293.

H. Gillespie, "Natural selection for variances in offspring numbers: a new evolutionary principle," *American Naturalist*, Vol. 111, 1977, pp. 1010–1014.

D. E. Goldberg, *Genetic Algorithms in Search, Optimization and Machine Learning*, Addison-Wesley, Reading, MA, 1989.

J. H. Holland, *Adaptation in Natural and Artificial Systems*, Univ. of Michigan Press, Ann Arbor, MI, 1975.

W. G. Macready and D. H. Wolpert, "Bandit problems and the exploration/exploitaton tradeoff," *IEEE Trans. Evolutionary Computation,* Vol. 2:1, 1998, pp. 2–22.

N. J. Radcliffe, "Non-linear genetic representations," *Parallel Problem Solving from Nature, 2,* R. Männer and B. Manderick (eds.), North-Holland, Amsterdam, pp. 259–268, 1992.

S. Rana, L. D. Whitley, and R. Cogswell, "Searching in the presence of noise," *Parallel Problem Solving from Nature — PPSN IV*, H.-M. Voigt, W. Ebeling, I. Rechenberg, and H.-P. Schwefel (eds.), Springier, Berlin, 1996, pp. 198–207.

G. Rudolph, "Reflections on bandit problems and selection methods in uncertain environments," In *Proc. of 7th Intern. Conf. on Genetic Algorithms*, T. Bäck (ed.), San Francisco: Morgan Kaufmann, 1997, pp. 166–173.

M. D. Vose and G. E. Liepins, "Schema disruption," *Proceedings of the Fourth International Conference on Genetic Algorithms*, R. K. Belew and L. B. Booker (eds.), Morgan Kaufmann, San Mateo, CA, pp. 237–240, 1991.

D. H. Wolpert and W. G. Macready, "No free lunch theorems for optimization," *IEEE Transactions on Evolutionary Computation*, vol. 1, no. 1, pp. 67–82, 1997.

5 Evolutionary Algorithms for Engineering Applications

Z. MICHALEWICZ[†], K. DEB[‡], M. SCHMIDT and TH. STIDSEN

Department of Computer Science, University of North Carolina
Charlotte, NC 28223, USA *and*
Institute of Computer Science, Polish Academy of Sciences
ul. Ordona 21, 01-237 Warsaw, Poland
E-mail: zbyszek@uncc.edu

Department of Mechanical Engineering, Indian Institute of Technology
Kanpur, Pin 208 016, India
E-mail: deb@iitk.ac.in

Department of Computer Science, Aarhus University
DK-8000 Aarhus C, Denmark
E-mail: marsch@daimi.au.dk

Department of Computer Science, Aarhus University
DK-8000 Aarhus C, Denmark
E-mail: stidsen@daimi.au.dk

5.1 INTRODUCTION

Evolutionary computation techniques have drawn much attention as optimization methods in the last two decades [1], [10], [21], [2]. They are are stochastic optimization methods which are conveniently presented using the metaphor of natural evolution: a randomly initialized *population of individuals* (set of points of the search space at hand) evolves following a crude parody of the Darwinian principle of *the survival of the fittest*. New individuals are generated using some variation operators (e.g., simulated genetic operations such as mutation and crossover). The probability of survival of the newly generated solutions depends on their

[†]Presently on leave at Aarhus University, Aarhus, Denmark
[‡]Presently on leave at University of Dortmund, Germany

fitness (how well they perform with respect to the optimization problem at hand): the best are kept with a high probability, the worst are rapidly discarded.

From the optimization point of view, one of the main advantages of evolutionary computation techniques is that they do not have much mathematical requirements about the optimization problem. They are 0–order methods (all they need is an evaluation of the objective function), they can handle nonlinear problems, defined on discrete, continuous or mixed search spaces, unconstrained or constrained. Moreover, the ergodicity of the evolution operators makes them global in scope (in probability).

Many engineering activities involve unstructured, real-life problems that are hard to model, since they require inclusion of unusual factors (from accident risk factors to esthetics). Other engineering problems are complex in nature: job shop scheduling, timetabling, traveling salesman or facility layout problems are examples of *NP-complete* problems. In both cases, evolutionary computation techniques represent a potential source of actual breakthroughs. Their ability to provide many near-optimal solutions at the end of an optimization run enables to choose the best solution afterwards, according to criteria that were either inarticulate from the expert, or badly modeled. Evolutionary algorithms can be made efficient because they are flexible, and relatively easy to hybridize with domain-dependent heuristics. Those features of evolutionary computation have already been acknowledged in the field of engineering, and many applications have been reported (see, for example, [6], [11], [34], [43]).

A vast majority of engineering optimization problems are constrained problems. The presence of constraints significantly affects the performance of any optimization algorithm, including evolutionary search methods [20]. This paper focuses on the issue of evaluation of constraints handling methods, as the advantages and disadvantages of various methods are not well understood. The general way of dealing with constraints — whatever the optimization method — is by penalizing infeasible points. However, there are no guidelines on designing penalty functions. Some suggestions for evolutionary algorithms are given in [37], but they do not generalize. Other techniques that can be used to handle constraints in are more or less problem dependent. For instance, the knowledge about linear constraints can be incorporated into specific operators [24], or a repair operator can be designed that projects infeasible points onto feasible ones [30].

Section 5.2 provides a general overview of constraints-handling methods for evolutionary computation techniques. The experimental results reported in many papers suggest that making an appropriate *a priori* choice of an evolutionary method for a nonlinear parameter optimization problem remains an open question. It seems that the most promising approach at this stage of research is experimental, involving a design of a *scalable* test suite of constrained optimization problems, in which many features could be easily tuned. Then it would be possible to evaluate merits and drawbacks of the available methods as well as test new methods efficiently. Section 5.3 discusses further the need for a such scalable test suite, whereas section 5.4 presents a particular test case generator proposed recently [23]. Section 5.5 provides an example of the use of this test case generator for evaluation of a particular constraint-handling method. Section 5.6 concludes the paper.

5.2 CONSTRAINT-HANDLING METHODS

The general nonlinear programming (NLP) problem is to find $\vec{x}$ so as to

$$\text{optimize } f(\vec{x}), \quad \vec{x} = (x_1, \ldots, x_n) \in \boldsymbol{R}^n, \tag{5.1}$$

where $\vec{x} \in \mathcal{F} \subseteq \mathcal{S}$. The *objective function* f is defined on the *search space* $\mathcal{S} \subseteq \boldsymbol{R}^n$ and the set $\mathcal{F} \subseteq \mathcal{S}$ defines the *feasible region*. Usually, the search space $\mathcal{S}$ is defined as a n-dimensional rectangle in $\boldsymbol{R}^n$ (domains of variables defined by their lower and upper bounds):

$$l_i \leq x_i \leq u_i, \quad 1 \leq i \leq n,$$

whereas the feasible region $\mathcal{F} \subseteq \mathcal{S}$ is defined by a set of p additional constraints ($p \geq 0$):

$$g_j(\vec{x}) \leq 0, \text{ for } j = 1, \ldots, q, \quad \text{and} \quad h_j(\vec{x}) = 0, \text{ for } j = q+1, \ldots, p.$$

At any point $\vec{x} \in \mathcal{F}$, the constraints g_j that satisfy $g_j(\vec{x}) = 0$ are called the *active* constraints at $\vec{x}$.

The NLP problem, in general, is intractable: it is impossible to develop a deterministic method for the NLP in the global optimization category, which would be better than the exhaustive search [12]. This makes a room for evolutionary algorithms, extended by some constraint-handling methods. Indeed, during the last few years, several evolutionary algorithms (which aim at complex objective functions (e.g., non differentiable or discontinuous) have been proposed for the NLP; a recent survey paper [28] provides an overview of these algorithms.

During the last few years several methods were proposed for handling constraints by genetic algorithms for parameter optimization problems. These methods can be grouped into five categories: (1) methods based on preserving feasibility of solutions, (2) methods based on penalty functions, (3) methods which make a clear distinction between feasible and infeasible solutions, (4) methods based on decoders, and (5) other hybrid methods. We discuss them briefly in turn.

5.2.1 Methods based on preserving feasibility of solutions

The best example of this approach is Genocop (for GEnetic algorithm for Numerical Optimization of COnstrained Problems) system [24], [25]. The idea behind the system is based on specialized operators which transform feasible individuals into feasible individuals, i.e., operators, which are closed on the feasible part $\mathcal{F}$ of the search space. The method assumes linear constraints only and a feasible starting point (or feasible initial population). Linear equations are used to eliminate some variables; they are replaced as a linear combination of remaining variables. Linear inequalities are updated accordingly. A closed set of operators maintains feasibility of solutions. For example, when a particular component x_i of a solution vector $\vec{x}$ is mutated, the system determines its current domain $dom(x_i)$ (which is a function of linear constraints and remaining values of the solution vector $\vec{x}$) and the new value of x_i is taken from this domain (either with flat probability distribution for uniform mutation, or other probability distributions for non-uniform and boundary mutations). In any case the offspring solution vector is always feasible. Similarly, arithmetic crossover, $a\vec{x} + (1-a)\vec{y}$, of two feasible solution vectors $\vec{x}$ and $\vec{y}$ yields always a feasible solution (for $0 \le a \le 1$) in convex search spaces (the system assumes linear constraints only which imply convexity of the feasible search space $\mathcal{F}$).

Recent work [27], [39], [40] on systems which search only the boundary area between feasible and infeasible regions of the search space, constitutes another example of the approach based on preserving feasibility of solutions. These systems are based on specialized boundary operators (e.g., sphere crossover, geometrical crossover, etc.): it is a common situation for many constrained optimization problems that some constraints are active at the target global optimum, thus the optimum lies on the boundary of the feasible space.

5.2.2 Methods based on penalty functions

Many evolutionary algorithms incorporate a constraint-handling method based on the concept of (exterior) penalty functions, which penalize infeasible solutions. Usually, the penalty function is based on the distance of a solution from the feasible region $\mathcal{F}$, or on the effort to "repair" the solution, i.e., to force it into $\mathcal{F}$. The former case is the most popular one; in many methods a set of functions f_j $(1 \le j \le m)$ is used to construct the penalty, where the function f_j measures the violation of the j-th constraint in the following way:

$$f_j(\vec{x}) = \begin{cases} \max\{0, g_j(\vec{x})\}, & \text{if } 1 \le j \le q \\ |h_j(\vec{x})|, & \text{if } q+1 \le j \le m. \end{cases}$$

However, these methods differ in many important details, how the penalty function is designed and applied to infeasible solutions. For example, a method of static penalties was proposed [14]; it assumes that for every constraint we establish a family of intervals which determine appropriate penalty coefficient. The method of dynamic penalties was examined [15], where individuals are evaluated (at the iteration t) by the following formula:

$$eval(\vec{x}) = f(\vec{x}) + (C \times t)^{\alpha} \sum_{j=1}^{m} f_j^{\beta}(\vec{x}),$$

where C, α and β are constants. Another approach (Genocop II), also based on dynamic penalties, was described [22]. In that algorithm, at every iteration active constraints only are considered, and the pressure on infeasible solutions is increased due to the decreasing values of temperature τ. In [9] a method for solving constraint satisfaction problems that changes the evaluation function based on the performance of a EA run was described: the penalties (weights) of those constraints which are violated by the best individual after termination are raised, and the new weights are used in the next run. A method based on adaptive penalty functions was was developed in [3], [13]: one component of the penalty function takes a feedback from the search process. Each individual is evaluated by the formula:

$$eval(\vec{x}) = f(\vec{x}) + \lambda(t) \sum_{j=1}^{m} f_j^2(\vec{x}),$$

where $\lambda(t)$ is updated every generation t with respect to the current state of the search (based on last k generations). The adaptive penalty function was also used in [42], where both the search length and constraint severity feedback was incorporated. It involves the estimation of a near-feasible threshold q_j for each constraint $1 \le j \le m$); such thresholds indicate distances from the feasible region $\mathcal{F}$ which are "reasonable" (or, in other words, which determine "interesting" infeasible solutions, i.e., solutions relatively close to the feasible region). Additional method (so-called segregated genetic algorithm) was proposed in [18] as yet another way to handle the problem of the robustness of the penalty level: two different penalized fitness functions with static penalty terms p_1 and p_2 were designed (smaller and larger, respectively). The main idea is that such an approach will result roughly in maintaining two subpopulations: the individuals selected on the basis of f_1 will more likely lie in the infeasible region while the ones selected on the basis of f_2 will probably stay in the feasible region; the overall process is thus allowed to reach the feasible optimum from both sides of the boundary of the feasible region.

5.2.3 Methods based on a search for feasible solutions

There are a few methods which emphasize the distinction between feasible and infeasible solutions in the search space $\mathcal{S}$. One method, proposed in [41] (called a "behavioral memory" approach) considers the problem constraints in a sequence; a switch from one constraint to another is made upon arrival of a sufficient number of feasible individuals in the population.

The second method, developed in [34] is based on a classical penalty approach with one notable exception. Each individual is evaluated by the formula:

$$eval(\vec{x}) = f(\vec{x}) + r\sum_{j=1}^{m} f_j(\vec{x}) + \theta(t, \vec{x}),$$

where r is a constant; however, the original component $\theta(t, \vec{x})$ is an additional iteration dependent function which influences the evaluations of infeasible solutions. The point is that the method distinguishes between feasible and infeasible individuals by adopting an additional heuristic rule (suggested earlier in [37]): for any feasible individual $\vec{x}$ and any infeasible individual $\vec{y}$: $eval(\vec{x}) < eval(\vec{y})$, i.e., any feasible solution is better than any infeasible one.[1] In a recent study [8], a modification to this approach is implemented with the tournament selection operator and with the following evaluation function:

$$eval(\vec{x}) = \begin{cases} f(\vec{x}), & \text{if } \vec{x} \text{ is feasible,} \\ f_{\max} + \sum_{j=1}^{m} f_j(\vec{x}), & \text{otherwise,} \end{cases}$$

where $f_{\max}$ is the function value of the worst feasible solution in the population. The main difference between this approach and Powell and Skolnick's approach is that in this approach the objective function value is not considered in evaluating an infeasible solution. Additionally, a niching scheme is introduced to maintain diversity among feasible solutions. Thus, initially the search focuses on finding feasible solutions and later when adequate number of feasible solutions are found, the algorithm finds better feasible solutions by maintaining

[1]For minimization problems.

a diversity in solutions in the feasible region. It is interesting to note that there is no need of the penalty coefficient r here, because the feasible solutions are always evaluated to be better than infeasible solutions and infeasible solutions are compared purely based on their constraint violations. However, normalization of constraints $f_j(\vec{x})$ is suggested. On a number of test problems and on an engineering design problem, this approach is better able to find constrained optimum solutions than Powell and Skolnick's approach.

The third method (Genocop III), proposed in [26] is based on the idea of repairing infeasible individuals. Genocop III incorporates the original Genocop system, but also extends it by maintaining two separate populations, where a development in one population influences evaluations of individuals in the other population. The first population P_s consists of so-called search points from $\mathcal{F}_l$ which satisfy linear constraints of the problem. The feasibility (in the sense of linear constraints) of these points is maintained by specialized operators. The second population P_r consists of so-called reference points from $\mathcal{F}$; these points are fully feasible, i.e., they satisfy *all* constraints. Reference points $\vec{r}$ from P_r, being feasible, are evaluated directly by the objective function (i.e., $eval(\vec{r}) = f(\vec{r})$). On the other hand, search points from P_s are "repaired" for evaluation.

5.2.4 Methods based on decoders

Decoders offer an interesting option for all practitioners of evolutionary techniques. In these techniques a chromosome "gives instructions" on how to build a feasible solution. For example, a sequence of items for the knapsack problem can be interpreted as: "take an item if possible"—such interpretation would lead always to a feasible solution.

However, it is important to point out that several factors should be taken into account while using decoders. Each decoder imposes a mapping T between a feasible solution and decoded solution. It is important that several conditions are satisfied: (1) for each solution $s \in \mathcal{F}$ there is an encoded solution d, (2) each encoded solution d corresponds to a feasible solution s, and (3) all solutions in $\mathcal{F}$ should be represented by the same number of encodings d.[2] Additionally, it is reasonable to request that (4) the transformation T is computationally fast and (5) it has locality feature in the sense that small changes in the coded solution result in small changes in the solution itself. An interesting study on coding trees in genetic algorithm was reported in [31], where the above conditions were formulated.

However, the use of decoders for continuous domains has not been investigated. Only recently [16], [17] a new approach for solving constrained numerical optimization problems was proposed. This approach incorporates a homomorphous mapping between n-dimensional cube and a feasible search space. The mapping transforms the constrained problem at hand into unconstrained one. The method has several advantages over methods proposed earlier (no additional parameters, no need to evaluate—or penalize—infeasible solutions, easiness of

[2] However, as observed by Davis [7], the requirement that all solutions in $\mathcal{F}$ should be represented by the same number of decodings seems overly strong: there are cases in which this requirement might be suboptimal. For example, suppose we have a decoding and encoding procedure which makes it impossible to represent suboptimal solutions, and which encodes the optimal one: this might be a good thing. (An example would be a graph coloring order-based chromosome, with a decoding procedure that gives each node its first legal color. This representation could not encode solutions where some nodes that could be colored were not colored, but this is a good thing!)

approaching a solution located on the edge of the feasible region, no need for special operators, etc).

5.2.5 Hybrid methods

It is relatively easy to develop hybrid methods which combine evolutionary computation techniques with deterministic procedures for numerical optimization problems. In [45] a combined evolutionary algorithm with the direction set method of Hooke-Jeeves is described; this hybrid method was tested on three (unconstrained) test functions. In [29] the authors considered a similar approach, but they experimented with constrained problems. Again, they combined evolutionary algorithm with some other method—developed in [19]. However, while the method of [45] incorporated the direction set algorithm as a problem-specific operator of his evolutionary technique, in [29] the whole optimization process was divided into two separate phases.

Several other constraint handling methods deserve also some attention. For example, some methods use of the values of objective function f and penalties f_j $(j = 1, \ldots, m)$ as elements of a vector and apply multi-objective techniques to minimize all components of the vector. For example, in [38], Vector Evaluated Genetic Algorithm (VEGA) selects $1/(m+1)$ of the population based on each of the objectives. Such an approach was incorporated by Parmee and Purchase [33] in the development of techniques for constrained design spaces. On the other hand, in the approach by [43], all members of the population are ranked on the basis of constraint violation. Such rank r, together with the value of the objective function f, leads to the two-objective optimization problem. This approach gave a good performance on optimization of gas supply networks.

Also, an interesting approach was reported in [32]. The method (described in the context of constraint satisfaction problems) is based on a co-evolutionary model, where a population of potential solutions co-evolves with a population of constraints: fitter solutions satisfy more constraints, whereas fitter constraints are violated by fewer solutions. There is some development connected with generalizing the concept of "ant colonies" [5] (which were originally proposed for order-based problems) to numerical domains [4]; first experiments on some test problems gave very good results [48]. It is also possible to incorporate the knowledge of the constraints of the problem into the belief space of cultural algorithms [35]; such algorithms provide a possibility of conducting an efficient search of the feasible search space [36].

5.3 A NEED FOR A TEST CASE GENERATOR

It is not clear what characteristics of a constrained problem make it difficult for an evolutionary technique (and, as a matter of fact, for any other optimization technique). Any problem can be characterized by various parameters; these may include the number of linear constraints, the number of nonlinear constraints, the number of equality constraints, the number of active constraints, the ratio $\rho = |\mathcal{F}|/|\mathcal{S}|$ of sizes of feasible search space to the whole, the type of the objective function (the number of variables, the number of local optima, the existence of derivatives, etc). In [28] eleven test cases for constrained numerical optimization problems were proposed ($G1$–$G11$). These test cases include objective functions of various types (linear, quadratic, cubic, polynomial, nonlinear) with various number of variables and different types (linear inequalities, nonlinear equations and inequalities) and numbers of constraints. The ratio ρ between the size of the feasible search space $\mathcal{F}$ and the size of the whole search space $\mathcal{S}$ for these test cases vary from 0% to almost 100%; the topologies of feasible search spaces are also quite different. These test cases are summarized in Table 5.1. For each test case the number n of variables, type of the function f, the relative size of the feasible region in the search space given by the ratio ρ, the number of constraints of each category (linear inequalities *LI*, nonlinear equations *NE* and inequalities *NI*), and the number a of active constraints at the optimum (including equality constraints) are listed.

Table 5.1 Summary of eleven test cases. The ratio $\rho = |\mathcal{F}|/|\mathcal{S}|$ was determined experimentally by generating 1,000,000 random points from $\mathcal{S}$ and checking whether they belong to $\mathcal{F}$ (for $G2$ and $G3$ we assumed $k = 50$). *LI*, *NE*, and *NI* represent the number of linear inequalities, and nonlinear equations and inequalities, respectively

Function	n	Type of f	ρ	*LI*	*NE*	*NI*	a
$G1$	13	quadratic	0.0111%	9	0	0	6
$G2$	k	nonlinear	99.8474%	0	0	2	1
$G3$	k	polynomial	0.0000%	0	1	0	1
$G4$	5	quadratic	52.1230%	0	0	6	2
$G5$	4	cubic	0.0000%	2	3	0	3
$G6$	2	cubic	0.0066%	0	0	2	2
$G7$	10	quadratic	0.0003%	3	0	5	6
$G8$	2	nonlinear	0.8560%	0	0	2	0
$G9$	7	polynomial	0.5121%	0	0	4	2
$G10$	8	linear	0.0010%	3	0	3	6
$G11$	2	quadratic	0.0000%	0	1	0	1

The results of many tests did not provide meaningful conclusions, as no single parameter (number of linear, nonlinear, active constraints, the ratio ρ, type of the function, number of variables) proved to be significant as a major measure of difficulty of the problem. For example, many tested methods approached the optimum quite closely for the test cases $G1$ and $G7$ (with $\rho = 0.0111\%$ and $\rho = 0.0003\%$, respectively), whereas most of the methods experienced difficulties for the test case $G10$ (with $\rho = 0.0010\%$). Two quadratic functions (the test cases $G1$ and $G7$) with a similar number of constraints (9 and 8, respectively) and an identical number (6) of active constraints at the optimum, gave a different challenge to most of these methods. Also, several methods were quite sensitive to the presence of a feasible solution in the initial population. Possibly a more extensive testing of various methods was required.

Not surprisingly, the experimental results of [28] suggested that making an appropriate *a priori* choice of an evolutionary method for a nonlinear optimization problem remained an open question. It seems that more complex properties of the problem (e.g., the characteristic of the objective function together with the topology of the feasible region) may constitute quite significant measures of the difficulty of the problem. Also, some additional measures of the problem characteristics due to the constraints might be helpful. However, this kind of information is not generally available. In [28] the authors wrote:

> "It seems that the most promising approach at this stage of research is experimental, involving the design of a scalable test suite of constrained optimization problems, in which many [...] features could be easily tuned. Then it should be possible to test new methods with respect to the corpus of all available methods."

Clearly, there is a need for a parameterized test-case generator which can be used for analyzing various methods in a systematic way (rather than testing them on a few selected test cases; moreover, it is not clear whether addition of a few extra test cases is of any help).

In this paper we propose such a test-case generator for constrained parameter optimization techniques. This generator is capable of creating various test cases with different characteristics:

- problems with different value of ρ: the relative size of the feasible region in the search space;
- problems with different number and types of constraints;
- problems with convex or non-convex objective function, possibly with multiple optima;
- problems with highly non-convex constraints consisting of (possibly) disjoint regions.

All this can be achieved by setting a few parameters which influence different characteristics of the optimization problem. Such test-case generator should be very useful for analyzing and comparing different constraint-handling techniques.

There were some attempts in the past to propose a test case generator for unconstrained parameter optimization [46], [47]. We are also aware of one attempt to do so for constrained cases; in [44] the author proposes so-called stepping-stones problem defined as:

$$\text{objective: maximize} \sum_{i=1}^{n}(x_i/\pi + 1),$$

where $-\pi \leq x_i \leq \pi$ for $i = 1, \ldots, n$ and the following constraints are satisfied:

$$e^{x_i/\pi} + cos(2x_i) \leq 1 \text{ for } i = 1, \ldots, n.$$

Note that the objective function is linear and that the feasible region is split into 2^n disjoint parts (called stepping-stones). As the number of dimensions n grows, the problem becomes more complex. However, as the stepping-stones problem has one parameter only, it can not be used to investigate some aspects of a constraint-handling method.

5.4 THE TEST CASE GENERATOR

As explained in the previous section, it is of great importance to have a parameterized generator of test cases for constrained parameter optimization problems. By changing values of some parameters it would be possible to investigate merits/drawbacks (efficiency, cost, etc) of many constraint-handling methods. Many interesting questions could be addressed:

- how the efficiency of a constraint-handling method changes as a function of the number of disjoint components of the feasible part of the search space?
- how the efficiency of a constraint-handling method changes as a function of the ratio between the sizes of the feasible part and the whole search space?
- what is a relationship between the number of constraints (or the number of dimensions, for example) of a problem and the computational effort of a method?

and many others. In [23] a parameterized test-case generator $\mathcal{TCG}$ was proposed:

$$\mathcal{TCG}(n, w, \lambda, \alpha, \beta, \mu);$$

the meaning of its six parameters is as follows:

n – the number of variables of the problem
w – a parameter to control the number of optima in the search space
λ – a parameter to control the number of constraints (inequalities)
α – a parameter to control the connectedness of the feasible search regions
β – a parameter to control the ratio of the feasible to total search space
μ – a parameter to influence the ruggedness of the fitness landscape

The ranges and types of the parameters are:

$n \geq 1$; integer	$w \geq 1$; integer	$0 \leq \lambda \leq 1$; float
$0 \leq \alpha \leq 1$; float	$0 \leq \beta \leq 1$; float	$0 \leq \mu \leq 1$; float

The general idea behind this test case generator was to divide the search space $\mathcal{S}$ into a number of disjoint subspaces $\mathcal{S}_k$ and to define a unimodal function f_k for every $\mathcal{S}_k$. Thus the objective function G is defined on the search space $\mathcal{S} = \prod_{i=1}^{n}[0, 1)$ as follows:

$$G(\vec{x}) = f_k(\vec{x}) \text{ iff } \vec{x} \in \mathcal{S}_k.$$

The total number of subspaces $\mathcal{S}_k$ is equal to w^n, as each dimension of the search space is divided into w equal length segments (for exact definition of a subspace $\mathcal{S}_k$, see [23]). This number indicates also the total number of local optima of function G. For each subspace $\mathcal{S}_k$ $(k = 0, \ldots, w^n - 1)$ a function f_k is defined:

$$f_k(x_1, \ldots, x_n) = a_k \left(\Pi_{i=1}^{n}(u_i^k - x_i)(x_i - l_i^k)\right)^{\frac{1}{n}} \tag{5.2}$$

where l_i^k and u_i^k are the boundaries of the k-th subspace for the i-th dimension. The constants a_k are defined as follows:

$$a_k = \begin{cases} 4w^2(1 - \alpha^2\beta^2)^{k'[(1-\mu)+\mu/\log_2(wn-n+1)]-1} & \text{if } \alpha\beta > 0 \\ 4w^2\left(\frac{(\mu-1)k''}{n(w-1)} + 1\right) & \text{if } \alpha\beta = 0, \end{cases} \tag{5.3}$$

where

$$k' = \log_2(\sum_{i=1}^{n} q_{i,k} + 1), \text{ and}$$
$$k'' = \sum_{i=1}^{n} q_{i,k},$$

where $(q_{1,k}, \ldots, q_{n,k})$ is a n-dimensional representation of the number k in w-ary alphabet.[3] Additionally, to remove this fixed pattern from the generated test cases, an additional mechanism: a permutation of subspaces $\mathcal{S}_k$, was used.

The third parameter λ of the test-case generator is related to the number m of constraints of the problem, as the feasible part of the search space $\mathcal{S}$ is defined by means on m double inequalities ($0 \leq m \leq w^n$):

$$r_1^2 \leq c_k(\vec{x}) \leq r_2^2, \quad k = 0, \ldots, m-1, \tag{5.4}$$

where $0 \leq r_1 \leq r_2$ and each $c_k(\vec{x})$ is a quadratic function:

$$c_k(\vec{x}) = (x_1 - p_1^k)^2 + \ldots + (x_n - p_n^k)^2,$$

where $p_i^k = (l_i^k + u_i^k)/2$. These m double inequalities define m feasible parts $\mathcal{F}_k$ of the search space:

$$\vec{x} \in \mathcal{F}_k \text{ iff } r_1^2 \leq c_k(\vec{x}) \leq r_2^2,$$

and the overall feasible search space $\mathcal{F} = \cup_{k=0}^{m-1} \mathcal{F}_k$. Note, the interpretation of constraints here is different than the one in the standard definition of the NLP problem. Here the feasible search space is defined as a *union* (not intersection) of all double constraints. In other words, a point $\vec{x}$ is feasible if and only if there exist an index $0 \leq k \leq m-1$ such that double inequality (5.4) is satisfied.

The parameter $0 \leq \lambda \leq 1$ determines the number m of constraints as follows:

$$m = \lfloor \lambda(w^n - 1) + 1 \rfloor. \tag{5.5}$$

Clearly, $\lambda = 0$ and $\lambda = 1$ imply $m = 1$ and $m = w^n$, i.e., minimum and maximum number of constraints, respectively.

The parameters α and β define the radii r_1 and r_2:

$$r_1 = \frac{\alpha\beta\sqrt{n}}{2w}, \quad r_2 = \frac{\alpha\sqrt{n}}{2w}. \tag{5.6}$$

These parameters determine the topology of the feasible part of the search space.

If $\alpha\beta > 0$, the function G has 2^n global maxima points, all in permuted $\mathcal{S}_0$. For any global solution $(x_1, \ldots, x_n)$, $x_i = 1/(2w) \pm \alpha\beta/(2w)$ for all $i = 1, 2, \ldots, n$. The function values

[3] For $w > 1$. If $w = 1$ the whole search space consists of one subspace $\mathcal{S}_0$ only. In this case $a_0 = 4/(1 - \alpha^2\beta^2)$ for all $0 \leq \alpha, \beta \leq 1$.

at these solutions are always equal to one. On the other hand, if $\alpha\beta = 0$, the function G has either one global maximum (if $\mu < 1$) or m maxima points (if $\mu = 1$), one in each of permuted $\mathcal{S}_0, \ldots, \mathcal{S}_{m-1}$. If $\mu < 1$, the global solution $(x_1, \ldots, x_n)$ is always at

$$(x_1, \ldots, x_n) = (1/(2w), 1/(2w), \ldots, 1/(2w)).$$

Fig. 5.1 displays the final landscape for test case $\mathcal{TCG}(2, 4, 1, \frac{1}{\sqrt{2}}, 0.8, 0)$.

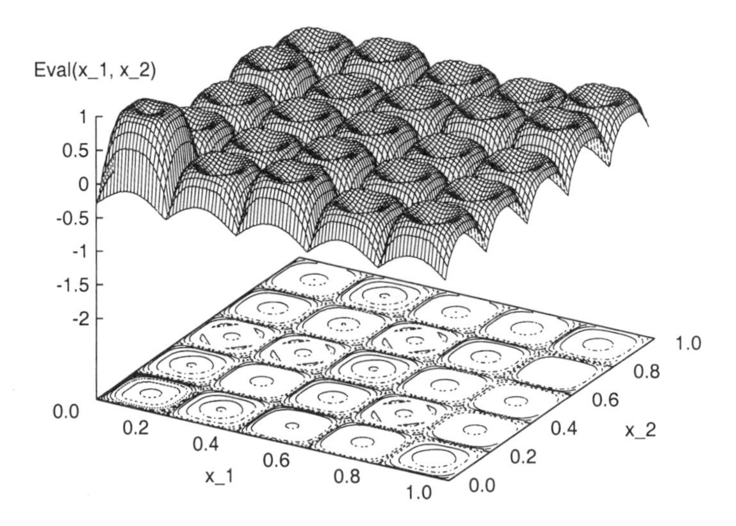

Fig. 5.1 Final landscape for the test case $\mathcal{TCG}(2, 5, 1, \frac{1}{\sqrt{2}}, 0.8, 0)$. Contours with function values ranging from -0.2 to 1.0 at a step of 0.2 are drawn at the base

The interpretation of the six parameters of the test-case generator $\mathcal{TCG}$ is as follows:

1. **Dimensionality:** By increasing the parameter n the dimensionality of the search space can be increased.

2. **Multimodality:** By increasing the parameter w the multimodality of the search space can be increased. For the unconstrained function, there are w^n locally maximum solutions, of which one is globally maximum. For the constrained test function with $\alpha\beta > 0$, there are $(2w)^n$ different locally maximum solutions, of which 2^n are globally maximum solutions.

3. **Constraints:** By increasing the parameter λ the number m of constraints is increased.

4. **Connectedness:** By reducing the parameter α (from 1 to $1/\sqrt{n}$ and smaller), the connectedness of the feasible subspaces can be reduced. When $\alpha < 1/\sqrt{n}$, the feasible subspaces $\mathcal{F}_k$ are completely disconnected.

5. **Feasibility:** By increasing the ratio β the proportion of the feasible search space to the complete search space (ratio ρ) can be reduced. For β values closer to one, the feasible search space becomes smaller and smaller. These test functions can be used to test an optimizer's ability to be find and maintain feasible solutions.

6. **Ruggedness:** By increasing the parameter μ the function ruggedness can be increased (for $\alpha\beta > 0$). A sufficiently rugged function will test an optimizer's ability to search for the globally constrained maximum solution in the presence of other almost equally significant locally maxima.

Increasing the each of the above parameters except α and decreasing α will cause difficulty to any optimizer. However, it is difficult to conclude which of these factors is most profoundly affect the performance of an optimizer. Thus, it is recommended that the user should first test his/her algorithm with simplest possible combination of the above parameters (small n, small w, small μ, large α, small β, and small λ). Thereafter, the parameters may be changed in a systematic manner to create difficult test functions. The most difficult test function is created when large values of parameters n, w, λ, β, and μ together with a small value of parameter α are used.

For full details of the test case generator the reader is referred to [23].

5.5 AN EXAMPLE

To test the usefulness of the $\mathcal{TCG}$, a simple steady-state evolutionary algorithm was developed. We have used a constant population size of 100; each individual is a vector $\vec{x}$ of n floating-point components. Parent selection was performed by a binary tournament selection; an offspring replaces the worse individual from a binary tournament. One of three operators was used in every generation (the selection of an operator was done accordingly to constant probabilities 0.5, 0.15, and 0.35, respectively):

- Gaussian mutation:

$$\vec{x} \leftarrow \vec{x} + N(0, \vec{\sigma}),$$

  where $N(0, \vec{\sigma})$ is a vector of independent random Gaussian numbers with a mean of zero and standard deviations $\vec{\sigma}$[4].

- uniform crossover:

$$\vec{z} \leftarrow (z_1, \ldots, z_n),$$

  where each z_i is either x_i or y_i (with equal probabilities), where $\vec{x}$ and $\vec{y}$ are two selected parents.

[4]In all experiments reported in this section, we have used a value of $1/(2\sqrt{n})$, which depends only on the dimensionality of the problem.

- heuristic crossover:

$$\vec{z} = r \cdot (\vec{x} - \vec{y}) + \vec{x},$$

where r is a random number between 0 and 0.25, and the parent $\vec{x}$ is not worse than $\vec{y}$.

The termination condition was to quit the evolutionary loop if an improvement in the last $N = 10,000$ generations was smaller than a predefined $\epsilon = 0.001$.

Now the $\mathcal{TCG}$ can be used for investigating the merits of any constraint-handling methods. For example, one of the simplest and the most popular constraint-handling method is based on static penalties; let us define a particular static penalty method as follows:

$$eval(\vec{x}) = f(\vec{x}) + W \cdot v(\vec{x}),$$

where f is an objective function, W is a constant (penalty coefficient) and function v measures the constraint violation. (Note that only one double constraint is taken into account as the $\mathcal{TCG}$ defines the feasible part of the search space as a union of all m double constraints.)

The penalty coefficient W was set to 10 and the value of v for any $\vec{x}$ is defined by the following procedure:

find k such that $\vec{x} \in \mathcal{S}_k$
set $C = (-2w)/\sqrt{n}$
if the whole $\mathcal{S}_k$ is infeasible **then** $v(\vec{x}) = -1$
else begin
  calculate distance $Dist$ between $\vec{x}$ and the center
  of the subspace $\mathcal{S}_k$
  if $Dist < r_1$ **then** $v(\vec{x}) = C \cdot (r_1 - Dist)$
  else if $Dist > r_2$ **then** $v(\vec{x}) = C \cdot (Dist - r_2)$
    else $v(\vec{x}) = 0$
end

Thus the constraint violation measure v returns -1, if the evaluated point is in infeasible subspace (i.e., subspace without a feasible ring); 0, if the evaluated point is feasible; or some number q from the range $[-1, 0)$, if the evaluated point is infeasible, but the corresponding subspace is partially feasible. It means that the point $\vec{x}$ is either inside the smaller ring or outside the larger one. In both cases q is a scaled negative distance of this point to the boundary of the closest ring. Note that the scaling factor C guarantees that $0 < q \leq 1$. Note, that the values of the objective function for feasible points of the search space stay in the range $[0, 1]$; the value at the global optimum is always 1. Thus, both the function and constraint violation values are normalized in [0,1].

The Figs 5.2 and 5.3 display[5] the results of typical runs of the system on two different test cases. Both test cases were generated for $n = 20$ (low dimensionality), $w = 2$ (i.e., the search space $\mathcal{S}$ was divided into 2^{20} subspaces), $\lambda = 1$ (i.e., there was a feasible ring in every subspace), $\alpha = 0.9/\sqrt{20} = 0.201246$ (the feasible search space consisted of disjoint feasible components of relatively large size), and $\beta = 0.1$ (there are small "inner rings", i.e., r_1 is one tenth of r_2). The only difference between these two test cases is in value of parameter μ, which is either 0.9 (Fig. 5.2) or 0.1 (Fig. 5.3).

[5] All figures of this section report averages of 10 runs.

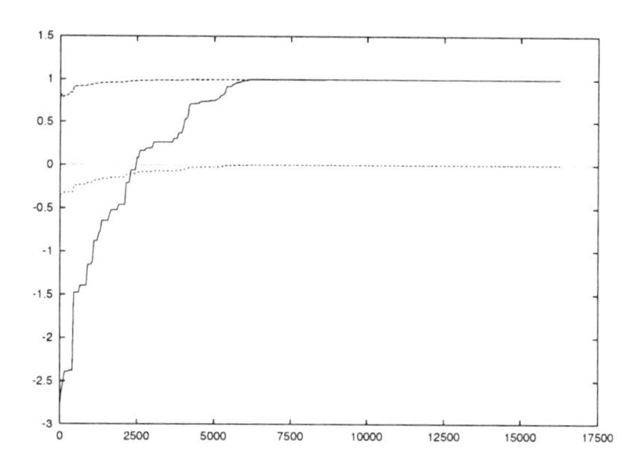

Fig. 5.2 A run of the system for the $\mathcal{TCG}(20, 2, 1, 0.201246, 0.1, 0.9)$; generation number versus objective value and constraint violation. The upper broken line represents the value of the objective function f, the lower broken line: the constraint violation v, and the continuous line: the final value $eval$

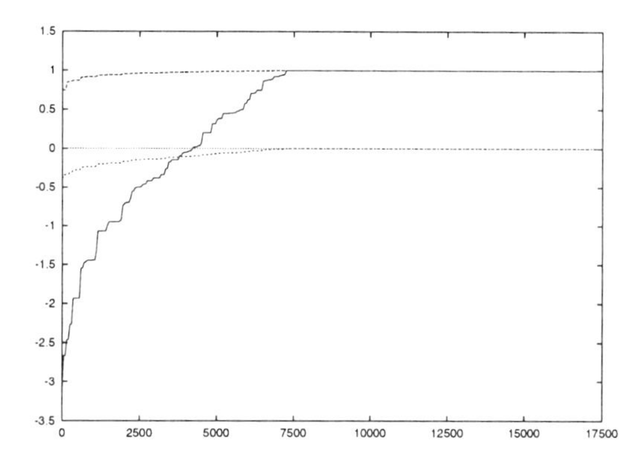

Fig. 5.3 A run of the system for the $\mathcal{TCG}(20, 2, 1, 0.201246, 0.1, 0.1)$; generation number versus objective value and constraint violation. The upper broken line represents the value of the objective function f, the lower broken line: the constraint violation v, and the continuous line: the final value $eval$

These results matched our expectations very well. The algorithm converges to a feasible solution (constraint violation is zero) in all runs. However, note that the height of the highest peak in the search space (i.e., the global feasible maximum) is 1, and the height of the lowest peak is a function of μ:

$$(1 - \alpha^2\beta^2)^{3.3923(1-\mu)}$$

for fixed values of other parameters (note that the total number of peaks is $w^n = 2^{20} = 1,048,576$). In both cases the algorithm converges to an "average" peak, which is slightly better for $\mu = 0.9$ than for $\mu = 0.1$. The algorithm failed to find the global peak in all runs.

Fig. 5.4 displays the performance of the algorithm for varying values of n (between 20 and 320). This is one of the most interesting (however, expected also) results. It indicates clearly that for a particular value of the penalty coefficient (which was $W = 10$), the system works well up to $n = 140$. For higher dimensions, the algorithm sometimes converges to an infeasible solution (which, due to a constant penalty factor) is better than a feasible one. For $n > 200$ the solutions from all 10 runs are infeasible. For different values of W we get different "breaking" points (i.e., values of n starting for which the algorithm converges to an infeasible solution). For example, if $W = 1$, the algorithm returns always feasible solutions up to $n = 54$. For $n > 54$ the algorithm may return infeasible solution, and for $n > 82$ all returned solutions are infeasible.

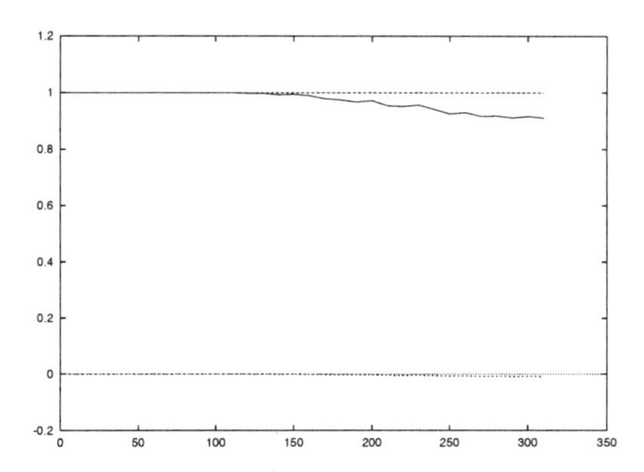

Fig. 5.4 The objective values f returned by the system for the $\mathcal{TCG}(n, 2, 1, 0.201246, 0.1, 0.1)$ with varying n between 20 and 320. The constraint violation is zero for $n < 120$ only

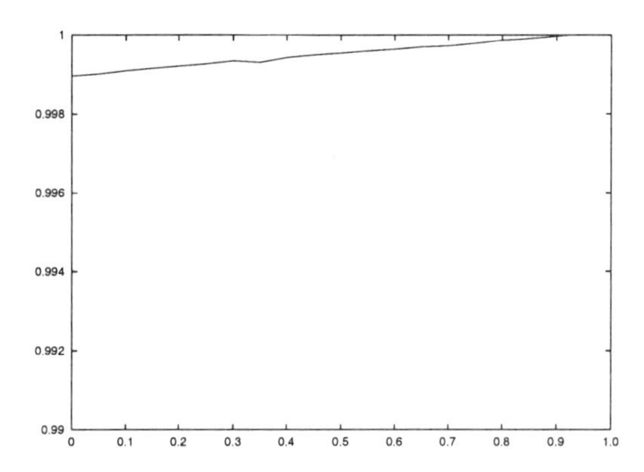

Fig. 5.5 The objective values f returned by the system for the $\mathcal{TCG}(20, 2, 1, 0.201246, 0.1, \mu)$ with varying μ between 0.0 and 1.0. The constraint violation is zero for all μ

Fig. 5.5 displays the performance of the algorithm for varying values of μ (between 0.0 and 1.0). These results indicate clearly that the final solution depends on the heights of the peaks in the search space: the algorithms always converges to a "slightly above average" peak. For larger values of μ, the average peaks are higher, hence the better final value.

It was also interesting to vary parameters α and β. Both changes have similar effect. With fixed $\beta = 0.1$, an increment of α decreases the performance of the system (Fig. 5.6). This seems rather counter-intuitive as the feasible area grows. However, it is important to remember that for each α the fitness landscape is different as we change the values of a_k's (equation (5.3)). Note that α varies between 0.011 and 0.224; the latter number equals to $1/\sqrt{20}$, so the feasible rings remain disjoint.

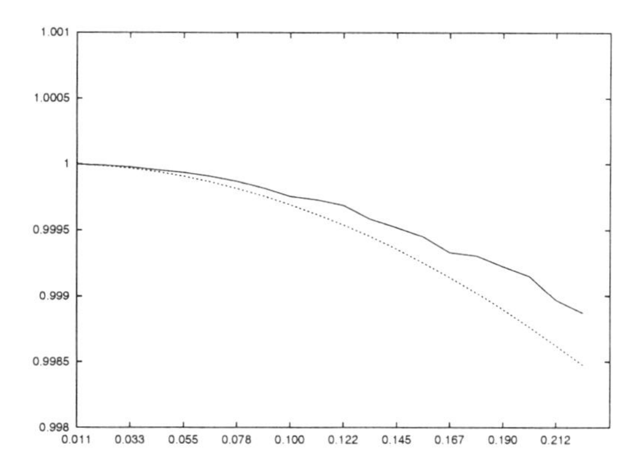

Fig. 5.6 The objective values f returned by the system for the $\mathcal{TCG}(20, 2, 1, \alpha, 0.1, 0.1)$ with varying α between 0.011 and 0.224. The constraint violation is zero for all α. The dotted line represents the height of the lowest peak in the search space

We get similar results while varying β. For $\beta = 0.05$ (the inner rings are very small), the system finds an avarage peak and converges there (for small values of β the value of the lowest peak is close to 1, thus the false impression about the quality of solution in that case). Increase of β triggers decrease of the lowest peak: this explains the negative slope of the dotted line (Fig. 5.7). For $\beta = 1$ there is a slight constraint violation as the feasible rings have the width zero.

A few observations can be made on the basis of the above experiments. Clearly, the results static penalty method depend on the penalty coefficient W, which should be selected carefully; in the case of the $\mathcal{TCG}$, the value of this coefficient depends on the dimension n of the problem. In the case of one active constraint (which is always the case), the system does not have any difficulties in locating a feasible solution (for selected $n = 20$); however, due to a large number of peaks (over one million for the case of $n = 20$ and $w = 2$), the method returns a feasible point from an average peak. For further experiments, modifications of the $\mathcal{TCG}$, and observations, see [23].

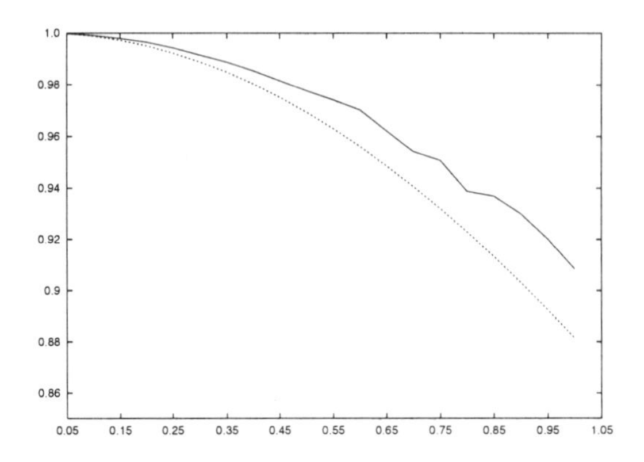

Fig. 5.7 The objective values f returned by the system for the $\mathcal{TCG}(20, 2, 1, 0.201246, \beta, 0.1)$ with varying β between 0.05 and 1.0. The constraint violation is zero for all β. The dotted line represents the height of the lowest peak in the search space

5.6 SUMMARY

We have discussed how a test case generator $\mathcal{TCG}(n, w, \lambda, \alpha, \beta, \mu)$ can be used for evaluation of a constraint-handling technique. As explained in section 5.4, the parameter n controls the dimensionality of the test function, the parameter w controls the modality of the function, the parameter λ controls the number of constraints in the search space, the parameter α controls the connectedness of the feasible search space, the parameter β controls the ratio of the feasible to total search space, and the parameter μ controls the ruggedness of the test function.

When the feasible region is disjointed, classical search and optimization methods may have difficulty finding the global optimal solution. Even for evolutionary algorithms, finding the feasible island containing the global optimal solution would be a challenging task. This task can be even more difficult if the fitness landscape is complex (e.g., for large values of μ).

We believe that such constrained test problem generator should serve the purpose of testing a constrained optimizer. In the previous section we have indicated how it can be used to evaluate merits and drawbacks of one particular constraint handling method (static penalties). Note that it is possible to analyse further the performance of a method by varying two or more parameters of the $\mathcal{TCG}$ (see [23] for a full discussion on these results).

The proposed test case generator is far from perfect. For example, it might be worthwhile to modify it further to parametrize the number of active constraints at the optimum. It seems necessary to introduce a possibility of a gradual increment of peaks (in the current version of the test case generator, $w = 1$ implies one peak, and $w = 2$ implies 2^n peaks). Also, the difference between the lowest and the highest values of the peaks in the search space are, in the present model, too small. However, the proposed $\mathcal{TCG}$ even in its current form is an important tool in analysing any constraint-handling method (for any algorithm: not necessarily evolutionary algorithm), and it represents an important step in the "right" direction.

Acknowledgments

The second author acknowledges the support provided by the Alexander von Humboldt Foundation, Germany. The research reported in this paper was partially supported by the ESPRIT Project 20288 Cooperation Research in Information Technology (CRIT-2): "Evolutionary Real-time Optimization System for Ecological Power Control".

REFERENCES

1. Bäck, T., *Evolutionary Algorithms in Theory and Practice*, New-York, Oxford University Press, 1995.

2. Bäck, T., Fogel, D.B., and Michalewicz, Z., *Handbook of Evolutionary Computation*, University Oxford Press, New York, 1996.

3. Bean, J. C. and A. B. Hadj-Alouane (1992). A dual genetic algorithm for bounded integer programs. Technical Report TR 92-53, Department of Industrial and Operations Engineering, The University of Michigan.

4. Bilchev, G. and I. Parmee (1995). Ant colony search vs. genetic algorithms. Technical report, Plymouth Engineering Design Centre, University of Plymouth.

5. Colorni, A., M. Dorigo, and V. Maniezzo (1991). Distributed optimization by ant colonies. In *Proceedings of the First European Conference on Artificial Life*, Paris. MIT Press/Bradford Book.

6. Dasgupta, D. and Michalewicz, Z., (Editors), *Evolutionary Algorithms in Engineering Applications*, Springer-Verlag, New York, 1997.

7. Davis, L. (1997). Private communication.

8. Deb, K. (in press). An efficient constraint handling method for genetic algorithms. *Computer Methods in Applied Mechanics and Engineering*.

9. Eiben, A. and Z. Ruttkay (1996). Self-adaptivity for Constraint Satisfaction: Learning Penalty Functions. In Proceedings of the 3rd IEEE Conference on Evolutionary Computation, IEEE Press, 1996, pp. 258–261.

10. Fogel, D. B., *Evolutionary Computation. Toward a New Philosophy of Machine Intelligence*, IEEE Press, 1995.

11. Goldberg, D.E., *Genetic Algorithms in Search, Optimization and Machine Learning*, Addison-Wesley, Reading, MA, 1989.

12. Gregory, J. (1995). Nonlinear Programming FAQ, Usenet sci.answers. Available at ftp://rtfm.mit.edu/pub/usenet/sci.answers/nonlinear-programming-faq.

13. Hadj-Alouane, A. B. and J. C. Bean (1992). A genetic algorithm for the multiple-choice integer program. Technical Report TR 92-50, Department of Industrial and Operations Engineering, The University of Michigan.

14. Homaifar, A., S. H.-Y. Lai, and X. Qi (1994). Constrained optimization via genetic algorithms. *Simulation 62*(4), 242–254.

15. Joines, J. and C. Houck (1994). On the use of non-stationary penalty functions to solve nonlinear constrained optimization problems with gas. In Z. Michalewicz, J. D. Schaffer, H.-P. Schwefel, D. B. Fogel, and H. Kitano (Eds.), *Proceedings of the First IEEE International Conference on Evolutionary Computation*, pp. 579–584. IEEE Press.

16. Kozieł, S. and Z. Michalewicz, (1998). A decoder-based evolutionary algorithm for constrained parameter optimization problems. In *Proceedings of the* 5th *Conference on Parallel Problems Solving from Nature*. Springer Verlag.

17. Kozieł, S. and Z. Michalewicz, (1999). Evolutionary Algorithms, Homomorphous Mappings, and Constrained Parameter Optimization. to appear in Evolutionary Computation.

18. Leriche, R. G., C. Knopf-Lenoir, and R. T. Haftka (1995). A segragated genetic algorithm for constrained structural optimization. In L. J. Eshelman (Ed.), *Proceedings of the* 6th *International Conference on Genetic Algorithms*, pp. 558–565.

19. Maa, C. and M. Shanblatt (1992). A two-phase optimization neural network. *IEEE Transactions on Neural Networks 3*(6), 1003–1009.

20. Michalewicz, Z. (1995b). Heuristic methods for evolutionary computation techniques. *Journal of Heuristics 1*(2), 177–206.

21. Michalewicz, Z. (1996). *Genetic Algorithms+Data Structures=Evolution Programs*. New-York: Springer Verlag. 3rd edition.

22. Michalewicz, Z. and N. Attia (1994). Evolutionary optimization of constrained problems. In *Proceedings of the* 3rd *Annual Conference on Evolutionary Programming*, pp. 98–108. World Scientific.

23. Michalewicz, Z., K. Deb, M. Schmidt, and T. Stidsen, *Test-case Generator for Constrained Parameter Optimization Techniques*, submitted for publication.

24. Michalewicz, Z. and C. Z. Janikow (1991). Handling constraints in genetic algorithms. In R. K. Belew and L. B. Booker (Eds.), *Proceedings of the* 4th *International Conference on Genetic Algorithms*, pp. 151–157. Morgan Kaufmann.

25. Michalewicz, Z., T. Logan, and S. Swaminathan (1994). Evolutionary operators for continuous convex parameter spaces. In *Proceedings of the* 3rd *Annual Conference on Evolutionary Programming*, pp. 84–97. World Scientific.

26. Michalewicz, Z. and G. Nazhiyath (1995). Genocop III: A co-evolutionary algorithm for numerical optimization problems with nonlinear constraints. In D. B. Fogel (Ed.), *Proceedings of the Second IEEE International Conference on Evolutionary Computation*, pp. 647–651. IEEE Press.

27. Michalewicz, Z., G. Nazhiyath, and M. Michalewicz (1996). A note on usefulness of geometrical crossover for numerical optimization problems. In P. J. Angeline and T. Bäck (Eds.), *Proceedings of the 5th Annual Conference on Evolutionary Programming*.

28. Michalewicz, Z. and M. Schoenauer (1996). Evolutionary computation for Constrained Parameter Optimization Problems. Evolutionary Computation, Vol.4, No.1, pp.1–32.

29. Myung, H., J.-H. Kim, and D. Fogel (1995). Preliminary investigation into a two-stage method of evolutionary optimization on constrained problems. In J. R. McDonnell, R. G. Reynolds, and D. B. Fogel (Eds.), *Proceedings of the 4th Annual Conference on Evolutionary Programming*, pp. 449–463. MIT Press.

30. Orvosh, D. and L. Davis (1993). Shall we repair? Genetic algorithms, combinatorial optimization, and feasibility constraints. In S. Forrest (Ed.), *Proceedings of the 5th International Conference on Genetic Algorithms*, pp. 650. Morgan Kaufmann.

31. Palmer, C.C. and A. Kershenbaum (1994). Representing Trees in Genetic Algorithms. In *Proceedings of the IEEE International Conference on Evolutionary Computation*, 27–29 June 1994, 379–384.

32. Paredis, J. (1994). Co-evolutionary constraint satisfaction. In Y. Davidor, H.-P. Schwefel, and R. Manner (Eds.), *Proceedings of the 3rd Conference on Parallel Problems Solving from Nature*, pp. 46–55. Springer Verlag.

33. Parmee, I. and G. Purchase (1994). The development of directed genetic search technique for heavily constrained design spaces. In *Proceedings of the Conference on Adaptive Computing in Engineering Design and Control*, pp. 97–102. University of Plymouth.

34. Powell, D. and M. M. Skolnick (1993). Using genetic algorithms in engineering design optimization with non-linear constraints. In S. Forrest (Ed.), *Proceedings of the 5th International Conference on Genetic Algorithms*, pp. 424–430. Morgan Kaufmann.

35. Reynolds, R. (1994). An introduction to cultural algorithms. In *Proceedings of the 3rd Annual Conference on Evolutionary Programming*, pp. 131–139. World Scientific.

36. Reynolds, R., Z. Michalewicz, and M. Cavaretta (1995). Using cultural algorithms for constraint handling in Genocop. In J. R. McDonnell, R. G. Reynolds, and D. B. Fogel (Eds.), *Proceedings of the 4th Annual Conference on Evolutionary Programming*, pp. 298–305. MIT Press.

37. Richardson, J. T., M. R. Palmer, G. Liepins, and M. Hilliard (1989). Some guidelines for genetic algorithms with penalty functions. In J. D. Schaffer (Ed.), *Proceedings of the 3rd International Conference on Genetic Algorithms*, pp. 191–197. Morgan Kaufmann.

38. Schaffer, D. (1985). Multiple objective optimization with vector evaluated genetic algorithms. In J. J. Grefenstette (Ed.), *Proceedings of the 1st International Conference on Genetic Algorithms*. Laurence Erlbaum Associates.

39. Schoenauer, M. and Z. Michalewicz (1996). Evolutionary computation at the edge of feasibility. W. Ebeling, and H.-M. Voigt (Eds.), *Proceedings of the 4th Conference on Parallel Problems Solving from Nature*, Springer Verlag.

40. Schoenauer, M. and Z. Michalewicz (1997). Boundary Operators for Constrained Parameter Optimization Problems. In T. Bäck (Ed.), *Proceedings of the* 7th *International Conference on Genetic Algorithms*, pp.320–329. Morgan Kaufmann.

41. Schoenauer, M. and S. Xanthakis (1993). Constrained GA optimization. In S. Forrest (Ed.), *Proceedings of the* 5th *International Conference on Genetic Algorithms*, pp. 573–580. Morgan Kaufmann.

42. Smith, A. and D. Tate (1993). Genetic optimization using a penalty function. In S. Forrest (Ed.), *Proceedings of the* 5th *International Conference on Genetic Algorithms*, pp. 499–503. Morgan Kaufmann.

43. Surry, P., N. Radcliffe, and I. Boyd (1995). A multi-objective approach to constrained optimization of gas supply networks. In T. Fogarty (Ed.), *Proceedings of the AISB-95 Workshop on Evolutionary Computing*, Volume 993, pp. 166–180. Springer Verlag.

44. van Kemenade, C.H.M. (1998). Recombinative evolutionary search. PhD Thesis, Leiden University, Netherlands, 1998.

45. Waagen, D., P. Diercks, and J. McDonnell (1992). The stochastic direction set algorithm: A hybrid technique for finding function extrema. In D. B. Fogel and W. Atmar (Eds.), *Proceedings of the* 1st *Annual Conference on Evolutionary Programming*, pp. 35–42. Evolutionary Programming Society.

46. Whitley, D., K. Mathias, S. Rana, and J. Dzubera (1995). Building better test functions. In L. Eshelmen (Editor), Proceedings of the 6th International Conference on Genetic Alforithms, Morgam Kaufmann, 1995.

47. Whitley, D., K. Mathias, S. Rana, and J. Dzubera (1995). Evaluating evolutionary algorithms. Artificial Intelligence Journal, Vol.85, August 1996, pp.245–276.

48. Wodrich, M. and G. Bilchev (1997). Cooperative distributed search: the ant's way. Control & Cybernetics, 26 (3).

6 Embedded Path Tracing and Neighbourhood Search Techniques in Genetic Algorithms

C. R. REEVES and T. YAMADA[†]

School of Mathematical and Information Sciences
Coventry University, UK
Email: C.Reeves@coventry.ac.uk

NTT Communication Science Laboratories
Kyoto, Japan
Email: yamada@cslab.kecl.ntt.co.jp

Abstract. Heuristic search methods have been increasingly applied to combinatorial optimization problems. However, it does not always seem to be recognized that there is an interaction between the type of search adopted (the representation, operators, and search strategy) and the nature of the problem. Thus, it sometimes appears as if such properties as local optimality are to be understood as an invariant of the *problem*. In fact, while a specific problem defines a unique search space, different 'landscapes' are created by the different heuristic search operators used to search it.

In this paper, after demonstrating this fact, we turn to the question of how we can exploit such effects in terms of a fruitful search strategy. Many approaches could be taken; the one described here embeds a development of neighbourhood search methods into a genetic algorithm. We present results where this approach is used for flowshop scheduling problems to generate very high-quality solutions.

6.1 INTRODUCTION

Genetic algorithms have become increasingly popular for finding near-optimal solutions to large combinatorial problems (COPs). At the same time other techniques (sometime called

[†]Much of the experimental work in this paper was undertaken when the second author was a visiting researcher in the School of Mathematical and Information Sciences, Coventry University, UK

'metaheuristics'—simulated annealing (SA) and tabu search (TS), for example) have also been found to perform very well, and some attempts have been made to combine them. For a review of some of these techniques see [1], [2].

Central to most heuristic search techniques is the concept of neighbourhood search. (GAs are apparently an exception to this, but in fact there are some points of contact that can be put to profitable use, as we shall discuss later.) If we assume that a solution is specified by a vector $\mathbf{x}$, that the set of all (feasible) solutions is denoted by $\mathbf{X}$ (which we shall also call the *search space*), and the cost of solution $\mathbf{x}$ is denoted by $c(\mathbf{x})$, then every solution $\mathbf{x} \in \mathbf{X}$ has an associated set of *neighbours*, $N(\mathbf{x}) \subset \mathbf{X}$, called the neighbourhood of $\mathbf{x}$. Each solution $\mathbf{x}' \in N(\mathbf{x})$ can be reached directly from $\mathbf{x}$ by an operation called a *move*. Many different types of move are possible in any particular case, and we can view a move as being generated by the application of an operator ω. For example, if $\mathbf{X}$ is the binary hypercube $\mathbb{Z}_2^l$, a simple operator is the bit flip $\phi(k)$, given by

$$\phi(k) : \mathbb{Z}_2^l \to \mathbb{Z}_2^l \qquad \begin{cases} z_k \mapsto 1 - z_k & \\ z_i \mapsto z_i & \text{if } k \neq i \end{cases}$$

where $\mathbf{z}$ is a binary vector of length l.

As another example, which we shall refer to extensively later, we can take the backward shift operator for the case where $\mathbf{X}$ is $\mathbf{\Pi}_n$—the space of permutations π of length n. The operator $\mathcal{BSH}(i,j)$ (where we assume $i < j$) is given by

$$\mathcal{BSH}(i,j) : \mathbf{\Pi}_n \to \mathbf{\Pi}_n \qquad \begin{cases} \pi_k \mapsto \pi_{k+1} & \text{if } i \le k < j \\ \pi_j \mapsto \pi_i & \\ \pi_k \mapsto \pi_k & \text{otherwise} \end{cases}$$

An analogous forward shift operator $\mathcal{FSH}(i,j)$ can similarly be described; the composite of $\mathcal{BSH}$ and $\mathcal{FSH}$ is denoted by $\mathcal{SH}$. Another alternative for such problems is an exchange operator $\mathcal{EX}(i,j)$ which simply exchanges the elements in the ith and jth positions.

A typical NS heuristic operates by generating neighbours in an iterative process where a move to a new solution is made whenever certain criteria are fulfilled. There are many ways in which candidate moves can be chosen for consideration, and in defining criteria for accepting candidate moves. Perhaps the most common case is that of *ascent*, in which the only moves accepted are to neighbours that improve the current solution. *Steepest* ascent corresponds to the case where all neighbours are evaluated before a move is made—that move being the best available. *Next* ascent is similar, but the next candidate (in some pre-defined sequence) that improves the current solution is accepted, without necessarily examining the complete neighbourhood. Normally, the search terminates when no moves can be accepted.

The trouble with NS is that the solution it generates is usually only a *local* optimum—a vector none of whose neighbours offer an improved solution. It should be stressed that the idea of a local optimum only has meaning with respect to a particular neighbourhood. Other associated concepts are those of 'landscapes', 'valleys', 'ridges', and 'basins of attraction' of a particular local optimum. However, these may alter in subtle ways, depending on the neighbourhood used, and the strategy employed for searching it.

Recent empirical analyses [3], [4], [5] have shown that, for many instances of (minimization) COPs, the landscapes induced by some commonly-used operators have a 'big valley' structure, where the local optima occur relatively close to each other, and to a global optimum. This obviously suggests that in developing algorithms, we should try to exploit this structure.

There is as yet no well-defined mathematical description of what it means for a landscape to possess a 'big valley'. The idea is a fairly informal one, based on the observation that in many combinatorial optimization problems local optima are indeed not distributed uniformly throughout the landscape. In the context of landscapes defined on binary strings, Kauffman has been the pioneer of such experiments, from which he suggested several descriptors of a 'big valley' landscape [6]. (Because he was dealing with fitness maximization, he used the term 'central massif', but it is clear that it is the same phenomenon.)

6.2 AVOIDING LOCAL OPTIMA

In recent years many techniques have been suggested for the avoidance of local optima. A survey of these methods can be found in [2]. For completeness, we refer here briefly to some of the most popular principles.

At the most basic level, we could use *iterative restarts* of NS from many different initial points, thus generating a collection of local optima from which the best can be selected.

Simulated annealing uses a controlled randomization strategy—inferior moves are accepted probabilistically, the chance of such acceptance decreasing slowly over the course of a search. By relaxing the acceptance criterion in this way, it becomes possible to move out of the basin of attraction of a local optimum.

Tabu search adopts a deterministic approach, whereby a 'memory' is implemented by the recording of previously-seen solutions. This record could be explicit, but is often an implicit one, making use of simple but effective data structures. These can be thought of as a 'tabu list' of moves which have been made in the recent past of the search, and which are 'tabu' or forbidden for a certain number of iterations. This prevents cycling, and also helps to promote a diversified coverage of the search space.

Perturbation methods improve the restart strategy: instead of retreating to an unrelated and randomly chosen initial solution, the current local optimum is perturbed in some way and the heuristic restarted from the new solution. Perhaps the most widely-known of such techniques is the 'iterated Lin-Kernighan' (ILK) method introduced by Johnson [7] for the travelling salesman problem.

Genetic algorithms differ in using a population of solutions rather than moving from one point to the next. Furthermore, new solutions are generated from two (or, rarely) more solutions by applying a 'crossover' operator. However, they can also be encompassed within an NS framework, as is shown in [8], and as we shall discuss later in this paper.

However, there are other possibilities besides a random perturbation of the population. Glover and Laguna [9] mention an idea called 'path relinking', which suggests an alternative means for exploring the landscape. The terminology does not describe exactly what is meant, at least in the context of this paper: it is points that are being linked, not paths; nor are the points being *re*-linked. For this reason we shall simply call it 'path tracing'.

6.3 PATH TRACING

Suppose we have 2 locally-optimal solutions to a COP. If the operators we are using induce a big valley structure, then it is a reasonable hypothesis that a local search that traces out a path from one such solution to another one will at some point enter the basin of attraction of a third local optimum. Even if no better solution is found on such a path, at least we have gained some knowledge about the relative size of the basins of attraction of the original local optima along one trajectory in the landscape.

Of course, there are many paths that could be taken, and many strategies that could be adopted to trace them out. One possibility is that used by [10] in investigating the idea of 'optima linking' in the context of a binary search space. Here a path was traced by finding at each step the *best* move among all those that would take the current point one step nearer (in the sense of Hamming distance) to the target solution.

If the representation space $\mathbf{X}$ is not the binary hypercube, distance between solutions may not be so easily measured, and several different approaches can be adopted. Later in this paper we shall describe some experimental work carried out in permutation spaces. First, however, we consider some links between the idea of path tracing and genetic algorithms.

6.4 LINKS TO GENETIC ALGORITHMS

If we consider the case of crossover of strings in $\mathbb{Z}_2^l$, any 'offspring' produced from two 'parents' will lie on a path that leads from one parent to another. In an earlier paper [11], we introduced the concept of an *intermediate vector* as follows:

Proposition The following definitions are equivalent:

1. A vector $\mathbf{y}$ in $\mathbb{Z}_2^l$ is *intermediate* between two vectors $\mathbf{x}$ and $\mathbf{z}$, written as $\mathbf{x} \diamond \mathbf{y} \diamond \mathbf{z}$, if and only if

$$\text{either } y_i = x_i \text{ or } y_i = z_i \text{ for } i = 1, \ldots, n.$$

2. Using Hamming distance as a distance measure between two vectors in $\mathbb{Z}_2^l$, given by $d(\mathbf{x}, \mathbf{y}) = \sum_{i=1}^{n} |x_i - y_i|$, $\mathbf{y}$ is *intermediate* between $\mathbf{x}$ and $\mathbf{z}$ if and only if

$$d(\mathbf{x}, \mathbf{z}) = d(\mathbf{x}, \mathbf{y}) + d(\mathbf{y}, \mathbf{z}).$$

The proof of this proposition is trivial, and the insight is not especially profound, but it helps to make clear the fact that every offspring produced by crossover is 'intermediate' between the parents, not merely in some analogical 'genetic' sense, but also in the sense of being an intermediate point on an underlying landscape—in this case, the Hamming landscape. (Whether the Hamming landscape is appropriate for GAs is actually another story—there is a discussion of its relevance in [12].)

The total number of paths that could be traced is very large. As an (extreme) example, *every* point clearly lies between $(00\ldots00)$ and $(11\ldots11)$. More generally, if the distance between two parents is d, there will be $d!$ possible paths that could be taken in the Hamming landscape.

In the case of non-binary strings, the distance measure may be more complicated, but the principle is still relevant. However, while crossover (of whatever variety) will generate a string that is intermediate between its parents, and is thus on such a path, it pays no attention to the cost function. It is also true that some crossover operators sample the set of all intermediate points in what might be seen as a rather eccentric way, as is discussed in [11]. In that paper it is suggested that some sort of local optimization should be attempted to find a 'good' intermediate point instead of making what is (in terms of the cost landscape) a rather arbitrary move. This led to the proposal for a 'neighbourhood search crossover' (NSX) that was found to produce good results in certain scheduling and graph problems [8]. Subsequently, Yamada and Nakano [13] extended this idea to 'multi-step crossover with fusion' (MSXF) which gave excellent results for some job-shop scheduling problems.

More recently, path tracing from one parent to another has also been suggested as a reasonable candidate for an 'intelligent' approach to recombination. This approach will be described in the rest of this paper.

6.5 EMBEDDED PATH TRACING IN SCHEDULING PROBLEMS

As remarked above, the problem of navigating around landscapes in scheduling problems is somewhat more interesting and challenging than in the case of binary strings. The problem we chose to study in detail was the permutation flowshop problem (PFSP). Earlier work [14] had shown that a fairly simple GA was able to obtain excellent results in a reasonable amount of computer time.

6.5.1 The flowshop problem

The PFSP is the following problem: if we have processing times $p(i, j)$ for job i on machine j, and a job permutation $\boldsymbol{\pi} = \{\pi_1, \pi_2, \cdots, \pi_n\}$, where there are n jobs and m machines, then we calculate the completion times $C(\pi_i, j)$ as follows:

$$
\begin{aligned}
C(\pi_1, 1) &= p(\pi_1, 1) \\
C(\pi_i, 1) &= C(\pi_{i-1}, 1) + p(\pi_i, 1) \quad for\ i = 2, \ldots, n \\
C(\pi_1, j) &= C(\pi_1, j-1) + p(\pi_1, j)\ for\ \ j = 2, \ldots, m \\
C(\pi_i, j) &= \max\{C(\pi_{i-1}, j), C(\pi_i, j-1)\} + p(\pi_i, j) \\
&\quad for\ i = 2, \ldots, n;\ j = 2, \ldots, m
\end{aligned}
$$

Finally, we define the *makespan* as

$$C_{max}(\boldsymbol{\pi}) = C(\pi_n, m).$$

The *makespan version* of the PFSP is then to find a permutation π^* such that

$$C_{max}(\pi^*) \leq C_{max}(\pi) \ \forall \, \pi \in \mathbf{\Pi}_n.$$

While it is the problem with the makespan objective that has received most attention, other objectives can also be defined. For example, we could seek to minimize the mean *flow-time* (the time a job spends in process), or the mean *tardiness* (assuming some deadline for each job). If the release dates of all jobs are zero (i.e., all jobs are available at the outset), the mean flow-time objective reduces to minimizing

$$C_{sum}(\pi) = \sum_{i=1}^{n} C(\pi_i, m).$$

We call this the C_{sum} *version* of the PFSP. Results obtained in both problems will be discussed later.

6.5.2 The meaning of distance

An immediate question that arises is how to measure distance in permutation problems. In the context of a permutation space, it would seem logical to measure distances on the landscape with respect to the operator that induces the landscape, but this is not always an easy task. It is possible for the $\mathcal{EX}$ operator to compute quite easily the minimum number of exchanges needed to transform one permutation π into another π'. But problems arise for $\mathcal{BSH}$—in fact for both $\mathcal{BSH}$ and $\mathcal{FSH}$ the induced distances are not even symmetric. In any case, in practice it would be more convenient to have an operator-independent measure of distance. In [4] 4 measures were investigated, and it was concluded that two of them were suitable:

- The **precedence-based measure** counts the number of times job j is *preceded* by job i in both π and π'; to obtain a 'distance', this quantity is subtracted from $n(n-1)/2$.
- The **position-based measure** compares the actual *positions* in the sequence of job j in each of π and π'. For a sequence π its inverse permutation σ gives the position of job π_i (i.e, $\sigma_{\pi_i} = i$). The position-based measure is then just

$$\sum_{j=1}^{n} |\sigma_j - \sigma'_j|.$$

These measures are closely related to each other, and in what follows we use only the precedence measure. It should also be mentioned, as a further justification for using these measures, that recent empirical investigations [15] have suggested that in fact the correlation of solution quality with operator-based distance measures is actually *less* than with these operator-independent measures.

Using such distance measures we were able to implement a very simple version of path tracing, as described in [16], which produced good-quality results for Taillard's benchmark instances [17] using a comparatively small amount of computation. However, the major goal of this research was to explore the potential of this approach for improving a traditional GA.

6.6 THE GENETIC ALGORITHM FRAMEWORK

Of course there is no such thing as **the** genetic algorithm, so the effect of embedded path tracing can only be measured relative to an existing example of the paradigm. In order to enable comparability with earlier work, the GA used was kept as close as possible to that used successfully in [14].

This was a steady-state GA in which parents were selected probabilistically, the probability of selection being based on a linear ranking of their makespan values. In the earlier work crossover was the PMX operator of [18] or a variant of Davis's order operator [19]. Mutation was implemented by an application of a $\mathcal{BSH}(i, j)$ or $\mathcal{FSH}(i, j)$ with i, j chosen randomly. An adaptive mutation rate was used to prevent premature convergence. A newly generated solution was inserted into the population only if its makespan is better than the worst in the current population (which is deleted).

However, the nature of the changes to be made were such that not all the prescriptions used previously were suitable. Both crossover and mutation have to be re-defined (as will be explained below), and the populations used were smaller. Also, in order to avoid premature convergence, no two individuals were allowed to have the same makespan.

6.6.1 'Crossover'

As proposed above, crossover in the conventional sense was replaced by a path tracing approach. The approach taken was a development of the MSXF operator mentioned above. MSXF carries out a short term local search starting from one of the parent solutions (the initial point) and moving in the direction of the other parent (the target point). It clearly, therefore, needs a neighbourhood, and a strategy for deciding whether or not to move to a candidate neighbour. The method adopted in [13], and also used here was as shown in Fig. 6.1, which describes the approach in a generic way. C_{max} can of course be replaced by whatever is appropriate to the problem given. The termination condition can be given by, for example, a fixed number of iterations L in the outer loop. The best solution $\mathbf{q}$ is used for the next generation.

The neighbourhood used was a variant of that generated by the $\mathcal{SH}$ operator. Many authors—for example, [20], [21]—have reported that $\mathcal{SH}$ seems on average to be better than $\mathcal{EX}$. As an additional reason for using $\mathcal{SH}$, it should be noted that it is possible to speed up the computation for $\mathcal{SH}$ by using Mohr's algorithm. This is described in [22] and adapted for NS in [21]. Nowicki and Smutnicki [20] have improved the effectiveness of $\mathcal{SH}$ still further by making use of the notion of a *critical path*: the set of *operations*—(job,machine) pairs—for which there is no idle time. The size of the $\mathcal{SH}$ based neighbourhood can be reduced by focusing on the *critical blocks*—subsequences of critical operations on the same machine. Only adjacent blocks were considered for the application of $\mathcal{SH}$, further reducing the size of the neighbourhood. The idea is pictured in Fig. 6.2. Further details can be found in [23].

Does this reduced neighbourhood still induce a big valley? The answer, from experiments carried out and reported in [23], seems to be 'yes'. Again, further details can be found in the paper cited.

- Let π_1, π_2 be the relevant solutions. Set $\mathbf{x} = \mathbf{q} = \pi_1$.

do

- For each neighbour $\mathbf{y}_i \in N(\mathbf{x})$, calculate $d(\mathbf{y}_i, \pi_2)$.
- Sort $\{\mathbf{y}_i\}$ in ascending order of $d(\mathbf{y}_i, \pi_2)$.

do

1. Select $\mathbf{y}_i$ from $N(\mathbf{x})$ with a probability inversely proportional to the index i.
2. Calculate $C_{max}(\mathbf{y}_i)$ if it is unknown.
3. Accept $\mathbf{y}_i$ with probability 1 if $C_{max}(\mathbf{y}_i) \leq C_{max}(\mathbf{x})$, and with probability $P_c(\mathbf{y}_i)$ otherwise.
4. Change the index of $\mathbf{y}_i$ from i to n, and the indices of $\mathbf{y}_k$ $(k \in \{i+1, \ldots, n\})$ from k to $k-1$.

until $\mathbf{y}_i$ is accepted.

- Set $\mathbf{x} = \mathbf{y}_i$.
- If $C_{max}(\mathbf{x}) < C_{max}(\mathbf{q})$ then set $\mathbf{q} = \mathbf{x}$.

until some termination condition is satisfied.

- $\mathbf{q}$ is used for the next generation.

Fig. 6.1 Path tracing crossover. The initial point on the path is π_1 and π_2 is the target. The neighbourhood of solution $\mathbf{x}$ is denoted by $N(\mathbf{x})$; neighbours that are closer to π_2 are probabilistically preferred; better neighbours (smaller C_{max} values) are always accepted, otherwise $\mathbf{y}_i$ may be accepted with probability $P_c(\mathbf{y}_i) = \exp(-\Delta C_{max}/c)$, where $\Delta C_{max} = C_{max}(\mathbf{y}_i) - C_{max}(\mathbf{x})$. This last prescription is similar to the approach of simulated annealing, corresponding to annealing at a constant temperature $T = c$.

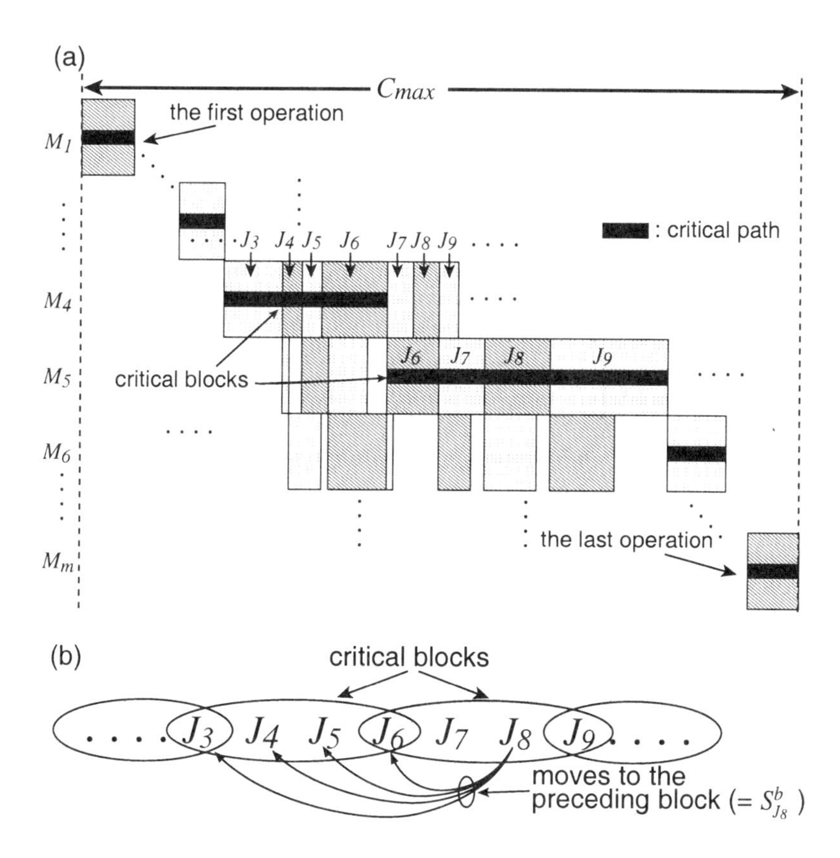

Fig. 6.2 An example of a critical path and blocks. Jobs $J_3, \ldots, J_6$ (on machine M_4) form one critical block and $J_6, \ldots, J_9$ another (on machine M_5); 4 possible moves are shown for shifting J_8 to some position in the preceding block

6.6.2 'Mutation'

It is also relevant to ask whether there is a path-tracing analogue for mutation. To answer this, we need to consider the purpose of mutation. Most would probably agree that if crossover promotes 'intensification' of the search, mutation promotes 'diversification' (where we use Glover's terminology [9]). That is, mutation tends to move the population away from the region to which crossover has brought it. In the analogy to path tracing, this corresponds to extrapolation of a path between parents, where crossover corresponds to interpolation.

This idea can be implemented by a simple modification in Fig. 6.1, where the neighbours of $\mathbf{x}$ are sorted in *descending* order of $d(\mathbf{y}_i, \pi_2)$ and the most *distant* solution is stored and used in the next generation instead of $\mathbf{q}$, if $\mathbf{q}$ does not improve the parent solutions. However, 'mutation' was only invoked if the distance between parents was less than a value d_{min}. This is reasonable, because if the parents are too close together, 'crossover' is of small value, since the potential neighbours and paths are very few—in the limit, where the distance is one unit, there is of course no path at all.

6.6.3 Local search

Early in the search, even the best members of the population are unlikely to be local optima. They may even be in quite the 'wrong' part of the landscape—some distance from the 'big valley'. This is in fact a criticism often made of GAs, and many authors have experimented with various ways of incorporating local neighbourhood search into a GA. (This is sometimes called a 'memetic' algorithm [5].)

There is good reason therefore for including NS explicitly in the algorithm. In other words, the path is used as a jumping-off point for an attempted improvement phase within the main GA loop. This has two benefits: it means that the big valley region can be found quickly, so it is important early on. But it is also important later, in searching the big valley more thoroughly than might be done by path tracing alone.

As it was not clear exactly how much local search should be done, we decided to make the choice a probabilistic one—with probability P_X the algorithm would use path tracing (either interpolatory or extrapolatory), otherwise NS was used on the first parent selected. However, a complete search to full local optimality could require considerable computation, so a balance was struck by making the NS stochastic (as in simulated annealing, but with a constant 'temperature' P_c), and fixing the number of iterations at a limit L. Fig. 6.3 gives an idealized picture of the concept.

6.7 RESULTS

6.7.1 The makespan version

The above ideas were incorporated into a C program and implemented on a DEC Alpha 600 5/226 computer. A template summarizing the final version of the algorithm is given in Fig. 6.4.

The parameter values used were as follows (no attempt was made to investigate whether these were in any sense 'optimal'). The population size $= 15$, constant temperature $c = 3$,

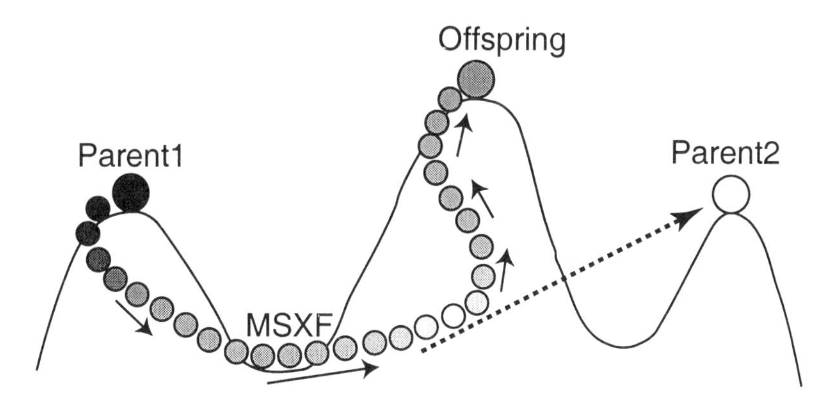

Fig. 6.3 Path tracing and local search: The search traces a path from the initial parent towards the target. In the 'middle' of the search, good solutions may be found somewhere between the parents. A local search can then exploit this new starting point by climbing towards the top of a hill (or the bottom of a valley, if it is a minimization problem)—a new local optimum.

number of iterations for each path tracing or local search phase $L = 1000$, and $d_{min} = n/2$. The probability $P_X = 0.5$.

Global optima are now known for all Taillard's smaller PFSP benchmarks (instances ranging from 20×5 to 50×10 in size). The algorithm was run on these, and in every case it quite quickly found a solution within about 1% of the global optimum.

Of more interest are the larger problem instances, where global optima are still unknown. Table 6.1 summarizes the performance statistics for a subset of Taillard's larger benchmarks, together with the results found by Nowicki and Smutnicki using their TS implementation [20] and the lower and upper bounds, taken from the OR-library [24]. (These upper bounds are the currently best-known makespans, most of them found by a branch and bound technique with computational time unknown).

6.7.2 The C_{sum} version

Th C_{sum} version of the PFSP has not been subject to the same intensity of investigation as the makespan version. Optimal solutions for the Taillard benchmarks are not known for this case. The problems are more difficult to optimize, mainly because the calculation of the objective function is more time consuming, good lower bounds are hard to generate, and problem specific knowledge such as critical blocks cannot be used. Some constructive algorithms based on heuristic rules have been suggested [25], and recently improved [26], but better quality solutions can be generated by the approach described in this paper.

Basically, this is the same as the one tested on the makespan version. However, some differences were inevitable—the inability to use critical blocks for example. The use of a reduced neighbourhood is however an important part of the algorithm suggested, since it prevents computation in the path tracing and local search stages from becoming too large. In

- Initialize population: randomly generate a set of permutation schedules. Sort the population members in descending order of their makespan values.

do

1. Select two sequences π_1, π_2 from the population with a probability inversely proportional to their ranks.
2. Do Step (a) with probability P_X, otherwise do Step (b).
   (a) **If** the precedence-based distance between π_1 and π_2 is less than d_{min}, extrapolate from π_1 and generate **q**. **Otherwise**, apply path tracing between π_1 and π_2 using the reduced neighbourhood and the precedence-based distance and generate a new schedule **q**.
   (b) Apply neighbourhood search to π_1 with acceptance probability P_c and the reduced neighbourhood. **q** is the best solution found during L iterations.
3. If **q**'s makespan is less than the worst in the population, and no member of the current population has the same makespan as **q**, replace the worst individual with **q**.

until some termination condition is satisfied.

- Output the best schedule in the population.

Fig. 6.4 The GA with embedded path tracing and neighbourhood search for the PFSP.

Table 6.1 Results of the Taillard benchmark problems: *best, mean, std* denote the best, average and standard deviation of our 30 makespan values; *nowi* denotes the results of Nowicki and Smutnicki; *lb, ub* are the theoretical lower bounds and currently best known makespan values taken from the OR-library; **bold** figures are those which are better than the current *ub*.

50 × 20	best	mean	std	nowi	lb – ub
1	**3861**	3880	9.3	3875	3771–3875
2	**3709**	3716	2.9	3715	3661–3715
3	**3651**	3668	6.2	3668	3591–3668
4	**3726**	3744	6.1	3752	3631–3752
5	**3614**	3636	8.9	3635	3551–3635
6	3690	3701	6.7	3698	3667–3687
7	3711	3723	5.6	3716	3672–3706
8	**3699**	3721	7.7	3709	3627–3700
9	3760	3769	5.2	3765	3645–3755
10	3767	3772	4.2	3777	3696–3767
100 × 20	best	mean	std	nowi	lb – ub
1	6242	6259	9.6	6286	6106–6228
2	6217	6234	8.9	6241	6183–6210
3	6299	6312	7.8	6329	6252–6271
4	6288	6303	2.7	6306	6254–6269
5	6329	6354	11.3	6377	6262–6319
6	**6380**	6417	12.7	6437	6302–6403
7	6302	6319	11.0	6346	6184–6292
8	6433	6466	17.3	6481	6315–6423
9	6297	6323	11.4	6358	6204–6275
0	6448	6471	10.6	6465	6404–6434
200 × 20	best	mean	std	nowi	lb – ub
1	11272	11316	20.8	11294	11152–11195
2	11299	11346	21.4	11420	11143–11223
3	11410	11458	25.2	11446	11281–11337
4	11347	11400	29.9	11347	11275–11299
5	11290	11320	16.6	11311	11259–11260
6	11250	11288	23.4	11282	11176–11189
7	11438	11455	9.3	11456	11337–11386
8	11395	11426	16.4	11415	11301–11334
9	11263	11306	21.5	11343	11145–11192
10	11335	11409	31.3	11422	11284–11313

this case, we used a reduced neighbourhood as follows: k members of the neighbourhood is chosen (at random) and tested. The best move in this candidate subset is then chosen. A mechanism for doing this efficiently is given in [21].

Of course, the selected move may not be an improving one, and in order to prevent the acceptance of a really bad move, the local search is made stochastic as before, but with a modification described in more detail in [27]. The other potential problem is that the search could cycle, revisiting points using a standard pattern of moves. This is classic tabu search (TS) territory, and this implementation enables us to make use of TS ideas of adaptive memory [9].

We make use of the simplest TS approach based on *recency*. If a solution is generated from the current solution π by moving a job $j = \pi_i$ to a new position, the pair (j, i), i.e. the job and its original position, is recorded as tabu for the next l iterations. Thus subsequent moves cannot generate a solution in which job j is in position i until at least l iterations have passed. The number of iterations of the local search or path tracing component was limited to a value L as before, while the GA (the 'outer loop') was allowed L' 'events' (i.e. Step 2 of Fig. 6.4).

The method was applied to some of Taillard's benchmark problems. For the relatively easy problems ($20 \times 5, 20 \times 10$ and 20×20) six runs were carried out for each problem with different random seeds. The parameters used in these experiments were: population size $= 5, L = 1000, L' = 700, P_X = 0.5$ and the tabu tenure $l = 7$.

Quite consistent results were obtained, i.e. almost all of the 6 runs converged to the same job sequence in a short time (from a few seconds to a few minutes) before the limit of $L_2 = 700$ was reached on a HP workstation. The best results (and they are also the average results in most cases) are reported in Table 6.2 together with the results obtained by the constructive method (NSPD) due to Liu [26].

Table 6.2 Taillard's benchmark results (20 jobs)

20 × 5			20 × 10			20 × 20		
prob	best	NSPD	prob	best	NSPD	prob	best	NSPD
1	14033	14281	11	20911	21520	21	33623	34119
2	15151	15599	12	22440	23094	22	31587	32706
3	13301	14121	13	19833	20561	23	33920	35290
4	15447	15925	14	18710	18867	24	31661	32717
5	13529	13829	15	18641	19580	25	34557	35367
6	13123	13420	16	19245	20010	26	32564	33153
7	13548	13953	17	18363	19069	27	32922	33763
8	13948	14235	18	20241	21048	28	32412	33234
9	14295	14552	19	20330	21138	29	33600	34416
10	12943	13054	20	21320	22212	30	32262	33045

The next group of problems (50×5 and 50×10) are much more difficult. In each run the best results were different. Ten runs were carried out for each problem with different random seeds. The parameters used in these experiments were: population size $= 30, L_1 = 10000, L_2 = 700, P_X = 0.5$. It takes 45 minutes per run for 50×5 instances and 90 minutes for 50×10.

Table 6.3 Taillard's benchmark results (50 jobs)

50×5				50×10			
prob	best	mean	NSPD	prob	best	mean	NSPD
31	64860	64934.8	66590	41	87430	87561.4	90373
32	68134	68247.2	68887	42	83157	83305.8	86926
33	63304	63523.2	64943	43	79996	80303.4	83213
34	68259	68502.7	70040	44	86725	86822.4	89527
35	69491	69619.6	71911	45	86448	86703.7	89190
36	67006	67127.6	68491	46	86651	86888.0	91113
37	66311	66450.0	67892	47	89042	89220.7	93053
38	64412	64550.1	66037	48	86924	87180.5	90614
39	63156	63223.8	64764	49	85674	85924.3	91289
40	68994	69137.4	69985	50	88215	88438.6	91622

It is not certain how good these solutions are. Certainly, they improve on the constructive method by a useful margin, but the lower bounds are some way off (on average around 30%—but they are probably not very good bounds). Even for the easier problems in Table 6.2, there is no guarantee that the best solutions obtained so far are optimal, although they are closer to their lower bounds. For the problems in Table 6.3, the best results could probably still be improved by increasing the amount of computation. For example, a solution to problem 31 was found with $C_{sum} = 64803$ by an overnight run.

6.8 CONCLUSIONS

The concept of path tracing has been introduced as a means of enhancing a genetic algorithm by extending the concept of crossover, and also, not entirely incidentally, mutation. In the cases examined the GA's performance was greatly enhanced. For both the makespan version and the C_{sum} version of the flowshop sequencing problem, the results for some established benchmarks are among some of the best known.

However, beating benchmarks is in some ways a mug's game. We would rather emphasize the methodological opportunity to bring GAs and neighbourhood search methods together. Using hybrid approaches consisting of GAs and other heuristics is not a new idea, but this perspective offers the opportunity for an almost seamless integration of their capabilities. While we cannot say whether these results would extend to all implementations of a GA, or to all combinatorial optimization problems, we would suggest that if the big valley conjecture holds for a particular landscape, the effect on a GA of embedding path tracing and neighbourhood search in this way is likely to be beneficial.

REFERENCES

1. C.R.Reeves (Ed.) (1993) *Modern Heuristic Techniques for Combinatorial Problems*, Blackwell Scientific Publications, Oxford, UK; re-issued by McGraw-Hill, London, UK (1995).

2. C.R.Reeves (1996) Heuristic search methods: A review. In D.Johnson and F.O'Brien (1996) *Operational Research: Keynote Papers 1996*, Operational Research Society, Birmingham, UK, 122-149.

3. K.D.Boese, A.B.Kahng and S.Muddu (1994) A new adaptive multi-start technique for combinatorial global optimizations. *Ops.Res. Letters*, **16**, 101-113.

4. C.R.Reeves (1997) Landscapes, operators and heuristic search. To appear in *Annals of Operations Research*.

5. P.Merz and B.Freisleben (1998) Memetic algorithms and the fitness landscape of the graph bi-partitioning problem. In A.E.Eiben, T.Bäck, M.Schoenauer, H-P.Schwefel (Eds.) (1998) *Parallel Problem-Solving from Nature—PPSN V*, Springer-Verlag, Berlin, 765-774.

6. S.Kauffman (1993) *The Origins of Order: Self-Organisation and Selection in Evolution*. Oxford University Press.

7. D.S.Johnson (1990) Local optimization and the traveling salesman problem. In G.Goos and J.Hartmanis (Eds.) (1990) *Automata, Languages and Programming*, Lecture Notes in Computer Science **443**, Springer-Verlag, Berlin, 446-461.

8. C.R.Reeves and C.Höhn (1996) Integrating local search into genetic algorithms. In V.J.Rayward-Smith, I.H.Osman, C.R.Reeves and G.D.Smith (Eds.) (1996) *Modern Heuristic Search Methods*, John Wiley & Sons, New York, 99-115.

9. F.Glover and M.Laguna (1993) Tabu Search. Chapter 3 in [1].

10. S.Rana and D.Whitley (1997) Bit representation with a twist. In T.Bäck (Ed.) (1997) *Proceedings of the 7th International Conference on Genetic Algorithms*, Morgan Kaufmann, San Francisco, CA, 188-195.

11. C.R.Reeves (1994) Genetic algorithms and neighbourhood search. In T.C.Fogarty (Ed.) (1994) *Evolutionary Computing: AISB Workshop, Leeds, UK, April 1994; Selected Papers*. Springer-Verlag, Berlin, 115-130.

12. C.Höhn and C.R.Reeves (1996) The crossover landscape for the *onemax* problem. In J.Alander (Ed.) (1996) *Proceedings of the 2nd Nordic Workshop on Genetic Algorithms and their Applications*, University of Vaasa Press, Vaasa, Finland, 27-43.

13. T.Yamada and R.Nakano (1996) Scheduling by genetic local search with multi-step crossover. In H-M Voigt, W.Ebeling, I.Rechenberg and H-P Schwefel (Eds.) (1996) *Parallel Problem-Solving from Nature-PPSN IV*, Springer-Verlag, Berlin, 960-969.

14. C.R.Reeves (1995) A genetic algorithm for flowshop sequencing. *Computers & Operations Research*, **22**, 5-13.

15. T.Yamada and C.R.Reeves (1998) Distance measures in the permutation flowshop problem. (In preparation.)

16. C.R.Reeves and T.Yamada (1998) Implicit tabu search methods for flowshop sequencing. In P.Borne, M.Ksouri and A.El Kamel (Eds.) *Proc. IMACS International Conference on Computational Engineering in Systems Applications*, 78-81, IEEE Publishing.

17. E.Taillard(1993) Benchmarks for basic scheduling problems. *European J. Operational Research*, **64**, 278-285.

18. D.E.Goldberg and R.Lingle (1985) Alleles, loci and the traveling salesman problem. In J.J.Grefenstette (Ed.) *Proceedings of an International Conference on Genetic Algorithms and Their Applications*, Lawrence Erlbaum Associates, Hillsdale, New Jersey, 154-159.

19. L.Davis (1985) Job shop scheduling with genetic algorithms. In J.J.Grefenstette (Ed.) *Proceedings of an International Conference on Genetic Algorithms and Their Applications*, Lawrence Erlbaum Associates, Hillsdale, New Jersey, 136-140.

20. E.Nowicki and C.Smutnicki (1996) A fast tabu search algorithm for the permutation flow-shop problem. *European J.Operational Research*, **91**, 160-175.

21. C.R.Reeves (1993) Improving the efficiency of tabu search in machine sequencing problems. *J.Opl.Res.Soc.*, **44**, 375-382.

22. E.Taillard(1990) Some efficient heuristic methods for the flow shop sequencing problem. *European J. Operational Research*, **47**, 65-74.

23. C.R.Reeves and T.Yamada (1998) Genetic algorithms, path relinking and the flowshop sequencing problem. *Evolutionary Computation*, **6**, 45-60.

24. J.E.Beasley (1990) OR-library: Distributing test problems by electronic mail. *European J.Operational Research*, **41**, 1069-1072.

25. C.Wang, C.Chu and J.Proth (1997) Heuristic approaches for $n/m/F/\sum C_i$ scheduling problems, *European J. Operational Research*, 636-644.

26. J. Liu (1997) A new heuristic algorithm for C_{sum} flowshop scheduling problems. Personal communication.

27. T.Yamada and C.R.Reeves (1998) Solving the C_{sum} permutation flowshop scheduling problem by genetic local search. In *Proc. of 1998 International Conference on Evolutionary Computation*, 230–234, IEEE Press.

7 Parallel and Distributed Evolutionary Algorithms: A Review

M. TOMASSINI

Institute of Computer Science, University of Lausanne, 1015 Lausanne, Switzerland.
E-mail: Marco.Tomassini@di.epfl.ch. Web: www-iis.unil.ch.

7.1 INTRODUCTION

Evolutionary algorithms (EAs) find their inspiration in the evolutionary processes occurring in Nature. The main idea is that in order for a population of individuals to adapt to some environment, it should behave like a natural system; survival, and therefore reproduction, is promoted by the elimination of useless or harmful traits and by rewarding useful behavior. Technically, evolutionary algorithms can be considered as a broad class of stochastic optimization techniques. They are particularly well suited for hard problems where little is known about the underlying search space. An evolutionary algorithm maintains a population of candidate solutions for the problem at hand, and makes it evolve by iteratively applying a (usually quite small) set of stochastic operators, known as *mutation*, *recombination* and *selection*. The resulting process tends to find globally satisfactory, if not optimal, solutions to the problem much in the same way as in Nature populations of organisms tend to adapt to their surrounding environment.

When applied to large hard problems evolutionary algorithms may become too slow. One way to overcome time and size constraints in a computation is to parallelize it. By sharing the workload, it is hoped that an N-processor system will do the job nearly N times faster than a uniprocessor system, thereby allowing researchers to treat larger and more interesting problem instances. Although an N-times speedup is difficult to achieve in practice, evolutionary algorithms are sufficiently regular in their space and time dimensions as to be suitable for parallel and distributed implementations. Furthermore, parallel evolutionary algorithms seem to be more in line with their natural counterparts and thus might yield algorithmic benefits besides the added computational power.

This chapter is organized as follows. The next section contains a brief introduction to genetic algorithms and genetic programming. This is followed by a short description of the main aspects of parallel and distributed computer architectures. Next, we review several models of parallel evolutionary algorithms, discussing their workings as well as their respective advantages and drawbacks with respect to the computer or network architecture on which they are executed. In the final sections, we review the state of the theory behind parallel evolutionary algorithms and s few unconventional models. A summary and conclusions section ends the chapter. Parallel genetic algorithms have been reviewed in the past by [6] which also includes an interesting account of the early studies in the field.

7.2 GENETIC ALGORITHMS AND GENETIC PROGRAMMING

Evolutionary algorithms are a class comprising several related techniques. Among them we find chiefly *genetic algorithms*, *genetic programming*, *evolution strategies* and *evolutionary programming*. From the point of view of parallel and distributed computation all such techniques roughly offer the same opportunities. We will therefore study parallel implementations of the two more widespread methodologies i.e., genetic algorithms and genetic programming. The following two sections give a brief introduction to both of these in a classical serial setting. For more details on evolutionary algorithms, the reader can consult the two recent reference texts [17], [4].

7.2.1 An introduction to genetic algorithms

Genetic algorithms (GAs) make use of a metaphor whereby an optimization problem takes the place of the environment; feasible solutions are viewed as individuals living in that environment. An individual's degree of adaptation to its surrounding environment is the counterpart of the objective function evaluated on a feasible solution. In the same way, a set of feasible solutions take the place of a population of organisms.

In genetic algorithms, individuals are just strings of binary digits or of some other set of symbols drawn from a finite set. As computer memory is made up of an array of bits, anything can be stored in a computer can also be encoded by a bit string of sufficient length. Each encoded individual in the population may be viewed as a representation, according to an appropriate encoding, of a particular solution to a problem,

A genetic algorithm starts with a population of randomly generated individuals, although it is also possible to use a previously saved population or a population of individuals encoding for solutions provided by a human expert or by another heuristic algorithm.

Once an initial population has been created, the genetic algorithm enters a loop. At the end of each iteration a new population will have been created by applying a certain number of stochastic operators to the previous population. One such iteration is referred to as a *generation*.

The first operator to be applied is *selection*: in order to create a new intermediate population of n "parents", n independent extractions of an individual from the old population are performed, where the probability for each individual of being extracted is linearly proportional to its fitness. Therefore, above average individuals will expectedly have more copies in the new population, while below average individuals will risk extinction.

Once the population of parents, that is of individuals that have been selected for reproduction, has been extracted, the individuals for the next generation will be produced through the application of a number of reproduction operators, which can involve just one parent (thus simulating asexual reproduction), in which case we speak of mutation, or more parents (thus simulating sexual reproduction), in which case we speak of recombination. In genetic algorithms two reproduction operators are used: crossover and mutation.

To apply crossover, couples are formed with all parent individuals; then, with a certain probability, called crossover rate p_{cross}, each couple actually undergoes crossover: the two bit strings are cut at the same random position and their second halves are swapped between the two individuals, thus yielding two novel individuals, each containing characters from both parents.

After crossover, all individuals undergo mutation. The purpose of mutation is to simulate the effect of transcription errors that can happen with a very low probability (p_{mut}) when a chromosome is duplicated. This is accomplished by flipping each bit in every individual with a very small probability, called mutation rate. In other words, each "0" has a small probability of being turned into a "1" and *vice versa*.

In principle, the above-described loop is infinite, but it can be stopped when a given termination condition specified by the user is met. Possible termination conditions are: a pre-determined number of generations or time has elapsed or a satisfactory solution has been found or no improvement in solution quality has been taking place for a pre-determined number of generations. At this point, it is useful to recall that evolutionary algorithms are stochastic iterative techniques that are not guaranteed to converge although they usually succeed in finding good enough solutions.

The evolutionary cycle can be summarized by the following pseudo-code:

```
generation = 0
seed population
while not termination condition do
    generation = generation + 1
    calculate fitness
    selection
    crossover(p_cross)
    mutation(p_mut)
end while
```

7.2.2 An introduction to genetic programming

Genetic programming (GP) is a more recent evolutionary approach which extends the genetic model of learning to the space of programs. It is a major variation of genetic algorithms in which the evolving individuals are themselves computer programs instead of fixed length strings from a limited alphabet of symbols. Genetic programming is a form of *program induction* that allows to automatically discover programs that solve or approximately solve a given task. The present form of GP is principally due to J. Koza [12].

Individual programs in GP might be expressed in principle in any current programming language. However, the syntax of most languages is such that GP operators would create a large percentage of syntactically incorrect programs. For this reason, Koza chose a syntax in prefix form analogous to LISP and a restricted language with an appropriate number of variables, constants and operators defined to fit the problem to be solved. In this way syntax constraints are respected and the program search space is limited. The restricted language is formed by a user-defined *function set* F and *terminal set* T. The functions chosen are those that are *a priori* believed to be useful for the problem at hand, and the terminals are usually either variables or constants. In addition, each function in the function set must be able to accept as arguments any other function return value and any data type in the terminal set T, a property that is called *syntactic closure*. Thus, the space of possible programs is constituted by the set of all possible compositions of functions that can be recursively formed from the elements of F and T.

As an example, suppose that we are dealing with simple arithmetic expressions in three variables. In this case suitable function and terminal sets might be defined as:

$$F = \{+, -, *, /\}$$

and

$$T = \{A, B, C\}$$

and the following are legal programs: $(+(*AB)(/CD)), and(*(-(+AC)B)A)$.

It is important to note that GP does not need to be implemented in the LISP language (though this was the original implementation). Any language that can represent programs internally as parse trees is adequate. Thus, most GP packages today are written in C, C++ or Java rather than LISP.

Programs are represented as trees with ordered branches in which the internal nodes are functions and the leaves are the terminals of the problem. Thus, the examples given above would give raise to the trees in Fig. 7.1.

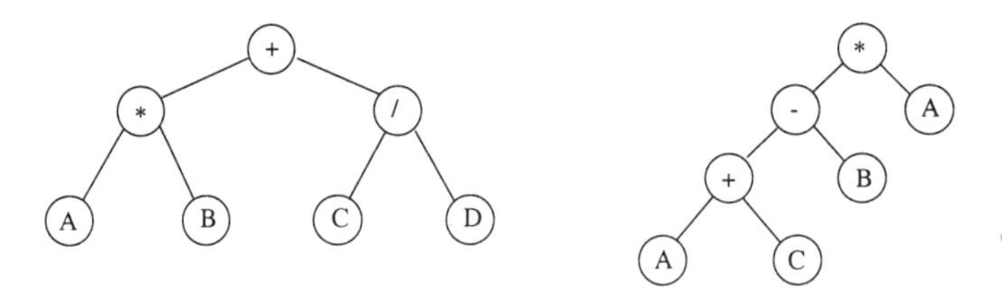

Fig. 7.1 Two GP trees corresponding to the LISP expressions in the text.

Evolution in GP is similar to GAs but different individual representation and genetic operators are used. Once suitable functions and terminals have been determined for the problem at hand, an initial random population of trees (programs) is constructed. From there on the population evolves as with a GA where fitness is assigned after actual execution of the program (individual) and with genetic operators adapted to the tree representation. Fitness calculation is a bit different for programs. In GP we would like to discover a program that satisfies a given number N of predefined input/output relations: these are called the *fitness cases*. For a given program p_i its fitness f_i on the i-th fitness case represents the difference between the output g_i produced by the program and the correct answer G_i for that case. The total fitness $F(p_i)$ is the sum over all N fitness cases of some norm of the cumulated difference:

$$F(p_i) = \sum_{k=1}^{N} \| g_k - G_k \| . \tag{7.1}$$

Obviously, a better program will have a lower fitness under this definition, and a perfect one will score 0 fitness.

The crossover operation starts by selecting a random crossover point in each parent tree and then exchanging the sub-trees, giving rise to two offspring trees, as shown in Fig. 7.2. The crossover site is usually chosen with non-uniform probability, in order to favor internal nodes with respect to leaves. Mutation, when used, is implemented by randomly removing a subtree at a selected point and replacing it with a randomly generated subtree.

One problematic step in GP is the choice of the appropriate language for a given problem. In general, the problem itself suggests a reasonable set of functions and terminals but this is not always the case. Although experimental evidence has shown that good results can be obtained with slightly different choices of F and T, it is clear that the choice of language has an influence on how hard the problem will be to solve with GP. For the time being, there is no guideline for estimating this dependence nor for choosing suitable terminal and function sets: the choice is left to the sensibility and knowledge of the problem solver.

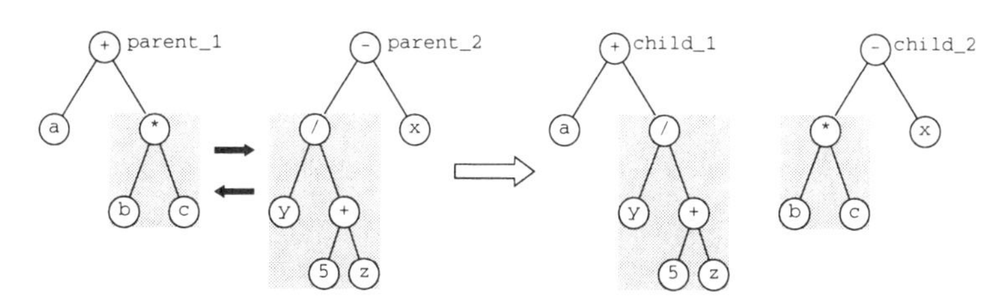

Fig. 7.2 Example of crossover of two genetic programs.

Plain GP works well for problems that are not too complex and that give rise to relatively short programs. To extend GP to more difficult problems some hierarchical principle has to be introduced. In any problem-solving activity hierarchical considerations are needed to produce economically viable solutions. Methods for automatically identifying and extracting useful modules within GP have been discussed by Koza under the name of Automatically Defined Functions (ADF) ([13]), by Angeline and Kinnear [10] and by Rosca [23].

GP is particularly useful for program discovery, i.e. the induction of programs that correctly solve a given problem with the assumption that the form of the program is unknown and that only the desired behavior is given, e.g. by specifying input-output relations. Genetic programming has been successfully applied to a wide variety of problems from many fields, described in [12], [13] and, more recently, in [2], [10], [4]. In conclusion, GP has been empirically shown to be quite a powerful automatic or semi-automatic program-induction and machine learning methodology.

7.3 PARALLEL AND DISTRIBUTED EVOLUTIONARY ALGORITHMS

Parallel and distributed computing is a key technology in the present days of networked and high-performance systems. The goal of increased performance can be met in principle by adding processors, memory and an interconnection network and putting them to work together on a given problem. By sharing the workload, it is hoped that an N-processor system will give rise to a *speedup* in the computation time. Speedup is defined as the the time it takes a single processor to execute a given problem instance with a given algorithm divided by the time on a N-processor architecture of the same type for the same problem instance and with the same algorithm. Sometimes, a different, more suitable algorithm is used in the parallel case. Clearly, in the ideal case the maximum speedup is equal to N. If speedup is indeed about linear in the number of processors, then time consuming problems can be solved in parallel in a fraction of the uniprocessor time or larger and more interesting problem instances can be tackled in the same amount of time. In reality things are not so simple since in most cases several overhead factors contribute to significantly lower the theoretical performance improvement expectations. Furthermore, general parallel programming models turn out to be difficult to design due to the large architectural space that they must span and to the resistance represented by current programming paradigms and languages. In any event, many important problems are sufficiently regular in their space and time dimensions as to be suitable for parallel or distributed computing and evolutionary algorithms are certainly among those.

In the next section we present a quick review of parallel and distributed architectures as an introduction to the discussion of parallel and distributed evolutionary algorithms.

7.3.1 Parallel and distributed computer architectures: an overview

Parallelism can arise at a number of levels in computer systems: task level, instruction level or at some lower machine level. Although there are several ways in which parallel architectures may be classified, the standard model of Flynn is widely accepted as a starting point. But the reader should be aware that it is a coarse-grain classification: for instance, even today's serial processors are in fact highly parallel in the way in which they execute instructions, as well as with respect to the memory interface. Even at a higher architectural level, many parallel architectures are in fact hybrids of the base classes. For a comprehensive treatement of the subject, the reader is referred to Hwang's text [11].

Flynn's taxonomy is based on the notion of instruction and data streams. There are four possible combinations conventionally called *SISD* (single instruction, single data stream), *SIMD* (single instruction, multiple data stream), *MISD* (multiple instruction, single data stream) and *MIMD* (multiple instruction, multiple data stream). Fig. 7.3 schematically depicts the three most important model architectures.

The SISD architecture corresponds to the classical mono-processor machine such as the typical PC or workstation. As stated above, there is already a great deal of parallelism in this architecture at the instruction level (pipelining, superscalar and very long instruction execution). This kind of parallelism is "invisible" to the programmer in the sense that it is built-in in the hardware or must be exploited by the compiler.

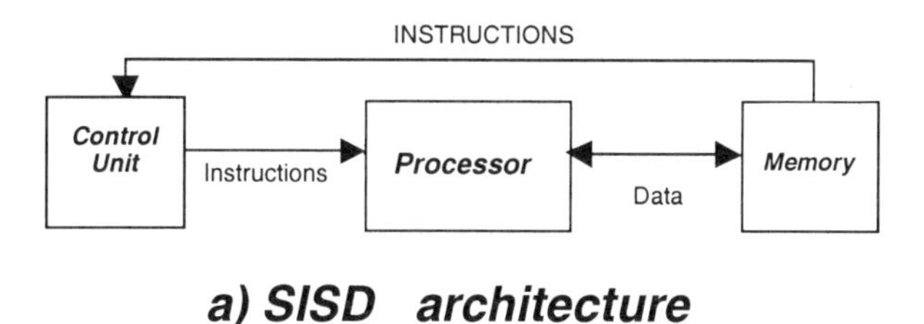

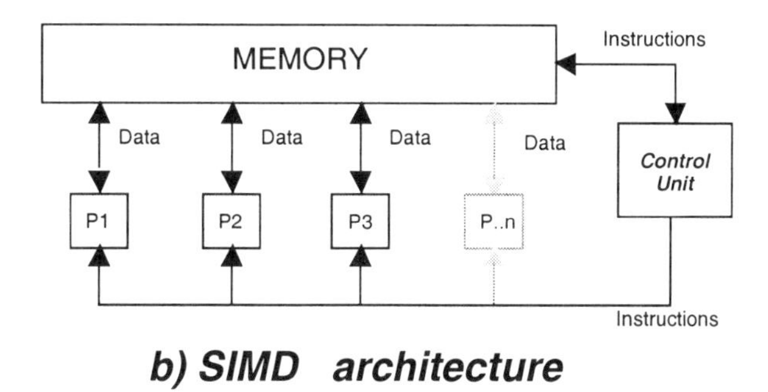

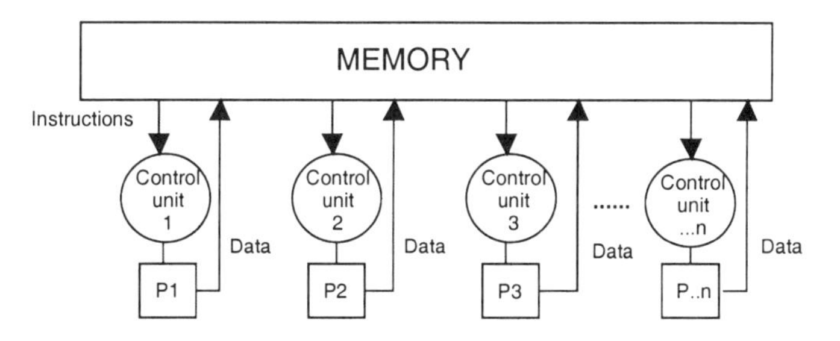

Fig. 7.3 Flynn's model of parallel architectures. The shared memory model is depicted; In the SIMD and MIMD case memory can also be distributed (see text).

In the SIMD architecture the same instruction is broadcast to all processors. The processors operate in lockstep executing the given instruction on different data stored in their local memories (hence the name: single instruction, multiple data). The processor can also remain idle, if this is appropriate. SIMD machines exploit *spatial* parallelism that may be present in a given problem and are suitable for large, regular data structures. If the problem domain is spatially or temporally irregular, many processors must remain idle at a given time step since different operations are called for in different regions. Obviously, this entails a serious loss in the amount of parallelism that can be exploited. Another type of SISD computer is the vector processor which is specialized in the pipelined execution of vector operations. This is the typical component of supercomputers but it is not a particularly useful architecture for evolutionary algorithms.

In the MIMD class of parallel architectures multiple processors work together through some form of interconnection. Different programs and data can be loaded into different processors which means that each processor can execute different instructions at any given point in time. Of course, usually the processors will require some form of synchronization and communication in order to cooperate on a given application. This class is the more generally useful and most commercial parallel and distributed architectures belong to it.

There has been little interest up to now in the MISD class since it does not lend itself to readily exploitable programming constructs. One type of architecture of this class that enjoys some popularity are the so-called *systolic arrays* which are used in specialized applications such as signal processing.

Another important design decision is whether the system memory spans a single address space or it is distributed into separated chunks that are addressed independently. The first type is called *shared memory* while the latter is known as *distributed memory*. This is only a logical subdivision independent of how the memory is physically built.

In shared memory multiprocessors all the data are accessible by all the processors. This poses some design problems for data integrity and for efficiency. Fast cache memories next to the processors are used in order to speedup memory access to often-used items. Cache coherency protocols are then needed to insure that all processors see the same value for a given piece of data.

Distributed memory multicomputers is also a popular architecture which is well suited to most parallel workloads. Since the address spaces of each processor are separate, communication between processors must be implemented through some form of message passing. To this class belong networked computers, sometimes called computer *clusters*. This kind of architecture is interesting for several reasons. First of all, it has a low cost since already existing local networks of workstations can be used just by adding a layer of communication software to implement message passing. Second, the machines in these networks usually feature up-to-date off-the-shelf processor and standard software environments which make program development easier. The drawbacks are that parallel computing performance is limited by comparatively high communication latencies and by the fact that the machines have different workloads at any given time and are possibly heterogeneous. Nevertheless, problems that do not need frequent communication are suitable for this architecture. Moreover, some of the drawbacks can be overcome by using networked computers in dedicated mode with a high-performance communication switch. Although this solution is more expensive, it can be cost-effective with respect to specialized parallel machines.

Finally, one should not forget that the World Wibe Web provides important infrastructures for distributed computation. As it implements a general distributed computing model, this Web technology can be used for parallel computing and for both computing and information related applications. Harnessing the Web or some other geographically distributed computer resource so that it looks like a single computer to the user is called *metacomputing*. The concept is very attractive but many challenges remain. In fact, in order to transparently and efficiently distribute a given computational problem over the available resources without the user taking notice requires advances in the field of user interfaces and in standardized languages, momitoring tools and protocols to cope with the problem of computer heterogeneity and uneven network load. The Java environment is an important step in this direction.

We will see in the next section how the different architectures can be used for distributed evolutionary computing.

7.3.2 Parallel and distributed evolutionary algorithms models

There are two main reasons for parallelizing an evolutionary algorithm: one is to achieve time savings by distributing the computational effort and the second is to benefit from a parallel setting from the algorithmic point of view, in analogy with the natural parallel evolution of spatially distributed populations.

A first type of parallel evolutionary algorithm makes use of the available processors or machines to run independent problems. This is trivial, as there is no communication between the different processes and for this reason it is sometimes called an *embarassingly parallel* algorithm. This extremely simple method of doing simultaneous work can be very useful. For example, this setting can be used to run several versions of the same problem with different initial conditions, thus allowing gathering statistics on the problem. Since evolutionary algorithms are stochastic in nature, being able to collect this kind of statistics is very important. This method is in general to be preferred with respect to a very long single run since improvements are more difficult at later stages of the simulated evolution. Other ways in which the model can be used is to solve N different versions of the same problem or to run N copies of the same problem but with different GA parameters, such as crossover or mutation rates. Neither of the above adds anything new to the nature of the evolutionary algorithms but the time savings can be large.

We now turn to genuine parallel evolutionary algorithms models. There are several possible levels at which an evolutionary algorithm can be parallelized: the population level, the individual level or the fitness evaluation level. Moreover, although genetic algorithms and genetic programming are similar in many respects, the differences in the individual representation make genetic programming a little bit different when implemented in parallel. The next section describes the parallelization of the fitness evaluation while the two following sections treat the population and the individual cases respectively.

7.3.3 Global parallel evolutionary algorithms

Parallelization at the fitness evaluation level does not require any change to the standard evolutionary algorithm since the fitness of an individual is independent of the rest of the population. Moreover, in many real-world problems, the calculation of the individual's fitness is by far the most time consuming step of the algorithm. This is also a necessary condition in order for the communication time to be small in comparison to the time spent in computations. In this case an obvious approach is to evaluate each individual fitness simultaneously on a different processor. A *master* process manages the population and hands out individuals to evaluate to a number of *slave* processes. After the evaluation, the master collects the results and applies the genetic operators to produce the next generations. Fig. 7.4 graphically depicts this architecture. If there are more individuals than processors, which is often the case, then the individuals to be evaluated are divided as evenly as possible among the available processors. This architecture can be implemented on both shared memory multiprocessors as well as distributed memory machines, including networked computers. For genetic algorithms it is assumed that fitness evaluation takes about the same time for any individual. The other parts of the algorithm are the same as in the sequential case and remain centralized. The following is an informal description of the algorithm:

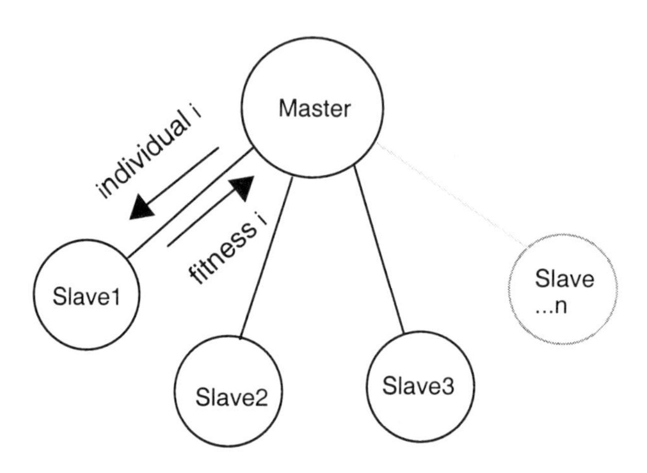

Fig. 7.4 A schematic view of the master-slave model.

```
produce an initial population of individuals
for all individuals do in parallel
   evaluate the individual's fitness
end parallel for
while not termination condition do
   select fitter individuals for reproduction
   produce new individuals
   mutate some individuals
   for all individuals do in parallel
      evaluate the individual's fitness
   end parallel for
   generate a new population by inserting some new good
   individuals and by discarding some old bad individuals
end while
```

In the GP case, due to the widely different sizes and complexities of the individual programs in the population different individuals may require different times to evaluate. This in turn may cause a load imbalance which decreases the utilization of the processors. Mouloud *et al* [21] implemented a simple method for load-balancing the system on a distributed memory parallel machine and observed nearly linear speedup. However, load balancing can be obtained for free if one gives up generational replacement and permits steady-state reproduction instead (see section 7.3.6).

7.3.4 Island distributed evolutionary algorithms

We now turn to individual or population-based parallel approaches for evolutionary algorithms. All these find their inspiration in the observation that natural populations tend to possess a *spatial* structure. As a result, so-called *demes* make their appearance. Demes are semi-independent groups of individuals or subpopulations having only a loose coupling to other neighboring demes. This coupling takes the form of the slow migration or diffusion of some individuals from one deme to another. A number of models based on spatial structure have been proposed. The two most important categories are the *island* and the *grid* models.

The *island* model [9] features geographically separated subpopulations of relatively large size. Subpopulations may exchange information from time to time by allowing some individuals to migrate from one subpopulation to another according to various patterns. The main reason for this approach is to periodically reinject diversity into otherwise converging subpopulations. As well, it is hoped that to some extent, different subpopulations will tend to explore different portions of the search space. When the migration takes place between nearest neighbor subpopulations the model is called *stepping stone*. Within each subpopulation a standard sequential evolutionary algorithm is executed between migration phases. Several migration topologies have been used: the ring structure, 2-d and 3-d meshes, hypercubes and random graphs being the most common. Fig. 7.5 schematically depicts this distributed model and the following is a high-level algorithmic description of the process:

```
initialize P subpopulations of size N each
generation number := 1
while not termination condition do
   for each subpopulation do in parallel
      evaluate and select individuals by fitness
      if generation number mod frequency = 0 then
         send K<N best individuals to
         a neighboring subpopulation
         receive K individuals from a
         neighboring population
         replace K individuals in
```

```
            the subpopulation
        end if
        produce new individuals
        mutate individuals
    end parallel for
    generation number := generation number + 1
end while
```

Here *frequency* is the number of generations before an exchange takes place. Several individual replacement policies have been described in the literature. One of the most common is for the migrating K individuals to displace the K worst individuals in the subpopulation. It is to be noted that the subpopulation size, the frequency of exchange, the number of migrating individuals, and the migration topology are all new parameters of the algorithm that have to be set in some way. At present there is no rigorous way for choosing them. However, several works have empirically arrived at rather similar topologies and parameter values [27], [9]. However, some new theoretical work is also shedding some light on these issues (see section 7.3.8 below).

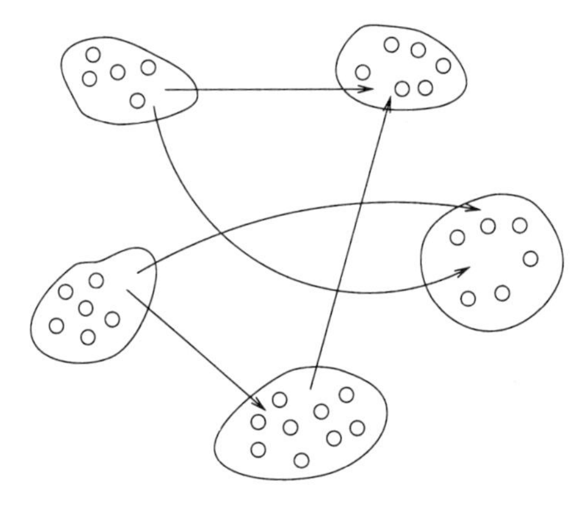

Fig. 7.5 The *island* model of semi-isolated populations.

7.3.5 Cellular genetic algorithms

In the *grid* or *fine-grained* model [16] individuals are placed on a large toroidal (the ends wrap around) one or two-dimensional grid, one individual per grid location (see Fig. 7.6). The model is also called *cellular* because of its similarity with cellular automata with stochastic transition rules [28], [29].

Fitness evaluation is done simultaneously for all individuals and selection, reproduction and mating take place locally within a small neighborhood. In time, semi-isolated niches of

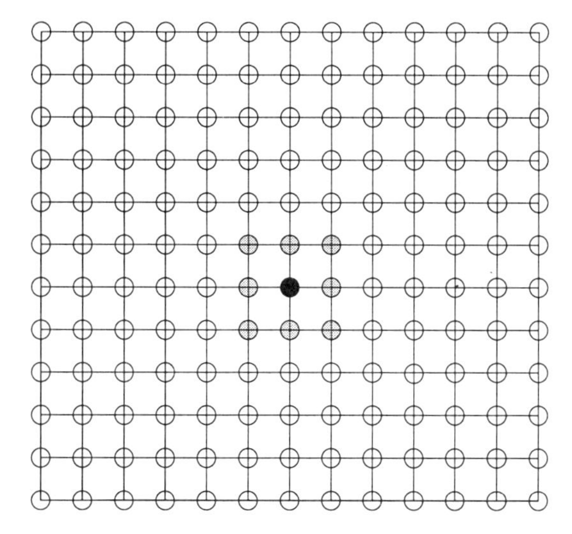

Fig. 7.6 A 2-D spatially extended population of individuals. A possible neighborhood of an individual (black) is marked in gray.

genetically homogeneous individuals emerge across the grid as a result of slow individual diffusion. This phenomenon is called *isolation by distance* and is due to the fact that the probability of interaction of two individuals is a fast-decaying function of their distance. The following is a pseudo-code description of a grid evolutionary algorithm.

for `each cell` i `in the grid` **do in parallel**

  `generate a random individual` i

end parallel for

while not `termination condition` **do**

  for `each cell` i **do in parallel**

    `evaluate individual` i

    `select a neighboring individual` k

    `produce offspring from` i `and` k

    `assign one of the offspring to` i

    `mutate` i `with probability` p_{mut}

  end parallel for

end while

In the preceding description the neighborhood is generally formed by the four or eight nearest neighbors of a given grid point (see Fig. 7.6). In the 1-D case a small number of cells on either side of the central one is taken into account. The selection of an individual in the neighborhood for mating with the central individual can be done in various ways. Tournament selection is commonly used since it matches nicely the spatial nature of the system. Local tournament selection extracts k individuals from the population with uniform probability but without re-insertion and makes them play a "tournament", which is won, in the deterministic case, by the fittest individual among the participants. The tournament may be probabilistic as well, in which case the probability for an individual to win it is generally proportional to its fitness. This makes full use of the available parallelism and is probably more appropriate if the biological methaphor is to be followed. Likewise, the replacement of the original individual can be done in several ways. For example, it can be replaced by the best among itself and the offspring or one of the offspring can replace it at random. The model can be made more dynamical by adding provisions for longer range individual movement through random walks, instead of having individuals interacting exclusively with their nearest neighbors [28]. A noteworthy variant of the cellular model is the so-called *cellular programming algorithm* [26]. Cellular programming has been extensively used to evolve cellular automata for performing computational tasks.

7.3.6 Implementation and experimental observations on parallel EAs

The master-slave global fitness parallelization model can be implemented on both shared memory multiprocessors and distributed memory multicomputers, including workstation clusters, and it is well adapted when fitness calculation takes up most of the computation time. However, it is well know that distributed memory machines have better scaling behavior. This means that performance is relatively unaffected if more processors (and memory) are added *and* larger problem instances are tackled. In this way, computation and communication times remain well balanced. Shared memory machines, on the other hand, suffer from memory acces contention, especially if global memory access is through a single path such as a bus. Cache memory alleviates the problem but it does not solve it.

Although both island and cellular models can be implemented on serial machines, thus comprising useful variants of the standard globally communicating GA, they are ideally suited for parallel computers. >From an implementation point of view, coarse-grained island models, where the ratio of computation to communication is high, are more adapted to distributed memory multicomputers and for clusters of workstations. Recently, some attempts have been made at using Web and Java-based computing for distributed evolutionary algorithms [20]. The advantages and the drawbacks of this approach have been discusses in section 7.3.1.

Massively parallel SIMD (Single Instruction Multiple Data) machines such as the Maspar and the Connection Machine CM-200 are appropriate for cellular models, since the necessary local communication operations, though frequent, are very efficiently implemented in hardware. Genetic programming is not suitable for cellular models since individuals may widely vary in size and complexity. This makes cellular implementations of GP difficult both because of the amount of local memory needed to store individuals as well as for efficiency reasons (e.g., sequential execution of different branches of code belonging to individuals stored on different processors).

In general, it has been found experimentally that parallel genetic algorithms, apart from being significantly faster, may help in alleviating the premature convergence problem and are effective for multimodal optimization. This is due to the larger total population size and to the relative isolation of the spatial regions where solutions start to co-evolve. Both of these factors help to preserve diversity while at the same time promoting local search. There are several results in the literature that support these conclusions, see for instance [14], [15], [27], [18]. As in the sequential case, the effectiveness of the search can be improved by permitting hill-climbing, i.e. local improvement around promising search points [18]. It has even been reported that for the island model [14], [1] *superlinear* speedup has been achieved in some cases. Superlinear speedup means getting more than N-fold acceleration with N processors with respect to the uniprocessor time. While superlinear speedup is strictly impossible for deterministic algorithms, it becomes possible when there is a random element in the ordering of the choices made by an algorithm. This has been shown to be the case in graph and tree search problems where, depending on the solution and the way in which the search space is subdivided, parallel search may be more than N-fold effective. The same phenomenon may occur in all kind of stochastic, Monte-Carlo type algorithms, including evolutionary algorithms. The net result is that in the multi-population case oftentimes *fewer* evaluations are needed to get the same solution quality than for the single population case with the same total number of individuals.

For coarse-grain, island-based parallel genetic programming the situation is somewhat controversial. Andre and Koza [1] reported excellent results on a GP problem implemented on a transputer network (the 5-parity problem). More recently, W. Punch [22] performed extensive experiments on two other problems: the "ant" problem and the "royal tree" problem. The first is a standard machine learning test problem [12], while the latter is a contructive benchmark for GP. Punch found that the multiple-population approach did not help in solving those problems. Although the two chosen problems were purposedly difficult for distributed GP approaches, these results point to the fact that the parallel, multi-population GP dynamics is not yet well understood. This is not surprising, given that there is no solid theoretical basis yet even for the standard sequential GP.

Until now only the spatial dimension entered into the picture. If we take into account the temporal dimension as well, we observe that parallel evolutionary algorithms can be *synchronous* or *asynchronous*. Island models are in general synchronous, using SPMD (Single Program Multiple Data), coarse-grain parallelism in which communication phases synchronize processes. This is not necessary and experiments have been carried out with asynchronous EAs in which subpopulations evolve at their own pace and exchange individuals only when some internally measured level of convergence has been attained [19]. This avoids constraining all co-evolving populations to swap individuals at the same time irrespective of subpopulation evolution and increases population diversity. Asynchronous behavior can be easily and conveniently obtained in island models by using *steady-state* reproduction instead of generational reproduction, although generational models can also be made to work asynchronously. In steady-state reproduction a small number of offspring replace other members of the population as soon as they are produced instead of replacing all the population at once after all the individuals have gone through a whole evaluation-selection-recombination mutation cycle[17]. For genetic programming the asynchronous setting also helps for the load-balancing problem caused by the variable size and complexity of GP programs. By independently and asynchronously evaluating each program, different nodes can proceed at different speeds without any need for synchronization [1].

Fine-grained parallel EAs are fully synchronous when they are implemented on SIMD machines and are an example of data-parallelism. Asynchronous behavior can be easily simulated on a sequential machine but I do not know of any example of asynchronous cellular evolutionary algorithms.

Fig. 7.7 schematically depicts the main parallel evolutionary algorithm classes according to their space and time dimensions. The reader should be aware that this is only a first cut at a general classification. Although the main classes are covered, a number of particular and hybrid cases exist in practice, some of which will be described in the next section.

	Coarse-grain	Fine-grain	
	Population	Individual	Fitness
Synchronous	**Island** GA and GP	**Cellular** GA Synchronous, stochastic CA	**Master-slave** GA and GP
Asynchronous	**Island** GA and GP	asynchronous, stochastic CA	**Master-slave** GA and GP

Fig. 7.7 Parallel and distributed algorithms classes according to their space and time behavior.

7.3.7 Non-standard parallel EAs models

There exist proposals for parallel evolutionary algorithms that do not fit into any of the classes that have been described in the previous sections. >From the topological and spatial point of view, the methods of parallelization can be combined to give *hybrid* models. Two-level hybrid algorithms are sometimes used. The island model can be used at the higher level, while another fine-grained model can be combined with it at the lower level. for example, one might consider an island model in which each island is structured as a grid of individuals interacting locally. Another possibility is to have an island model at the higher level where each island is a global parallel EA in which fitness calculation is parallelized. Other combinations are also possible and a more detailed description can be found in [6].

Although these combinations may give rise to interesting and efficient new algorithms, they have the drawback that even more new parameters have to be introduced to account for the more complex topological structure.

So far, we have made the implicit hypothesis that the genetic material as well as the evolutionary conditions such as selection and crossover methods were the same for the different subpopulations in a multi-population algorithm. If one gives up these constraints and allows different subpopulations to evolve with different parameters and/or with different individual representations for the same problem then new distributed algorithms may arise. We will name these algorithms *non-uniform* parallel EAs. One recent example of this class is the *Injection island GA* (iiGA) of Lin *et al* [14]. In an iiGA there are multiple populations that encode the same problem using a different representation size and thus different resolutions in different islands. The migration rules are also special in that migration is only one-way, going from a low-resolution to a high resolution node. According to Lin *et al*, such a hierarchy has a number of advantages with respect to a standard island algorithm.

In biological terms, having different semi-isolated populations in which evolution takes place at different stages and with different, but compatible genetic material makes sense from a biological point of view: it might also be advantageous for evolutionary algorithms. Genetic programming in particular might benefit from these ideas since the choice of the function and terminal sets is often a fuzzy and rather subjective issue.

7.3.8 Theoretical work

Theoretical studies of evolutionary algorithms have been performed mostly for the case of binary-coded, generational genetic algorithms with standard genetic operators and selection [3]. The standard form of evolution strategies and evolutionary programming is also sufficiently well understood. However, the dynamical behavior of genetic programming and non-standard GAs is much less clear, although progress is being made.

Parallel evolutionary algorithms are even more difficult to analyze because they introduce a number of new parameters such as migration rates and frequencies, subpopulation topology, grid neighborhood shape and size among others. As a consequence, there seems to be lacking a better understanding of their working. Nevertheless, performance evaluation models have been built for the simpler case of the master-slave parallel fitness evaluation in the GA case [7] and for GP [21]. As well, some limiting simplified cases of the island approach have been modeled: a set of isolated demes and a set of fully connected demes [5]. These are only first steps towords establishing more principled ways for choosing suitable parameters for parallel EAs.

In another study, Whitley *et al* [30] have presented an abstract model and made experiments of when one might expect the island model to out-perform single population GAs on separable problems. Linear separable problems are those problems in which the evaluation function can be expressed as a sum of independent nonlinear contributions. Although the results are not clear-cut, as the authors themselves acknowledge, there are indications that partitioning the population may be advantageous.

There have also been some attempts at studying the behavior of cellular genetic algorithms. For instance, Sarma and De Jong [25] investigated the effects of the size and shape of the cellular neighborhood on the selection mechanism. They found that the size of the neighborhood is a parameter that can be used to control the selection pressure over the population. In another paper [24] the convergence properties of a typical cellular algorithm were studied using the formalism of probabilistic cellular automata and Markov chains. The authors reached some conclusions concerning global convergence and the influence of the neighborhood size on the quality and speed at which solutions can be found. Unfortunately, their algorithm is not wholly standard since it includes a special recipe for introducing a form of elitism into the population. Recently, Capcarrere *et al* [8] presented a number of statistical measures that can be applied to cellular algorithms in order to understand their dynamics. The proposed statistics were demonstrated on the specific example of the evolution of non-uniform cellular automata but they can be applied generally.

7.4 SUMMARY AND CONCLUSIONS

If one is to follow the biological methaphor, parallel and distributed evolutionary algorithms seem more natural than their sequential counterparts. When implemented on parallel computer architectures they offer the advantage of increased performance due to the execution of simultaneous work on different machines. Moreover, genuine evolutionary parallel models such as the island or the cellular model constitute new useful varieties of evolutionary algorithms and there is experimental evidence that they can be more effective as problem solvers. This may be due to several factors such as independent and simultaneous exploration of different portions of the search space, larger total population sizes, and the possibility of delaying the uniformisation of a population by migration and diffusion of individuals. These claims have not been totally and unambiguously confirmed by the few theoretical studies that have been carried out to date. It is also unclear what is the kind of problem that can benefit more from a parallel/distributed setting, although separable and multimodal problems seem good candidates.

One drawback of parallel and evolutionary algorithms is that they are more complicated than the sequential versions. Since a number of new parameters must be introduced such as communication topologies and migration/diffusion policies, the mathematical modeling of the dynamics of these algorithms is very difficult. Only a few studies have been performed to date and even those have been mostly applied to simplified, more tractable models. Clearly, a large amount of work remains to be done on these issues before a clearer picture starts to emerge.

The effectiveness of parallel and distributed EAs has also been shown in practice in a number of industrial and commercial applications. Because of lack of space I cannot review these applications here, but it is clear that for some hard problems parallelism can be effective. Some real-life problems may need days or weeks of computing time to solve on serial machines. Although the intrinsic complexity of a problem cannot be lowered by using a finite amount of computing resources in parallel, the use of parallelism often allows to reduce these times to reasonable amounts. This can be very important in an industrial or commercial setting where the time to solution is instrumental for decision making and competitivity. Parallel and evolutionary algorithms have been used successfully in operations research, engineering and manufacturing, finance, VLSI and telecommunication problemes among others. The interested reader can find out more about these applications by consulting the proceedings of the major conferences in the field.

Acknowledgments

I am indebted to my colleagues B. Chopard, M. Sipper and A. Tettamanzi for stimulating discussions and to R. Gomez for his help with some of the figures appearing in this work.

REFERENCES

1. D. Andre and J. R. Koza. Parallel genetic programming: A scalable implementation using the transputer network architecture. In P. Angeline and K. Kinnear, editors, *Advances in Genetic Programming 2*, Cambridge, MA, 1996. The MIT Press.

2. P.J. Angeline and K.E. Kinnear Jr. (Eds.). *Advances in Genetic Programming 2*. The MIT Press, Cambridge, Massachusetts, 1996.

3. T. Bäck. *Evolutionary algorithms in theory and practice*. Oxford University Press, Oxford, 1996.

4. W. Banzhaf, P. Nordin, R. E. Keller, and F. D. Francone. *Genetic programming, An Introduction*. Morgan Kaufmann, San Francisco CA, 1998.

5. E. Cantú-Paz. Designing efficient master-slave parallel genetic algorithms. Technical Report 95004, Illinois Genetic Algorithms Laboratory, University of Illinois at Urbana-Champaign, Urbana, IL, 1997.

6. E. Cantú-Paz. A survey of parallel genetic algorithms. Technical Report 97003, Illinois Genetic Algorithms Laboratory, University of Illinois at Urbana-Champaign, Urbana, IL, 1997.

7. E. Cantú-Paz and D. E. Goldberg. Modeling idealized bounding cases of parallel genetic algorithms. In J. R. Koza, K. Deb, M. Dorigo, D. B. Fogel, M. Garzon, H. Iba, and R. L. Riolo, editors, *Genetic Programming 1997: Proceedings of the Second Annual Conference*, pages 456–462. Morgan Kaufmann, San Francisco, CA, 1997.

8. M. Capcarrere, A. Tettamanzi, M. Tomassini, and M. Sipper. Studying parallel evolutionary algorithms: the cellular programming case. In A. Eiben, T. Bäck, M. Shoenauer, and H.P. Schwefel, editors, *Parallel Problem Solving from Nature*, pages 573–582. Springer, 1998.

9. J. P. Cohoon, S. U. Hedge, W. N. Martin, and D. Richards. Punctuated equilibria: A parallel genetic algorithm. In J. J. Grefenstette, editor, *Proceedings of the Second International Conference on Genetic Algorithms*, page 148. Lawrence Erlbaum Associates, 1987.

10. K.E. Kinnear Jr. (Ed.). *Advances in Genetic Programming*. The MIT Press, Cambridge, Massachusetts, 1994.

11. K. Hwang. *Advanced Computer Architecture*. Mc Graw-Hill, New York, 1993.

12. J. R. Koza. *Genetic Programming*. The MIT Press, Cambridge, Massachusetts, 1992.

13. J. R. Koza. *Genetic Programming II*. The MIT Press, Cambridge, Massachusetts, 1994.

14. S. C. Lin, W. F. Punch, and E. D. Goodman. Coarse-grain parallel genetic algorithms: Categorization and a new approach. In *Sixth IEEE SPDP*, pages 28–37, 1994.

15. A. Loraschi, A. Tettamanzi, M. Tomassini, and P. Verda. Distributed genetic algorithms with an application to portfolio selection problems. In *Proceedings of the International Conference on Artificial Neural Networks and Genetic Algorithms*, pages 384–387, Wien, New York, 1995. Springer-Verlag.

16. B. Manderick and P. Spiessens. Fine-grained parallel genetic algorithms. In J. D. Schaffer, editor, *Proceedings of the Third International Conference on Genetic Algorithms*, page 428. Morgan Kaufmann, 1989.

17. Z. Michalewicz. *Genetic Algorithms + Data Structures = Evolution Programs*. Springer-Verlag, Heidelberg, third edition, 1996.

18. H. Mühlenbein, M. Schomish, and J. Born. The parallel genetic algorithm as a function optimizer. *Parallel Computing*, 17:619–632, 1991.

19. M. Munetomo, Y. Takai, and Y. Sato. An efficient migration scheme for subpopulation-based ansynchronously parallel genetic algorithms. In S. Forrest, editor, *Proceedings of the Fifth International Conference on Genetic Algorithms*, page 649. Morgan Kaufmann Publishers, San Mateo, California, 1993.

20. P. Nangsue and S. E. Conry. An agent-oriented, massively distributed parallelization model of evolutionary algorithms. In J. Koza, editor, *Late Breaking Papers, Genetic Programming 1998*, pages 160–168. Stanford University, 1998.

21. M. Oussaidene, B. Chopard, O. Pictet, and M. Tomassini. Parallel genetic programming and its application to trading model induction. *Parallel Computing*, 23:1183–1198, 1997.

22. W. Punch. How effective are multiple poplulations in genetic programming. In J. R. Koza, W. Banzhaf, K. Chellapilla, K. Deb, M. Dorigo, D. B. Fogeland M. Garzon, D. Goldberg, H. Iba, and R. L. Riolo, editors, *Genetic Programming 1998: Proceedings of the Third Annual Conference*, pages 308–313, San Francisco, CA, 1998. Morgan Kaufmann.

23. J. Rosca and D. Ballard. Discovery of subroutines in genetic programming. In P.J. Angeline and K.E. Kinnear Jr. (Eds.), editors, *Advances in Genetic Programming 2*. Cambridge, Massachusetts, 1996.

24. G. Rudolph and J. Sprave. A cellular genetic algorithm with self-adjusting acceptance thereshold. In *In First IEE/IEEE International Conference on Genetic Algorithms in*

Engineering Systems: Innovations and Applications, pages 365–372, London, 1995. IEE.

25. J. Sarma and K. De Jong. An analysis of the effect of the neighborhood size and shape on local selection algorithms. In H.-M. Voigt, W. Ebeling, I. Rechenberg, and H.-P. Schwefel, editors, *Parallel Problem Solving from Nature - PPSN IV*, volume 1141 of *Lecture Notes in Computer Science*, pages 236–244. Springer-Verlag, Heidelberg, 1996.

26. M. Sipper. *Evolution of Parallel Cellular Machines: The Cellular Programming Approach*. Springer-Verlag, Heidelberg, 1997.

27. T. Starkweather, D. Whitley, and K. Mathias. Optimization using distributed genetic algorithms. In H.-P. Schwefel and R. Männer, editors, *Parallel Problem Solving from Nature*, volume 496 of *Lecture Notes in Computer Science*, page 176, Heidelberg, 1991. Springer-Verlag.

28. M. Tomassini. The parallel genetic cellular automata: Application to global function optimization. In R. F. Albrecht, C. R. Reeves, and N. C. Steele, editors, *Proceedings of the International Conference on Artificial Neural Networks and Genetic Algorithms*, pages 385–391. Springer-Verlag, 1993.

29. D. Whitley. Cellular genetic algorithms. In S. Forrest, editor, *Proceedings of the Fifth International Conference on Genetic Algorithms*, page 658. Morgan Kaufmann Publishers, San Mateo, California, 1993.

30. D. Whitley, S. Rana, and R. B. Heckendorn. Island model genetic algorithms and linearly separable problems. In D. Corne and J. L. Shapiro, editors, *Evolutionary Computing: Proceedings of the AISB Workshop, Lecture notes in computer science, vol. 1305*, pages 109–125. Springer-Verlag, Berlin, 1997.

8 Evolutionary Algorithms for Multi-Criterion Optimization in Engineering Design

K. DEB†

Kanpur Genetic Algorithms Laboratory (KanGAL)
Department of Mechanical Engineering
Indian Institute of Technology Kanpur
Kanpur, PIN 208 016, India
E-mail: deb@iitk.ernet.in

8.1 INTRODUCTION

Many real-world engineering design or decision making problems involve simultaneous optimization of multiple objectives. The principle of multi-criterion optimization is different from that in a single-objective optimization. In single-objective optimization, the goal is to find the *best* design solution, which corresponds to the minimum or maximum value of the objective function [9], [34]. On the contrary, in a multi-criterion optimization with *conflicting* objectives, there is no single optimal solution. The interaction among different objectives gives rise to a set of compromised solutions, largely known as the Pareto-optimal solutions [40], [2]. Since none of these Pareto-optimal solutions can be identified as better than others without any further consideration, the goal in a multi-criterion optimization is to find as many Pareto-optimal solutions as possible. Once such solutions are found, it usually requires a higher-level decision-making with other considerations to choose one of them for implementation. Here, we address the first task of finding multiple Pareto-optimal solutions. There are two objectives in a multi-criterion optimization: (i) find solutions close to the true Pareto-optimal solutions and (ii) find solutions that are widely different from each other. The first task is desired to satisfy optimality conditions in the obtained solutions. The second task is desired to have no bias towards any particular objective function.

†Currently at the Computer Science Department, University of Dortmund, Germany

In dealing with multi-criterion optimization problems, classical search and optimization methods are not efficient, simply because (i) most of them cannot find multiple solutions in a single run, thereby requiring them to be applied as many times as the number of desired Pareto-optimal solutions, (ii) multiple application of these methods do not guarantee finding widely different Pareto-optimal solutions, and (iii) most of them cannot efficiently handle problems with discrete variables and problems having multiple optimal solutions [9]. On the contrary, the studies on evolutionary search algorithms, over the past few years, have shown that these methods can be efficiently used to eliminate most of the above difficulties of classical methods [38], [19], [26]. Since they use a population of solutions in their search, multiple Pareto-optimal solutions can, in principle, be found in one single run. The use of diversity-preserving mechanisms can be added to the evolutionary search algorithms to find widely different Pareto-optimal solutions.

In this paper, we briefly outline the principles of multi-objective optimization. Thereafter, we discuss why classical search and optimization methods are not adequate for multi-criterion optimization by discussing the working of two popular methods. We then outline several evolutionary methods for handling multi-criterion optimization problems. Of them, we discuss one implementation (non-dominated sorting GA or NSGA [38]) in somewhat greater details. Thereafter, we demonstrate the working of the evolutionary methods by applying NSGA to three test problems having constraints and discontinuous Pareto-optimal region. We also show the efficacy of evolutionary algorithms in engineering design problems by solving a welded beam design problem. The results show that evolutionary methods can find widely different yet near-Pareto-optimal solutions in such problems. Based on the above studies, this paper also suggests a number of immediate future studies which would make this emerging field more mature and applicable in practice.

8.2 PRINCIPLES OF MULTI-CRITERION OPTIMIZATION

The principles of multi-criterion optimization are different from that in a single-objective optimization. The main goal in a single-objective optimization is to find the global optimal solution, resulting in the optimal value for the single objective function. However, in a multi-criterion optimization problem, there are more than one objective function, each of which may have a *different* individual optimal solution. If there is sufficient difference in the optimal solutions corresponding to different objectives, the objective functions are often known as *conflicting* to each other. Multi-criterion optimization with such conflicting objective functions gives rise to a set of optimal solutions, instead of one optimal solution. The reason for the optimality of many solutions is that no one can be considered to be better than any other with respect to all objective functions. These optimal solutions have a special name—Pareto-optimal solutions. Let us illustrate this aspect with a hypothetical example problem shown in Fig. 8.1. The figure considers two objectives—cost and accident rate—both of which are to be minimized. The point A represents a solution which incurs a near-minimal cost, but is highly accident-prone. On the other hand, the point B represents a solution which is costly, but is near least accident-prone. If both objectives (cost and accident rate) are important goals of design, one cannot really say whether solution A is better than solution B, or vice versa. One solution is better than other in one objective, but is worse in the other. In fact, there exist

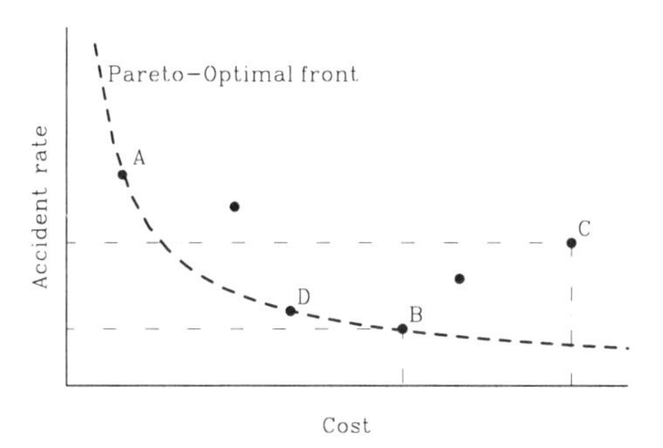

Fig. 8.1 The concept of Pareto-optimal solutions is illustrated.

many such solutions (like solution D) which also belongs to the Pareto-optimal set and one cannot conclude about an absolute hierarchy of solutions A, B, D, or any other solution in the set. All these solutions (in the front marked by the dashed line) are known as Pareto-optimal solutions.

Looking at the figure, we also observe that there exists non-Pareto-optimal solutions, like the point C. If we compare solution C with solution A, we again are in a fix and cannot say whether one is better than the other in both objectives. Does this mean that solution C is also a member of the Pareto-optimal set. The answer is no. This is because there exists another solution D in the search space, which is better than solution C in *both* objectives. That is why solutions like C are known as *dominated* solutions or *inferior* solutions.

It is now clear that the concept of optimality in multi-criterion optimization deals with a number (or a set) of solutions, instead of one solution. Based on the above discussions, we first define conditions for a solution to become dominated with respect to another solution and then present conditions for a set of solutions to become a Pareto-optimal set.

For a problem having more than one objective function (say, $f_j, j = 1, \ldots, M$ and $M > 1$), any two solutions $x^{(1)}$ and $x^{(2)}$ can have one of two possibilities—one dominates the other or none dominates the other. A solution $x^{(1)}$ is said to dominate the other solution $x^{(2)}$, if both the following conditions are true:

1. The solution $x^{(1)}$ is no worse (say the operator $\prec$ denotes worse and $\succ$ denotes better) than $x^{(2)}$ in all objectives, or $f_j(x^{(1)}) \not\prec f_j(x^{(2)})$ for all $j = 1, 2, \ldots, M$ objectives.

2. The solution $x^{(1)}$ is strictly better than $x^{(2)}$ in at least one objective, or $f_{\bar{j}}(x^{(1)}) \succ f_{\bar{j}}(x^{(2)})$ for at least one $\bar{j} \in \{1, 2, \ldots, M\}$.

If any of the above condition is violated, the solution $x^{(1)}$ does not dominate the solution $x^{(2)}$. If $x^{(1)}$ dominates the solution $x^{(2)}$, it is also customary to write $x^{(2)}$ is dominated by $x^{(1)}$, or $x^{(1)}$ is non-dominated by $x^{(2)}$, or, simply, among the two solutions, $x^{(1)}$ is the non-dominated solution.

The following definitions ensure whether a set of solutions belong to a local or global Pareto-optimal set, similar to the definitions of local and global optimal solutions in single-objective optimization problems:

Local Pareto-optimal Set: If for every member x in a set $\underline{P}$, there exist no solution y satisfying $||y - x||_\infty \leq \epsilon$, where ϵ is a small positive number (in principle, y is obtained by perturbing x in a small neighborhood), which dominates any member in the set $\underline{P}$, then the solutions belonging to the set $\underline{P}$ constitute a local Pareto-optimal set.

Global Pareto-optimal Set: If there exists no solution in the search space which dominates any member in the set $\bar{P}$, then the solutions belonging to the set $\bar{P}$ constitute a global Pareto-optimal set.

We would like to highlight here that there exists a difference between a non-dominated set and a Pareto-optimal set. A non-dominated set is defined in the context of a sample of the search space. In a sample of search points, solutions that are not dominated (according to the above definition) by any other solutions in the sample space are non-dominated solutions. A Pareto-optimal set is a non-dominated set, when the sample is the entire search space.

From the above discussions, we observe that there are primarily two goals that a multi-criterion optimization algorithm must achieve:

1. Guide the search towards the global Pareto-optimal region, and
2. Maintain population diversity in the Pareto-optimal front.

The first task is a natural goal of any optimization algorithm. The second task is unique to multi-criterion optimization[1]. Since no one solution in the Pareto-optimal set can be said to be better than the other, what an algorithm can do best is to find as many different Pareto-optimal solutions as possible.

We now review a couple of popular classical search and optimization methods briefly and discuss why there is a need for better algorithms for multi-criterion optimization.

8.3 CLASSICAL METHODS

Here, we shall discuss two popular classical optimization methods used for solving multi-criterion optimization problems.

8.3.1 Weighted sum method

Multiple objective functions are combined into one overall objective function, F, as follows:

$$\begin{aligned} \text{Minimize} \quad & F = \textstyle\sum_{j=1}^{M} w_j f_j(\vec{x}), \\ & \vec{x} \in \mathcal{F}, \end{aligned} \tag{8.1}$$

[1] In multi-modal optimization problems, often, the goal is also to find multiple global optimal solutions simultaneously [14], [22].

where w_j is the weight used for the j-th objective function $f_j(\vec{x})$. Usually, non-zero fractional weights are used so that sum of all weights $\sum_{j=1}^{M} w_j$ is equal to one. All Pareto-optimal solutions must lie in the feasible region $\mathcal{F}$. The procedure is simple. Choose a random weight vector and optimize the single-objective function F to get an optimal solution. Hopefully, the obtained optimal solution belongs to the set of the desired Pareto-optimal set. In order to get different Pareto-optimal solutions, choose different random vectors and optimize the resulting F in each case. We illustrate the working of this method in a hypothetical problem shown in Fig. 8.2. The figure shows the feasible search space in the function space, having two objectives (which are to be minimized). Each point inside the feasible region represents a solution ($\vec{x}$) having two objective function values (such as cost and accident rate). Fixing a weight vector and optimizing equation 8.1 means finding a hyper-plane (a line for two objective functions) with a fixed orientation in the function space. The optimal solution is the point where a hyperplane having this fixed orientation has a common tangent with the feasible search space. We show this solution as the point A in the figure, where the line with intercepts at the f_1 and f_2 axes in proportions of w_2 and w_1, respectively, is tangent to the feasible region. One can now imagine finding other solutions such as B or C by choosing a different weight vector and finding the corresponding common tangent point again. A collection of such solutions (A, B, C, and others) constitute the Pareto-optimal set (shown by a thick line). Although such a simple strategy is intuitively a computationally expensive method, there is a major difficulty with such a method.

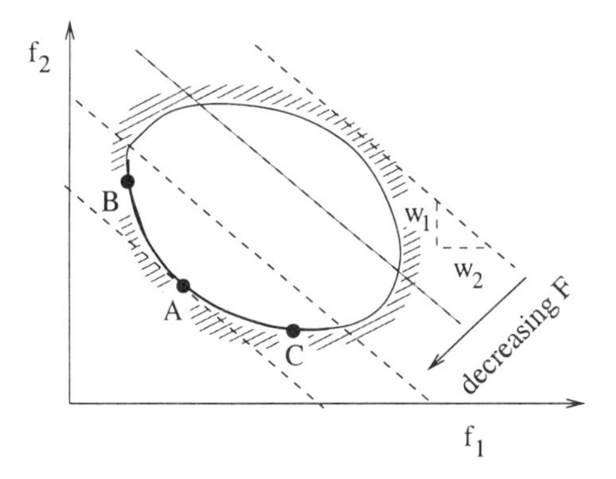

Fig. 8.2 A convex Pareto-optimal front is shown

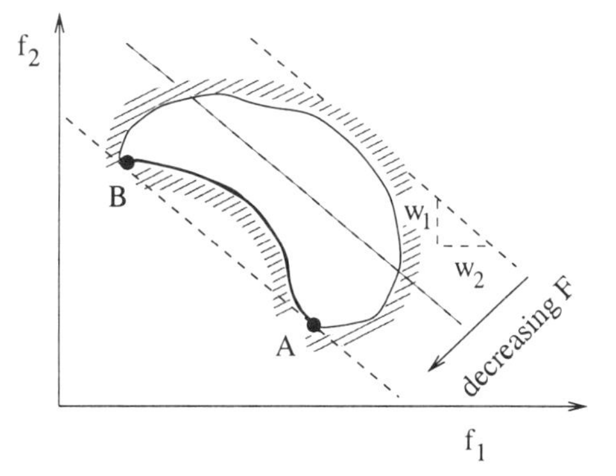

Fig. 8.3 A non-convex Pareto-optimal front is shown

This weighted sum method cannot be used to find Pareto-optimal solutions in multi-criterion optimization problems having a non-convex Pareto-optimal front. In Fig. 8.2, the Pareto-optimal front is convex. But, one can also think of multi-criterion optimization problems having a non-convex Pareto-optimal front (Fig. 8.3). In this figure, fixing a different weight vector and finding the tangents closest to the origin do not give rise to finding different Pareto-optimal solutions. For every weight vector, only the solutions A or B will be found, and all other Pareto-optimal solutions within A and B cannot be found.

8.3.2 The ϵ-perturbation method

In order to remedy the above difficulty, a single-objective optimization problem is constructed in which all but one objectives are used as constraints and only one is used as the objective function:

$$\begin{array}{rl} \text{Minimize} & f_k(\vec{x}), \\ \text{Subject to} & f_j(\vec{x}) \leq \epsilon_j \quad \forall j \neq k, \\ & \vec{x} \in \mathcal{F}. \end{array} \tag{8.2}$$

To find one Pareto-optimal solution, a suitable value of ϵ_j is chosen for the j-th objective function (where $j \neq k$). Thereafter, the above single-objective constraint optimization problem is solved to find the solution A. This procedure is continued with different values of ϵ_j to find different Pareto-optimal solutions. Fig. 8.4 shows that this method can find non-convex Pareto-optimal solutions. However, there also exists a difficulty with this method. A knowledge of an appropriate range of ϵ_j values for each objective function is required to be known a priori.

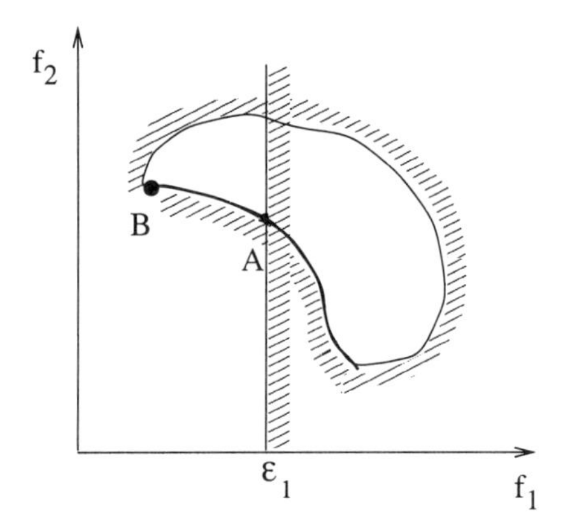

Fig. 8.4 The ϵ-perturbation method is illustrated. The first objective function is used as a constraint $f_1(\vec{x}) \leq \epsilon_1$.

Although there exists a few other methods such as goal programming, min-max method, and others [40], all methods require some kind of problem knowledge. The most profound difficulty with all these methods is that all need to be applied many times, hopefully finding one different Pareto-optimal solution each time. This makes the methods unattractive and this is one of the reasons why multi-criterion optimization problems are mostly avoided in practice. Based on the above discussions, we summarize the difficulties with the classical optimization methods:

1. An algorithm is needed to be applied many times to find multiple Pareto-optimal solutions,
2. Most algorithms demand some knowledge about the problem being solved,
3. Some algorithms are sensitive to the shape of the Pareto-optimal front,

4. The spread of Pareto-optimal solutions depends on efficiency of the single-objective optimizer,

5. In problems involving uncertainties or stochasticities, classical methods are not reliable,

6. Since classical single-objective optimizers are not efficient in handling discrete search space problems [9], [13], they will also not be efficient for multi-criterion optimization problems having discrete search space.

The first difficulty can be eliminated if an appropriate population-based search algorithm is used and special operators are used to emphasize maintenance of multiple Pareto-optimal solutions in the population. However, all the above difficulties can be handled by using an evolutionary search algorithm. In the following, we briefly describe a number of evolutionary-based implementations and describe one such method in details.

8.4 EVOLUTIONARY METHODS

As early as in 1967, Rosenberg suggested, but did not simulate, a genetic search method for finding the chemistry of a population of single-celled organisms with multiple properties or objectives [35]. However, the first practical implementation was suggested by David Schaffer in the year 1984 [36]. Thereafter, no significant study was performed for almost a decade, except a revolutionary 10-line sketch of a new non-dominated sorting procedure outlined in David Goldberg's book [20]. The book came out in the year 1989. Getting a clue for an efficient multi-objective optimization technique, many researchers developed different versions of multi-objective optimization algorithms [38], [19], [26] based on their interpretations of the 10-line sketch. The idea was so sound and appropriate, that almost any such implementation has been successful in many test problems. The publication of the above-mentioned algorithms showed the superiority of evolutionary multi-criterion optimization techniques over classical methods, and since then there has been no looking back. Many researchers have modified the above-mentioned approaches and developed their own versions. Many researchers have also applied these techniques to more complex test problems and to real-world engineering design problems. Till to date, most of the successful evolutionary implementations for multi-criterion optimization rely on the concept of non-domination. Although there are other concepts that can be used to develop a search algorithm [28], the concept of non-domination is simple to use and understand. However, a recent study [7] has shown that search algorithms based on non-domination need not always lead to the true Pareto-optimal front. The algorithm can get stuck to a non-dominated front which is different from the true Pareto-optimal front. This exception is in the details of the implementational issues and we would not like to belabor this here, simply because in most test problems tried so far most of the search algorithms based on non-domination concept has found the true Pareto-optimal front.

With the development of many new algorithms, many researchers have attempted to summarize the studies in the field from different perspectives [18], [25], [42], [3]. These reviews list many different techniques of multi-criterion optimization that exist to date.

We now present a brief summary of a few salient evolutionary multi-criterion optimization algorithms.

8.4.1 Schaffer's VEGA

Schaffer [36] modified the simple tripartite genetic algorithm by performing independent selection cycles according to each objective. He modified the public-domain GENESIS software by creating a loop around the traditional selection procedure so that the selection method is repeated for each individual objective to fill up a portion of the mating pool. Then the entire population is thoroughly shuffled to apply crossover and mutation operators. This is performed to achieve the mating of individuals of different subpopulation groups.

The algorithm worked efficiently for some generations but in some cases suffered from its bias towards some individuals or regions. The independent selection of specialists resulted in speciation in the population. The outcome of this effect is the convergence of the entire population towards the individual optimum regions after a large number of generations. From a designer's point of view, it is not desirable to have any bias towards such middling individuals, rather it is of interest to find as many non-dominated solutions as possible. Schaffer tried to minimize this speciation by developing two heuristics—the non-dominated selection heuristic (a wealth redistribution scheme), and the mate selection heuristic (a cross breeding scheme) [36], [37]. In the non-dominated selection heuristic, dominated individuals are penalized by subtracting a small fixed penalty from their expected number of copies during selection. Then the total penalty for dominated individuals was divided among the non-dominated individuals and was added to their expected number of copies during selection. But this algorithm failed when the population has very few non-dominated individuals, resulting in a large fitness value for those few non-dominated points, eventually leading to a high selection pressure. The mate selection heuristic was intended to promote the cross breeding of specialists from different subgroups. This was implemented by selecting an individual, as a mate to a randomly selected individual, which has the maximum Euclidean distance in the performance space from its mate. But it failed too to prevent the participation of poorer individuals in the mate selection. This is because of random selection of the first mate and the possibility of a large Euclidean distance between a champion and a mediocre. Schaffer concluded that the random mate selection is far superior than this heuristic.

8.4.2 Fonseca amd Fleming's multi-objective GA

Fonesca and Fleming [19] implemented Goldberg's suggestion in a different way. In this study, the multi-objective optimization GA (MOGA) uses a non-dominated sorting procedure. In MOGA, the whole population is checked and all non-dominated individuals are assigned rank '1'. Other individuals are ranked by checking the non-dominance of them with respect to the rest of the population in the following way. For an individual solution, the number of solutions that strictly dominate it in the population is first found. Thereafter, the rank of that individual is assigned to be one more than that number. Therefore, at the end of this ranking procedure, there may exist many solutions having the same rank. The selection procedure then uses these ranks to select or delete blocks of points to form the mating pool. As discussed elsewhere [21], this type of blocked fitness assignment is likely to produce a large selection pressure which might cause premature convergence. MOGA also uses a niche-formation method to distribute the population over the Pareto-optimal region. But instead of performing sharing on the parameter values, they have used sharing on objective function values. Even though this maintain diversity in the objective function values, this may not maintains diversity in the parameter set, a matter which is important for a decision maker. Moreover, MOGA may not be able to find multiple solutions in problems where different Pareto-optimal points correspond to the same objective function value [7]. However, the ranking of individuals according to their non-dominance in the population is an important aspect of this work.

8.4.3 Horn, Nafploitis, and Goldberg's niched Pareto GA

Horn, Nafploitis, and Goldberg [26] used a *Pareto domination tournaments* instead of non-dominated sorting and ranking selection method in solving multi-objective optimization problems. In this method, a *comparison set* comprising of a specific number (t_{dom}) of individuals is picked at random from the population at the beginning of each selection process. Two random individuals are picked from the population for selecting a winner in a tournament selection according to the following procedure. Both individuals are compared with the members of the comparison set for domination with respect to objective functions. If one of them is non-dominated and the other is dominated, then the non-dominated point is selected. On the other hand, if both are either non-dominated or dominated, a niche count is found for each individual in the entire population. The niche count is calculated by simply counting the number of points in the population within a certain distance (σ_{share}) from an individual. The individual with least niche count is selected. The effect of multiple objectives is taken into the non-dominance calculation. Since this non-dominance is computed by comparing an individual with a randomly chosen population set of size t_{dom}, the success of this algorithm highly depends on the parameter t_{dom}. If a proper size is not chosen, true non-dominated (Pareto-optimal) points may not be found. If a small t_{dom} is chosen, this may result in a few non-dominated points in the population. Instead, if a large t_{dom} is chosen, premature convergence may result. This aspect is also observed by the authors. They have presented some empirical results with various t_{dom} values. Nevertheless, the concept of niche formation among the non-dominated points is an important aspect of this work.

8.4.4 Zitzler and Theile's strength Pareto approach (SPEA)

Zitzler and Theile [44] have recently suggested an elitist multi-criterion EA with the concept of non-domination. They suggested maintaining an external population at every generation storing a set of non-dominated solutions discovered so far beginning from the initial population. This external population participates in genetic operations. The fitness of each individual in the current population and in the external population is decided based on the number of dominated solutions. Specifically, the following procedure is adopted. A combined population with the external and the current population is first constructed. All non-dominated solutions in the combined population are assigned a fitness based on the number of solutions they dominate. To maintain diversity and in the context of minimizing the fitness function, they assigned more fitness to a non-dominated solution having more dominated solutions in the combined population. On the other hand, more fitness is also assigned to solutions dominated by more solutions in the combined population. Care is taken to assign no non-dominated solution a fitness worse than that of the best dominated solution. This assignment of fitness makes sure that the search is directed towards the non-dominated solutions and simultaneously diversity among dominated and non-dominated solutions are maintained. On knapsack problems, they have reported better results than any other method used for comparison in that study. However, such comparisons of algorithms is not appropriate, simply because SPEA approach uses a inherent elitism mechanism of using best non-dominated solutions discovered up to the current generation, whereas other algorithms do not use any such mechanism. Nevertheless, an interesting aspect of that study is that it shows the importance of introducing elitism in evolutionary multi-criterion optimization. Similar effect of elitism in multi-criterion optimization was also observed elsewhere [30].

8.4.5 Srinivas and Deb's non-dominated sorting genetic algorithm (NSGA)

Srinivas and Deb [38] have implemented Goldberg's idea most directly. The idea behind NSGA is that a ranking selection method is used to emphasize current non-dominated points and a niching method is used to maintain diversity in the population. We describe the NSGA procedure in somewhat more details.

NSGA varies from a simple genetic algorithm only in the way the selection operator in used. The crossover and mutation operators remain as usual. Before the selection is performed, the population is first ranked on the basis of an individual's non-domination level, which is found by the following procedure, and then a fitness is assigned to each population member.

8.4.5.1 Fitness assignment Consider a set of N population members, each having M (> 1) objective function values. The following procedure can be used to find the non-dominated set of solutions:

Step 0: Begin with $i = 1$.

Step 1: For all $j = 1, \ldots, N$ and $j \neq i$, compare solutions $x^{(i)}$ and $x^{(j)}$ for domination using two conditions (page 137) for all M objectives.

Step 2: If for any j, $x^{(i)}$ is dominated by $x^{(j)}$, mark $x^{(i)}$ as 'dominated'.

Step 3: If all solutions (that is, when $i = N$ is reached) in the set are considered, Go to Step 4, else increment i by one and Go to Step 1.

Step 4: All solutions that are not marked 'dominated' are non-dominated solutions.

All these non-dominated solutions are assumed to constitute the first non-dominated front in the population and assigned a large dummy fitness value (we assign a fitness N). The same fitness value is assigned to give an equal reproductive potential to all these non-dominated individuals. In order to maintain diversity in the population, these non-dominated solutions are then *shared* with their dummy fitness values. Sharing methods are discussed elsewhere in details [11], [22]; we give a brief description of the sharing procedure in the next subsection. Sharing is achieved by dividing the dummy fitness value of an individual by a quantity (called the niche count) proportional to the number of individuals around it. This procedure causes multiple optimal points to co-exist in the population. The worst shared fitness value in the solutions of the first non-dominated front is noted for further use.

After sharing, these non-dominated individuals are ignored temporarily to process the rest of population members. The above step-by-step procedure is used to find the second level of non-dominated solutions in the population. Once they are identified, a dummy fitness value which is a little smaller than the worst shared fitness value observed in solutions of first non-dominated set is assigned. Thereafter, the sharing procedure is performed among the solutions of second non-domination level and shared fitness values are found as before. This process is continued till all population members are assigned a shared fitness value.

The population is then reproduced with the shared fitness values. A stochastic remainder proportionate selection [20] is used in this study. Since individuals in the first front have better fitness values than solutions of any other front, they always get more copies than the rest of population. This was intended to search for non-dominated regions, which will finally lead to the Pareto-optimal front. This results in quick convergence of the population towards non-dominated regions and sharing procedure helps to distribute it over this region.

Another aspect of this method is that practically any number of objectives can be used. Both minimization and maximization problems can also be handled by this algorithm. The only place a change is required for the above two cases is the way the non-dominated points are identified (according to the definition outlined on page 137).

8.4.5.2 ***Sharing procedure*** Given a set of n_k solutions in the k-th non-dominated front each having a dummy fitness value f_k, the sharing procedure is performed in the following way for each solution $i = 1, 2, \ldots, n_k$:

Step 1: Compute a normalized Euclidean distance measure with another solution j in the k-th non-dominated front, as follows:

$$d_{ij} = \sqrt{\sum_{p=1}^{P} \left(\frac{x_p^{(i)} - x_p^{(j)}}{x_p^u - x_p^l} \right)^2},$$

where P is the number of variables in the problem. The parameters x_p^u and x_p^l are the upper and lower bounds of variable x_p.

Step 2: This distance d_{ij} is compared with a pre-specified parameter σ_{share} and the following *sharing function* value is computed [22]:

$$Sh(d_{ij}) = \begin{cases} 1 - \left(\frac{d_{ij}}{\sigma_{\text{share}}} \right)^2, & \text{if } d_{ij} \leq \sigma_{\text{share}}, \\ 0, & \text{otherwise.} \end{cases}$$

Step 3: Increment j. If $j \leq n_k$, go to Step 1 and calculate $Sh(d_{ij})$. If $j > n_k$, calculate niche count for i-th solution as follows:

$$m_i = \sum_{j=1}^{n_k} Sh(d_{ij}).$$

Step 4: Degrade the dummy fitness f_k of i-th solution in the k-th non-domination front to calculate the shared fitness, f'_i, as follows:

$$f'_i = \frac{f_k}{m_i}.$$

This procedure is continued for all $i = 1, 2, \ldots, n_k$ and a corresponding f'_i is found. Thereafter, the smallest value $f_k^{\min}$ of all f'_i in the k-th non-dominated front is found for further processing. The dummy fitness of the next non-dominated front is assigned to be $f_{k+1} = f_k^{\min} - \epsilon_k$, where ϵ_k is a small positive number.

The above sharing procedure requires a pre-specified parameter σ_{share}, which can be calculated as follows [11]:

$$\sigma_{\text{share}} \approx \frac{0.5}{\sqrt[p]{q}}, \tag{8.3}$$

where q is the desired number of distinct Pareto-optimal solutions. Although the calculation of σ_{share} depends on this parameter q, it has been been shown elsewhere [7], [38] that the use of above equation with $q \approx 10$ works in many test problems. Moreover, the performance of NSGAs is not very sensitive to this parameter near σ_{share} values calculated using $q \approx 10$.

It may be mentioned here that the above sharing procedure can also be implemented with a distance measure defined in the gene space. That is, instead of the Euclidean distance measure as given above, d_{ij} can also be calculated using the number of bit differences (or the Hamming distance) between solutions i and j [11]. Such distance measures will be useful in problems where the problem is defined in the gene space [7]. The calculation of σ_{share} is different in this case and interested readers may refer to the original study for more details [11].

8.5 PROOF-OF-PRINCIPLE RESULTS

In this section, we apply NSGA to three test problems: (i) a single-variable simple problem [36], (ii) a two-variable constrained problem [2], and (iii) a two-variable problem having discontinuous Pareto-optimal region [7]. In all simulations, we use binary-coded genetic algorithms [20], [24] with a single-point crossover operator with probability one. Mutation probability is kept zero in order to observe the effectiveness of NSGA alone. The parameters are held constant across all runs. Unbiased initial population is generated by randomly spreading solutions over the entire variable space in consideration.

8.5.1 Problem F1

This is a single-variable problem which has been widely used in the multi-criterion optimization literature [36], [38]:

$$\begin{aligned} \text{Minimize} \quad & f_1(x) = x^2, \\ \text{Minimize} \quad & f_2(x) = (x-2)^2. \end{aligned} \tag{8.4}$$

Initial range for the design variable used in simulations is $(-100.0, 100.0)$. The Pareto-optimal solutions lie in $x \in [0, 2]$. The variable is coded using binary strings of size 30. We use a

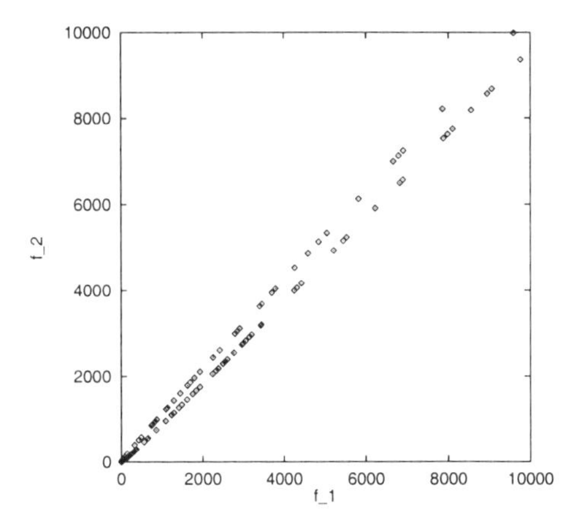

Fig. 8.5 Initial population for function F1.

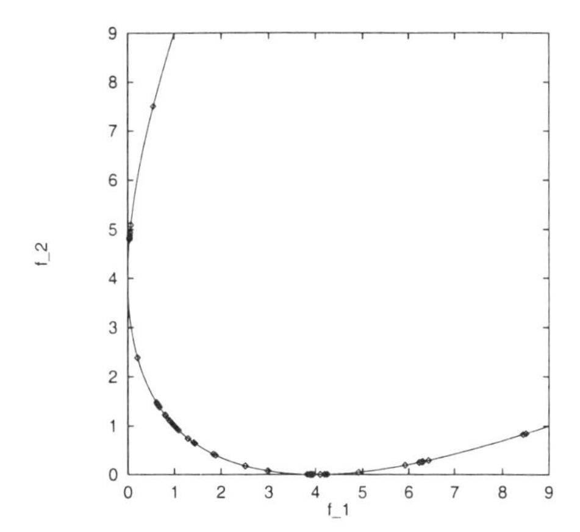

Fig. 8.6 Population at generation 10 for function F1.

population size[2] of 100. The parameter σ_{share} is set to be 0.0002 here. This value induces about N or 100 niches in the Pareto-optimal region. Fig. 8.5 shows the population distribution in the initial generation. The figure shows that the population is widely distributed. With the above parameter settings, it is expected to have only one (out of 100) population member in the Pareto-optimal region. Thereafter, we show the population of 100 members after 10 generations in Fig. 8.6. The figure shows how quickly the NSGA has managed to get the population from function values of about four order of magnitudes to one order of magnitude. However, most of the population members are now inside the Pareto-optimal region. Fig. 8.7 shows the population history at generation 100. All population members are now inside the Pareto-optimal region and notice how the population gets uniformly distributed over the Pareto-optimal region. Finally, Fig. 8.8 shows the population at generation 500. NSGAs are run so long just to show that they have the capabilities of maintaining a sustainable wide-spread population over the entire Pareto-optimal region.

8.5.2 Problem F2

Next, we consider a two-variable problem with constraints [2], [15]:

$$\begin{array}{ll} \text{Minimize} & f_1(\vec{x}) = 2 + (x_1 - 2)^2 + (x_2 - 1)^2, \\ \text{Minimize} & f_2(\vec{x}) = 9x_1 - (x_2 - 1)^2, \\ \text{Subject to} & g_1(\vec{x}) \equiv 225 - x_1^2 - x_2^2 \geq 0, \\ & g_2(\vec{x}) \equiv 3x_2 - x_1 - 10 \geq 0. \end{array} \tag{8.5}$$

[2] Although a much smaller population size will also be able to find Pareto-optimal solutions, this rather large population size is chosen for a specific purpose. This will allow creation of around 100 niches in the Pareto-optimal region, demonstrating that a large number of different niches can be found and maintained in the population using the sharing procedure.

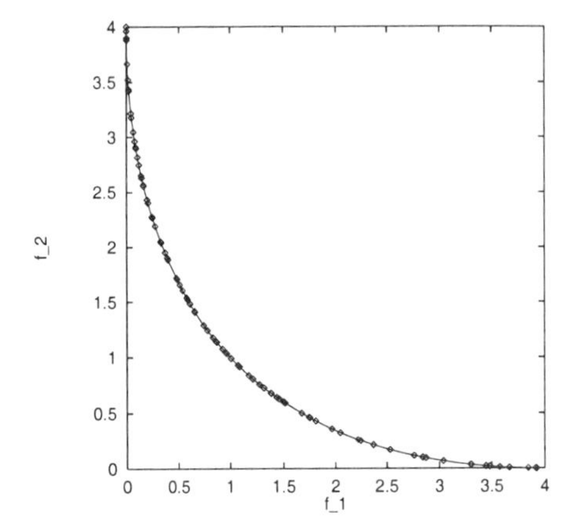

Fig. 8.7 Population at generation 100 for function F1.

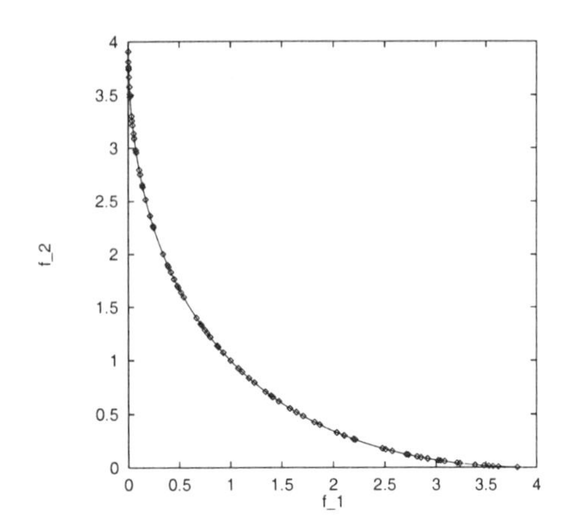

Fig. 8.8 Population at generation 500 for function F1.

Both variables are initialized in the range $[-20.0, 20.0]$. Both variables are coded using binary strings of size 15. We handle constraints by first normalizing them and then using the bracket-operator penalty function [9] with a penalty parameter 10^3. It was shown in an earlier study [38], that resulting Pareto-optimal region in this problem lies where both functions f_1 and f_2 are tangent to each other. This happens for $x_1 = -2.5$ in the unconstrained problem. The addition of two constraints makes only a portion of the search space feasible. Thus, the resulting Pareto-optimal region is as follows:

$$x_1 = -2.5, \quad 2.5 \leq x_2 \leq 14.79.$$

Fig. 8.9 shows the initial population of 100 solutions in the x_1-x_2 space. The figure shows that although some of the solutions are feasible, most solutions are infeasible in the initial population. Fig. 8.10 shows the population at generation 10. This figure shows that most population members are now in the feasible region. At generation 20, Fig. 8.11 shows that most population members are close to the true Pareto-optimal front. Finally, Fig. 8.12 shows that even up to 200 generations the population is able to maintain solutions in the true Pareto-optimal region.

8.5.3 Problem F3

We construct a multi-objective problem having a Pareto-optimal front which is discontinuous [7]:

$$\begin{aligned} &\text{Minimize} \quad f_1(\vec{x}) = x_1, \\ &\text{Minimize} \quad f_2(\vec{x}) = (1 + 10x_2)\left(1 - \left(\frac{x_1}{1+10x_2}\right)^2 - \frac{x_1}{1+10x_2}\sin(8\pi x_1)\right). \end{aligned} \tag{8.6}$$

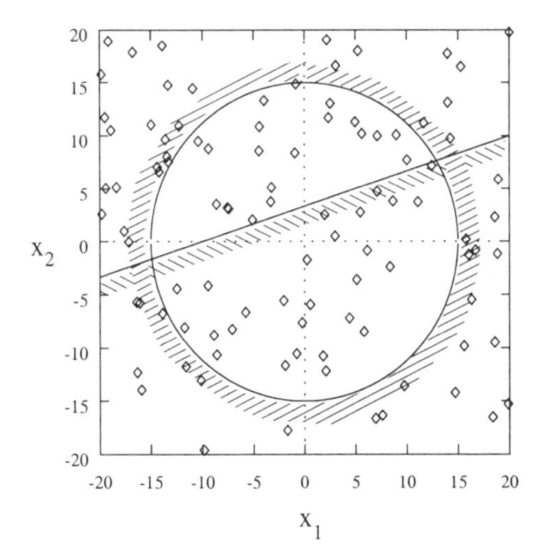

Fig. 8.9 Initial population for function F2.

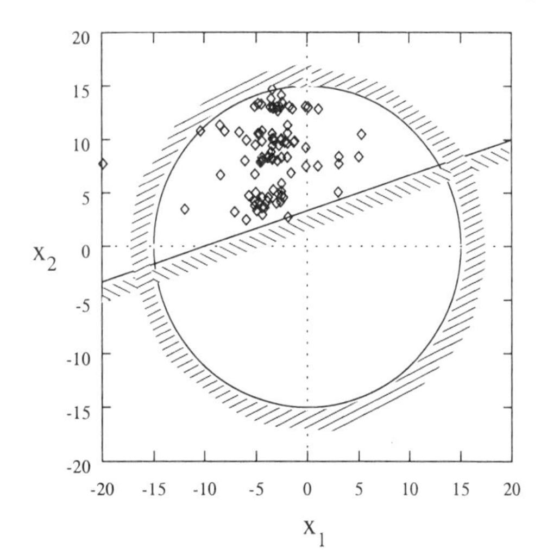

Fig. 8.10 Population at generation 10 for function F2.

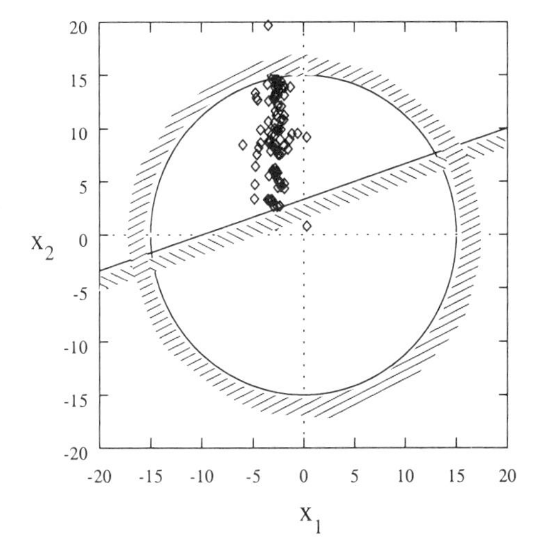

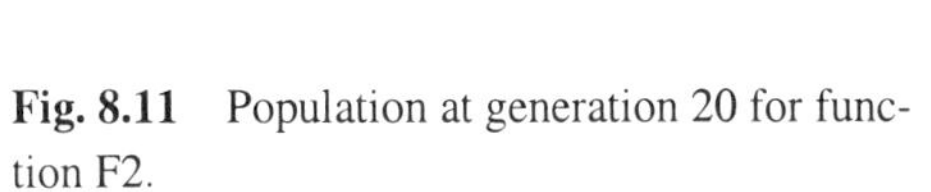

Fig. 8.11 Population at generation 20 for function F2.

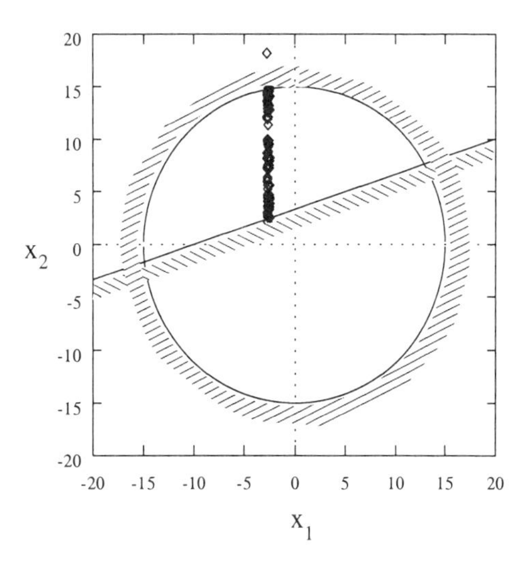

Fig. 8.12 Population at generation 200 for function F2.

Both x_1 and x_2 varies in the interval $[0,1]$ and are coded in binary strings of size 15 each. The discontinuity in the Pareto-optimal region comes due to the periodicity in function f_2. This function tests an algorithm's ability to maintain subpopulations in all discontinuous Pareto-optimal regions.

Fig. 8.13 shows the 50,000 random solutions in f_1-f_2 space. The discontinuous Pareto-optimal regions are also shown by showing solid lines. When NSGAs (population size of 200 and σ_{share} of 0.1) are tried on this problem, the resulting population at generation 300 is shown in Fig. 8.14. The plot shows that if reasonable GA parameter values are chosen,

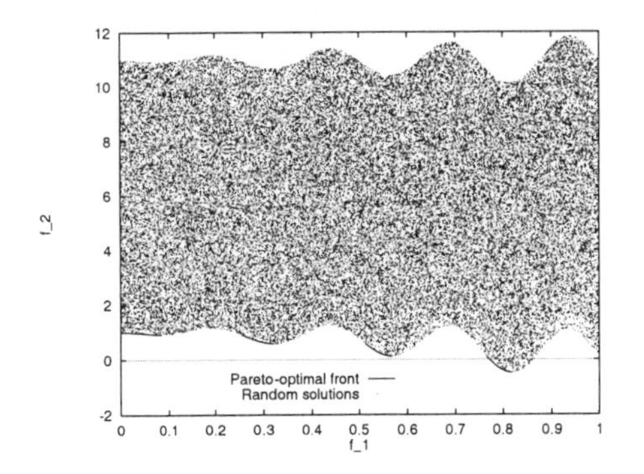

Fig. 8.13 50,000 random solutions are shown on a f_1-f_2 plot of a multi-objective problem having discontinuous Pareto-optimal front.

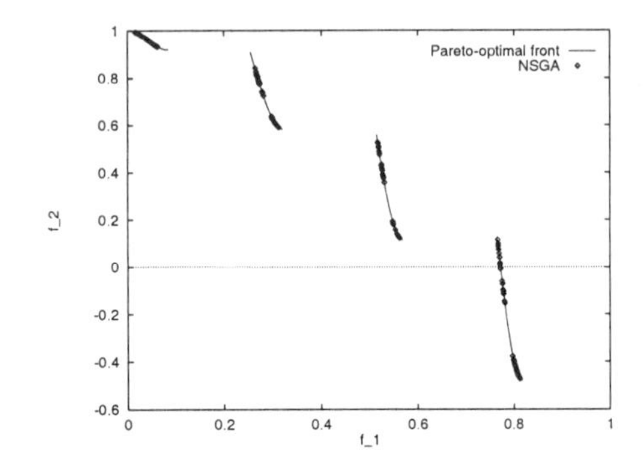

Fig. 8.14 The population at generation 300 for a NSGA run is shown to have found solutions in all four discontinuous Pareto-optimal regions.

NSGAs can find Pareto-optimal solutions in discontinuous regions. A population size of 200 is deliberately used to have a wide distribution of solutions in all discontinuous regions. A little thought will reveal that Pareto-optimal region comprises of solutions $0 \leq x_1 \leq 1$ and $x_2 = 0$. It is interesting to note how NSGAs avoid creating the non-Pareto-optimal solutions, although the corresponding x_2 value is same (and equal to zero) as that of the Pareto-optimal solutions.

8.6 AN ENGINEERING DESIGN

This problem has been well studied in the context of single-objective optimization [34]. A beam needs to be welded on another beam and must carry a certain load F (Fig. 8.15). In the context of single-objective optimal design, it is desired to find four design parameters (thickness of the beam, b, width of the beam t, length of weld ℓ, and weld thickness h) for which the cost of the beam is minimum. The overhang portion of the beam has a length of 14 inch and $F = 6,000$ lb force is applied at the end of the beam. It is intuitive that an optimal design for cost will make all four design variables to take small values. When the beam dimensions are small, it is likely that the deflection at the end of the beam is going to be large. In the parlance of mechanics of materials, this means that the rigidity of the beam is smaller for smaller dimensions of the beam. In mechanical design activities, optimal design for maximum rigidity is also common. Again, a little thought will reveal that a design

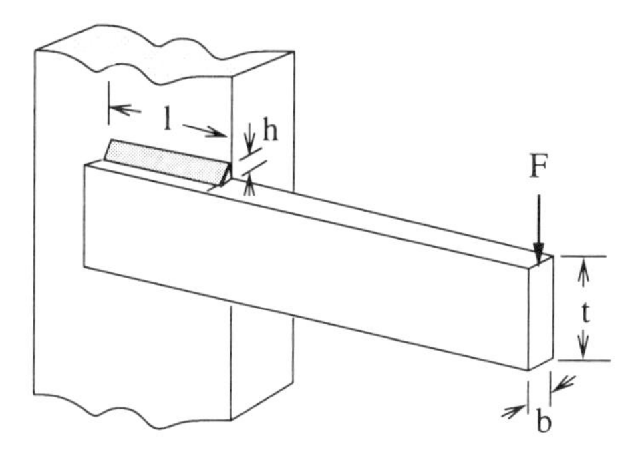

Fig. 8.15 The welded beam design problem. Minimizations of cost and end deflection are two objectives.

for maximum rigidity of the above beam will make all four design dimensions to take large dimensions. Thus, the design solutions for minimum cost and maximum rigidity (or minimum end deflection) are conflicting to each other. In other words, a design that is near-optimal from cost consideration is not near-optimal from rigidity consideration and vice versa. This kind of conflicting objective functions leads to Pareto-optimal solutions. In the following, we present the mathematical formulation of the two-objective optimization problem of minimizing cost and the end deflection [15], [8]:

$$\begin{array}{ll}\text{Minimize} & f_1(\vec{x}) = 1.10471h^2\ell + 0.04811tb(14.0+\ell),\\ \text{Minimize} & f_2(\vec{x}) = \delta(\vec{x}),\\ \text{Subject to} & g_1(\vec{x}) \equiv 13,600 - \tau(\vec{x}) \geq 0,\\ & g_2(\vec{x}) \equiv 30,000 - \sigma(\vec{x}) \geq 0,\\ & g_3(\vec{x}) \equiv b - h \geq 0,\\ & g_4(\vec{x}) \equiv P_c(\vec{x}) - 6,000 \geq 0.\end{array} \tag{8.7}$$

The deflection term $\delta(\vec{x})$ is given as follows:

$$\delta(\vec{x}) = \frac{2.1952}{t^3 b}.$$

There are four constraints. The first constraint makes sure that the shear stress developed at the support location of the beam is smaller than the allowable shear strength of the material (13,600 psi). The second constraint makes sure that normal stress developed at the support location of the beam is smaller than the allowable yield strength of the material (30,000 psi). The third constraint makes sure that thickness of the beam is not smaller than the weld thickness from a practical standpoint. The fourth constraint makes sure that the allowable buckling load (along t direction) of the beam is more than the applied load F. A violation of any of the above four constraints will make the design unacceptable. The stress and buckling terms are given as follows [34]:

$$\tau(\vec{x}) = \sqrt{(\tau')^2 + (\tau'')^2 + (\ell\tau'\tau'')/\sqrt{0.25(\ell^2+(h+t)^2)}},$$

$$\tau' = \frac{6,000}{\sqrt{2}h\ell},$$
$$\tau'' = \frac{6,000(14+0.5\ell)\sqrt{0.25(\ell^2+(h+t)^2)}}{2\{0.707h\ell(\ell^2/12+0.25(h+t)^2)\}},$$
$$\sigma(\vec{x}) = \frac{504,000}{t^2 b},$$
$$P_c(\vec{x}) = 64,746.022(1-0.0282346t)tb^3.$$

The variables are initialized in the following range: $0.125 \leq h, b \leq 5.0$ and $0.1 \leq \ell, t \leq 10.0$. Constraints are handled using the bracket-operator penalty function [9]. Penalty parameters of 100 and 0.1 are used for the first and second objective functions, respectively. We use real-parameter GAs with simulated binary crossover (SBX) operator [10] to solve this problem. Unlike in the binary-coded GAs, variables are used directly and a crossover operator that creates two real-valued children solutions from two parent solutions is used. For details of this crossover implementation, refer to original studies [10], [15], [13]. In order to investigate the search space, we plot about 230,000 random feasible solutions in f_1-f_2 space in Fig. 8.16. The figure clearly shows the Pareto-optimal front near the origin. It can be seen that the Pareto-optimal solutions have a maximum end deflection value of around 0.01 inch, beyond which the an increase in end deflection also causes an increase in cost. It is also interesting to observe that a large density of solutions lie near the Pareto-optimal region, rather than away from it. An initial population of 100 solutions are shown in the figure. We use a σ_{share} of 0.281 (refer to equation 8.3 with $P = 4$ and $q = 10$). Fig. 8.17 shows the populations at different generations. The figure clearly shows that NSGA progresses towards the Pareto-optimal front with generation. Finally, Fig. 8.18 shows that the population after 500 generations has truly come near the Pareto-optimal front. Note here that the deflection axis is now plotted for a reduced range compared to that in Fig. 8.16. These figures demonstrate the efficacy of NSGAs in converging close to the Pareto-optimal front with a wide variety of solutions.

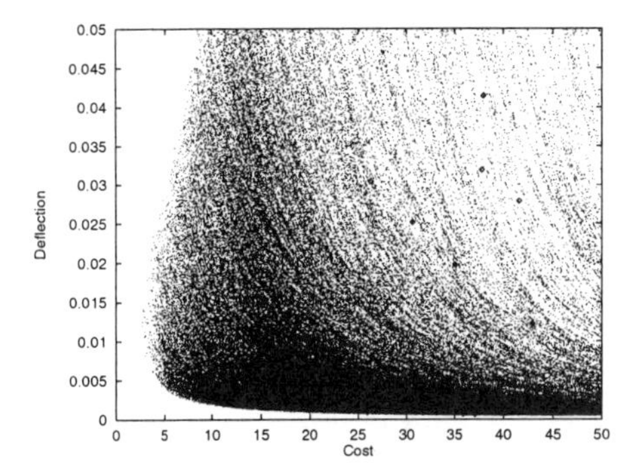

Fig. 8.16 About 230,000 random feasible solutions are shown in a function space plot.

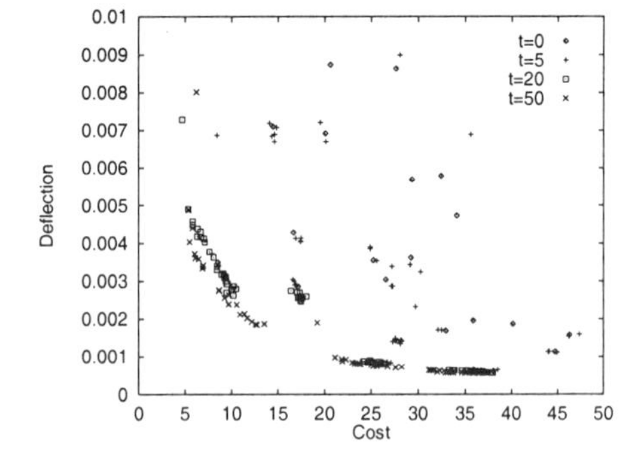

Fig. 8.17 Population history for a few generations are shown.

Before we end the discussion on this problem, we would like to highlight an important by-product of multi-criterion optimization. If we investigate the two extreme solutions (the

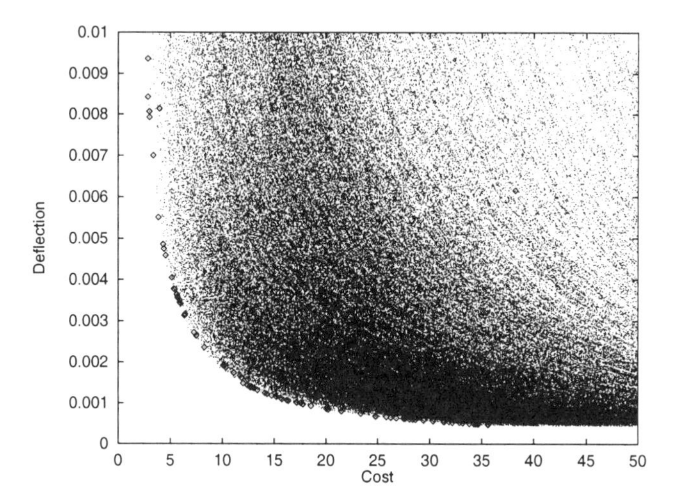

Fig. 8.18 Population at generation 500 shows that a wide range of Pareto-optimal solutions are found.

smallest cost solution and the smallest deflection solution), we observe an interesting property among the two solutions (Table 8.1). It is remarkable to note that both optimal solutions have identical values for three of four variables (h, ℓ, and t). The only way the solutions differ is in variable b. As long as the first three variables are set to the above values, any value of b (within lower and upper limits satisfying all constraints) would produce a Pareto-optimal or a near-Pareto-optimal solution. For a smaller value of b, the resulting design is a closer to a smaller cost design. This information is crisp and very useful for designers and is not apparent how else such an information could be found. It is not clear whether such a property exists in Pareto-optimal solutions of other multi-criterion optimal design problems, but if there exists such a property, it is obvious that any designer would like to know it at any cost. Evolutionary approaches to multi-criterion optimization problems allow such a study to be carried out in other engineering design problems as well.

Table 8.1 Two extreme Pareto-optimal solutions are shown.

Cost	Defl.	h	ℓ	t	b
3.944	0.005	0.422	2.465	9.990	0.439
20.677	0.001	0.422	2.465	9.990	2.558

8.7 FUTURE DIRECTIONS FOR RESEARCH

As mentioned earlier, evolutionary multi-criterion optimization techniques are getting growing interests among researchers and practitioners. As the interests grow, there also exists a need of a directed research and application. In the following, we describe some of the immediate research directions that may help develop better algorithms for multi-criterion optimization.

1. Develop constrained test problems for multi-objective optimization
2. Comparison of existing multi-objective GA implementations
3. Understand dynamics of GA populations with generations
4. Scalability of multi-objective GAs with number of objectives
5. Convergence to Pareto-optimal front
6. Define appropriate multi-objective GA parameters (such as elitism)
7. Metrics for comparing two populations
8. Hybrid multi-objective GAs
9. Real-world applications
10. Multi-objective scheduling and other optimization problems

Because of page restrictions, we only discuss here the first issue by presenting a number of test problems. Interested readers may refer to [7] for details on other issues.

With the development of multi-objective optimization algorithms, there is an obvious need of a good set of test problems, which will test and compare different multi-criterion optimization algorithms with each other. Many test problems that are used in the literature are listed in [42]. However, it is not clear what aspect of an algorithm is tested by these test problems. Recently, a systematic procedure of constructing test problems have been suggested [7]. In that study, test problems are constructed from single-objective optimization problems to test two main features a multi-criterion optimization technique must have: (i) convergence ability to the true Pareto-optimal front and (ii) ability to find diverse Pareto-optimal solutions. To test these two features, three different functions (g, f_1, and h) are used to construct a two-objective optimization problem:

$$\begin{array}{ll} \text{Minimize} & f_1(\vec{x}) = f_1(x_1, x_2, \ldots, x_m), \\ \text{Minimize} & f_2(\vec{x}) = g(x_{m+1}, \ldots, x_N) h(f_1(x_1, \ldots, x_m), g(x_{m+1}, \ldots, x_N)). \end{array} \tag{8.8}$$

Since there are no variables which are common to both functions g and f_1, the Pareto-optimal solutions take those values of x_{m+1} to x_N variables which correspond to the global optimum of g. Since the function f_1 is also to be minimized, Pareto-optimal solutions take all possible values of x_1 to x_m variables. The above construction procedure allows the function g to control the search space lateral to the Pareto-optimal front. Thus, just by choosing a multi-modal or a *deceptive* g function [12], a multi-modal or deceptive two-objective optimization problem can be constructed. The function f_1 controls the search space along the Pareto-optimal front. Thus, a non-linear function of f_1 can construct a two-objective optimization problem having bias against some regions in the Pareto-optimal region. Finally, the function h controls the shape of the Pareto-optimal front. By choosing a convex, concave, or periodic function for h, a two-objective optimization problem having convex, concave, or discontinuous Pareto-optimal region can be constructed. In this study, we have already shown one such function (problem F3) which has a discontinuous Pareto-optimal region. This construction method allows each of two important features of a multi-criterion optimization algorithm to be tested in a controlled manner. In the following, we present a few test problems which can be used to test multi-criterion optimization algorithms.

Multi-modal multi-objective problem: Construct the problem in equation 8.8 using following functions:

$$\begin{aligned} f_1(x_1) &= x_1, \\ g(x_2, \ldots, x_N) &= 1 + 10(N-1) + \sum_{i=2}^{N} x_i^2 - 10\cos(2\pi x_i), \\ h(f_1, g) &= \begin{cases} 1 - \left(\frac{f_1}{g}\right)^{0.5}, & \text{if } f_1 \le g, \\ 0, & \text{otherwise.} \end{cases} \end{aligned}$$

The first variable lies in $x_1 \in [0, 1]$ and all others in $[-30, 30]$. Global Pareto-optimal solutions are as follows: $0 \le x_1 \le 1$ and $x_i = 0.0$ for $i = 2, \ldots, N$. A $N = 10$ or more is recommended. Multi-criterion optimization face difficulty with this problem because of the existence of $(61^{N-1} - 1)$ other local Pareto-optimal solutions each corresponding to an integer value of x_2 to x_N variables.

Deceptive multi-objective problem: The function is defined in *unitation*[3] u of N substrings of different lengths ℓ_i $(i = 1, \ldots, N)$:

$$\begin{aligned} f_1 &= 1 + u(\ell_1), \\ f_2 &= \frac{\sum_{i=2}^{N} g(u(\ell_i))}{1 + u(\ell_1)}, \end{aligned}$$

where $g(u(\ell_i)) = \begin{cases} 2 + u(\ell_i), & \text{if } u(\ell_i) < \ell_i, \\ 1, & \text{if } u(\ell_i) = \ell_i. \end{cases}$

The global Pareto-optimal front has the following solutions: $0 \le u(\ell_1) \le \ell_1$ and $u(\ell_i) = \ell_i$ for all other $i = 2, \ldots, N$. Note that this function is defined in gene the space. This problem is deceptive because the subfunction $g(u(\ell_i))$ shows that the function values of unitation $u = 0$ to $(\ell_i - 1)$ directs the search towards the deceptive attractor $u = 0$, whereas the solution $u = \ell_i$ is the true minimum of subfunction g.

[3]Unitation is the number of 1s in the substring.

Biased Pareto-optimal front: Construct the problem using equation 8.8 and following functions:

$$
\begin{aligned}
f_1(x_1) &= 1 - \exp(-4x_1)\sin^6(5\pi x_1), \\
g(x_2, \ldots, x_N) &= g_{\min} + (g_{\max} - g_{\min}) \left(\frac{\sum_{i=2}^{N} x_i - \sum_{i=2}^{N} x_i^{\min}}{\sum_{i=2}^{N} x_i^{\max} - \sum_{i=2}^{N} x_i^{\min}} \right)^{\gamma}, \\
h(f_1, g) &= \begin{cases} 1 - \left(\frac{f_1}{g}\right)^2, & \text{if } f_1 \leq g, \\ 0, & \text{otherwise.} \end{cases}
\end{aligned}
$$

Use $\gamma = 0.25$, $g_{\min} = 1$, and $g_{\max} = 10$. All variables vary in $[0, 1]$, thus $x_i^{\min} = 0$ and $x_i^{\max} = 1$. The true Pareto-optimal front has following values of variables: $0 \leq x_1 \leq 1$ and $x_i = x_i^{\min} = 0$ for $i = 2, \ldots, N$. Multi-criterion optimization problems may face difficulty in this problem because of the non-linearity in f_1 function. There are two aspects of this problem: (i) There are more solutions having f_1 close to one than solutions having f_1 close to zero, and (ii) there will be a tendency for algorithms to find x_1 values within $[0, 0.2]$ than the other space ($x_1 \in [0.2, 1]$), although all solutions $x_1 \in [0, 1]$ are Pareto-optimal solutions. Thus, this problem tests an algorithm's ability to spread solutions over the entire Pareto-optimal region, although the problem introduces a bias against certain portion of the search space.

Concave Pareto-optimal front: Equation 8.8 with following functions can be used to construct this problem:

$$
\begin{aligned}
f_1(x_1) &= x_1, \\
g(x_2, \ldots, x_N) &= 1 + 10 \frac{\sum_{i=2}^{N} x_i}{N - 1}, \\
h(f_1, g) &= \begin{cases} 1 - \left(\frac{f_1}{g}\right)^{\alpha}, & \text{if } f_1 \leq g, \\ 0, & \text{otherwise.} \end{cases}
\end{aligned}
$$

All variables lie in $[0, 1]$. The true Pareto-optimal front has following variable values: $0 \leq x_1 \leq 1$ and $x_i = 0$ for $i = 2, \ldots, N$. By setting $\alpha = 0.5$, we obtain a problem having convex Pareto-optimal front (like in Fig. 8.2) and by setting $\alpha = 2$, we shall obtain a problem having non-convex Pareto-optimal front (line in Fig. 8.3). Thus, multi-objective algorithms can be tested for the effect of non-convexity in the Pareto-optimal front just by changing the α parameter.

Discontinuous Pareto-optimal front: We construct the problem using equation 8.8 and the following functions:

$$
\begin{aligned}
f_1(x_1) &= x_1, \\
g(x_2, \ldots, x_N) &= 1 + 10 \frac{\sum_{i=2}^{N} x_i}{N - 1}, \\
h(f_1, g) &= 1 - \left(\frac{f_1}{g}\right)^{0.25} - \frac{f_1}{g} \sin(10\pi f_1).
\end{aligned}
$$

All variables vary in $[0, 1]$ and true Pareto-optimal solutions are constituted with $x_i = 0$ for $i = 2, \ldots, N$ and discontinuous values of x_1 in the range $[0, 1]$. This problem tests an algorithm's ability to find Pareto-optimal solutions which are discontinuously spread in the search space.

More complex test problems can be created by combining more than one aspects described above. Difficulties can also be increased by introducing parameter interactions by first rotating the variables by a transformation matrix and then using above equations. For more details, refer to the original study [7].

In Table 8.2, we outline some of the engineering design problems where an evolutionary multi-criterion optimization algorithm has been successfully applied.

Table 8.2 Applications of evolutionary multi-criterion optimization algorithms.

Researcher(s)	Year	Application area
P. Hajela and C.-Y. Lin	1992	Multi-criterion structure design [23]
J. W. Eheart, S. E. Cieniawski and S. Ranjithan	1993	Ground-water quality monitoring system [16]
C. M. Fonseca and P. J. Fleming	1993	Gas turbine engine design [17]
A. D. Belegundu et al.	1994	Laminated ceramic composites [1]
T. J. Stanley and T. Mudge	1995	Microprocessor chip design [39]
D. S. Todd and P. Sen	1997	Containership loading design [41]
D. S. Weile, E. Michielssen, and D. E. Goldberg	1996	Broad-band microwave absorber design [43]
A. G. Cunha, P. Oliviera, and J. A. Covas	1997	Extruder screw design [4], [6]
D. H. Loughlin and S. Ranjithan	1997	Air pollution management [27]
C. Poloni et al.	1997	Aerodynamic shape design [33], [32]
E. Zitzler and L. Thiele	1998	Synthesis of digital hardware-software multi-processor system [45]
G. T. Parks and I. Miller	1998	Pressurized water reactor reload design [31]
S. Obayashi, S. Takahashi, and Y. Takeguchi	1998	Aircraft wing planform shape design [30]
K. Mitra, K. Deb, and S. K. Gupta	1998	Dynamic optimization of an industrial nylon 6 semibatch reactor [29]
D. Cvetkovic and I. Parmee	1998	Airframe design [5]

8.8 SUMMARY

In this paper, we have discussed evolutionary algorithms for multi-criterion optimization. By reviewing a couple of popular classical algorithms, it has been argued that there is a need for more efficient search algorithms for multi-criterion optimization. A number of evolutionary algorithm implementations have been outlined and one particular implementation (called non-dominated sorting GA (NSGA)) has been discussed in details. The efficacy of NSGA has been demonstrated by showing simulation results on three test problems and on one engineering design problem. The results show that evolutionary algorithms are effective tools for doing multi-criterion optimization in that multiple Pareto-optimal solutions can be found in one simulation run. The study also suggests a number of immediate future studies including suggesting a few test problems for multi-criterion optimization, which should help develop better algorithms and maintain a focused research direction for the development of this fast growing field.

Acknowledgments

The author acknowledges the support provided by Alexander von Humboldt Foundation, Germany during the course of this study.

REFERENCES

1. Belegundu, A. D., Murthy, D. V., Salagame, R. R., and Constants, E. W. (1994). Multi-objective optimization of laminated ceramic composites using genetic algorithms. *Proceedings of the Fifth AIAA/USAF/NASA Symposium on Multidisciplinary Analysis and Optimization*, 1015–1022.

2. Chankong, V., and Haimes, Y. Y. (1983). *Multiobjective decision making theory and methodology*. New York:North-Holland.

3. Coello, C. A. C. (1998). A comprehensive survey of evolutionary based multiobjective optimization techniques. *Unpublished document.*

4. Cunha, A. G., Oliveira, P., and Covas, J. A. (1997). Use of genetic algorithms in multicriteria optimization to solve industrial problems. *Proceedings of the Seventh International Conference on Genetic Algorithms*. 682–688.

5. Cvetkovic, D. and Parmee, I. (1998). Evolutionary design and multi-objective optimisation. *Proceedings of the Sixth European Congress on Intelligent Techniques and Soft Computing (EUFIT)*, 397–401.

6. Covas, J. A., Cunha, A. G., and Oliveira, P. (in press). Optimization of single screw extrusion: Theoretical and experimental results. *International Journal of Forming Processes.*

7. Multi-objective genetic algorithms: Problem difficulties and construction of test problems. Technical Report No: CI-49/98. Dortmund: Department of Computer Science/LS11, University of Dortmund, Germany.

8. Deb, K. (in press). An efficient constraint handling method for genetic algorithms. *Computer Methods in Applied Mechanics and Engineering*.

9. Deb, K. (1995). *Optimization for engineering design: Algorithms and examples.* New Delhi: Prentice-Hall.

10. Deb, K. and Agrawal, R. B. (1995) Simulated binary crossover for continuous search space. *Complex Systems, 9* 115–148.

11. Deb, K. and Goldberg, D. E. (1989). An investigation of niche and species formation in genetic function optimization. *Proceedings of the Third International Conference on Genetic Algorithms* (pp. 42–50).

12. Deb, K. and Goldberg, D. E. (1994). Sufficient conditions for arbitrary binary functions. *Annals of Mathematics and Artificial Intelligence, 10*, 385–408.

13. Deb, K. and Goyal, M. (1998). A robust optimization procedure for mechanical component design based on genetic adaptive search. *Transactions of the ASME: Journal of Mechanical Design, 120*(2), 162–164.

14. Deb, K., Horn, J., and Goldberg, D. E. (1993). Multi-Modal deceptive functions. *Complex Systems, 7*, 131–153.

15. Deb, K. and Kumar, A. (1995). Real-coded genetic algorithms with simulated binary crossover: Studies on multi-modal and multi-objective problems. *Complex Systems, 9*(6), 431–454.

16. Eheart, J. W., Cieniawski, S. E., and Ranjithan, S. (1993). Genetic-algorithm-based design of groundwater quality monitoring system. *WRC Research Report No. 218.* Urbana: Department of Civil Engineering, The University of Illinois at Urbana-Champaign.

17. Fonseca, C. M. and Fleming, P. J. (1998). Multiobjective optimization and multiple constraint handling with evolutionary algorithms–Part II: Application example. *IEEE Transactions on Systems, Man, and Cybernetics: Part A: Systems and Humans*. 38–47.

18. Fonseca, C. M. and Fleming, P. J. (1995). An overview of evolutionary algorithms in multi-objective optimization. *Evolutionary Computation, 3*(1). 1–16.

19. Fonseca, C. M. and Fleming, P. J. (1993). Genetic algorithms for multiobjective optimization: Formulation, discussion, and generalization. *Proceedings of the Fifth International Conference on Genetic Algorithms*. 416–423.

20. Goldberg, D. E. (1989). *Genetic algorithms for search, optimization, and machine learning*. Reading, MA: Addison-Wesley.

21. Goldberg, D. E. and Deb, K. (1991). A comparison of selection schemes used in genetic algorithms, *Foundations of Genetic Algorithms*, 69–93.

22. Goldberg, D. E. and Richardson, J. (1987). Genetic algorithms with sharing for multimodal function optimization. *Proceedings of the First International Conference on Genetic Algorithms and Their Applications*. 41–49.

23. Hajela, P. and Lin, C.-Y. (1992). Genetic search strategies in multi-criterion optimal design. *Structural Optimization, 4*. 99–107.

24. Holland, J. H. (1975). *Adaptation in natural and artificial systems*. Ann Arbor, Michigan: MIT Press.

25. Horn, J. (1997). Multicriterion decision making. In Eds. (T. Bäck et al.) *Handbook of Evolutionary Computation*.

26. Horn, J. and Nafploitis, N., and Goldberg, D. E. (1994). A niched Pareto genetic algorithm for multi-objective optimization. *Proceedings of the First IEEE Conference on Evolutionary Computation*. 82–87.

27. Loughlin, D. H. and Ranjithan, S. (1997). The neighborhood constraint-method: A genetic algorithm-based multiobjective optimization technique. *Proceedings of the Seventh International Conference on Genetic algorithms*, 666–673.

28. Laumanns, M., Rudolph, G., and Schwefel, H.-P. (1998). A spatial predator-prey approach to multi-objective optimization: A preliminary study. *Proceedings of the Parallel Problem Solving from Nature, V*. 241–249.

29. Mitra, K., Deb, K., and Gupta, S. K. (1998). Multiobjective dynamic optimization of an industrial Nylon 6 semibatch reactor using genetic algorithms. *Journal of Applied Polymer Science, 69*(1), 69–87.

30. Obayashi, S., Takahashi, S., and Takeguchi, Y. (1998). Niching and elitist models for MOGAs. *Parallel Problem Solving from nature, V*, 260–269.

31. Parks, G. T. and Miller, I. (1998). Selective breeding in a multi-objective genetic algorithm. *Proceedings of the Parallel Problem Solving from Nature, V*, 250–259.

32. Poloni, C. (1997). Hybrid GA for multi-objective aerodynamic shape optimization. In G. Winter, J. Periaux, M. Galan, and P. Cuesta (Eds.), *Genetic algorithms in engineering and computer science*. Chichester: Wiley (Pages 397–414).

33. Poloni, C., Giurgevich, A., Onesti, L., and Pediroda, V. (in press). Hybridisation of multiobjective genetic algorithm, neural networks and classical optimiser for complex design problems in fluid dynamics. *Computer Methods in Applied Mechanics and Engineering*.

34. Reklaitis, G. V., Ravindran, A. and Ragsdell, K. M. (1983). *Engineering optimization methods and applications*. New York: Wiley.

35. Rosenberg, R. S. (1967). *Simulation of genetic populations with biochemical properties*. PhD dissertation. University of Michigan.

36. Schaffer, J. D. (1984). Some experiments in machine learning using vector evaluated genetic algorithms. (Doctoral Dissertation). Nashville, TN: Vanderbilt University.

37. Schaffer, J. D. (1985). Multiple objective optimization with vector evaluated genetic algorithms. *Proceedings of the First International Conference on Genetic Algorithms*, 93–100.

38. Srinivas, N. and Deb, K. (1994). Multi-Objective function optimization using non-dominated sorting genetic algorithms, *Evolutionary Computation, 2*(3), 221–248.

39. Stanley, T. J. and Mudge, T. (1995). A parallel genetic algorithm for multiobjective microprocessor design. *Proceedings of the Sixth International Conference on Genetic algorithms*, 597–604.

40. Steuer, R. E. (1986). *Multiple criteria optimization: Theory, computation, and application*. New York: Wiley.

41. Todd, D. S. and Sen, P. (1997). A multiple criteria genetic algorithm for containership loading. *Proceedings of the Seventh International Conference on Genetic algorithms*, 674–681.

42. van Veldhuizen, D. and Lamont, G. B. (1998). Multiobjective evolutionary algorithm research: A history and analysis. *Report Number TR-98-03*. Wright-Patterson AFB, Ohio: Department of Electrical and Computer Engineering, Air Force Institute of Technology.

43. Weile, D. S., Michielssen, E., and Goldberg, D. E. (1996). Genetic algorithm design of Pareto-optimal broad band microwave absorbers. *IEEE Transactions on Electromagnetic Compatibility, 38*(4).

44. Zitzler, E. and Thiele, L. (1998). Multiobjective optimization using evolutionary algorithms—A comparative case study. *Parallel Problem Solving from Nature, V*, 292–301.

45. Zitzler, E. and Thiele, L. (1998). An evolutionary algorithm for multiobjective optimization: The strength Pareto approach. *Technical Report No. 43 (May 1998)*. Zürich: Computer Engineering and Networks Laboratory, Switzerland.

9 ACO Algorithms for the Traveling Salesman Problem

TH. STÜTZLE[†] and M. DORIGO

IRIDIA, Université Libre de Bruxelles, Belgium
E-mail: {tstutzle,mdorigo}@ulb.ac.be

9.1 INTRODUCTION

Ant algorithms [18], [14], [19] are a recently developed, population-based approach which has been successfully applied to several $\mathcal{NP}$-hard combinatorial optimization problems [6], [13], [17], [23], [34], [40], [49]. As the name suggests, ant algorithms have been inspired by the behavior of real ant colonies, in particular, by their foraging behavior. One of the main ideas of ant algorithms is the indirect communication of a colony of agents, called (artificial) ants, based on pheromone trails (pheromones are also used by real ants for communication). The (artificial) pheromone trails are a kind of distributed numeric information which is modified by the ants to reflect their experience while solving a particular problem. Recently, the Ant Colony Optimization (ACO) meta-heuristic has been proposed which provides a unifying framework for most applications of ant algorithms [15], [16] to combinatorial optimization problems. In particular, all the ant algorithms applied to the TSP fit perfectly into the ACO meta-heuristic and, therefore, we will call these algorithms also ACO algorithms.

The first ACO algorithm, called Ant System (AS) [18], [14], [19], has been applied to the Traveling Salesman Problem (TSP). Starting from Ant System, several improvements of the basic algorithm have been proposed [21], [22], [17], [51], [53], [7]. Typically, these improved algorithms have been tested again on the TSP. All these improved versions of AS have in common a stronger exploitation of the best solutions found to direct the ants' search process; they mainly differ in some aspects of the search control. Additionally, the best performing ACO algorithms for the TSP [17], [49] improve the solutions generated by the ants using local search algorithms.

In this paper we give an overview on the available ACO algorithms for the TSP. We first introduce, in Section 9.2, the TSP. In Section 9.3 we outline how ACO algorithms can be applied to that problem and present the available ACO algorithms for the TSP. Section 9.4

[†]Presently on leave from FG Intellektik, TU Darmstadt, Germany

briefly discusses local search for the TSP, while Section 9.5 presents experimental results which have been obtained with $\mathcal{MAX}$–$\mathcal{MIN}$ Ant System, one of the improved versions of Ant System. Since the first application of ACO algorithms to the TSP, they have been applied to several other combinatorial optimization problems. On many important problems ACO algorithms have proved to be among the best available algorithms. In Section 9.6 we give a concise overview of these other applications of ACO algorithms.

9.2 THE TRAVELING SALESMAN PROBLEM

The TSP is extensively studied in literature [29], [31], [45] and has attracted since a long time a considerable amount of research effort. The TSP also plays an important role in Ant Colony Optimization since the first ACO algorithm, called Ant System [18], [14], [19], as well as many of the subsequently proposed ACO algorithms [21], [17], [52], [53], [7] have initially been applied to the TSP. The TSP was chosen for many reasons: (i) it is a problem to which ACO algorithms are easily applied, (ii) it is an $\mathcal{NP}$-hard [26] optimization problem, (iii) it is a standard test-bed for new algorithmic ideas and a good performance on the TSP is often taken as a proof of their usefulness, and (iv) it is easily understandable, so that the algorithm behavior is not obscured by too many technicalities.

Intuitively, the TSP is the problem of a salesman who wants to find, starting from his home town, a shortest possible trip through a given set of customer cities and to return to its home town. More formally, it can be represented by a complete weighted graph $G = (N, A)$ with N being the set of nodes, representing the cities, and A the set of arcs fully connecting the nodes N. Each arc is assigned a value d_{ij}, which is the length of arc $(i, j) \in A$, that is, the distance between cities i and j, with $i, j \in N$. The TSP is the problem of finding a minimal length Hamiltonian circuit of the graph, where an Hamiltonian circuit is a closed tour visiting exactly once each of the $n = |N|$ nodes of G. For symmetric TSPs, the distances between the cities are independent of the direction of traversing the arcs, that is, $d_{ij} = d_{ji}$ for every pair of nodes. In the more general asymmetric TSP (ATSP) at least for one pair of nodes i, j we have $d_{ij} \neq d_{ji}$.

In case of symmetric TSPs, we will use *Euclidean* TSP instances in which the cities are points in the Euclidean space and the inter-city distances are calculated using the Euclidean norm. All the TSP instances we use are taken from the TSPLIB Benchmark library [44] which contains a large collection of instances; these have been used in many other studies or stem from practical applications of the TSP. TSPLIB is accessible on the WWW at the address http://www.iwr.uni-heidelberg.de/iwr/comopt/soft/TSPLIB95/TSPLIB.html

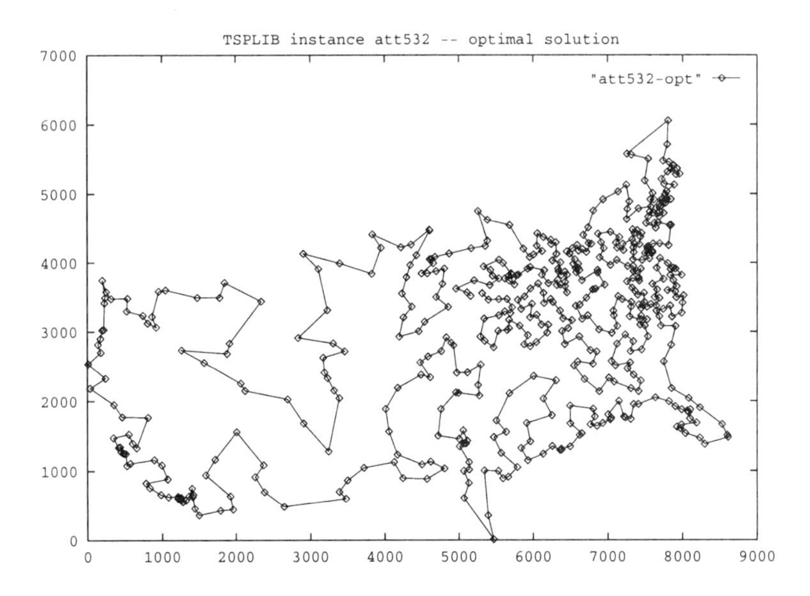

Fig. 9.1 TSP instance `att532` from TSPLIB; it comprises 532 cities in the USA. The given tour is of length 27686, which in fact is an optimal tour.

9.3 AVAILABLE ACO ALGORITHMS FOR THE TSP

9.3.1 Applying ACO algorithms to the TSP

In ACO algorithms ants are simple agents which, in the TSP case, construct tours by moving from city to city on the problem graph. The ants' solution construction is guided by (artificial) pheromone trails and an a priori available heuristic information. When applying ACO algorithm to the TSP, a pheromone strength $\tau_{ij}(t)$ is associated to each arc (i, j), where $\tau_{ij}(t)$ is a numerical information which is modified during the run of the algorithm and t is the iteration counter. If an ACO algorithm is applied to symmetric TSP instances, we always have $\tau_{ij}(t) = \tau_{ji}(t)$; in applications to asymmetric TSPs (ATSPs), we will possibly have $\tau_{ij}(t) \neq \tau_{ji}(t)$.

Initially, each of the m ants is placed on a randomly chosen city and then iteratively applies at each city a state transition rule. An ant constructs a tour as follows. At a city i, the ant chooses a still unvisited city j probabilistically, biased by the pheromone trail strength $\tau_{ij}(t)$ on the arc between city i and city j and a locally available heuristic information, which is a function of the arc length. Ants probabilistically prefer cities which are close and are connected by arcs with a high pheromone trail strength. To construct a feasible solution each ant has a limited form of memory, called *tabu list*, in which the current partial tour is stored. The memory is used to determine at each construction step the set of cities which still has to be visited and to guarantee that a feasible solution is built. Additionally, it allows the ant to retrace its tour, once it is completed.

After all ants have constructed a tour, the pheromones are updated. This is typically done by first lowering the pheromone trail strengths by a constant factor and then the ants are allowed to deposit pheromone on the arcs they have visited. The trail update is done in such a form that arcs contained in shorter tours and/or visited by many ants receive a higher amount of pheromone and are therefore chosen with a higher probability in the following iterations of the algorithm. In this sense the amount of pheromone $\tau_{ij}(t)$ represents the learned desirability of choosing next city j when an ant is at city i.

The best performing ACO algorithms for the TSP improve the tours constructed by the ants applying a local search algorithm [17], [49], [53]. Hence, these algorithms are in fact hybrid algorithms combining probabilistic solution construction by a colony of ants with standard local search algorithms. Such a combination may be very useful since constructive algorithms for the TSP often result in a relatively poor solution quality compared to local search algorithms [29]. Yet, it has been noted that repeating local search from randomly generated initial solutions results in a considerable gap to the optimal solution [29]. Consequently, on the one hand local search algorithms may improve the tours constructed by the ants, while on the other hand the ants may guide the local search by constructing promising initial solutions. The initial solutions generated by the ants are promising because the ants use with higher probability those arcs which, according to the search experience, have more often been contained in shorter tours.

In general, all the ACO algorithms for the TSP follow a specific algorithmic scheme, which is outlined in Algorithm 9.3.1.1. After the initialization of the pheromone trails and some parameters a main loop is repeated until a termination condition, which may be a certain number of solution constructions or a given CPU-time limit, is met. In the main loop, first, the ants construct feasible tours, then the generated tours are improved by applying local search, and finally the pheromone trails are updated. In fact, most of the best performing ACO algorithms for $\mathcal{NP}$-hard combinatorial optimization problems follow this algorithmic scheme [17], [23], [34], [49], [53]. It must be noted that the ACO meta-heuristic [15], [16] is more general than the algorithmic scheme given above. For example, the later does not capture the application of ACO algorithms to network routing problems.

9.3.1.1 Algorithm Algorithmic skeleton for ACO algorithm applied to the TSP.

```
procedure ACO algorithm for TSPs
    Set parameters, initialize pheromone trails
    while (termination condition not met) do
        ConstructSolutions
        ApplyLocalSearch                    % optional
        UpdateTrails
    end
end ACO algorithm for TSPs.
```

9.3.2 Ant System

When Ant System (AS) was first introduced, it was applied to the TSP. Initially, three different versions of AS were proposed [14], [9], [18]; these were called *ant-density*, *ant-quantity*, and *ant-cycle*. While in *ant-density* and *ant-quantity* the ants updated the pheromone directly after a move from a city to an adjacent one, in *ant-cycle* the pheromone update was only done after all the ants had constructed the tours and the amount of pheromone deposited by each ant was set to be a function of the tour quality. Because *ant-cycle* performed much better than the other two variants, here we only present the ant-cycle algorithm, referring to it as Ant System in the following. In AS each of m (artificial) ants builds a solution (tour) of the TSP, as described before. In AS no local search is applied. (Obviously, it would be straightforward to add local search to AS.)

Tour construction Initially, each ant is put on some randomly chosen city. At each construction step, ant k applies a probabilistic action choice rule. In particular, the probability with which ant k, currently at city i, chooses to go to city j at the tth iteration of the algorithm is:

$$p_{ij}^k(t) = \frac{[\tau_{ij}(t)]^\alpha \cdot [\eta_{ij}]^\beta}{\sum_{l \in \mathcal{N}_i^k} [\tau_{il}(t)]^\alpha \cdot [\eta_{il}]^\beta} \qquad \text{if } j \in \mathcal{N}_i^k \tag{9.1}$$

where $\eta_{ij} = 1/d_{ij}$ is an a priori available heuristic value, α and β are two parameters which determine the relative influence of the pheromone trail and the heuristic information, and $\mathcal{N}_i^k$ is the feasible neighborhood of ant k, that is, the set of cities which ant k has not yet visited. The role of the parameters α and β is the following. If $\alpha = 0$, the closest cities are more likely to be selected: this corresponds to a classical stochastic greedy algorithm (with multiple starting points since ants are initially randomly distributed on the cities). If $\beta = 0$, only pheromone amplification is at work: this method will lead to the rapid emergence of a *stagnation* situation with the corresponding generation of tours which, in general, are strongly suboptimal [14]. Search stagnation is defined in [19] as the situation where all the ants follow the same path and construct the same solution. Hence, a tradeoff between the influence of the heuristic information and the pheromone trails exists.

Pheromone update After all ants have constructed their tours, the pheromone trails are updated. This is done by first lowering the pheromone strength on *all* arcs by a constant factor and then allowing each ant to add pheromone on the arcs it has visited:

$$\tau_{ij}(t+1) = (1-\rho) \cdot \tau_{ij}(t) + \sum_{k=1}^{m} \Delta\tau_{ij}^k(t) \tag{9.2}$$

where $0 < \rho \leq 1$ is the pheromone trail evaporation. The parameter ρ is used to avoid unlimited accumulation of the pheromone trails and it enables the algorithm to "forget" previously done bad decisions. If an arc is not chosen by the ants, its associated pheromone strength decreases exponentially. $\Delta\tau_{ij}^k(t)$ is the amount of pheromone ant k puts on the arcs it has visited; it is defined as follows:

$$\Delta\tau_{ij}^{k}(t) = \begin{cases} 1/L^k(t) & \text{if arc } (i,j) \text{ is used by ant } k \\ 0 & \text{otherwise} \end{cases} \tag{9.3}$$

where $L^k(t)$ is the length of the kth ant's tour. By Equation (9.3), the better the ant's tour is, the more pheromone is received by arcs belonging to the tour. In general, arcs which are used by many ants and which are contained in shorter tours will receive more pheromone and therefore are also more likely to be chosen in future iterations of the algorithm.

AS has been compared with other general purpose heuristics on some relatively small TSP instances with maximally 75 cities. Despite encouraging initial results, for larger instances AS gives a rather poor solution quality. Therefore, a substantial amount of recent research in ACO has focused on improvements over AS. A first improvement on the initial form of AS, called the *elitist strategy* for Ant System, has been introduced in [14], [19]. The idea is to give a strong additional reinforcement to the arcs belonging to the best tour found since the start of the algorithm; this tour is denoted as T^{gb} (global-best tour) in the following. This is achieved by adding to the arcs of tour T^{gb} a quantity $e \cdot 1/L^{gb}$, where e is the number of elitist ants, when the pheromone trails are updated according to Equation (9.2). Some limited results presented in [14], [19] suggest that the use of the elitist strategy with an appropriate number of elitist ants allows AS: (i) to find better tours, and (ii) to find them earlier in the run. Yet, for too many elitist ants the search concentrates early around suboptimal solutions leading to an early stagnation of the search.

9.3.3 Ant Colony System

Ant Colony System (ACS) has been introduced in [17], [22] to improve the performance of AS. ACS is based on an earlier algorithm proposed by the same authors, called Ant-Q, which exploited connections of ACO algorithms with Q-learning [55], a specific type of reinforcement learning algorithm. ACS and Ant-Q differ only in one specific aspect, which is described below. We concentrate on ACS, since it is somewhat simpler and it is preferred by its authors.

ACS differs in three main aspects from ant system. First, ACS uses a more aggressive action choice rule than AS. Second, the pheromone is added only to arcs belonging to the global-best solution. Third, each time an ant uses an arc (i,j) to move from city i to city j it removes some pheromone from the arc. In the following we present these modifications in more detail.

Tour construction In ACS ants choose the next city using the *pseudo-random-proportional action choice rule*: When located at city i, ant k moves, with probability q_0, to city l for which $\tau_{il}(t) \cdot [\eta_{il}]^{\beta}$ is maximal, that is, with probability q_0 the best possible move as indicated by the learned pheromone trails and the heuristic information is made (exploitation of learned knowledge). With probability $(1 - q_0)$ an ant performs a biased exploration of the arcs according to Equation (9.1).

Global pheromone trail update In ACS only the global best ant is allowed to add pheromone after each iteration. Thus, the update according to Equation (9.2) is modified to

$$\tau_{ij}(t+1) = (1-\rho) \cdot \tau_{ij}(t) + \rho \cdot \Delta\tau_{ij}^{gb}(t), \tag{9.4}$$

where $\Delta\tau_{ij}^{gb}(t) = 1/L^{gb}$. It is important to note that the trail update only applies to the arcs of the global-best tour, not to all the arcs like in AS. The parameter ρ again represents the pheromone evaporation. In ACS only the global best solution receives feedback. Initially, also using the iteration best solution was considered for the pheromone update. Although for smaller TSP instances the difference in solution quality between using the global-best solution or the iteration-best solution is minimal, for larger instances the use of the global-best tour gives by far better results.

Local pheromone trail update Additionally to the global updating rule, in ACS the ants use a local update rule that they apply immediately after having crossed an arc during the tour construction:

$$\tau_{ij} = (1-\xi) \cdot \tau_{ij} + \xi \cdot \tau_0 \tag{9.5}$$

where ξ, $0 < \xi < 1$, and τ_0 are two parameters. The effect of the local updating rule is to make an already chosen arc less desirable for a following ant. In this way the exploration of not yet visited arcs is increased.

The only difference between ACS and Ant-Q is the definition of the term τ_0. Ant-Q uses a formula for τ_0 which was inspired by Q-learning [55]. In Ant-Q the term τ_0 corresponds to the discounted evaluation of the next state and is set to $\gamma \cdot \max_{l \in \mathcal{N}_j^k}\{\tau_{jl}\}$ [21]. It was found that replacing the discounted evaluation of the next state by a small constant resulted in approximately the same performance and therefore, due to its simplicity, ACS is preferably used.

9.3.4 $\mathcal{MAX}$–$\mathcal{MIN}$ Ant System

$\mathcal{MAX}$–$\mathcal{MIN}$ Ant System ($\mathcal{MM}$AS) [52], [51] is a direct improvement over AS. The solutions in $\mathcal{MM}$AS are constructed in exactly the same way as in AS, that is, the selection probabilities are calculated as in Equation (9.1). Additionally, in [53] a variant of $\mathcal{MM}$AS is considered which uses the pseudo-random-proportional action choice rule of ACS. Using that action choice rule, very good solutions could be found faster, but the final solution quality achieved was worse.

The main modifications introduced by $\mathcal{MM}$AS with respect to AS are the following. First, to exploit the best solutions found, after each iteration only one ant (like in ACS) is allowed to add pheromone. Second, to avoid search stagnation, the allowed range of the pheromone trail strengths is limited to the interval $[\tau_{\min}, \tau_{\max}]$, that is, $\forall\tau_{ij}\ \tau_{\min} \leq \tau_{ij} \leq \tau_{\max}$. Last, the pheromone trails are initialized to the upper trail limit, which causes a higher exploration at the start of the algorithm [49].

Update of pheromone trails After all ants have constructed a solution, the pheromone trails are updated according to

$$\tau_{ij}(t+1) = (1-\rho) \cdot \tau_{ij}(t) + \Delta\tau_{ij}^{best} \tag{9.6}$$

where $\Delta\tau_{ij}^{best} = 1/L^{best}$. The ant which is allowed to add pheromone may be the *iteration-best* solution T^{ib}, or the *global-best* solution T^{gb}. Hence, if specific arcs are often used in the best solutions, they will receive a larger amount of pheromone. Experimental results have shown that the best performance is obtained by gradually increasing the frequency of choosing T^{gb} for the trail update [49].

Trail limits In $\mathcal{MM}$AS lower and upper limits on the possible pheromone strengths on any arc are imposed to avoid search stagnation. In particular, the imposed trail limits have the effect of indirectly limiting the probability p_{ij} of selecting a city j when an ant is in city i to an interval $[p_{\min}, p_{\max}]$, with $0 < p_{\min} \leq p_{ij} \leq p_{\max} \leq 1$. Only if an ant has one single possible choice for the next city, then $p_{\min} = p_{\max} = 1$. Experimental results [49] suggest that the lower trail limits used in $\mathcal{MM}$AS are the more important ones, since the maximum possible trail strength on arcs is limited in the long run due to pheromone evaporation.

Trail initialization The pheromone trails in $\mathcal{MM}$AS are initialized to their upper pheromone trail limits. Doing so the exploration of tours at the start of the algorithms is increased, since the relative differences between the pheromone trail strengths are less pronounced.

9.3.5 Rank-based version of Ant System

Another improvement over Ant System is the *rank-based* version of Ant System (AS$_{rank}$) [7]. In AS$_{rank}$, always the global-best tour is used to update the pheromone trails, similar to the elitist strategy of AS. Additionally, a number of the best ants of the current iteration are allowed to add pheromone. To this aim the ants are sorted by tour length ($L^1(t) \leq L^2(t) \leq \ldots \leq L^m(t)$), and the quantity of pheromone an ant may deposit is weighted according to the rank r of the ant. Only the $(w-1)$ best ants of each iteration are allowed to deposit pheromone. The global best solution, which gives the strongest feedback, is given weight w. The rth best ant of the current iteration contributes to pheromone updating with a weight given by $\max\{0, w-r\}$. Thus the modified update rule is:

$$\tau_{ij}(t+1) = (1-\rho) \cdot \tau_{ij}(t) + \sum_{r=1}^{w-1} (w-r) \cdot \Delta\tau_{ij}^{r}(t) + w \cdot \Delta\tau_{ij}^{gb}(t), \tag{9.7}$$

where $\Delta\tau_{ij}^{r}(t) = 1/L^{r}(t)$ and $\Delta\tau_{ij}^{gb}(t) = 1/L^{gb}$.

In [7] AS$_{rank}$ has been compared to AS, AS with elitist strategy, to a genetic algorithm, and to a simulated annealing algorithm. For the larger TSP instances (the largest instance considered had 132 cities) the AS-based approaches showed to be superior to the genetic algorithm and the simulated annealing procedure. Among the AS-based algorithms, both,

AS_{rank} and elitist AS performed significantly better than AS, with AS_{rank} giving slightly better results than elitist AS.

9.3.6 Synopsis

Obviously, the proposed ACO algorithms for the TSP share many common features. Ant System can mainly be seen as a first study to demonstrate the viability of ACO algorithms to attack $\mathcal{NP}$-hard combinatorial optimization problems, but its performance compared to other approaches is rather poor. Therefore, several ACO algorithms have been proposed which strongly increase the performance of Ant System. The main common feature among the proposed improvements is that they exploit more strongly the best solutions found during the ants' search. This is typically achieved by giving a higher weight to better solutions for the pheromone update and often allowing to deposit additional pheromone trail on the arcs of the global-best solution. In particular, in elitist AS and in AS_{rank} the global best solution adds a strong additional reinforcement (in AS_{rank} additionally some of the best ants of an iteration add pheromone); in ACS and $\mathcal{MM}$AS only one single ant, either the global-best or the iteration-best ant, deposit pheromone. Obviously, by exploiting the best solutions more strongly, arcs contained in the best tours receive a strong additional feedback and therefore are chosen with higher probability in subsequent algorithm iterations.

Yet, a problem associated with the stronger exploitation of search experience may be search stagnation, that is, the situation in which all ants follow the same path. Hence, some of the proposed ACO algorithms, in particular ACS and $\mathcal{MM}$AS, introduce additional features to avoid search stagnation. In ACS stagnation is avoided by the local updating rule which decreases the amount of pheromone on an arc and makes this arc less and less attractive for following ants. In this way the exploration of not yet visited arcs is favored. In $\mathcal{MM}$AS search stagnation is avoided by imposing limits on the allowed pheromone strength associated with an arc. Hence, since the pheromone trail limits influence directly the selection probabilities given by Equation (9.1), the selection probability for an arc cannot fall below some lower value. By appropriate choices for the trail limits, it is very unlikely that all ants follow the same path.

All proposed improvements over AS show significantly better performance on the TSP [19], [7], [21], [49] and they all make use of a stronger exploitation of the best solutions found by the ants. It is therefore reasonable to think that the concentration of the search around the best solutions found during the search is the key to the improved performance shown by these algorithms.

The characteristics of the TSP search space may explain this phenomenon. In particular, recent research has addressed the issue of the correlation between the solution cost and the distance from very good or from the optimal solution [4], [3] (an obvious distance measure for the TSP is given by the number of different arcs in two tours). Similarly, the Fitness-Distance Correlation (FDC) has been introduced in the context of genetic algorithm research [30]. The FDC measures the correlation between the distance of a solution from the closest optimal solution and the solution costs. Plots of the solution cost versus the distance from optimal solutions have been particularly useful to visualize the correlation. In Fig. 9.2 we present such a plot for a randomly generated Euclidean TSP instance with 500 cities. The plot is based on 5000 local optima which have been obtained by a `3-opt` local search algorithm (see next section). The x-axis gives the distance from the optimal solution, while on the y-axis the tour length is given. Obviously, a significant correlation between the solution cost and the distance from the global optimum exists (the correlation coefficient is 0.52); the closer a local optimum is to the known global optimum, the better, on average, is the solution.

The task of algorithms like ACO algorithms is to guide the search towards regions of the search space containing high quality solutions and, possibly, the global optimum. Clearly, the notion of *search space region* is tightly coupled to the notion of distance, defined by an appropriate distance measure, between solutions. The most important guiding mechanism of ACO algorithms is the objective function value of solutions; the better a solution the more feedback it is allowed to give. This guiding mechanism relies on the general intuition that the better a solution is, the more likely it is to find even better solutions close to it. The fitness-distance correlation describes this intuitive relation between the fitness (cost) of solutions and their distance from very good solutions or from the global best solution. Since for the TSP such a correlation obviously exists, it appears to be a good idea to concentrate the search around the best solutions found.

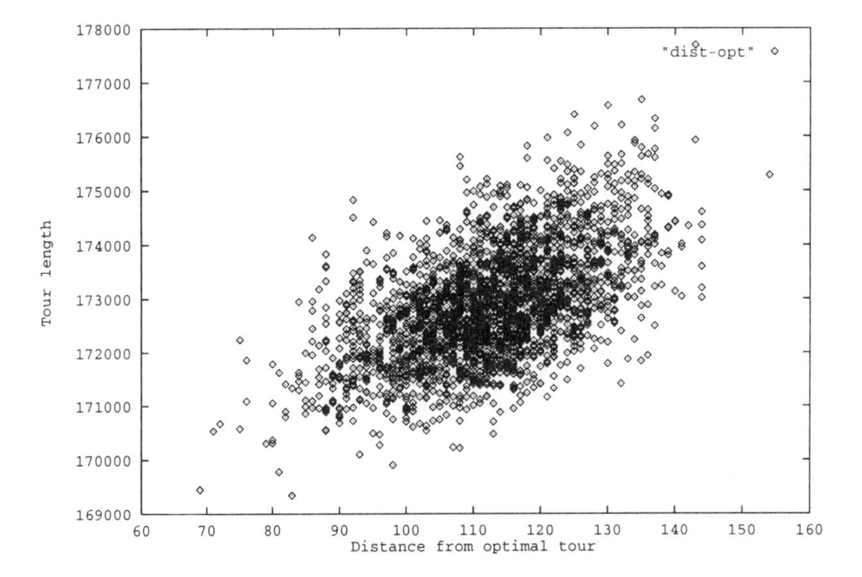

Fig. 9.2 Fitness distance plot for a Euclidean TSP instances with 500 cities distributed randomly according to an uniform distribution on a square. The plot is based on 5000 `3-opt` tours. The distance from the optimal tour is measured by the number of different arcs.

9.4 LOCAL SEARCH FOR THE TSP

Local Search starts from some initial assignment and repeatedly tries to improve the current assignment by local changes. If in the neighborhood of the current tour T a better tour T' is found, it replaces the current tour and the local search is continued from T'. The most widely known iterative improvement algorithms for the TSP are certainly `2-opt` [12] and `3-opt` [32]. They proceed by systematically testing whether the current tour can be improved by replacing 2 or at most 3 arcs, respectively. Both local search algorithms are widely studied in the literature [45], [29] and have been shown empirically to yield good solution quality. Local search algorithms using $k > 3$ arcs to be exchanged are not used commonly, due to the high computation times involved and the low additional increase in solution quality. Already for `2-opt` and `3-opt` a straightforward implementation would require $\mathcal{O}(n^2)$ or $\mathcal{O}(n^3)$ exchanges to be examined. This is clearly infeasible when trying to solve instances with several hundreds of cities in reasonable computation time.

Fortunately, there exist quite a few speed-up techniques [1], [45], [29] achieving, in practice, run-times which grow sub-quadratically. This effect is obtained by examining only a small part of the whole neighborhood. We use three techniques to reduce the run-time of `2-opt` and `3-opt` implementations. One, consists in restricting the set of moves which are examined to those contained in a candidate list of the nearest neighbors ordered according to nondecreasing distances [1], [33], [45]. Using candidate lists, for a given starting node i we only consider moves which add a new arc between i and one of the nodes in its candidate list. Hence, by using a neighborhood list of bounded length, an improving move can be found in constant time.

An additional speed-up is achieved by performing a fixed radius nearest neighbor search [1]. For `2-opt` at least one newly introduced arc has to be shorter than any of the two removed arcs (i, j) and (k, l). Due to symmetry reasons, it suffices to check whether $d_{ij} > d_{ik}$. A similar argument also holds for `3-opt` [1].

To yield a further reduction in run-time we use *don't look bits* associated with each node. Initially, all don't look bits are turned off (set to 0). If for a node no improving move can be found, the don't look bit is turned on (set to 1) and the node is not considered as a starting node for finding an improving move in the next iteration. In case an arc incident to a node is changed by a move, the node's don't look bit is turned off again.

For asymmetric TSPs, `2-opt` is not directly applicable because one part of the tour has to be traversed in the opposite direction. In case a sub-tour is reversed, the length of this sub-tour has to be computed from scratch. Yet, one of the three possible `3-opt` moves does not lead to a reversal of a sub-tour. We call `reduced 3-opt` this special form of `3-opt`. The implementation of `reduced 3-opt` uses the same speed-up techniques as described before.

The local search algorithm producing the highest quality solutions for symmetric TSPs is the Lin-Kernighan heuristic (`LK`) [33]. The Lin-Kernighan heuristic considers in each step a variable number of arcs to be exchanged. Yet, a disadvantage of the `LK` heuristic is that its implementation is more difficult than that of `2-opt` and `3-opt`, and careful fine-tuning is necessary to get it to run fast and produce high quality solutions [45], [29].

9.5 EXPERIMENTAL RESULTS

In this section we present some computational results obtained with $\mathcal{MMAS}$ on some symmetric and asymmetric TSP instances from TSPLIB. Since we apply $\mathcal{MMAS}$ to larger TSP instances with several hundreds of cities, we first present two techniques which decrease the complexity of the ACO algorithm iterations.

9.5.1 Additional techniques

Candidate lists Each tour construction has complexity $O(n^2)$, which would lead to substantial run-times for larger instances. A standard technique used when solving large TSPs is the use of candidate lists [29], [38], [45]. Typically, a candidate list comprises a fixed number of nearest neighbors for each city in order of nondecreasing distances. The use of candidate lists for the tour construction by the ants has first been proposed for the application of ACS [22] to the TSP. When constructing a tour an ant chooses the next city among those of the candidate list, if possible. Only if all the members of the candidate list of a city have already been visited, one of the remaining cities is chosen. Note that using nearest neighbor lists is a reasonable approach when trying to solve TSPs. For example, in many TSP instances the optimal tour can be found within a surprisingly low number of nearest neighbors. For example, an optimal solution is found for instance `pcb442.tsp`, which has 442 cities, within a subgraph of the 6 nearest neighbors and for instance `pr2392.tsp` (2392 cities) within a subgraph of the 8 nearest neighbors [45].

Fastening the trail update When applying ACO algorithms to the TSP, pheromone trails are stored in a matrix with $\mathcal{O}(n^2)$ entries (one for each arc). All the entries of this matrix should be updated at every iteration because of pheromone trail evaporation implemented by Formula (9.2) (the update of the whole trail matrix does not happen in ACS). Obviously, this is a very expensive operation if large TSP instances should be solved. To avoid this, in $\mathcal{MM}$AS pheromone evaporation is applied only to arcs connecting a city i to cities belonging to i's candidate list. Hence, the pheromone trails can be updated in $\mathcal{O}(n)$.

9.5.2 Parameter settings

Suitable parameter settings were determined in some preliminary experiments. We use $m = 25$ ants (all ants apply a local search to their solution), $\rho = 0.2$, $\alpha = 1$, and $\beta = 2$. During the tour construction the ants use a candidate list of size 20. The most important aspect concerning the allowed interval of the trail values is that they have to be in some reasonable interval around the expected amount of pheromone deposited by the trail update. Hence, the interval for the allowed values of the pheromone trail strength is determined to reflect this intuition. We set $\tau_{\max} = \frac{1}{\rho} \cdot \frac{1}{T^{gb}}$, where T^{gb} is the tour length of the global best tour found (this setting corresponds to the maximum possible trail level for longer runs). The lower trail limit is chosen as $\tau_{\min} = \tau_{\max}/2 \cdot n$, where n is the number of cities of an instance.

As indicated in Section 9.3.4, the best results with $\mathcal{MM}$AS are obtained when the frequency with which the global-best solution is used to update pheromone trails increases at run-time. To do so, we apply a specific schedule to alternate the pheromone trail update between T^{gb} and T^{ib}. Let u^{gb} indicate that every u^{gb} iterations T^{gb} is chosen to deposit pheromone. In the first 25 iterations only T^{ib} is used to update the trails; we set u^{gb} to 5 if $25 < t < 75$ (where t is the iteration counter), to 3 if $75 < t < 125$ to 3 (where t is the iteration counter), to 2 if $125 < t < 250$ to 2, and from then on to 1. By shifting the emphasis from the iteration-best to the global-best solution for the trail update, we in fact are shifting from a stronger exploration of the search space to an exploitation of the best solution found so far. Additionally, we reinitialize the pheromone trails to $\tau_{\max}$ if the pheromone trail strengths on almost all arcs not contained in T^{gb} is very close to $\tau_{\min}$ (as indicated by the average branching factor [21]). After a reinitialization of the pheromone trails, the schedule is applied from the start again.

For the local search we use `3-opt`, enhanced by the speed-up techniques described in Section 9.4. The size of the candidate list was set to 40 for the local search, following recommendations of [29]. Initially, all the don't look bits are turned off.

9.5.3 Experimental results for symmetric TSPs

In this section we report on experimental results obtained with $\mathcal{MM}$AS, on some symmetric TSP instances from TSPLIB. Most of the instances were proposed as benchmark instances in the First International Contest on Evolutionary Computation (1st ICEO) [2]. We compare the computational results obtained with $\mathcal{MM}$AS to a standard iterated local search algorithm [38], [37], [29] for the TSP using the same `3-opt` implementation and the same maximal computation time t_{max} for each trial as $\mathcal{MM}$AS.

Iterated local search (ILS) [38], [37], [29] is well known to be among the best algorithms for the TSP. In ILS a local search is applied iteratively to starting solutions which are obtained by mutations of a previously found local optimal solution, typically the best solution found so far. In particular, the ILS algorithm we use for comparison first locally optimizes an initial tour produced by the nearest neighbor heuristic. Then it applies iteratively `3-opt` to initial solutions which are obtained by the application of random, sequential 4-opt moves (called double-bridge moves in [38]) to the best solution found so far. The runs are performed on a Sun UltraSparc II Workstation with two UltraSparc I 167MHz processors with 0.5MB external cache. Due to the sequential implementation of the algorithm only one processor is used.

The computational results show that $\mathcal{MM}$AS is able to find very high quality solutions for all instances and on most instances $\mathcal{MM}$AS achieves a better average solution quality than ILS. This result is very encouraging, since ILS is cited in the literature as a very good algorithm for the TSP [37], [29].

9.5.4 Experimental results for asymmetric TSPs

We applied the same algorithms also to the more general asymmetric TSP; the computational results are given in Table 9.2. On the asymmetric instances the performance difference between $\mathcal{MM}$AS and ILS becomes more pronounced. While $\mathcal{MM}$AS solves all instances in all runs to optimality, ILS gets stuck at suboptimal solutions in all the four instances tested. Among the instances tested, `ry48p` and `kro124p` are easily solved by $\mathcal{MM}$AS, while on these the relatively poor performance of ILS is most striking. Only the two instances `ft70` and `ftv170` are somewhat harder. Computational results from other researchers suggest that in particular instance `ft70` is, considering its small size, relatively hard to solve [39], [54], [17].

Table 9.1 Comparison of $\mathcal{MM}$AS with iterated 3-opt (ILS) applied to some symmetric TSP instances (available in TSPLIB). Given are the instance name (the number in the name gives the problem dimension, that is, the number of cities), the algorithm used, the best solution, the average solution quality (its percentage deviation from the optimum in parentheses), the worst solution generated, the average time t_{avg} to find the best solution in a run, and the maximally allowed computation time t_{max}. Averages are taken over 25 trials for $n < 1000$, over 10 trials on the larger instances. Best average results are in boldface.

Instance	Algorithm	Best	Average	Worst	t_{avg}	t_{max}
d198	$\mathcal{MM}$AS	15780	15780.4 (0.0)	15784	61.7	170
	ILS	15780	**15780.2 (0.0)**	15784	18.6	
lin318	$\mathcal{MM}$AS	42029	**42029.0 (0.0)**	42029	94.2	450
	ILS	42029	42064.6 (0.09)	42163	55.6	
pcb442	$\mathcal{MM}$AS	50778	**50911.2 (0.26)**	51047	308.9	600
	ILS	50778	50917.7 (0.28)	51054	180.7	
att532	$\mathcal{MM}$AS	27686	**27707.9 (0.08)**	27756	387.3	1250
	ILS	27686	27709.7 (0.09)	27759	436.2	
rat783	$\mathcal{MM}$AS	8806	**8814.4 (0.10)**	8837	965.2	2100
	ILS	8806	8828.4 (0.34)	8850	1395.2	
u1060	$\mathcal{MM}$AS	224455	224853.5 (0.34)	225131	2577.6	3600
	ILS	224152	**224408.4 (0.14)**	224743	1210.4	
pcb1173	$\mathcal{MM}$AS	56896	**56956.0 (0.11)**	57120	3219.5	5400
	ILS	56897	57029.5 (0.24)	57251	1892.0	
d1291	$\mathcal{MM}$AS	50801	**50821.6 (0.04)**	50838	1894.4	5400
	ILS	50825	50875.7 (0.15)	50926	1102.9	
fl1577	$\mathcal{MM}$AS	22286	**22311.0 (0.28)**	22358	3001.8	7200
	ILS	22311.0	22394.6 (0.65)	22505	3447.6	

9.5.5 Related work on the TSP

Because the TSP is a standard benchmark problem for meta-heuristic algorithms, it has received considerable attention from the research community. Here, we only mention some of the most recent work, for a discussion of earlier work we refer to the overview article by Johnson and McGeoch [29].

Currently, the iterated LK heuristic (ILK), that is, ILS using the LK heuristic for the local search, is the most efficient approach to symmetric TSPs for short to medium run-times [29]. Recently, several new approaches and improved implementations have been presented which appear to perform as well or better than ILK for larger run-times. Among these algorithms we find the genetic local search approach of Merz and Freisleben [20], [39], an improved implementation of ASPARAGOS (one of the first genetic local search approaches to the TSP) of Gorges-Schleuter [27], a new genetic local search approach using a repair-based crossover operator and brood selection by Walters [54], a genetic algorithm using a specialized crossover operator, called edge assembly crossover, due to Nagata and Kobayashi [42], and finally a specialized local search algorithm for the TSP called Iterative Partial Transcription by Möbius et.al. [41]. Some of these algorithms [39], [41], [42], [54] achieve better computational results

Table 9.2 Comparison of $\mathcal{MM}$AS with iterated `3-opt` (ILS) applied to some asymmetric TSP instances (available in TSPLIB). Given are the instance name (the number in the name gives the problem dimension, that is, the number of cities; an exception is instance kro124p with 100 cities), the algorithm used, the best solution, the average solution quality (its percentage deviation from the optimum in parentheses), the worst solution generated, the average time t_{avg} to find the best solution in a run, and the maximally allowed computation time t_{max}. Averages are taken at least over 25 trials. Best average results are in boldface.

Instance	Algorithm	Best	Average	Worst	t_{avg}	t_{max}
ry48p	$\mathcal{MM}$AS	14422	**14422.0 (0.0)**	14422	2.3	120
	ILS	14422	14425.6 (0.03)	14507	24.6	
ft70	$\mathcal{MM}$AS	38673	**38673.0 (0.0)**	38673	37.2	300
	ILS	38673	38687.9 (0.04)	38707	3.1	
kro124p	$\mathcal{MM}$AS	36230	**36230.0 (0.0)**	36230	7.3	300
	ILS	36230	36542.5 (0.94)	37114	23.4	
ftv170	$\mathcal{MM}$AS	2755	**2755.0 (0.0)**	2755	56.2	600
	ILS	2755	2756.8 (0.07)	2764	21.7	

than the ones presented here. For example, the genetic local search approach presented in [39], which uses the `LK` heuristic for the local search, reaches on average a solution of 8806.2 on instance `rat783` in 424 seconds on a DEC Alpha station 255 MHz. Yet, the computational results with $\mathcal{MM}$AS also would benefit from the application of a more powerful local search algorithm like the `LK` heuristic. In fact, some initial experimental results presented in [49] confirm this conjecture, but still the computation times are relatively high compared to, for example, the results given in [39].

Applied to asymmetric TSP instances, our computational results with respect to solution quality compare more favorably to these approaches. For example, the solution quality we obtain with $\mathcal{MM}$AS is better than that of the genetic local search approach of [39] and the same as reported in [27], [54], but at the cost of higher run-times.

9.6 OTHER APPLICATIONS OF ACO ALGORITHMS

Research on ACO algorithms, and on ant algorithms more in general, has led to a number of other successful applications to combinatorial optimization problems. We call ant algorithm those algorithms that take inspiration by some aspects of social insects behavior. In general, ant algorithms are characterized by being multi-agent systems using only local information to make stochastic decisions and interacting with each other via a form of indirect communication called stigmergy [28].. A general introduction to ant algorithms can be found in Bonabeau, Dorigo, and Theraulaz [5]. In this paper we limit our attention to ACO algorithms, a particular class of ant algorithms which follow the meta-heuristic described in Dorigo, Di Caro, and Gambardella [16] and Dorigo and Di Caro [15].

The most widely studied problems using ACO algorithms are the traveling salesman problem and the quadratic assignment problem (QAP). Applications of ACO algorithms to the

TSP have been reviewed in this paper. As in the TSP case, the first application of an ACO algorithm to the QAP has been that of Ant System [36]. In the last two years several ant and ACO algorithms for the QAP have been presented by Maniezzo and Colorni [35], Maniezzo [34], Gambardella, Taillard, and Dorigo [25], and Stützle [48]. Currently, ant algorithms are among the best algorithms for attacking real-life QAP instances. We refer to [50] for an overview of these approaches.

The sequential ordering problem is closely related to the asymmetric TSP, but additional precedence constraints between the nodes have to be satisfied. Gambardella and Dorigo have tackled this problem by an extension of ACS enhanced by a local search algorithm [23]. They obtained excellent results and were able to improve the best known solutions for many benchmark instances.

A first application of AS to the job-shop scheduling problem has been presented in [10]. Despite showing the viability of the approach, the computational results are not competitive with state-of-the-art algorithms. More recently, $\mathcal{MM}$AS has been applied to the flow shop scheduling problem (FSP) in [47]. The computational results have shown that $\mathcal{MM}$AS outperforms earlier proposed Simulated Annealing algorithms and performs comparably to Tabu Search algorithms applied to the FSP.

Costa and Herz [11] have proposed an extension of AS to assignment type problems and present an application of their approach to the graph coloring problem obtaining results comparable to those obtained by other heuristic approaches.

Applications of AS_{rank} to vehicle routing problems are presented by Bullnheimer, Hartl, and Strauss [8], [6]. They obtained good computational results for standard benchmark instances, slightly worse than the best performing Tabu Search algorithms. A recent application by Gambardella, Taillard, and Agazzi to vehicle routing problems with time windows, improves the best known solutions for some benchmark instances [24].

A further application of AS to the shortest common supersequence problem has been proposed by Michel and Middendorf [40]. They introduce the novel aspect of using a lookahead function during the solution construction by the ants. Additionally, they present a parallel implementation of their algorithm based on an island model often also used in parallel genetic algorithms.

$\mathcal{MM}$AS has recently been applied to the generalized assignment problem by Ramalhinho Lorençou and Serra [43], obtaining very promising computational results. In particular, their algorithm is shown to find optimal and near optimal solutions faster than a GRASP algorithm which was used for comparison.

Applications of ACO algorithms to telecommunication networks, in particular to network routing problems, have recently received a strongly increased interest in the ACO community (see, for example, Schoonderwoerd et al. [46] and Di Caro and Dorigo [13]). The application of ACO algorithms to network optimization problems is appealing, since these problems have characteristics like distributed information, non-stationary stochastic dynamics, and asynchronous evolution of the network status which well match some of the properties of ACO algorithms like the use of local information to generate solutions, indirect communication via the pheromone trails and stochastic state transitions. A detailed overview of routing applications of ACO algorithms can be found in Dorigo, Di Caro and Gambardella [16].

Acknowledgments

This work was supported by a Madame Curie Fellowship awarded to Thomas Stützle (CEC-TMR Contract No. ERB4001GT973400). Marco Dorigo acknowledges support from the Belgian FNRS, of which he is a Research Associate.

REFERENCES

1. J.L. Bentley. Fast Algorithms for Geometric Traveling Salesman Problems. *ORSA Journal on Computing*, 4(4):387–411, 1992.

2. H. Bersini, M. Dorigo, S. Langerman, G. Seront, and L. Gambardella. Results of the First International Contest on Evolutionary Optimisation. In *Proceedings of the IEEE International Conference on Evolutionary Computation (ICEC'96)*, pages 611–615. IEEE Press, 1996.

3. K.D. Boese. *Models for Iterative Global Optimization*. PhD thesis, University of California, Computer Science Department, Los Angeles, 1996.

4. K.D. Boese, A.B. Kahng, and S. Muddu. A New Adaptive Multi-Start Technique for Combinatorial Global Optimization. *Operations Research Letters*, 16:101–113, 1994.

5. E. Bonabeau, M. Dorigo, and G. Theraulaz. *From Natural to Artificial Swarm Intelligence*. Oxford University Press, 1999.

6. B. Bullnheimer, R.F. Hartl, and C. Strauss. An Improved Ant System Algorithm for the Vehicle Routing Problem. *Annals of Operations Research*. To appear.

7. B. Bullnheimer, R.F. Hartl, and C. Strauss. A New Rank Based Version of the Ant System — A Computational Study. *Central European Journal for Operations Research and Economics*. To appear.

8. B. Bullnheimer, R.F. Hartl, and C. Strauss. Applying the Ant System to the Vehicle Routing Problem. In S. Voss, S. Martello, I.H. Osman, and C. Roucairol, editors, *Meta-Heuristics: Advances and Trends in Local Search Paradigms for Optimization*, pages 285–296. Kluwer, Boston, 1999.

9. A. Colorni, M. Dorigo, and V. Maniezzo. Distributed Optimization by Ant Colonies. In *Proceedings of the First European Conference on Artificial Life (ECAL 91)*, pages 134–142. Elsevier, 1991.

10. A. Colorni, M. Dorigo, V. Maniezzo, and M. Trubian. Ant System for Job-Shop Scheduling. *Belgian Journal of Operations Research, Statistics and Computer Science*, 34(1):39–53, 1994.

11. D. Costa and A. Hertz. Ants Can Colour Graphs. *Journal of the Operational Research Society*, 48:295–305, 1997.

12. G.A. Croes. A Method for Solving Traveling Salesman Problems. *Operations Research*, 6:791–812, 1958.

13. G. Di Caro and M. Dorigo. AntNet: Distributed Stigmergetic Control for Communications Networks. *Journal of Artificial Intelligence Research (JAIR)*, 9:317–365, December 1998. Available at http://www.jair.org.

14. M. Dorigo. *Optimization, Learning, and Natural Algorithms (in Italian)*. PhD thesis, Dip. Elettronica, Politecnico di Milano, 1992.

15. M. Dorigo and G. Di Caro. The Ant Colony Optimization Meta-Heuristic. In D. Corne, M. Dorigo, and F. Glover, editors, *New Ideas in Optimization*. McGraw-Hill, 1999.

16. M. Dorigo, G. Di Caro, and L.M. Gambardella. Ant Algorithms for Distributed Discrete Optimization. *Artificial Life*. To appear.

17. M. Dorigo and L.M. Gambardella. Ant Colony System: A Cooperative Learning Approach to the Traveling Salesman Problem. *IEEE Transactions on Evolutionary Computation*, 1(1):53–66, 1997.

18. M. Dorigo, V. Maniezzo, and A. Colorni. Positive Feedback as a Search Strategy. Technical Report 91-016, Dip. Elettronica, Politecnico di Milano, 1991.

19. M. Dorigo, V. Maniezzo, and A. Colorni. The Ant System: Optimization by a Colony of Cooperating Agents. *IEEE Transactions on Systems, Man, and Cybernetics – Part B*, 26(1):29–42, 1996.

20. B. Freisleben and P. Merz. A Genetic Local Search Algorithm for Solving Symmetric and Asymmetric Traveling Salesman Problems. In *Proceedings of the IEEE International Conference on Evolutionary Computation (ICEC'96)*, pages 616–621. IEEE Press, 1996.

21. L.M. Gambardella and M. Dorigo. Ant-Q: A Reinforcement Learning Approach to the Traveling Salesman Problem. In *Proceedings of the Eleventh International Conference on Machine Learning*, pages 252–260. Morgan Kaufmann, 1995.

22. L.M. Gambardella and M. Dorigo. Solving Symmetric and Asymmetric TSPs by Ant Colonies. In *Proceedings of the IEEE International Conference on Evolutionary Computation (ICEC'96)*, pages 622–627. IEEE Press, 1996.

23. L.M. Gambardella and M. Dorigo. HAS-SOP: Hybrid Ant System for the Sequential Ordering Problem. Technical Report IDSIA 11-97, IDSIA, Lugano, Switzerland, 1997.

24. L.M. Gambardella, É. Taillard, and G. Agazzi. MACS-VRPTW: A Multiple Ant Colony System for Vehicle Routing Problems with Time Windows. In D. Corne, M. Dorigo, and F. Glover, editors, *New Ideas in Optimization*. McGraw-Hill, 1999.

25. L.M. Gambardella, É.D. Taillard, and M. Dorigo. Ant Colonies for the QAP. *Journal of the Operational Research Society*. To appear.

26. M.R. Garey and D.S. Johnson. *Computers and Intractability: A Guide to the Theory of $\mathcal{NP}$-Completeness*. Freeman, San Francisco, CA, 1979.

27. M. Gorges-Schleuter. Asparagos96 and the Travelling Salesman Problem. In T. Baeck, Z. Michalewicz, and X. Yao, editors, *Proceedings of the IEEE International Conference on Evolutionary Computation (ICEC'97)*, pages 171–174, 1997.

28. P.P. Grassé. La Reconstruction du Nid et les Coordinations Interindividuelles chez *bellicositermes natalensis et cubitermes sp.* La Théorie de la Stigmergie: Essai d'interprétation du Comportement des Termites Constructeurs. *Insectes Sociaux*, 6:41–81, 1959.

29. D.S. Johnson and L.A. McGeoch. The Travelling Salesman Problem: A Case Study in Local Optimization. In E.H.L. Aarts and J.K. Lenstra, editors, *Local Search in Combinatorial Optimization*, pages 215–310. John Wiley & Sons, 1997.

30. T. Jones and S. Forrest. Fitness Distance Correlation as a Measure of Problem Difficulty for Genetic Algorithms. In L.J. Eshelman, editor, *Proceedings of the 6th International Conference on Genetic Algorithms*, pages 184–192. Morgan Kaufman, 1995.

31. E.L. Lawler, J.K. Lenstra, A.H.G. Rinnooy Kan, and D.B. Shmoys. *The Travelling Salesman Problem*. John Wiley & Sons, 1985.

32. S. Lin. Computer Solutions for the Traveling Salesman Problem. *Bell Systems Technology Journal*, 44:2245–2269, 1965.

33. S. Lin and B.W. Kernighan. An Effective Heuristic Algorithm for the Travelling Salesman Problem. *Operations Research*, 21:498–516, 1973.

34. V. Maniezzo. Exact and Approximate Nondeterministic Tree-Search Procedures for the Quadratic Assignment Problem. Technical Report CSR 98-1, Scienze dell'Informazione, Universitá di Bologna, Sede di Cesena, 1998.

35. V. Maniezzo and A. Colorni. The Ant System Applied to the Quadratic Assignment Problem. *IEEE Transactions on Knowledge and Data Engineering*. To appear.

36. V. Maniezzo, M. Dorigo, and A. Colorni. The Ant System Applied to the Quadratic Assignment Problem. Technical Report IRIDIA/94-28, Université de Bruxelles, Belgium, 1994.

37. O. Martin and S.W. Otto. Combining Simulated Annealing with Local Search Heuristics. *Annals of Operations Research*, 63:57–75, 1996.

38. O. Martin, S.W. Otto, and E.W. Felten. Large-Step Markov Chains for the Traveling Salesman Problem. *Complex Systems*, 5(3):299–326, 1991.

39. P. Merz and B. Freisleben. Genetic Local Search for the TSP: New Results. In *Proceedings of the IEEE International Conference on Evolutionary Computation (ICEC'97)*, pages 159–164. IEEE Press, 1997.

40. R. Michel and M. Middendorf. An Island Based Ant System with Lookahead for the Shortest Common Supersequence Problem. In *Proceedings of the Fifth International Conference on Parallel Problem Solving from Nature*, volume 1498, pages 692–708. Springer Verlag, 1998.

41. A. Möbius, B. Freisleben, P. Merz, and M. Schreiber. Combinatorial Optimization by Iterative Partial Transcription. Submitted to *Physical Review E*, 1998.

42. Y. Nagata and S. Kobayashi. Edge Assembly Crossover: A High-power Genetic Algorithm for the Traveling Salesman Problem. In *Proc. of ICGA'97*, pages 450–457, 1997.

43. H. Ramalhinho Lourenço and D. Serra. Adaptive Approach Heuristics for the Generalized Assignment Problem. Technical Report Economic Working Papers Series No.304, Universitat Pompeu Fabra, Dept. of Economics and Management, Barcelona, May 1998.

44. G. Reinelt. TSPLIB — A Traveling Salesman Problem Library. *ORSA Journal on Computing*, 3:376–384, 1991.

45. G. Reinelt. *The Traveling Salesman: Computational Solutions for TSP Applications*, volume 840 of *LNCS*. Springer Verlag, 1994.

46. R. Schoonderwoerd, O. Holland, J. Bruten, and L. Rothkrantz. Ant-based Load Balancing in Telecommunications Networks. *Adaptive Behavior*, 5(2):169–207, 1996.

47. T. Stützle. An Ant Approach to the Flow Shop Problem. In *Proceedings of the 6th European Congress on Intelligent Techniques & Soft Computing (EUFIT'98)*, volume 3, pages 1560–1564. Verlag Mainz, Aachen, 1997.

48. T. Stützle. $\mathcal{MAX}$–$\mathcal{MIN}$ Ant System for the Quadratic Assignment Problem. Technical Report AIDA–97–4, FG Intellektik, TU Darmstadt, July 1997.

49. T. Stützle. *Local Search Algorithms for Combinatorial Problems — Analysis, Improvements, and New Applications*. PhD thesis, Darmstadt University of Technology, Department of Computer Science, 1998.

50. T. Stützle. ACO Algorithms for the Quadratic Assignment Problem. In D. Corne, M. Dorigo, and F. Glover, editors, *New Ideas in Optimization*. McGraw-Hill, 1999.

51. T. Stützle and H.H. Hoos. The $\mathcal{MAX}$–$\mathcal{MIN}$ Ant System and Local Search for the Traveling Salesman Problem. In T. Baeck, Z. Michalewicz, and X. Yao, editors, *Proceedings of the IEEE International Conference on Evolutionary Computation (ICEC'97)*, pages 309–314, 1997.

52. T. Stützle and H.H. Hoos. Improvements on the Ant System: Introducing the $\mathcal{MAX}$–$\mathcal{MIN}$ Ant System. In R.F. Albrecht G.D. Smith, N.C. Steele, editor, *Artificial Neural Networks and Genetic Algorithms*, pages 245–249. Springer Verlag, Wien New York, 1998.

53. T. Stützle and H.H. Hoos. $\mathcal{MAX}$–$\mathcal{MIN}$ Ant System and Local Search for Combinatorial Optimization Problems. In S. Voss, S. Martello, I.H. Osman, and C. Roucairol, editors, *Meta-Heuristics: Advances and Trends in Local Search Paradigms for Optimization*, pages 313–329. Kluwer, Boston, 1999.

54. T. Walters. Repair and Brood Selection in the Traveling Salesman Problem. In A.E. Eiben, T. Bäck, M. Schoenauer, and H.-P. Schwefel, editors, *Proc. of Parallel Problem*

Solving from Nature – PPSN V, volume 1498 of *LNCS*, pages 813–822. Springer Verlag, 1998.

55. C.J.C.H. Watkins and P. Dayan. Q-Learning. *Machine Learning*, 8:279–292, 1992.

10 Genetic Programming: Turing's Third Way to Achieve Machine Intelligence

J. R. KOZA, F. H BENNETT III, D. ANDRE and M. A. KEANE

Stanford University, Stanford, California, USA

Genetic Programming Inc., Los Altos, California, USA

University of California, Berkeley, California, USA

Econometrics Inc., Chicago, Illinois, USA

One of the central challenges of computer science is to get a computer to solve a problem without explicitly programming it. In particular, it would be desirable to have a problem-independent system whose input is a high-level statement of a problem's requirements and whose output is a working computer program that solves the given problem. The challenge is to make computers do what needs to be done, without having to tell the computer exactly how to do it.

Alan Turing recognized that machine intelligence may be realized using a biological approach. In his 1948 essay "Intelligent Machines" [9], Turing made the connection between searches and the challenge of getting a computer to solve a problem without explicitly programming it.

> Further research into intelligence of machinery will probably be very greatly concerned with "searches" ...

Turing then identified three broad approaches by which search techniques might be used to automatically create an intelligent computer program.

One approach that Turing identified is a search through the space of integers representing candidate computer programs. This approach reflects the orientation of much of Turing's own work on the logical basis for computer algorithms.

A second approach that Turing identified is the "cultural search" which relies on knowledge and expertise acquired over a period of years from others. This approach is akin to present-day knowledge-based systems and expert systems.

The third approach that Turing identified is "genetical or evolutionary search." Turing said,

> There is the genetical or evolutionary search by which a combination of genes is looked for, the criterion being the survival value. The remarkable success of this search confirms to some extent the idea that intellectual activity consists mainly of various kinds of search.

Turing did not specify in this essay how to conduct the "genetical or evolutionary search" for a computer program. However, his 1950 paper "Computing Machinery and Intelligence" [24] suggested how natural selection and evolution might be incorporated into the search for intelligent machines.

> We cannot expect to find a good child-machine at the first attempt. One must experiment with teaching one such machine and see how well it learns. One can then try another and see if it is better or worse. There is an obvious connection between this process and evolution, by the identifications
>
> Structure of the child machine = Hereditary material
>
> Changes of the child machine = Mutations
>
> Natural selection = Judgment of the experimenter

Since the 1940s, an enormous amount of research effort has been expended on trying to achieve machine intelligence using what Turing called the "logical" approach and the "cultural" approach. In contrast, Turing's third way to achieve machine intelligence – namely, "genetical or evolutionary" approach – has received comparatively less effort.

This paper is about genetic programming – a way to implement Turing's third way to achieve machine intelligence. Genetic programming is a "genetical or evolutionary" technique that automatically creates a computer program from a high-level statement of a problem's requirements. In particular, genetic programming is an extension of the genetic algorithm described in John Holland's pioneering 1975 book *Adaptation in Natural and Artificial Systems* [7]. Starting with a primordial ooze of thousands of randomly created computer programs, genetic programming progressively breeds a population of computer programs over a series of generations. Genetic programming employs the Darwinian principle of survival of the fittest, analogs of naturally occurring operations such as sexual recombination (crossover), mutation, gene duplication, and gene deletion, and certain mechanisms of developmental biology [2, 3, 4, 10, 12–19, 23].

When we talk about a computer program (Fig. 10.1), we mean an entity that receives inputs, performs computations, and produces outputs. Computer programs perform basic arithmetic and conditional computations on variables of various types (including integer, floating-point, and Boolean variables), perform iterations and recursions, store intermediate results in memory, contain reusable groups of operations that are organized into subroutines, pass information to subroutines in the form of dummy variables (formal parameters), receive information from subroutines in the form of return values (or through side effects). The subroutines and main program are typically organized into a hierarchy.

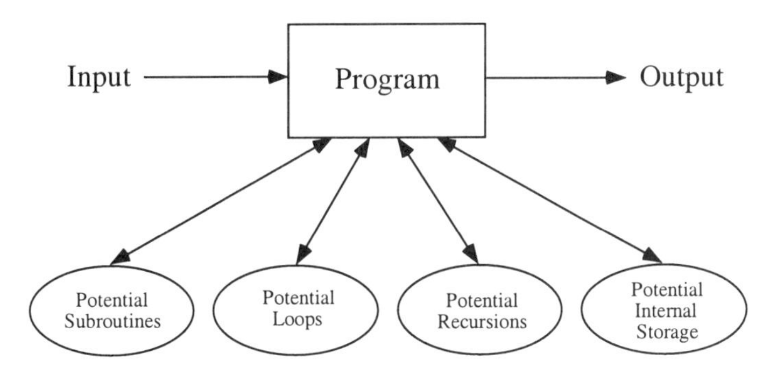

Fig. 10.1 A computer program.

We think that it is reasonable to expect that a system for automatically creating computer programs should be able to create entities that possess most or all of the above capabilities (or reasonable equivalents of them). A list of attributes for a system for automatically creating computer programs might include the following 16 items:

- **Attribute No. 1 (Starts with "What needs to be done"):** It starts from a high-level statement specifying the requirements of the problem.

- **Attribute No. 2 (Tells us "How to do it"):** It produces a result in the form of a sequence of steps that satisfactorily solves the problem.

- **Attribute No. 3 (Produces a computer program):** It produces an entity that can run on a computer.

- **Attribute No. 4 (Automatic determination of program size):** It has the ability to automatically determine the number of steps that must be performed and thus does not require the user to prespecify the exact size of the solution.

- **Attribute No. 5 (Code reuse):** It has the ability to automatically organize useful groups of steps so that they can be reused.

- **Attribute No. 6 (Parameterized reuse):** It has the ability to reuse groups of steps with different instantiations of values (formal parameters or dummy variables).

- **Attribute No. 7 (Internal storage):** It has the ability to use internal storage in the form of single variables, vectors, matrices, arrays, stacks, queues, lists, relational memory, and other data structures.

- **Attribute No. 8 (Iterations, loops, and recursions):** It has the ability to implement iterations, loops, and recursions.

- **Attribute No. 9 (Self-organization of hierarchies):** It has the ability to automatically organize groups of steps into a hierarchy.

- **Attribute No. 10 (Automatic determination of program architecture):** It has the ability to automatically determine whether to employ subroutines, iterations, loops, recursions, and internal storage, and the number of arguments possessed by each subroutine, iteration, loop, and recursion.

- **Attribute No. 11 (Wide range of programming constructs):** It has the ability to implement analogs of the programming constructs that human computer programmers find useful, including macros, libraries, typing, pointers, conditional operations, logical functions, integer functions, floating-point functions, complex-valued functions, multiple inputs, multiple outputs, and machine code instructions.

- **Attribute No. 12 (Well-defined):** It operates in a well-defined way. It unmistakably distinguishes between what the user must provide and what the system delivers.

- **Attribute No. 13 (Problem-independent):** It is problem-independent in the sense that the user does not have to modify the system's executable steps for each new problem.

- **Attribute No. 14 (Wide applicability):** It produces a satisfactory solution to a wide variety of problems from many different fields.

- **Attribute No. 15 (Scalability):** It scales well to larger versions of the same problem.

- **Attribute No. 16 (Competitive with human-produced results):** It produces results that are competitive with those produced by human programmers, engineers, mathematicians, and designers.

As shown in detail in *Genetic Programming III: Darwinian Invention and Problem Solving* [16], genetic programming currently unconditionally possesses 13 of the 16 attributes that can reasonably be expected of a system for automatically creating computer programs and that genetic programming at least partially possesses the remaining three attributes.

Attribute No. 16 is especially important because it reminds us that the ultimate goal of a system for automatically creating computer programs is to produce useful programs – not merely programs that solve "toy" or "proof of principle" problems. As Samuel [22] said,

> The aim [is] ... to get machines to exhibit behavior, which if done by humans, would be assumed to involve the use of intelligence.

There are 14 specific instances reported in *Genetic Programming III* where genetic programming automatically created a computer program that is competitive with a human-produced result.

What do we mean when we say that an automatically created solution to a problem is competitive with a result produced by humans? We are not referring to the fact that a computer can rapidly print ten thousand payroll checks or that a computer can compute π to a million decimal places. As Fogel, Owens, and Walsh [6] said,

> Artificial intelligence is realized only if an inanimate machine can solve problems ... not because of the machine's sheer speed and accuracy, but because it can discover for itself new techniques for solving the problem at hand.

We think it is fair to say that an automatically created result is competitive with one produced by human engineers, designers, mathematicians, or programmers if it satisfies any of the following eight criteria (or any other similarly stringent criterion):

A. The result was patented as an invention in the past, is an improvement over a patented invention, or would qualify today as a patentable new invention.

B. The result is equal to or better than a result that was accepted as a new scientific result at the time when it was published in a peer-reviewed journal.

C. The result is equal to or better than a result that was placed into a database or archive of results maintained by an internationally recognized panel of scientific experts.

D. The result is publishable in its own right as a new scientific result (independent of the fact that the result was mechanically created).

E. The result is equal to or better than the most recent human-created solution to a long-standing problem for which there has been a succession of increasingly better human-created solutions.

F. The result is equal to or better than a result that was considered an achievement in its field at the time it was first discovered.

G. The result solves a problem of indisputable difficulty in its field.

H. The result holds its own or wins a regulated competition involving human contestants (in the form of either live human players or human-written computer programs).

Table 10.1 shows 14 instances reported in *Genetic Programming III* [16], where we claim that genetic programming has produced results that are competitive with those produced by human engineers, designers, mathematicians, or programmers. Each claim is accompanied by the particular criterion that establishes the basis for the claim. The instances in Table 10.1 include classification problems from the field of computational molecular biology, a long-standing problem involving cellular automata, a problem of synthesizing the design of a minimal sorting network, and several problems of synthesizing the design of analog electrical circuits. As can be seen, 10 of the 14 instances in the table involve previously patented inventions. Several of these items are discussed in a companion paper in this volume.

Engineering design offers a practical yardstick for evaluating a method for automatically creating computer programs because the design process is usually viewed as requiring human intelligence. Design is a major activity of practicing engineers. The process of design involves the creation of a complex structure to satisfy user-defined requirements. For example, the design process for analog electrical circuits begins with a high-level description of the circuit's desired behavior and characteristics and entails the creation of both the circuit's topology and the values of each of the circuit's components. The design process typically entails tradeoffs between competing considerations. The design (synthesis) of analog electrical circuits is especially challenging because there is no previously known general automated technique for creating both the topology and sizing of an analog circuit from a high-level statement of the circuit's desired behavior and characteristics.

Table 10.1 Fourteen instances where genetic programming has produced results that are competitive with human-produced results.

	Claimed instance	Basis for the claim	Section in *Genetic Programming III* [16]
1	Creation of four different algorithms for the transmembrane segment identification problem for proteins	B, E	16.6
2	Creation of a sorting network for seven items using only 16 steps	A, D	21.4.4
3	Rediscovery of recognizable ladder topology for filters	A, F	25.15.1
4	Rediscovery of "*M*-derived half section" and "constant K" filter sections	A, F	26.1.3
5	Rediscovery of the Cauer (elliptic) topology for filters	A, F	27.3.7
6	Automatic decomposition of the problem of synthesizing a crossover filter	A, F	32.3
7	Rediscovery of a recognizable voltage gain stage and a Darlington emitter-follower section of an amplifier and other circuits	A, F	42.3
8	Synthesis of 60 and 96 decibel amplifiers	A, F	45.3
9	Synthesis of analog computational circuits for squaring, cubing, square root, cube root, logarithm, and Gaussian functions	A, D, G	47.5.3
10	Synthesis of a real-time analog circuit for time-optimal control of a robot	G	48.3
11	Synthesis of an electronic thermometer	A, G	49.3
12	Synthesis of a voltage reference circuit	A, G	50.3
13	Creation of a cellular automata rule for the majority classification problem that is better than the Gacs-Kurdyumov-Levin (GKL) rule and better than all other known rules written by humans over the past 20 years	D, E	58.4
14	Creation of motifs that detect the D–E–A-D box family of proteins and the manganese superoxide dismutase family as well as, or slightly better than, the human-written motifs archived in the PROSITE database of protein motifs	C	59.8

There are, of course, additional instances where others have used genetic programming to evolve programs that are competitive with human-produced results. Other examples include Howley's use of genetic programming to control a spacecraft's attitude maneuvers [8]. Two additional noteworthy recent examples are Sean Luke and Lee Spectors' genetically evolved entry at the 1997 Robo Cup competition at the International Joint Conference on Artificial Intelligence in Nagoya, Japan [21] and the genetically evolved entry at the 1998 Robo Cup competition in Paris [1]. Both of these entries consisted of soccer-playing programs that were evolved entirely by genetic programming. The other entries in this annual competition consisted of programs written by teams of human programmers. Most were entirely human-written; however, minor elements of a few of the human-written programs were produced by machine learning. Both of the genetically evolved entries held their own against the human-written programs and placed in the middle of the field (consisting of 16 and 34 total entries, respectively).

Of course, we do not claim that genetic programming is the only possible approach to the challenge of getting computers to solve problems without explicitly programming them. However, we are not aware at this time of any other method of artificial intelligence, machine learning, neural networks, adaptive systems, reinforcement learning, or automated logic that can be said to possess more than a few of the above 16 attributes.

Conspicuously, the above list of 16 attributes does not preordain that formal logic or an explicit knowledge base be the method used to achieve the goal of automatically creating computer programs. Many artificial intelligence researchers and computer scientists unquestioningly assume that formal logic must play a preeminent role in any system for automatically creating computer programs. Similarly, the vast majority of contemporary researchers in artificial intelligence believe that a system for automatically creating computer programs must employ an explicit knowledge base. Indeed, over the past four decades, the field of artificial intelligence has been dominated by the strongly asserted belief that the goal of getting a computer to solve problems automatically can be achieved *only* by means of formal logic inference methods and knowledge. This approach typically entails the selection of a knowledge representation, the acquisition of the knowledge, the codification of the knowledge into a knowledge base, the depositing of the knowledge base into a computer, and the manipulation of the knowledge in the computer using the inference methods of formal logic. As Lenat [20] stated,

> All our experiences in AI research have led us to believe that for automatic programming, the answer lies in *knowledge*, in adding a collection of expert rules which guide code synthesis and transformation. (Emphasis in original).

However, the existence of a strenuously asserted belief for four decades does not, in itself, validate the belief. Moreover, the popularity of a belief does not preclude the possibility that there might be an alternative way of achieving a particular goal. In particular, the popularity over the past four decades of the "logical" and the "cultural" (knowledge-based) approach does not preclude the possibility that Turing's third way to achieve machine intelligence may prove to be more fruitful.

Genetic programming is different from all other approaches to artificial intelligence, machine learning, neural networks, adaptive systems, reinforcement learning, or automated logic in all (or most) of the following seven ways:

1. **Representation:** Genetic programming overtly conducts its search for a solution to the given problem in program space.

2. **Role of point-to-point transformations in the search:** Genetic programming does not conduct its search by transforming a single point in the search space into another single point, but instead transforms a set of points into another set (population) of points.

3. **Role of hill climbing in the search:** Genetic programming does not rely exclusively on greedy hill climbing to conduct its search, but instead allocates a certain number of trials, in a principled way, to choices that are known to be inferior.

4. **Role of determinism in the search:** Genetic programming conducts its search probabilistically.

5. **Role of an explicit knowledge base:** None.

6. **Role of in the inference methods of formal logic in the search:** None.

7. **Underpinnings of the technique:** Biologically inspired.

First, consider the issue of representation. Most techniques of artificial intelligence, machine learning, neural networks, adaptive systems, reinforcement learning, or automated logic employ specialized structures in lieu of ordinary computer programs. These surrogate structures include if-then production rules, Horn clauses, decision trees, Bayesian networks, propositional logic, formal grammars, binary decision diagrams, frames, conceptual clusters, concept sets, numerical weight vectors (for neural nets), vectors of numerical coefficients for polynomials or other fixed expressions (for adaptive systems), genetic classifier system rules, fixed tables of values (as in reinforcement learning), or linear chromosome strings (as in the conventional genetic algorithm).

Tellingly, except in unusual situations, the world's several million computer programmers do not use any of these surrogate structures for writing computer programs. Instead, for five decades, human programmers have persisted in writing computer programs that intermix a multiplicity of types of computations (e.g., arithmetic and logical) operating on a multiplicity of types of variables (e.g., integer, floating-point, and Boolean). Programmers have persisted in using internal memory to store the results of intermediate calculations in order to avoid repeating the calculation on each occasion when the result is needed. They have persisted in using iterations and recursions. They have similarly persisted for five decades in organizing useful sequences of operations into reusable groups (subroutines) so that they avoid reinventing the wheel on each occasion when they need a particular sequence of operations. Moreover, they have persisted in passing parameters to subroutines so that they can reuse those subroutines with different instantiations of values. And, they have persisted in organizing their subroutines and main program into hierarchies.

All of the above tools of ordinary computer programming have been in use since the beginning of the era of electronic computers in the 1940s. Significantly, none has fallen into disuse by human programmers. Yet, in spite of the manifest utility of these everyday tools of computer programming, these tools are largely absent from existing techniques of automated machine learning, neural networks, artificial intelligence, adaptive systems, reinforcement learning, and automated logic. On one of the relatively rare occasions when one or two of these everyday tools of computer programming is available within the context of one of these automated techniques, it is usually available only in a hobbled and barely recognizable form. In contrast, genetic programming draws on the full arsenal of tools that human programmers have found useful for five decades. It conducts its search for a solution to a problem overtly in the space of computer programs. Our view is that computer programs are the best representation of computer programs. We believe that the search for a solution to the challenge of getting computers to solve problems without explicitly programming them should be conducted in the space of computer programs.

Of course, once one realizes that the search should be conducted in program space, one is immediately faced with the task of finding the desired program in the enormous space of possible programs. As will be seen, genetic programming performs this task of program discovery. It provides a problem-independent way to productively search the space of possible computer programs to find a program that satisfactorily solves the given problem.

Second, another difference between genetic programming and almost every other automated technique concerns the nature of the search conducted in the technique's chosen search space. Almost all of these non-genetic methods employ a point-to-point strategy that transforms a single point in the search space into another single point. Genetic programming is different in that it operates by explicitly cultivating a diverse population of often-inconsistent and often-contradictory approaches to solving the problem. Genetic programming performs a beam search in program space by iteratively transforming one population of candidate computer programs into a new population of programs.

Third, consider the role of hill climbing. When the trajectory through the search space is from one single point to another single point, there is a nearly irresistible temptation to extend the search only by moving to a point that is known to be superior to the current point. Consequently, almost all automated techniques rely exclusively on greedy hill climbing to make the transformation from the current point in the search space to the next point. The temptation to rely on hill climbing is reinforced because many of the toy problems in the literature of the fields of machine learning and artificial intelligence are so simple that they can, in fact, be solved by hill climbing. However, popularity cannot cure the innate tendency of hill climbing to become trapped on a local optimum that is not a global optimum. Interesting and non-trivial problems generally have high-payoff points that are inaccessible to greedy hill climbing. In fact, the existence of points in the search space that are not accessible to hill climbing is a good working definition of non-triviality. The fact that genetic programming does not rely on a point-to-point search strategy helps to liberate it from the myopia of hill climbing. Genetic programming is free to allocate a certain measured number of trials to points that are known to be inferior. This allocation of trials to known-inferior individuals is not motivated by charity, but in the expectation that it will often unearth an unobvious trajectory through the search space leading to points with an ultimately higher payoff. The fact that genetic programming operates from a population enables it to make a small number of adventurous moves while simultaneously pursuing the more immediately gratifying avenues of advance through the search space. Of course, genetic programming is not the only search technique that avoids mere hill climbing. For example, both simulated annealing [11] and genetic algorithms [7] allocate a certain number of trials to inferior points in a similar principled way. However, most of the techniques currently used in the fields of artificial intelligence, machine learning, neural networks, adaptive systems, reinforcement learning, or automated logic are trapped on the local optimum of hill climbing.

Fourth, another difference between genetic programming and almost every other technique of artificial intelligence and machine learning is that genetic programming conducts a probabilistic search. Again, genetic programming is not unique in this respect. For example, simulated annealing and genetic algorithms are also probabilistic. However, most existing techniques in the fields of artificial intelligence and machine learning are deterministic.

Fifth, consider the role of a knowledge base in the pursuit of the goal of automatically creating computer programs. In genetic programming, there is no explicit knowledge base. While there are numerous optional ways to incorporate domain knowledge into a run of genetic programming, genetic programming does not require (or usually use) an explicit knowledge base to guide its search.

Sixth, consider the role of the inference methods of formal logic. Many computer scientists unquestioningly assume that every problem-solving technique must be logically sound, deterministic, logically consistent, and parsimonious. Accordingly, most conventional methods of artificial intelligence and machine learning possess these characteristics. However, logic does not govern two of the most important types of complex problem solving processes, namely the invention process performed by creative humans and the evolutionary process occurring in nature.

A new idea that can be logically deduced from facts that are known in a field, using transformations that are known in a field, is not considered to be an invention. There must be what the patent law refers to as an "illogical step" (i.e., an unjustified step) to distinguish a putative invention from that which is readily deducible from that which is already known. Humans supply the critical ingredient of “illogic” to the invention process. Interestingly, everyday usage parallels the patent law concerning inventiveness: People who mechanically apply existing facts in well-known ways are summarily dismissed as being uncreative. Logical thinking is unquestionably useful for many purposes. It usually plays an important role in setting the stage for an invention. But, at the end of the day, logical thinking is the antithesis of invention and creativity.

Recalling his invention in 1927 of the negative feedback amplifier, Harold S. Black of Bell Laboratories [5] said,

> Then came the morning of Tuesday, August 2, 1927, when the concept of the negative feedback amplifier came to me in a flash while I was crossing the Hudson River on the Lackawanna Ferry, on my way to work. For more than 50 years, I have pondered how and why the idea came, and I can't say any more today than I could that morning. All I know is that after several years of hard work on the problem, I suddenly realized that if I fed the amplifier output back to the input, in reverse phase, and kept the device from oscillating (singing, as we called it then), I would have exactly what I wanted: a means of canceling out the distortion of the output. I opened my morning newspaper and on a page of *The New York Times* I sketched a simple canonical diagram of a negative feedback amplifier plus the equations for the amplification with feedback.

Of course, inventors are not oblivious to logic and knowledge. They do not thrash around using blind random search. Black did not try to construct the negative feedback amplifier from neon bulbs or doorbells. Instead, "several years of hard work on the problem" set the stage and brought his thinking into the proximity of a solution. Then, at the critical moment, Black made his “illogical” leap. This unjustified leap constituted the invention.

The design of complex entities by the evolutionary process in nature is another important type of problem-solving that is not governed by logic. In nature, solutions to design problems are discovered by the probabilistic process of evolution and natural selection. There is nothing logical about this process. Indeed, inconsistent and contradictory alternatives abound. In fact, such genetic diversity is necessary for the evolutionary process to succeed. Significantly, the solutions evolved by evolution and natural selection almost always differ from those created by conventional methods of artificial intelligence and machine learning in one very important respect. Evolved solutions are not brittle; they are usually able to easily grapple with the perpetual novelty of real environments.

Similarly, genetic programming is not guided by the inference methods of formal logic in its search for a computer program to solve a given problem. When the goal is the automatic creation of computer programs, all of our experience has led us to conclude that the non-logical approach used in the invention process and in natural evolution are far more fruitful than the logic-driven and knowledge-based principles of conventional artificial intelligence. In short, "logic considered harmful."

Seventh, the biological metaphor underlying genetic programming is very different from the underpinnings of all other techniques that have previously been tried in pursuit of the goal of automatically creating computer programs. Many computer scientists and mathematicians are baffled by the suggestion biology might be relevant to their fields. In contrast, we do not view biology as an unlikely well from which to draw a solution to the challenge of getting a computer to solve a problem without explicitly programming it. Quite the contrary – we view biology as a most likely source. Indeed, genetic programming is based on the only method that has ever produced intelligence – the time-tested method of evolution and natural selection. As Stanislaw Ulam said in his 1976 autobiography [25],

> [Ask] not what mathematics can do for biology, but what biology can do for mathematics.

REFERENCES

1. Andre, David and Teller, Astro. 1998. Evolving team Darwin United. In Asada, Minoru (editor). *RoboCup-98: Robot Soccer World Cup II*. Lecture Notes in Computer Science. Berlin: Springer Verlag. In press.

2. Angeline, Peter J. and Kinnear, Kenneth E. Jr. (editors). 1996. *Advances in Genetic Programming 2*. Cambridge, MA: The MIT Press.

3. Banzhaf, Wolfgang, Nordin, Peter, Keller, Robert E., and Francone, Frank D. 1998. *Genetic Programming – An Introduction*. San Francisco, CA: Morgan Kaufmann and Heidelberg: dpunkt.

4. Banzhaf, Wolfgang, Poli, Riccardo, Schoenauer, Marc, and Fogarty, Terence C. 1998. *Genetic Programming: First European Workshop. EuroGP'98. Paris, France, April 1998 Proceedings. Paris, France. April 1998*. Lecture Notes in Computer Science. Volume 1391. Berlin, Germany: Springer-Verlag.

5. Black, Harold S. 1977. Inventing the negative feedback amplifier. *IEEE Spectrum*. December 1977. Pages 55 – 60.

6. Fogel, Lawrence J., Owens, Alvin J., and Walsh, Michael. J. 1966. *Artificial Intelligence through Simulated Evolution*. New York: John Wiley.

7. Holland, John H. 1975. *Adaptation in Natural and Artificial Systems*. Ann Arbor, MI: University of Michigan Press.

8. Howley, Brian. 1996. Genetic programming of near-minimum-time spacecraft attitude maneuvers. In Koza, John R., Goldberg, David E., Fogel, David B., and Riolo, Rick L. (editors). 1996. *Genetic Programming 1996: Proceedings of the First Annual Conference, July 28-31, 1996, Stanford University*. Cambridge, MA: MIT Press. Pages 98–106.

9. Ince, D. C. (editor). 1992. *Mechanical Intelligence: Collected Works of A. M. Turing*. Amsterdam: North Holland.

10. Kinnear, Kenneth E. Jr. (editor). 1994. *Advances in Genetic Programming*. Cambridge, MA: The MIT Press.

11. Kirkpatrick, S., Gelatt, C. D., and Vecchi, M. P. 1983. Optimization by simulated annealing. *Science* 220, pages 671-680.

12. Koza, John R. 1992. *Genetic Programming: On the Programming of Computers by Means of Natural Selection.* Cambridge, MA: MIT Press.

13. Koza, John R. 1994a. *Genetic Programming II: Automatic Discovery of Reusable Programs.* Cambridge, MA: MIT Press.

14. Koza, John R. 1994b. *Genetic Programming II Videotape: The Next Generation.* Cambridge, MA: MIT Press.

15. Koza, John R., Banzhaf, Wolfgang, Chellapilla, Kumar, Deb, Kalyanmoy, Dorigo, Marco, Fogel, David B., Garzon, Max H., Goldberg, David E., Iba, Hitoshi, and Riolo, Rick L. (editors). *Genetic Programming 1998: Proceedings of the Third Annual Conference, July 22-25, 1998, University of Wisconsin, Madison, Wisconsin.* San Francisco, CA: Morgan Kaufmann.

16. Koza, John R., Bennett III, Forrest H, Andre, David, and Keane, Martin A. 1999. *Genetic Programming III: Darwinian Invention and Problem Solving.* San Francisco, CA: Morgan Kaufmann.

17. Koza, John R., Deb, Kalyanmoy, Dorigo, Marco, Fogel, David B., Garzon, Max, Iba, Hitoshi, and Riolo, Rick L. (editors). 1997. *Genetic Programming 1997: Proceedings of the Second Annual Conference* San Francisco, CA: Morgan Kaufmann.

18. Koza, John R., Goldberg, David E., Fogel, David B., and Riolo, Rick L. (editors). 1996. *Genetic Programming 1996: Proceedings of the First Annual Conference.* Cambridge, MA: The MIT Press.

19. Koza, John R., and Rice, James P. 1992. *Genetic Programming: The Movie.* Cambridge, MA: MIT Press.

20. Lenat, Douglas B. 1983. The role of heuristics in learning by discovery: Three case studies. In Michalski, Ryszard S., Carbonell, Jaime G., and Mitchell, Tom M. (editors) *Machine Learning: An Artificial Intelligence Approach, Volume I.* Los Altos, CA: Morgan Kaufmann. Pages 243-306.

21. Luke, Sean and Spector, Lee. 1998. Genetic programming produced competitive soccer softbot teams for RoboCup97. In Koza, John R., Banzhaf, Wolfgang, Chellapilla, Kumar, Deb, Kalyanmoy, Dorigo, Marco, Fogel, David B., Garzon, Max H., Goldberg, David E., Iba, Hitoshi, and Riolo, Rick. (editors). *Genetic Programming 1998:*

Proceedings of the Third Annual Conference, July 22-25, 1998, University of Wisconsin, Madison, Wisconsin. San Francisco, CA: Morgan Kaufmann. Pages 214 – 222.

22. Samuel, Arthur L. 1983. AI: Where it has been and where it is going. *Proceedings of the Eighth International Joint Conferènce on Artificial Intelligence*. Los Altos, CA: Morgan Kaufmann. Pages 1152 – 1157.

23. Spector, Lee, Langdon, William B., O'Reilly, Una-May, and Angeline, Peter (editors). 1999. *Advances in Genetic Programming 3*. Cambridge, MA: The MIT Press.

24. Turing, Alan M. 1950. Computing machinery and intelligence. *Mind*. 59(236) 433 – 460. Reprinted in Ince, D. C. (editor). 1992. *Mechanical Intelligence: Collected Works of A. M. Turing*. Amsterdam: North Holland. Pages 133 – 160.

25. Ulam, Stanislaw M. 1991. *Adventures of a Mathematician*. Berkeley, CA: University of California Press.

11 Automatic Synthesis of the Topology and Sizing for Analog Electrical Circuits Using Genetic Programming

F. H BENNETT III, M. A. KEANE, D. ANDRE and J. R. KOZA

Genetic Programming Inc.
Los Altos, California, USA

Econometrics Inc.
Chicago, Illinois, USA

University of California
Berkeley, California, USA

Stanford University
Stanford, California, USA

11.1 INTRODUCTION

Design is a major activity of practicing engineers. The design process entails creation of a complex structure to satisfy user-defined requirements. Since the design process typically entails tradeoffs between competing considerations, the end product of the process is usually a satisfactory and compliant design as opposed to a perfect design. Design is usually viewed as requiring creativity and human intelligence. Consequently, the field of design is a source of challenging problems for automated techniques of machine intelligence. In particular, design problems are useful for determining whether an automated technique can produce results that are competitive with human-produced results.

The design (synthesis) of analog electrical circuits is especially challenging. The design process for analog circuits begins with a high-level description of the circuit's desired behavior and characteristics and includes creation of both the topology and the sizing of a satisfactory circuit. The *topology* of a circuit involves specification of the gross number of components in the circuit, the type of each component (e.g., a capacitor), and a *netlist* specifying where each of a component's leads are to be connected. *Sizing* involves specification of the values (typically numerical) of each component. Since the design process typically involves tradeoffs between competing considerations, the end product of the process is usually a satisfactory and compliant design as opposed to a perfect design.

Although considerable progress has been made in automating the synthesis of certain categories of purely digital circuits, the synthesis of analog circuits and mixed analog-digital circuitshas not proved to be as amenable to automation [30]. Until recently, there has been no general technique for automatically creating the topology and sizing for an entire analog electrical circuit from a high-level statement of the design goals of the circuit. Describing "the analog dilemma," O. Aaserud and I. Ring Nielsen [1] noted

> "Analog designers are few and far between. In contrast to digital design, most of the analog circuits are still handcrafted by the experts or so-called 'zahs' of analog design. The design process is characterized by a combination of experience and intuition and requires a thorough knowledge of the process characteristics and the detailed specifications of the actual product. "

> "Analog circuit design is known to be a knowledge-intensive, multiphase, iterative task, which usually stretches over a significant period of time and is performed by designers with a large portfolio of skills. It is therefore considered by many to be a form of art rather than a science."

There has been extensive previous work (surveyed in [23]) on the problem of automated circuit design (synthesis) using simulated annealing, artificial intelligence, and other techniques, including work employing genetic algorithms [11, 27, 34].

This paper presents a uniform approach to the automatic synthesis of both the topology and sizing of analog electrical circuits. Section 11.2 presents seven design problems involving prototypical analog circuits. Section 11.3 describes genetic programming. Section 11.4 details the circuit-constructing functions used in applying genetic programming to the problem of analog circuit synthesis. Section 11.5 presents the preparatory steps required for applying genetic programming to a particular design problem. Section 11.6 shows the results for the seven problems. Section 11.7 cites other circuits that have been designed by genetic programming.

11.2 SEVEN PROBLEMS OF ANALOG DESIGN

This paper applies genetic programming to an illustrative suite of seven problems of analog circuit design. The circuits comprise a variety of types of components, including transistors, diodes, resistors, inductors, and capacitors. The circuits have varying numbers of inputs and outputs. They circuits encompass both passive and active circuits.

1. Design a one-input, one-output lowpass filter composed of capacitors and inductors that passes all frequencies below 1,000 Hz and suppresses all frequencies above 2,000 Hz.

2. Design a one-input, one-output highpass filter composed of capacitors and inductors that suppresses all frequencies below 1,000 Hz and passes all frequencies above 2,000 Hz.

3. Design a one-input, one-output bandpass filter composed of capacitors and inductors that passes all frequencies between 500 Hz and 1,000 Hz and suppresses frequencies that are less than 250 Hz and greater than 2,000 Hz.

4. Design a one-input, one-output frequency-measuring circuit that is composed of capacitors and inductors whose output in millivolts (from 1 millivolt to 1,000 millivolts) is proportional to the frequency of an incoming signal (between 1 Hz and 100,000 Hz).

5. Design a one-input, one-output computational circuit that is composed of transistors, diodes, resistors, and capacitors and that produces an output voltage equal to the square root of its input voltage.

6. Design a two-input, one-output time-optimal robot controller circuit that is composed of the above components and that navigates a constant-speed autonomous mobile robot (with nonzero turning radius) to an arbitrary destination in minimal time.

7. Design a one-input, one-output amplifier composed of the above components and that delivers amplification of 60 dB (i.e., 1,000 to 1) with low distortion and low bias.

The above seven prototypical circuits are representative of analog circuits that are in widespread use. Filters extract specified ranges of frequencies from electrical signals and amplifiers enhance the amplitude of signal. Amplifiers are used to increase the amplitude of an incoming signal. Analog computational circuits are used to perform real-time mathematical calculations on signals. Embedded controllers are used to control the operation of numerous automatic devices.

11.3 BACKGROUND ON GENETIC PROGRAMMING

Genetic programming is a biologically inspired, domain-independent method that automatically creates a computer program from a high-level statement of a problem's requirements. Genetic programming is an extension of the genetic algorithm described in John Holland's pioneering book *Adaptation in Natural and Artificial Systems* [13]. In genetic programming, the genetic algorithm operates on a population of computer programs of varying sizes and shapes [18, 26].

Starting with a primordial ooze of thousands of randomly created computer programs, genetic programming progressively breeds a population of computer programs over a series of generations. Genetic programming applies the Darwinian principle of survival of the fittest, analogs of naturally occurring operations such as sexual recombination (crossover), mutation, gene duplication, and gene deletion, and certain mechanisms of developmental biology. The computer programs are compositions of functions (e.g., arithmetic operations, conditional operators, problem-specific functions) and terminals (e.g., external inputs, constants, zero-argument functions). The programs may be thought of as trees whose points are labeled with the functions and whose leaves are labeled with the terminals.

Genetic programming breeds computer programs to solve problems by executing the following three steps:

1. Randomly create an initial population of individual computer programs.

2. Iteratively perform the following substeps (called a *generation*) on the population of programs until the termination criterion has been satisfied:

(a) Assign a fitness value to each individual program in the population using the fitness measure.

(b) Create a new population of individual programs by applying the following three genetic operations. The genetic operations are applied to one or two individuals in the population selected with a probability based on fitness (with reselection allowed).

(i) Reproduce an existing individual by copying it into the new population.

(ii) Create two new individual programs from two existing parental individuals by genetically recombining subtrees from each program using the crossover operation at randomly chosen crossover points in the parental individuals.

(iii) Create a new individual from an existing parental individual by randomly mutating one randomly chosen subtree of the parental individual.

3. Designate the individual computer program that is identified by the method of result designation (e.g., the *best-so-far* individual) as the result of the run of genetic programming. This result may represent a solution (or an approximate solution) to the problem.

The hierarchical character and the dynamic variability of the computer programs that are developed along the way to a solution are important features of genetic programming. It is often difficult and unnatural to try to specify or restrict the size and shape of the eventual solution in advance.

Automated programming requires some hierarchical mechanism to exploit, *by reuse* and *parameterization*, the regularities, symmetries, homogeneities, similarities, patterns, and modularities inherent in problem environments. Subroutines do this in ordinary computer programs. Automatically defined functions [19, 20] can be implemented within the context of genetic programming by establishing a constrained syntactic structure for the individual programs in the population. Each multi-part program in the population contains one (or more) function-defining branches and one (or more) main result-producing branches. The result-producing branch usually has the ability to call one or more of the automatically defined functions. A function-defining branch may have the ability to refer hierarchically to other already-defined automatically defined functions.

Since each individual program in the population of this example consists of function-defining branch(es) and result-producing branch(es), the initial random generation is created so that every individual program in the population has this particular constrained syntactic structure. Since a constrained syntactic structure is involved, crossover is performed so as to preserve this syntactic structure in all offspring.

Architecture-altering operations enhance genetic programming with automatically defined functions by providing a way to automatically determine the number of such automatically defined functions, the number of arguments that each automatically defined function possesses, and the nature of the hierarchical references, if any, among such automatically defined functions [21]. These operations include branch duplication, argument duplication, branch creation, argument creation, branch deletion, and argument deletion. The architecture-altering operations are motivated by the naturally occurring mechanism of gene duplication that creates new proteins (and hence new structures and new behaviors in living things) [28].

Genetic programming has been applied to numerous problems in fields such as system identification, control, classification, design, optimization, and automatic programming. Current research is described in [3, 4, 5, 16, 22, 23, 24, 25, 32] and on the World Wide Web at `www.genetic-programming.org`.

11.4 APPLYING GENETIC PROGRAMMING TO ANALOG CIRCUIT SYNTHESIS

Genetic programming can be applied to the problem of synthesizing circuits if a mapping is established between the program trees (rooted, point-labeled trees – that is, acyclic graphs – with ordered branches) used in genetic programming and the labeled cyclic graphs germane to electrical circuits. The principles of developmental biology and work on applying genetic algorithms and genetic programming to evolve neural networks [12, 17] provide the motivation for mapping trees into circuits by means of a developmental process that begins with a simple embryo. The initial circuit typically includes fixed wires that connect the inputs and outputs of the particular circuit being designed and certain fixed components (such as source and load resistors). Until these wires are modified, the initial circuit does not produce interesting output. An electrical circuit is developed by progressively applying the functions in a circuit-constructing program tree to the modifiable wires of the embryo (and, during the developmental process, to new components and modifiable wires).

An electrical circuit is created by executing the functions in a circuit-constructing program tree. The functions are progressively applied in a developmental process to the embryo and its successors until all of the functions in the program tree are executed. That is, the functions in the circuit-constructing program tree progressively side-effect the embryo and its successors until a fully developed circuit eventually emerges. The functions are applied in a breadth-first order.

The functions in the circuit-constructing program trees are divided into five categories: (1) topology-modifying functions that alter the circuit topology, (2) component-creating functions that insert components into the circuit, (3) development-controlling functions that control the development process by which the embryo and its successors is changed into a fully developed circuit, (4) arithmetic-performing functions that appear in subtrees as argument(s) to the component-creating functions and specify the numerical value of the component, and (5) automatically defined functions that appear in the function-defining branches and potentially enable certain substructures of the circuit to be reused (with parameterization).

Each branch of the program tree is created in accordance with a constrained syntactic structure. Each branch is composed of topology-modifying functions, component-creating functions, development-controlling functions, and terminals. Component-creating functions typically have one arithmetic-performing subtree, while topology-modifying functions, and development-controlling functions do not. Component-creating functions and topology-modifying functions are internal points of their branches and possess one or more arguments (construction-continuing subtrees) that continue the developmental process. The syntactic validity of this constrained syntactic structure is preserved using structure-preserving crossover with point typing. For details, see [23].

11.4.1 The initial circuit

An electrical circuit is created by executing a circuit-constructing program tree that contains various component-creating, topology-modifying, and development-controlling functions. Each circuit-constructing program tree in the population creates one circuit.

Fig. 11.1 shows a one-input, one-output initial circuit in which VSOURCE is the input signal and VOUT is the output signal (the probe point). The circuit is driven by an incoming alternating circuit source VSOURCE. There is a fixed load resistor RLOAD and a fixed source resistor RSOURCE. All of the foregoing items are called the *test fixture*. The test fixture is fixed. It provides a means for testing individual evolved circuits. An embryo is embedded within the test fixture. The embryo contains one or more modifiable wires or modifiable components. In the figure, there a modifiable wire Z0 between nodes 2 and 3. All development originates from modifiable wires or modifiable components.

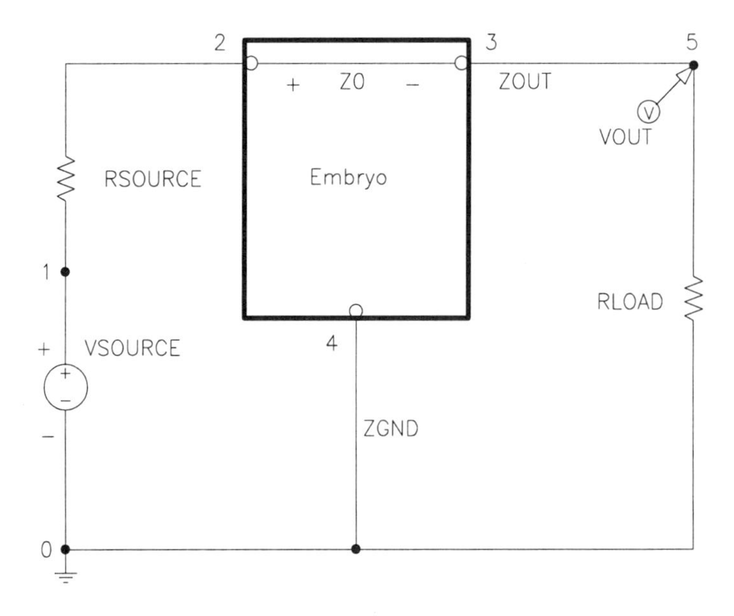

Fig. 11.1 One-input, one-output embryo.

11.4.2 Component-creating functions

The component-creating functions insert a component into the developing circuit and assign component value(s) to the component.

Each component-creating function has a writing head that points to an associated highlighted component in the developing circuit and modifies that component in a specified manner. The construction-continuing subtree of each component-creating function points to a successor function or terminal in the circuit-constructing program tree.

The arithmetic-performing subtree of a component-creating function consists of a composition of arithmetic functions (addition and subtraction) and random constants (in the range -1.000 to +1.000). The arithmetic-performing subtree specifies the numerical value of a component by returning a floating-point value that is interpreted on a logarithmic scale as the value for the component in a range of 10 orders of magnitude (using a unit of measure that is appropriate for the particular type of component).

The two-argument resistor-creating R function causes the highlighted component to be changed into a resistor. The value of the resistor in kilo Ohms is specified by its arithmetic-performing subtree.

Fig. 11.2 shows a modifiable wire Z0 connecting nodes 1 and 2 of a partial circuit containing four capacitors (C2, C3, C4, and C5). The circle indicates that Z0 has a writing head (i.e., is the highlighted component and that Z0 is subject to subsequent modification). Fig. 11.3 shows the result of applying the R function to the modifiable wire Z0 of Fig. 11.2. The circle indicates that the newly created R1 has a writing head so that R1 remains subject to subsequent modification.

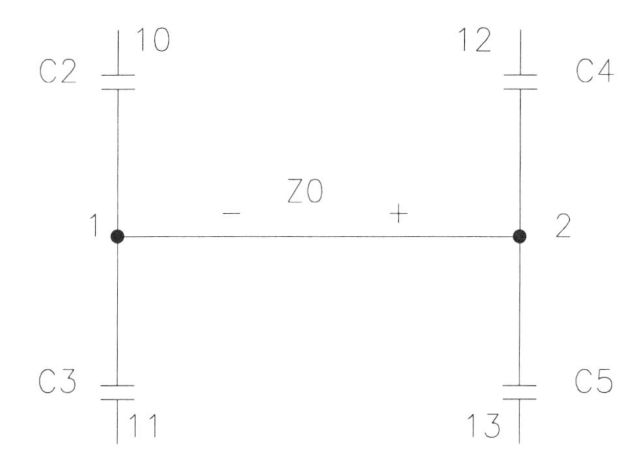

Fig. 11.2 Modifiable wire Z0.

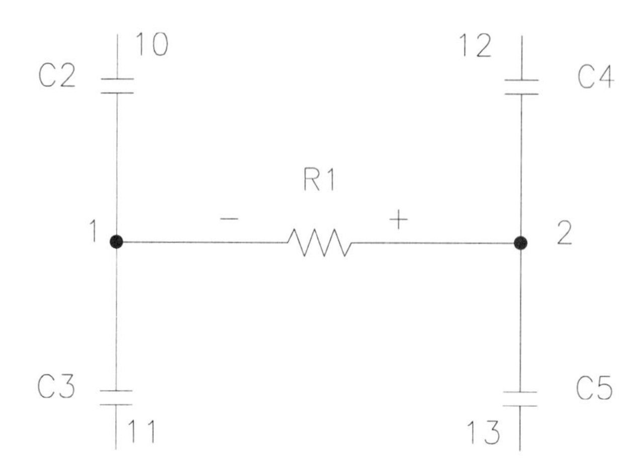

Fig. 11.3 Result of applying the R function.

Similarly, the two-argument capacitor-creating C function causes the highlighted component to be changed into a capacitor whose value in micro-Farads is specified by its arithmetic-performing subtree. In addition, the two-argument inductor-creating L function causes the highlighted component to be changed into an inductor whose value in micro-Henrys is specified by its arithmetic-performing subtree.

The one-argument Q_D_PNP diode-creating function causes a diode to be inserted in lieu of the highlighted component. This function has only one argument because there is no numerical value associated with a diode and thus no arithmetic-performing subtree. In practice, the diode is implemented here using a pnp transistor whose collector and base are connected to each other. The Q_D_NPN function inserts a diode using an npn transistor in a similar manner.

There are also six one-argument transistor-creating functions (Q_POS_COLL_NPN, Q_GND_EMIT_NPN, Q_NEG_EMIT_NPN, Q_GND_EMIT_PNP, Q_POS_EMIT_PNP, Q_NEG_COLL_PNP) that insert a bipolar junction transistor in lieu of the highlighted component and that directly connect the collector or emitter of the newly created transistor to a fixed point of the circuit (the positive power supply, ground, or the negative power supply). For example, the Q_POS_COLL_NPN function inserts a bipolar junction transistor whose collector is connected to the positive power supply.

Each of the functions in the family of six different three-argument transistor-creating Q_3_NPN functions causes an npn bipolar junction transistor to be inserted in place of the highlighted component and one of the nodes to which the highlighted component is connected. The Q_3_NPN function creates five new nodes and three modifiable wires. There is no writing head on the new transistor, but there is a writing head on each of the three new modifiable wires. There are 12 members (called Q_3_NPN0, ..., Q_3_NPN11) in this family of functions because there are two choices of nodes (1 and 2) to be bifurcated and then there are six ways of attaching the transistor's base, collector, and emitter after the bifurcation. Similarly the family of 12 Q_3_PNP functions causes a pnp bipolar junction transistor to be inserted.

11.4.3 Topology-modifying functions

Each topology-modifying function in a program tree points to an associated highlighted component and modifies the topology of the developing circuit.

The three-argument SERIES division function creates a series composition of the highlighted component (with a writing head), a copy of it (with a writing head), one new modifiable wire (with a writing head), and two new nodes.

The four-argument PARALLEL0 parallel division function creates a parallel composition consisting of the original highlighted component (with a writing head), a copy of it (with a writing head), two new modifiable wires (each with a writing head), and two new nodes. There are potentially two topologically distinct outcomes of a parallel division. Since we want the outcome of all circuit-constructing functions to be deterministic, there are two members (called PARALLEL0 and PARALLEL1) in the PARALLEL family of topology-modifying functions. The two functions operate differently depending on degree and numbering of the preexisting components in the developing circuit. The use of the two functions breaks the symmetry between the potentially distinct outcomes. Fig. 11.4 shows the result of applying PARALLEL0 to the resistor R1 from Fig. 11.3. Modifiable resistors R1 and R7 and modifiable wires Z6 and Z8 are each linked to the top-most function in one of the four construction-continuing subtrees of the PARALLEL0 function.

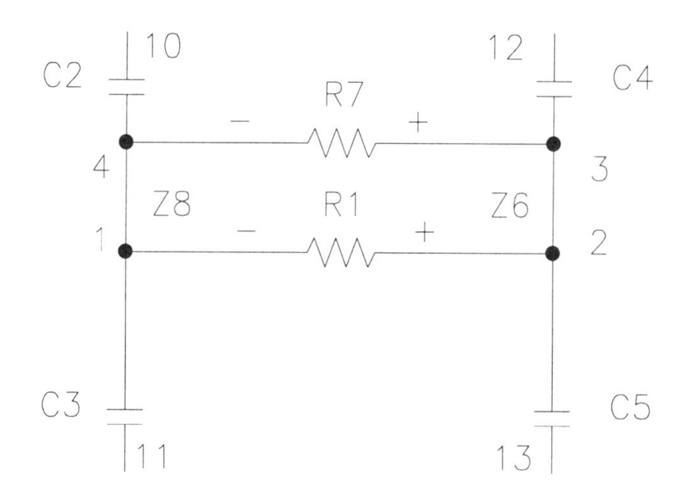

Fig. 11.4 Result of the PARALLEL0 function.

The reader is referred to [23] for a detailed description of the operation of the PARALLEL0 and PARALLEL1 functions (and other functions mentioned herein).

If desired, other topology-modifying functions may be defined to create the Y-shaped divisions and Δ-shaped divisions that are frequently seen in human-designed circuits.

The one-argument polarity-reversing FLIP function reverses the polarity of the highlighted component.

There are six three-argument functions (T_GND_0, T_GND_1, T_POS_0, T_POS_1, T_NEG_0, T_NEG_1) that insert two new nodes and two new modifiable wires, and then make a connection to ground, positive power supply, or negative power supply, respectively.

There are two three-argument functions (PAIR_CONNECT_0 and PAIR_CONNECT_1) that enable distant parts of a circuit to be connected together. The first PAIR_CONNECT to occur in the development of a circuit creates two new wires, two new nodes, and one temporary port. The next PAIR_CONNECT creates two new wires and one new node, connects the temporary port to the end of one of these new wires, and then removes the temporary port.

If desired, numbered vias can be created to provide connectivity between distant points of the circuit by using a three-argument VIA function.

The zero-argument SAFE_CUT function causes the highlighted component to be removed from the circuit provided that the degree of the nodes at both ends of the highlighted component is three (i.e., no dangling components or wires are created).

11.4.4 Development-controlling functions

The one-argument NOOP ("No Operation") function has no effect on the modifiable wire or modifiable component with which it is associated; however, it has the effect of delaying activity on the developmental path on which it appears in relation to other developmental paths in the overall circuit-constructing program tree.

The zero-argument END function makes the modifiable wire or modifiable component with which it is associated into a non-modifiable wire or component (thereby ending a particular developmental path).

11.4.5 Example of developmental process

Fig. 11.5 is an illustrative circuit-constructing program tree shown as a rooted, point-labeled tree with ordered branches. The overall program consists of two main result-producing branches joined by a connective LIST function (labeled 1 in the figure). The first (left) result-producing branch is rooted at the capacitor-creating C function (labeled 2). The second result-producing branch is rooted at the polarity-reversing FLIP function (labeled 3). This figure also contains four occurrences of the inductor-creating L function (at 17, 11, 20, and 12). The figure contains two occurrences of the topology-modifying SERIES function (at 5 and 10). The figure also contains five occurrences of the development-controlling END function (at 15, 25, 27, 31, and 22) and one occurrence of the development-controlling "no operation" NOP function (at 6). There is a seven-point arithmetic-performing subtree at 4 under the capacitor-creating C function at 4. Similarly, there is a three-point arithmetic-performing subtree at 19 under the inductor-creating L function at 11. There are also one-point arithmetic-performing subtrees (i.e., constants) at 26, 30, and 21. Additional details can be found in [23].

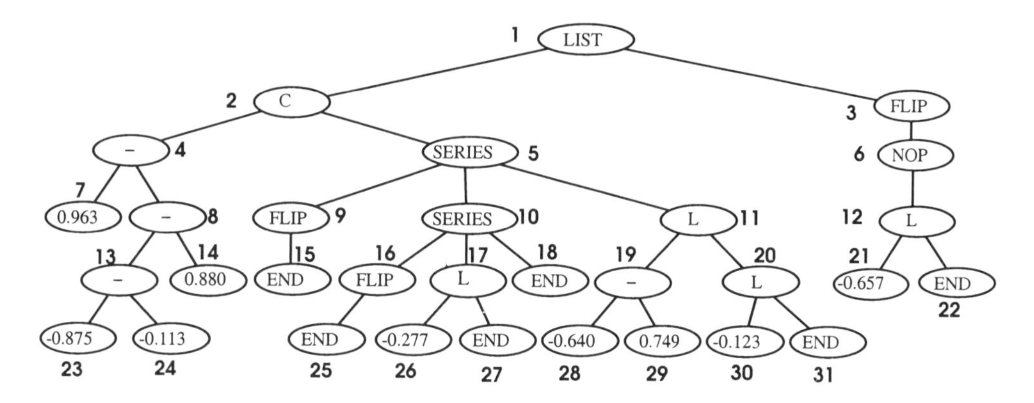

Fig. 11.5 Illustrative circuit-constructing program tree.

11.5 PREPARATORY STEPS

Before applying genetic programming to a problem of circuit design, seven major preparatory steps are required: (1) identify the initial circuit, (2) determine the architecture of the circuit-constructing program trees, (3) identify the primitive functions of the program trees, (4) identify the terminals of the program trees, (5) create the fitness measure, (6) choose control parameters for the run, and (7) determine the termination criterion and method of result designation.

11.5.1 Initial circuit

The initial circuit used on a particular problem depends on the circuit's number of inputs and outputs. All development originates from the modifiable wires of the embryo within the initial circuit.

An embryo with two modifiable wires (Z0 and Z1) was used for the one-input, one-output lowpass filter, highpass filter, bandpass filter, and frequency-measuring circuits.

The robot controller circuit has two inputs (VSOURCE1 and VSOURCE2) representing the two-dimensional position of the target point. Therefore, this problem requires a different initial circuit than that used above. Both voltage inputs require their own separate source resistor (RSOURCE1 and RSOURCE2). The embryo for the robot controller circuit has three modifiable wires (Z0, Z1, and Z2) in order to provide full connectivity between the two inputs and the one output.

In some problems, such as the amplifier, the embryo contains additional fixed components (as detailed in [23]).

For historical reasons, an embryo with one modifiable wire was used for the computational circuit. However, an embryo with two modifiable wires could have been used on these problems and an embryo with one modifiable wire could have been used on the lowpass filter, highpass filter, and bandpass filter, and frequency-measuring circuits.

11.5.2 Program architecture

Since there is one result-producing branch in the program tree for each modifiable wire in the embryo, the architecture of each circuit-constructing program tree depends on the embryo. One result-producing branch was used for the computational circuit; two were used for lowpass, highpass, and bandpass filter, problems; and three were used for the robot controller and amplifier.

The architecture of each circuit-constructing program tree also depends on the use, if any, of automatically defined functions. Automatically defined functions provide a mechanism enabling certain substructures to be reused and are described in detail in [23]. Automatically defined functions and architecture-altering operations were used in the robot controller, and amplifier. For these problems, each program in the initial population of programs had a uniform architecture with no automatically defined functions. In later generations, the number of automatically defined functions, if any, emerged as a consequence of the architecture-altering operations (also described in [23]).

11.5.3 Function and terminal sets

The function set for each design problem depends on the type of electrical components that are to be used for constructing the circuit.

For the problems of synthesizing a lowpass, highpass, and bandpass filter and the problem of synthesizing a frequency-measuring circuit, the function set included two component-creating functions (for inductors and capacitors), topology-modifying functions (for series and parallel divisions and for flipping components), one development-controlling function ("no operation"), functions for creating a via to ground, and functions for connecting pairs of points. That is, the function set, $\mathcal{F}_{\text{ccs-initial}}$, for each construction-continuing subtree was

$\mathcal{F}_{\text{ccs-initial}}$ = {L, C, SERIES, PARALLEL0, PARALLEL1, FLIP, NOOP, T_GND_0, T_GND_1, PAIR_CONNECT_0, PAIR_CONNECT_1}.

Capacitors, resistors, diodes, and transistors were used for the computational circuit, the robot controller, and the amplifier. In addition, the function set also included functions to provide connectivity to the positive and negative power supplies (in order to provide a source of energy for the transistors). Thus, the function set, $\mathcal{F}_{\text{ccs-initial}}$, for each construction-continuing subtree was

$\mathcal{F}_{\text{ccs-initial}}$ = {R, C, SERIES, PARALLEL0, PARALLEL1, FLIP, NOOP, T_GND_0, T_GND_1, T_POS_0, T_POS_1, T_NEG_0, T_NEG_1, PAIR_CONNECT_0, PAIR_CONNECT_1, Q_D_NPN, Q_D_PNP, Q_3_NPN0, ..., Q_3_NPN11, Q_3_PNP0, ..., Q_3_PNP11, Q_POS_COLL_NPN, Q_GND_EMIT_NPN, Q_NEG_EMIT_NPN, Q_GND_EMIT_PNP, Q_POS_EMIT_PNP, Q_NEG_COLL_PNP}.

For the *npn* transistors, the Q2N3904 model was used. For *pnp* transistors, the Q2N3906 model was used.

The initial terminal set, $\mathcal{T}_{\text{ccs-initial}}$, for each construction-continuing subtree was

$\mathcal{T}_{\text{ccs-initial}}$ = {END, SAFE_CUT}.

The initial terminal set, $\mathcal{T}_{\text{aps-initial}}$, for each arithmetic-performing subtree consisted of

$$\mathcal{T}_{\text{aps-initial}} = \{\Re\},$$

where $\Re$ represents floating-point random constants from –1.0 to +1.0.

The function set, $\mathcal{F}_{\text{aps}}$, for each arithmetic-performing subtree was,

$$\mathcal{F}_{\text{aps}} = \{+, -\}.$$

The terminal and function sets were identical for all result-producing branches for a particular problem.

For the lowpass filter, highpass filter, and bandpass filter, there was no need for functions to provide connectivity to the positive and negative power supplies.

For the robot controller and the amplifier, the architecture-altering operations were used and the set of potential new functions, $\mathcal{F}_{\text{potential}}$, was

$$\mathcal{F}_{\text{potential}} = \{\texttt{ADF0}, \texttt{ADF1}, \ldots\}.$$

The set of potential new terminals, $\mathcal{T}_{\text{potential}}$, for the automatically defined functions was

$$\mathcal{T}_{\text{potential}} = \{\texttt{ARG0}\}.$$

The architecture-altering operations change the function set, $\mathcal{F}_{\text{ccs}}$ for each construction-continuing subtree of all three result-producing branches and the function-defining branches, so that

$$\mathcal{F}_{\text{ccs}} = \mathcal{F}_{\text{ccs-initial}} \approx \mathcal{F}_{\text{potential}}.$$

The architecture-altering operations generally change the terminal set for automatically defined functions, $\mathcal{T}_{\text{aps-adf}}$, for each arithmetic-performing subtree, so that

$$\mathcal{T}_{\text{aps-adf}} = \mathcal{T}_{\text{aps-initial}} \approx \mathcal{T}_{\text{potential}}.$$

11.5.4 Fitness measure

The evolutionary process is driven by the *fitness measure*. Each individual computer program in the population is executed and then evaluated, using the fitness measure. The nature of the fitness measure varies with the problem. The high-level statement of desired circuit behavior is translated into a well-defined measurable quantity that can be used by genetic programming to guide the evolutionary process. The evaluation of each individual circuit-constructing program tree in the population begins with its execution. This execution progressively applies the functions in each program tree to the initial circuit, thereby creating a fully developed circuit. A netlist is created that identifies each component of the developed circuit, the nodes to which each component is connected, and the value of each component. The netlist becomes the input to our modified version of the 217,000-line SPICE (Simulation Program with Integrated Circuit Emphasis) simulation program [29]. SPICE then determines the behavior of the circuit. It was necessary to make considerable modifications in SPICE so that it could run as a submodule within the genetic programming system.

11.5.4.1 Fitness measure for the lowpass filter A simple *filter* is a one-input, one-output electronic circuit that receives a signal as its input and passes the frequency components of the incoming signal that lie in a specified range (called the *passband*) while suppressing the frequency components that lie in all other frequency ranges (the *stopband*).

The desired lowpass LC filter has a passband below 1,000 Hz and a stopband above 2,000 Hz. The circuit is driven by an incoming AC voltage source with a 2 volt amplitude.

The *attenuation* of the filter is defined in terms of the output signal relative to the reference voltage (half of 2 volt here). A *decibel* is a unitless measure of relative voltage that is defined as 20 times the common (base 10) logarithm of the ratio between the voltage at a particular probe point and a reference voltage.

In this problem, a voltage in the passband of exactly 1 volt and a voltage in the stopband of exactly 0 volts is regarded as ideal. The (preferably small) variation within the passband is called the *passband ripple*. Similarly, the incoming signal is never fully reduced to zero in the stopband of an actual filter. The (preferably small) variation within the stopband is called the *stopband ripple*. A voltage in the passband of between 970 millivolts and 1 volt (i.e., a passband ripple of 30 millivolts or less) and a voltage in the stopband of between 0 volts and 1 millivolts (i.e., a stopband ripple of 1 millivolts or less) is regarded as acceptable. Any voltage lower than 970 millivolts in the passband and any voltage above 1 millivolts in the stopband is regarded as unacceptable.

A fifth-order *elliptic* (*Cauer*) *filter* with a modular angle Θ of 30 degrees (i.e., the arcsin of the ratio of the boundaries of the passband and stopband) and a reflection coefficient ρ of 24.3% is required to satisfy these design goals [35].

Since the high-level statement of behavior for the desired circuit is expressed in terms of frequencies, the voltage VOUT is measured in the frequency domain. SPICE performs an AC small signal analysis and reports the circuit's behavior over five decades (between 1 Hz and 100,000 Hz) with each decade being divided into 20 parts (using a logarithmic scale), so that there are a total of 101 fitness cases.

Fitness is measured in terms of the sum over these cases of the absolute weighted deviation between the actual value of the voltage that is produced by the circuit at the probe point VOUT and the target value for voltage. The smaller the value of fitness, the better. A fitness of zero represents an (unattainable) ideal filter.

Specifically, the standardized fitness is

$$F(t) = \sum_{i=0}^{100} (W(d(f_i), f_i)d(f_i))$$

where f_i is the frequency of fitness case i; $d(x)$ is the absolute value of the difference between the target and observed values at frequency x; and $W(y,x)$ is the weighting for difference y at frequency x.

The fitness measure is designed to not penalize ideal values, to slightly penalize every acceptable deviation, and to heavily penalize every unacceptable deviation. Specifically, the procedure for each of the 61 points in the 3-decade interval between 1 Hz and 1,000 Hz for the intended passband is as follows:

- If the voltage equals the ideal value of 1.0 volt in this interval, the deviation is 0.0.

- If the voltage is between 970 millivolts and 1 volt, the absolute value of the deviation from 1 volt is weighted by a factor of 1.0.

- If the voltage is less than 970 millivolts, the absolute value of the deviation from 1 volt is weighted by a factor of 10.0.

The acceptable and unacceptable deviations for each of the 35 points from 2,000 Hz to 100,000 Hz in the intended stopband are similarly weighed (by 1.0 or 10.0) based on the amount of deviation from the ideal voltage of 0 volts and the acceptable deviation of 1 millivolts.

For each of the five "don't care" points between 1,000 and 2,000 Hz, the deviation is deemed to be zero.

The number of "hits" for this problem (and all other problems herein) is defined as the number of fitness cases for which the voltage is acceptable or ideal or that lie in the "don't care" band (for a filter).

Many of the random initial circuits and many that are created by the crossover and mutation operations in subsequent generations cannot be simulated by SPICE. These circuits receive a high penalty value of fitness (10^8) and become the worst-of-generation programs for each generation. For details, see [23].

11.5.4.2 Fitness measure for the highpass filter The fitness cases for the highpass filter are the same 101 points in the five decades of frequency between 1 Hz and 100,000 Hz as for the lowpass filter. The fitness measure is substantially the same as that for the lowpass filter problem above, except that the locations of the passband and stopband are reversed. Notice that the only difference in the seven preparatory steps for a highpass filter versus a lowpass filter is this change in the fitness measure.

11.5.4.3 Fitness measure for the bandpass filter The fitness cases for the bandpass filter are the same 101 points in the five decades of frequency between 1 Hz and 100,000 Hz as for the lowpass filter. The acceptable deviation in the desired passband between 500 Hz and 1,000 Hz is 30 millivolts (i.e., the same as for the passband of the lowpass and highpass filters above). The acceptable deviation in the two stopbands (i.e., between 1 Hz and 250 Hz and between 2,000 Hz and 100,000 Hz is 1 millivolt1 (i.e., the same as for the stopband of the lowpass and highpass filters above). Again, notice that the only difference in the seven preparatory steps for a bandpass filter versus a lowpass or highpass filter is a change in the fitness measure.

11.5.4.4 Fitness measure for frequency-measuring circuit The fitness cases for the frequency-measuring circuit are the same 101 points in the five decades of frequency (on a logarithmic scale) between 1 Hz and 100,000 Hz as for the lowpass and lowpass filters. The circuit's output in millivolts (from 1 millivolt to 1,000 millivolts) is intended to be proportional to the frequency of an incoming signal (between 1 Hz and 100,000 Hz). Fitness is the sum, over the 101 fitness cases, of the absolute value of the difference between the circuit's actual output and the desired output voltage.

11.5.4.5 Fitness measure for the computational circuit SPICE is called to perform a DC sweep analysis at 21 equidistant voltages between –250 millivolts and +250 millivolts.

Fitness is the sum, over these 21 fitness cases, of the absolute weighted deviation between the actual value of the voltage that is produced by the circuit and the target value for voltage. For details, see [23].

11.5.4.6 Fitness measure for the robot controller circuit The fitness of a robot controller was evaluated using 72 randomly chosen fitness cases each representing different two-dimensional target points. Fitness is the sum, over the 72 fitness cases, of the travel times of the robot to the target point. If the robot came within a capture radius of 0.28 meters of its target point before the end of the 80 time steps allowed for a particular fitness case, the contribution to fitness for that fitness case was the actual time. However, if the robot failed to come within the capture radius during the 80 time steps, the contribution to fitness was a penalty value of 0.160 hours (i.e., double the worst possible time).

The two voltage inputs to the circuit represents the two-dimensional location of the target point. SPICE performs a nested DC sweep, which provides a way to simulate the DC behavior of a circuit with two inputs. The nested DC sweep resembles a nested pair of `FOR` loops in a computer program in that both of the loops have a starting value for the voltage, an increment, and an ending value for the voltage. For each voltage value in the outer loop, the inner loop simulates the behavior of the circuit by stepping through its range of voltages. Specifically, the starting value for voltage is –4 volt, the step size is 0.2 volt, and the ending value is +4 volt. These values correspond to the dimensions of the robot's world of 64 square meters extending 4 meters in each of the four directions from the origin of a coordinate system (i.e., 1 volt equals 1 meter). For details, see [23].

11.5.4.7 Fitness measure for the 60 dB amplifier SPICE was requested to perform a DC sweep analysis to determine the circuit's response for several different DC input voltages. An ideal inverting amplifier circuit would receive the DC input, invert it, and multiply it by the amplification factor. A circuit is flawed to the extent that it does not achieve the desired amplification, the output signal is not perfectly centered on 0 volts (i.e., it is biased), or the DC response is not linear. Fitness is calculated by summing an amplification penalty, a bias penalty, and two non-linearity penalties – each derived from these five DC outputs. For details, see [6].

11.5.5 Control parameters

The population size, M, was 1,400,000 for the bandpass filter problem and 640,000 for all other problems.

11.5.6 Implementation on parallel computer

All problems except the bandpass filter were run on a medium-grained parallel Parsytec computer system consisting of 64 80-MHz PowerPC 601 processors arranged in an 8 by 8 toroidal mesh with a host PC Pentium type computer. The bandpass filter problem was run on a medium-grained parallel Beowulf-style [33] arranged in an 7 by 10 toroidal mesh with a host Alpha computer. The distributed genetic algorithm [2] was used. When the population size was 640,000, Q = 10,000 at each of the D = 64 demes (semi-isolated subpopulations). When the population size was 1,400,000 Q = 20,000 at each of the D = 70 demes. On each generation, four boatloads of emigrants, each consisting of B = 2% (the migration rate) of the node's subpopulation (selected on the basis of fitness) were dispatched to each of the four adjacent processing nodes.

11.6 RESULTS

In all seven problems, fitness was observed to improve from generation to generation during the run. Satisfactory results were generated on the first or second run of each of the seven problems. Most of the seven problems were solved on the very first run. When a second run was required (i. e., a run with different random number seeds), the first run always produced a nearly satisfactory result. The fact that each of these seven illustrative problems were solved after only one or two runs suggests that the ability of genetic programming to evolve analog electrical circuits was not severely challenged by any of these seven problems. Thus augers well for handling more challenging problems in the future.

11.6.1 Lowpass filter

Genetic programming has evolved numerous lowpass filters having topologies similar to that devised by human engineers. For example, a circuit (Fig. 11.6) was evolved in generation 49 of one run with a near-zero fitness of 0.00781. The circuit was 100% compliant with the design requirements in the sense that it scored 101 hits (out of 101). As can be seen, this evolved circuit consists of seven inductors (L5, L10, L22, L28, L31, L25, and L13) arranged horizontally across the top of the figure "in series" with the incoming signal VSOURCE and the source resistor RSOURCE. It also contains seven capacitors (C12, C24, C30, C3, C33, C27, and C15) that are each shunted to ground. This circuit is a classical ladder filter with seven rungs [35].

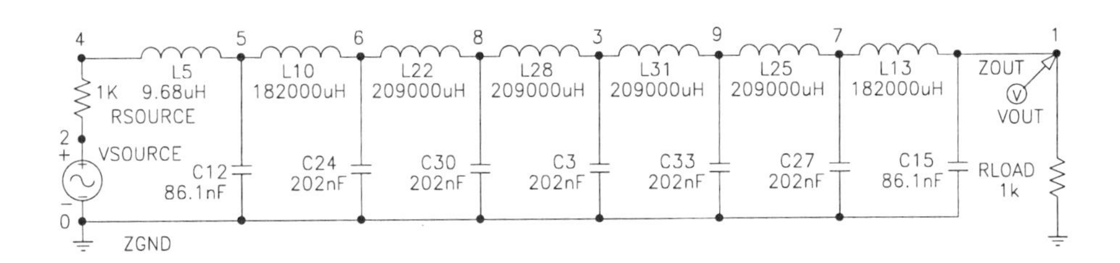

Fig. 11.6 Evolved seven-rung ladder lowpass filter.

After the run, this evolved circuit (and all other evolved circuits herein) were simulated anew using the commercially available MicroSim circuit simulator to verify performance. Fig. 11.7 shows the behavior in the frequency domain of this evolved lowpass filter. As can be seen, the evolved circuit delivers about 1 volt for all frequencies up to 1,000 Hz and about 0 volts for all frequencies above 2,000 Hz. There is a sharp drop-off in voltage in the transition region between 1,000 Hz and 2,000 Hz.

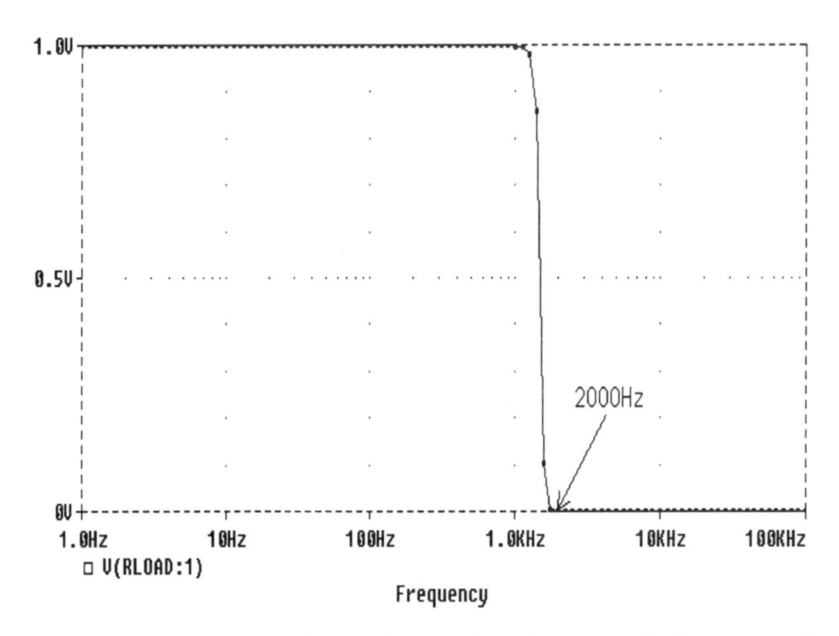

Fig. 11.7 Frequency domain behavior of genetically evolved 7-rung ladder lowpass filter.

The circuit of Fig. 11.6 has the recognizable features of the circuit for which George Campbell of American Telephone and Telegraph received U. S. patent 1,227,113 [7]. Claim 2 of Campbell's patent covered,

> "An electric wave filter consisting of a connecting line of negligible attenuation composed of a plurality of sections, each section including a capacity element and an inductance element, one of said elements of each section being in series with the line and the other in shunt across the line, said capacity and inductance elements having precomputed values dependent upon the upper limiting frequency and the lower limiting frequency of a range of frequencies it is desired to transmit without attenuation, the values of said capacity and inductance elements being so proportioned that the structure transmits with practically negligible attenuation sinusoidal currents of all frequencies lying between said two limiting frequencies, while attenuating and approximately extinguishing currents of neighboring frequencies lying outside of said limiting frequencies."

An examination of the evolved circuit of Fig. 11.6 shows that it indeed consists of "a plurality of sections." (specifically, seven). In the figure, "Each section include[es] a capacity element and an inductance element." Specifically, the first of the seven sections consists of inductor L5 and capacitor C12; the second section consists of inductor L10 and capacitor C24; and so forth. Moreover, "one of said elements of each section [is] in series with the line and the other in shunt across the line." Inductor L5 of the first section is indeed "in series with the line" and capacitor C12 is indeed "in shunt across the line." This is also true for the circuit's remaining six sections. Moreover, Fig. 11.6 herein matches Figure 7 of Campbell's 1917 patent. In addition, this circuit's 100% compliant behavior in the frequency domain (Fig. 11.7 herein) confirms the fact that the values of the inductors and capacitors are such as to transmit "with practically negligible attenuation sinusoidal currents" of the passband frequencies "while attenuating and approximately extinguishing currents" of the stopband frequencies.

In short, genetic programming evolved an electrical circuit that infringes on the claims of Campbell's now-expired patent.

Moreover, the evolved circuit of Fig. 11.6 also approximately possesses the numerical values recommended in Campbell's 1917 patent. After making several very minor adjustments and approximations (detailed in [23]), the evolved lowpass filter circuit of Fig. 11.6 can be viewed as what is now known as a cascade of six identical symmetric π-sections [15]. Such π-sections are characterized by two key parameters. The first parameter is the characteristic resistance (impedance) of the π-section. This characteristic resistance should match the circuit's fixed load resistance RLOAD (1,000 Ω). The second parameter is the nominal cutoff frequency which separates the filter's passband from its stopband. This second parameter should lie somewhere in the transition region between the end of the passband (1,000 Hz) and the beginning of the stopband (2,000 Hz). The characteristic resistance, R, of each of the π-sections is given by the formula $\sqrt{L/C}$. Here L equals 200,000 μH and C equals 197 nF when employing this formula after making the minor adjustments and approximations detailed in [23]. This formula yields a characteristic resistance, R, of 1,008 Ω. This value is very close to the value of the 1,000 Ω load resistance of this problem. The nominal cutoff frequency, f_c, of each of the π-sections of a lowpass filter is given by the formula $1/\pi\sqrt{LC}$. This formula yields a nominal cutoff frequency, f_c, of 1,604 Hz (i.e., roughly in the middle of the transition region between the passband and stopband of the desired lowpass filter).

The legal criteria for obtaining a U. S. patent are that the proposed invention be "new" and "useful" and

> ... the differences between the subject matter sought to be patented and the prior art are such that the subject matter as a whole would [not] have been obvious at the time the invention was made to a person having ordinary skill in the art to which said subject matter pertains. (35 *United States Code* 103a).

George Campbell was part of the renowned research team of the American Telephone and Telegraph Corporation. He received a patent for his filter in 1917 because his idea was new in 1917, because it was useful, and because satisifed the above statutory test for unobviousness. The fact that genetic programming rediscovered an electrical circuit that was unobvious "to a person having ordinary skill in the art" establishes that this evolved result satisfies Arthur Samuel's criterion [31] for artificial intelligence and machine learning, namely

> "The aim [is] ... to get machines to exhibit behavior, which if done by humans, would be assumed to involve the use of intelligence."

In another run, a 100% compliant recognizable "bridged T" arrangement was evolved. The "bridged T" filter topology was invented and patented by Kenneth S. Johnson of Western Electric Company in 1926 [14]. In yet another run of this same problem using automatically defined functions, a 100% compliant circuit emerged with the recognizable elliptic topology that was invented and patented by Wilhelm Cauer [8, 9, 10]. The Cauer filter was a significant advance (both theoretically and commercially) over the Campbell, Johnson, Butterworth, Chebychev, and other earlier filter designs. Details are found in [23].

It is important to note that when we performed the preparatory steps for applying genetic programming to the problem of synthesizing a lowpass filter, we did not employ any significant domain knowledge about filter design. We did not, for example, incorporate knowledge of Kirchhoff's laws, integro-differential equations, Laplace transforms, poles, zeroes, or the other mathematical techniques and insights about circuits that are known to electrical engineers who design filters. We did, of course, specify the basic ingredients from which a circuit is composed, such as appropriate electrical components (e.g., inductors and capacitors). We also specified various generic methods for constructing the topology of electrical circuits (e.g., series divisions, parallel divisions, and vias). Genetic programming then proceeded to evolve a satisfactory circuit under the guidance of the fitness measure.

The choices of electrical components in the preparatory steps are, of course, important. If, for example, we had included an insufficient set of components (e.g., only resistors and neon bulbs), genetic programming would have been incapable of evolving a satisfactory solution to any of the seven problems herein. On other hand, if we had included transistor-creating functions in the set of component-creating functions for this problem (instead of functions for creating inductors and capacitors), genetic programming would have evolved an active filter composed of transistors, instead of a passive filter composed of inductors and capacitors. An example of the successful evolution of an active filter satisfying the design requirements of this problem is found in [23].

There are various ways of incorporating problem-specific domain knowledge into a run of genetic programming if a practicing engineer desires to bring such additional domain knowledge to bear on a particular problem. For example, subcircuits that are known (or believed) to be necessary (or helpful) in solving a particular problem may be provided as primitive components. Also, a particular subcircuit may be hard-wired into an embryo (so that it is not subject to modification during the developmental process). In addition, a circuit may be divided into a prespecified number of distinct stages. A constrained syntactic structure can be used to mandate certain desired circuit features. Details and examples are found in [23].

11.6.2 Highpass filter

In generation 27 of one run, a 100% compliant circuit (Fig. 11.8) was evolved with a near-zero fitness of 0.213. This circuit has four capacitors and five inductors (in addition to the fixed components of the embryo). As can be seen, capacitors appear in series horizontally across the top of the figure, while inductors appear vertically as shunts to ground.

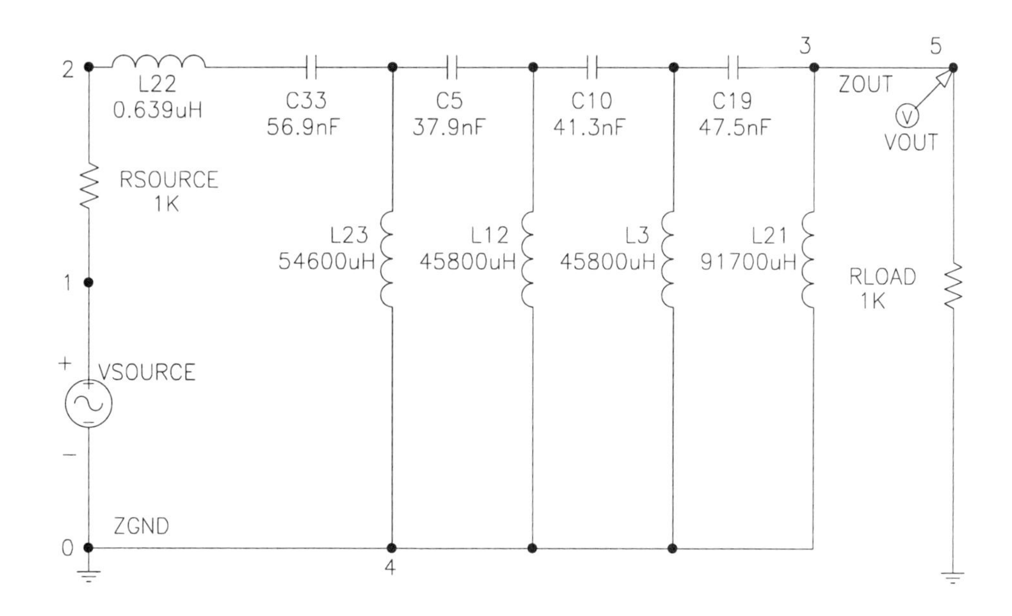

Fig. 11.8 Evolved four-rung ladder highpass filter.

Fig. 11.9 shows the behavior in the frequency domain of this evolved highpass filter. As desired, the evolved highpass delivers about 0 volts for all frequencies up to 1,000 Hz and about 1 volt for all frequencies above 2,000 Hz.

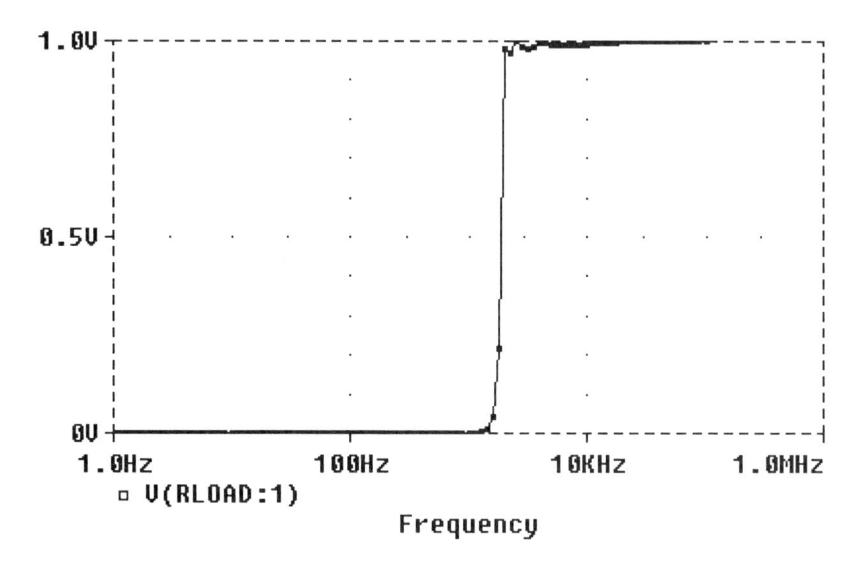

Fig. 11.9 Frequency domain behavior of evolved four-rung ladder highpass filter.

The reversal of roles for the capacitors and inductors in lowpass and highpass ladder filters is well known to electrical engineers. It arises because of the duality of the single terms (derivatives versus integrals) in the integro-differential equations that represent the voltages and currents of the inductors and capacitors in the loops and nodes of a circuit. However, genetic programming was not given any domain knowledge concerning integro-differential equations or this duality. In fact, the only difference in the preparatory steps for the problem of synthesizing the highpass filter versus the problem of synthesizing the lowpass filter was the fitness measure. The fitness measure was merely a high-level statement of the goals of the problem (i.e., suppression of the low frequencies, instead of the high frequencies, and passage at full voltage of the high frequencies, instead of the low frequencies). In spite of the absence of explicit domain knowledge about integro-differential equations or this duality, genetic programming evolved a 100% compliant highpass filter embodying the well-known highpass ladder topology. Using the altered fitness measure appropriate for highpass filters, genetic programming searched the same space (i.e., the space of circuit-constructing program trees composed of the same component-creating functions, the same topology-modifying functions, and the same development-controlling functions) and discovered a circuit-constructing program tree that yielded a 100%-complaint highpass filter.

11.6.3 Bandpass filter

The 100%-compliant evolved bandpass filter circuit (Fig. 11.10) from generation 718 scores 101 hits (out of 101).

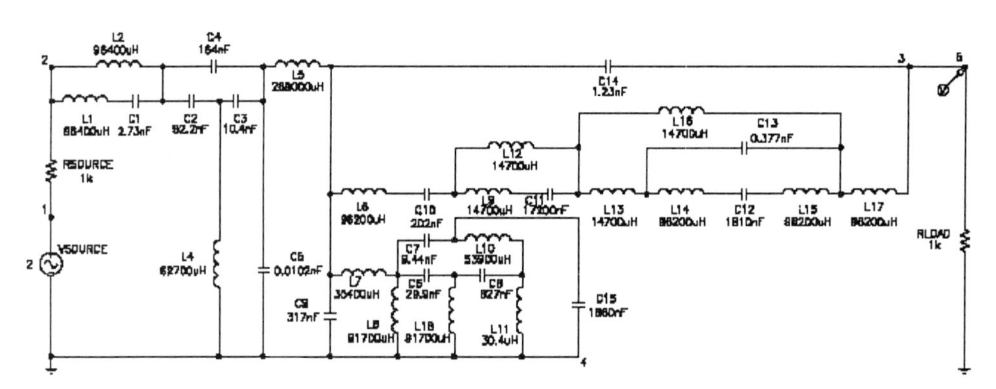

Fig. 11.10 Evolved bandpass filter.

Fig. 11.11 shows the behavior in the frequency domain of this evolved bandpass filter. The evolved circuit satisfies all of the stated requirements.

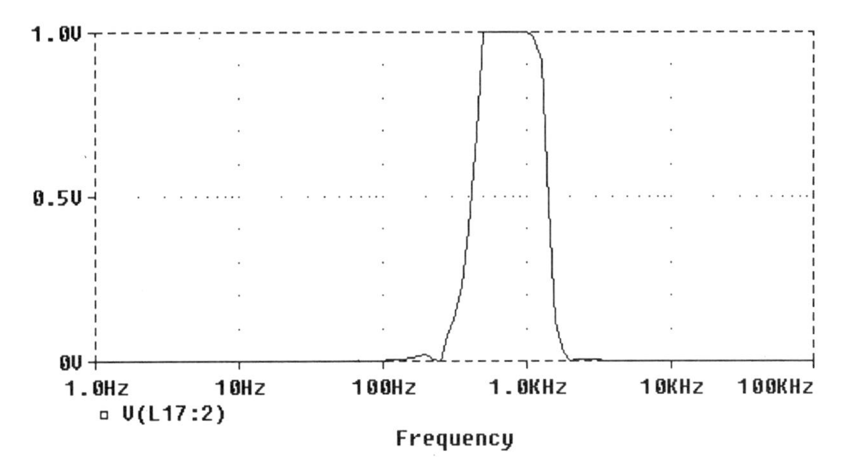

Fig. 11.11 Frequency domain behavior of evolved bandpass filter.

11.6.4 Frequency-measuring circuit

The 100%-compliant evolved frequency-measuring circuit (Fig. 11.12) from generation 101 scores 101 hits (out of 101).

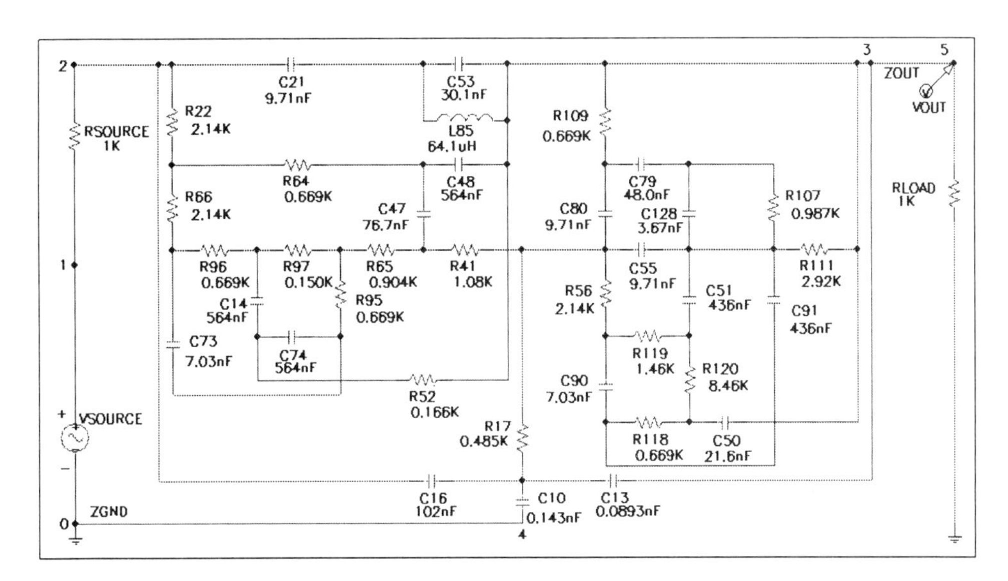

Fig. 11.12 Evolved frequency-measuring circuit.

Fig. 11.13 shows that the output of the circuit varies linearly with the frequency (on a logarithmic scale) of the incoming signal from 1 Hz to 1,000 Hz.

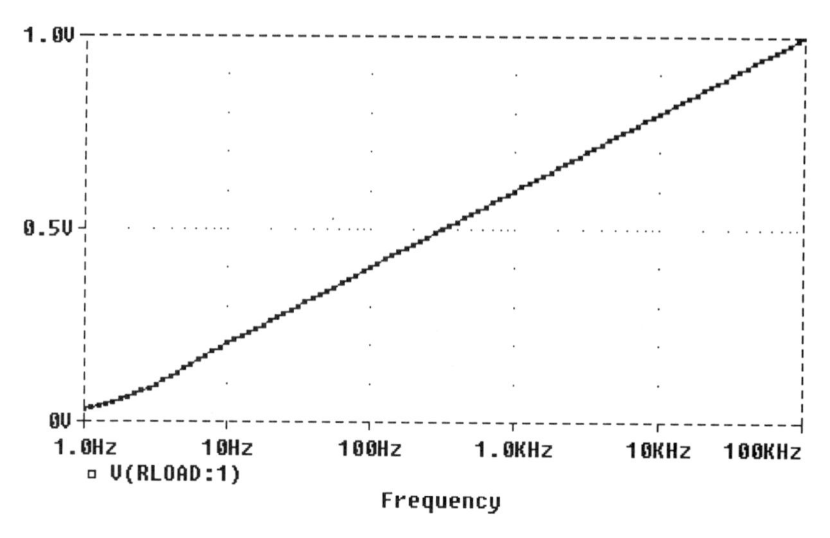

Fig. 11.13 Frequency domain behavior of evolved frequency-measuring circuit.

11.6.5 Computational circuit

The genetically evolved computational circuit for the square root from generation 57 (Fig. 11.14), achieves a fitness of 1.19, and has 38 transistors, seven diodes, no capacitors, and 18 resistors (in addition to the source and load resistors in the embryo). The output voltages produced by this best-of-run circuit are almost exactly the required values.

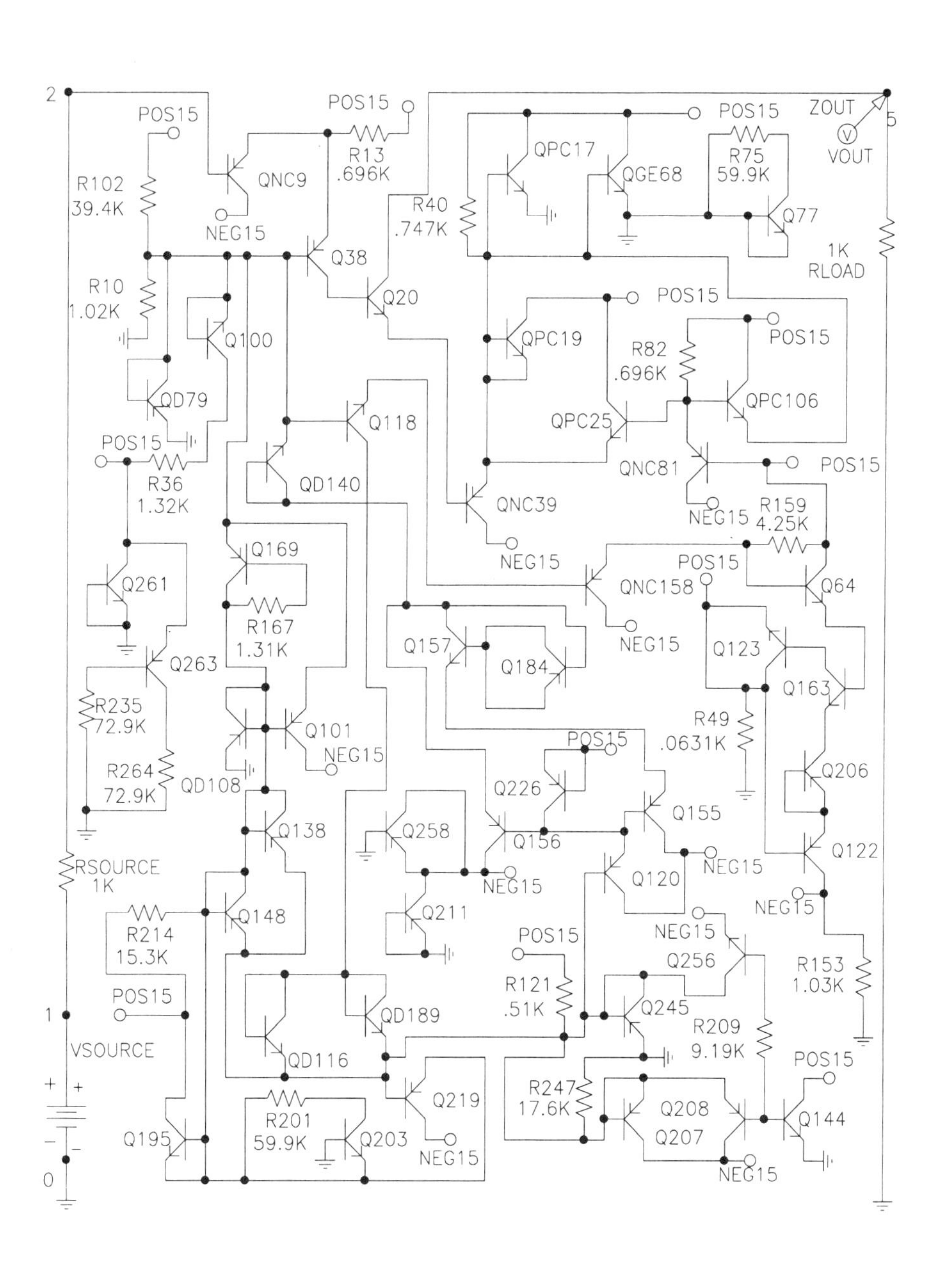

Fig. 11.14 Evolved square root circuit.

11.6.6 Robot controller circuit

The best-of-run time-optimal robot controller circuit (Fig. 11.15) appeared in generation 31, scores 72 hits, and achieves a near-optimal fitness of 1.541 hours. In comparison, the optimal value of fitness for this problem is known to be 1.518 hours. This best-of-run circuit has 10 transistors and 4 resistors. The program has one automatically defined function that is called twice (incorporated into the figure).

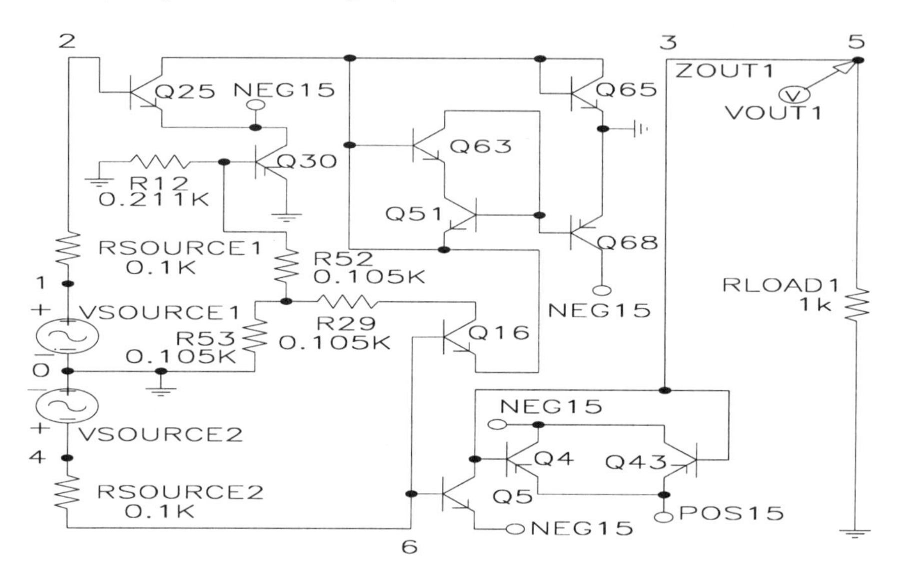

Fig. 11.15 Evolved robot controller.

This problem entails navigating a robot to a destination in minimum time, so its fitness measure (Section 11.5.4.6) is expressed in terms of elapsed time. The fitness measure is a high-level description of "what needs to be done" – namely, get the robot to the destination in a time-optimal way. However, the fitness measure does not specify "how to do it." In particular, the fitness measure conveys no hint about the critical (and counterintuitive) tactic needed to minimize elapsed time in time-optimal control problem – namely, that it is sometimes necessary to veer away from the destination in order to reach it in minimal time. Nonetheless, the evolved time-optimal robot controller embodies this counterintuitive tactic. For example, Fig. 11.16 shows the trajectory for the fitness case where the destination is (0.409, –0.892). Correct time-optimal handling of this difficult destination point requires a trajectory that begins by veering away from the destination (thereby increasing the distance to the destination) followed by a circular trajectory to the destination. The small circle in the figure represents the capture radius of 0.28 meters around the destination point.

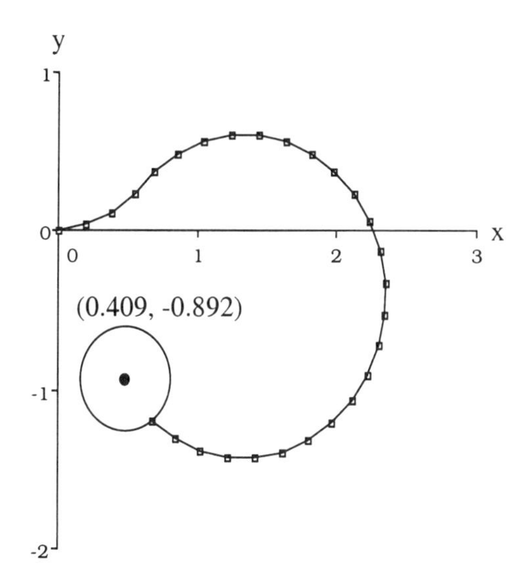

Fig. 11.16 Evolved time-optimal trajectory to destination point (0.409, –0.892).

The evolved time-optimal robot controller generalizes so as to correctly handle all other possible destinations in the plane.

11.6.7 60 dB amplifier

The best circuit from generation 109 (Fig. 11.17) achieves a fitness of 0.178. Based on a DC sweep, the amplification is 60 dB here (i.e., 1,000-to-1 ratio) and the bias is 0.2 volt. Based on a transient analysis at 1,000 Hz, the amplification is 59.7 dB; the bias is 0.18 volts; and the distortion is very low (0.17%). Based on an AC sweep, the amplification at 1,000 Hz is 59.7 dB; the flatband gain is 60 dB; and the 3 dB bandwidth is 79,333 Hz. Thus, a high-gain amplifier with low distortion and acceptable bias has been evolved.

11.7 OTHER CIRCUITS

Numerous other analog electrical circuits have been designed using the techniques described herein, including a difficult-to-design asymmetric bandpass filter, a crossover filter, a double passband filter, other amplifiers, a temperature-sensing circuit, and a voltage reference circuit [23]. Eleven of the circuit described in [23] are subjects of U. S. patents.

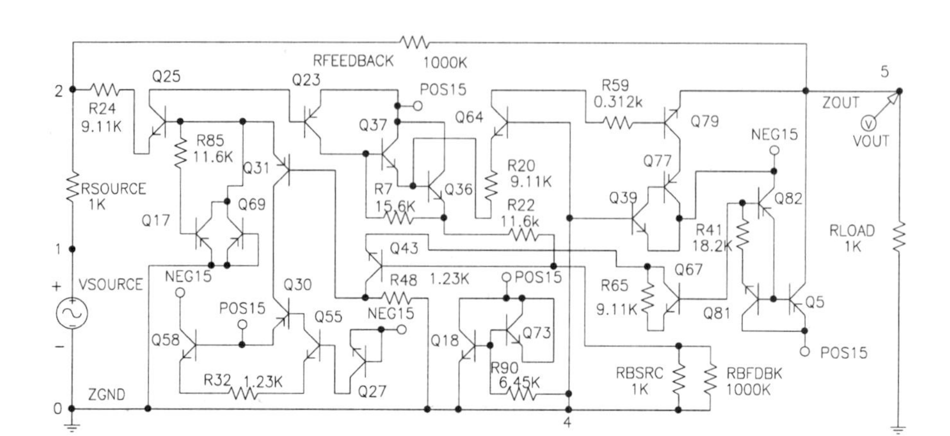

Fig. 11.17 Genetically evolved amplifier.

11.8 CONCLUSION

The design (synthesis) of an analog electrical circuit entails the creation of both the topology and sizing (numerical values) of all of the circuit's components. There has previously been no general automated technique for automatically creating the design for an analog electrical circuit from a high-level statement of the circuit's desired behavior. We have demonstrated how genetic programming can be used to automate the design of seven prototypical analog circuits, including a lowpass filter, a highpass filter, a passband filter, a bandpass filter, a frequency-measuring circuit, a 60 dB amplifier, a differential amplifier, a computational circuit for the square root function, and a time-optimal robot controller circuit. All seven of these genetically evolved circuits constitute instances of an evolutionary computation technique solving a problem that is usually thought to require human intelligence. The approach described herein can be directly applied to many other problems of analog circuit synthesis.

REFERENCES

1. Aaserud, O. and Nielsen, I. Ring. 1995. Trends in current analog design: A panel debate. *Analog Integrated Circuits and Signal Processing*. 7(1) 5-9.

2. Andre, David and Koza, John R. 1996. Parallel genetic programming: A scalable implementation using the transputer architecture. In Angeline, P. J. and Kinnear, K. E. Jr. (editors). 1996. *Advances in Genetic Programming 2*. Cambridge: MIT Press.

3. Angeline, Peter J. and Kinnear, Kenneth E. Jr. (editors). 1996. *Advances in Genetic Programming 2*. Cambridge, MA: The MIT Press.

4. Banzhaf, Wolfgang, Nordin, Peter, Keller, Robert E., and Francone, Frank D. 1998. *Genetic Programming – An Introduction*. San Francisco, CA: Morgan Kaufmann and Heidelberg: dpunkt.

5. Banzhaf, Wolfgang, Poli, Riccardo, Schoenauer, Marc, and Fogarty, Terence C. 1998. *Genetic Programming: First European Workshop. EuroGP'98. Paris, France, April 1998 Proceedings. Paris, France. April 1998*. Lecture Notes in Computer Science. Volume 1391. Berlin, Germany: Springer-Verlag.

6. Bennett III, Forrest H, Koza, John R., Andre, David, and Keane, Martin A. 1996. Evolution of a 60 Decibel op amp using genetic programming. In Higuchi, Tetsuya, Iwata, Masaya, and Lui, Weixin (editors). *Proceedings of International Conference on Evolvable Systems: From Biology to Hardware (ICES-96)*. Lecture Notes in Computer Science, Volume 1259. Berlin: Springer-Verlag. Pages 455-469.

7. Campbell, George A. 1917. *Electric Wave Filter*. Filed July 15, 1915. U. S. Patent 1,227,113. Issued May 22, 1917.

8. Cauer, Wilhelm. 1934. *Artificial Network*. U. S. Patent 1,958,742. Filed June 8, 1928 in Germany. Filed December 1, 1930 in United States. Issued May 15, 1934.

9. Cauer, Wilhelm. 1935. *Electric Wave Filter*. U. S. Patent 1,989,545. Filed June 8, 1928. Filed December 6, 1930 in United States. Issued January 29, 1935.

10. Cauer, Wilhelm. 1936. *Unsymmetrical Electric Wave Filter*. Filed November 10, 1932 in Germany. Filed November 23, 1933 in United States. Issued July 21, 1936.

11. Grimbleby, J. B. 1995. Automatic analogue network synthesis using genetic algorithms. *Proceedings of the First International Conference on Genetic Algorithms in Engineering Systems: Innovations and Applications*. London: Institution of Electrical Engineers. Pages 53–58.

12. Gruau, Frederic. 1992. *Cellular Encoding of Genetic Neural Networks*. Technical report 92-21. Laboratoire de l'Informatique du Parallélisme. Ecole Normale Supérieure de Lyon. May 1992.

13. Holland, John H. 1975. *Adaptation in Natural and Artificial Systems*. Ann Arbor, MI: University of Michigan Press.

14. Johnson, Kenneth S. 1926. *Electric-Wave Transmission*. Filed March 9, 1923. U. S. Patent 1,611,916. Issued December 28, 1926.

15. Johnson, Walter C. 1950. *Transmission Lines and Networks*. New York: NY: McGraw-Hill.

16. Kinnear, Kenneth E. Jr. (editor). 1994. *Advances in Genetic Programming*. Cambridge, MA: The MIT Press.

17. Kitano, Hiroaki. 1990. Designing neural networks using genetic algorithms with graph generation system. *Complex Systems*. 4(1990) 461–476.

18. Koza, John R. 1992. *Genetic Programming: On the Programming of Computers by Means of Natural Selection*. Cambridge, MA: MIT Press.

19. Koza, John R. 1994a. *Genetic Programming II: Automatic Discovery of Reusable Programs.* Cambridge, MA: MIT Press.

20. Koza, John R. 1994b. *Genetic Programming II Videotape: The Next Generation.* Cambridge, MA: MIT Press.

21. Koza, John R. 1995. Evolving the architecture of a multi-part program in genetic programming using architecture-altering operations. In McDonnell, John R., Reynolds, Robert G., and Fogel, David B. (editors). 1995. *Evolutionary Programming IV: Proceedings of the Fourth Annual Conference on Evolutionary Programming.* Cambridge, MA: The MIT Press. Pages 695–717.

22. Koza, John R., Banzhaf, Wolfgang, Chellapilla, Kumar, Deb, Kalyanmoy, Dorigo, Marco, Fogel, David B., Garzon, Max H., Goldberg, David E., Iba, Hitoshi, and Riolo, Rick L. (editors). *Genetic Programming 1998: Proceedings of the Third Annual Conference, July 22-25, 1998, University of Wisconsin, Madison, Wisconsin.* San Francisco, CA: Morgan Kaufmann.

23. Koza, John R., Bennett III, Forrest H, Andre, David, and Keane, Martin A. 1999. *Genetic Programming III: Darwinian Invention and Problem Solving.* San Francisco, CA: Morgan Kaufmann.

24. Koza, John R., Deb, Kalyanmoy, Dorigo, Marco, Fogel, David B., Garzon, Max, Iba, Hitoshi, and Riolo, Rick L. (editors). 1997. *Genetic Programming 1997: Proceedings of the Second Annual Conference* San Francisco, CA: Morgan Kaufmann.

25. Koza, John R., Goldberg, David E., Fogel, David B., and Riolo, Rick L. (editors). 1996. *Genetic Programming 1996: Proceedings of the First Annual Conference.* Cambridge, MA: The MIT Press.

26. Koza, John R., and Rice, James P. 1992. *Genetic Programming: The Movie.* Cambridge, MA: MIT Press.

27. Kruiskamp Marinum Wilhelmus and Leenaerts, Domine. 1995. DARWIN: CMOS opamp synthesis by means of a genetic algorithm. *Proceedings of the 32nd Design Automation Conference.* New York, NY: Association for Computing Machinery. Pages 433–438.

28. Ohno, Susumu. *Evolution by Gene Duplication.* New York: Springer-Verlag 1970.

29. Quarles, Thomas, Newton, A. R., Pederson, D. O., and Sangiovanni-Vincentelli, A. 1994. *SPICE 3 Version 3F5 User's Manual.* Department of Electrical Engineering and Computer Science, University of California, Berkeley, CA. March 1994.

30. Rutenbar, R. A. 1993. Analog design automation: Where are we? Where are we going? *Proceedings of the l5th IEEE CICC.* New York: IEEE. 13.1.1-13.1.8.

31. Samuel, Arthur L. 1983. AI: Where it has been and where it is going. *Proceedings of the Eighth International Joint Conference on Artificial Intelligence.* Los Altos, CA: Morgan Kaufmann. Pages 1152 – 1157.

32. Spector, Lee, Langdon, William B., O'Reilly, Una-May, and Angeline, Peter (editors). 1999. *Advances in Genetic Programming 3.* Cambridge, MA: The MIT Press.

33. Sterling, Thomas L., Salmon, John, and Becker, Donald J., and Savarese. 1999. *How to Build a Beowulf: A Guide to Implementation and Application of PC Clusters*. Cambridge, MA: MIT Press.

34. Thompson, Adrian. 1996. Silicon evolution. In Koza, John R., Goldberg, David E., Fogel, David B., and Riolo, Rick L. (editors). 1996. *Genetic Programming 1996: Proceedings of the First Annual Conference*. Cambridge, MA: MIT Press.

35. Williams, Arthur B. and Taylor, Fred J. 1995. *Electronic Filter Design Handbook*. Third Edition. New York, NY: McGraw-Hill.

Part II

Application-oriented approaches

12 Multidisciplinary Hybrid Constrained GA Optimization

G.S. DULIKRAVICH, T.J. MARTIN, B.H. DENNIS and N.F. FOSTER

Department of Aerospace Engineering, 233 Hammond Building
The Pennsylvania State University, University Park, PA 16802, USA

12.1 INTRODUCTION

Realistic engineering problems always involve interaction of several disciplines like fluid dynamics, heat transfer, elasticity, electromagnetism, dynamics, etc. Thus, realistic problems are always multidisciplinary and the geometric space is typically arbitrarily shaped and three-dimensional. Each of the individual disciplines is governed by its own system of governing partial differential equations or integral equations of different degree of non-linearity and based on often widely disparate time scales and length scales. All of these factors make a typical multidisciplinary optimization problem highly non-linear and interconnected. Consequently, an objective function space for a typical multidisciplinary problem could be expected to have a number of local minimums. A typical multidisciplinary optimization problem therefore requires the use of optimization algorithms that can either avoid the local minimums or escape from the local minimums. Non-gradient based optimizers have these capabilities. On the other hand, once the neighborhood of the global minimum has been found, the non-gradient based optimizers have difficulty converging to the global minimum. For this purpose, it is more appropriate to use gradient-based optimizers.

Addition of constraints of both equality and inequality type to a typical multidisciplinary optimization problem reduces significantly the feasible domain of the objective function space. To find such often-small feasible function space, the optimizer should be able to search as large portion of the objective function space as possible. Again, non-gradient based optimizers are capable of performing this task. When constraints of equality type are to be enforced, the gradient-based optimizers can perform this task very accurately.

One of the primary concerns of any optimization algorithm is the computational effort required to achieve convergence. Except in the case of certain sensitivity based optimization algorithms and genetic algorithms with extremely large populations, the computer memory is not an issue. Typical constrained optimization problems in engineering require large number of objective function evaluations. Each function evaluation involves a very time-consuming computational analysis of the physical processes involved. The real issue is the reduction of the overall computing time required to perform the optimization. Therefore, an efficient optimization algorithm should require the least number of objective function evaluations and should utilize the most efficient disciplinary (or multidisciplinary) analysis algorithms for the objective function evaluations.

This suggests that it might be beneficial to utilize several different optimization algorithms during different phases of the overall optimization process. The use of several optimization algorithms can be viewed as a backup strategy (Dulikravich, 1997) so that, if one optimization method fails, another optimization algorithm can automatically take over. This strategy of using a hybrid optimization approach and performing the switching among the optimizers will be demonstrated and discussed in this text.

In addition, it might be beneficial to perform disciplinary (or multidisciplinary) analysis by using simplified or surrogate physical models during the initial stages of optimization. Because of their reduced degree of non-linearity, such models typically require significantly less computing time to evaluate than the full non-linear model of the physical process. This strategy of using progressively more accurate (and computationally expensive) objective function evaluation models will be demonstrated and discussed.

Finally, the flexibility of the parameterization of the design variable space (for example, parameterization of geometry in the case of shape optimization) can affect the convergence rate of the optimization process and the quality of the final result. The effects of the degree of inherent flexibility of the discretization algorithm on the optimization convergence rate will be discussed in this text.

12.2 HYBRID CONSTRAINED OPTIMIZATION

Various optimization algorithms have been known to provide faster convergence over others depending upon the size and shape of the mathematical design space, the nature of the constraints, and where they are during the optimization process. This is why we created a hybrid constrained optimization software. Our hybrid optimizer incorporates four of the most popular optimization modules; the Davidon-Fletcher-Powell (DFP) (Davidon, 1959; Fletcher and Powell, 1963) gradient search method, a genetic algorithm (GA) (Goldberg, 1989), the Nelder-Mead (NM) (Nelder and Mead, 1965) simplex method, and simulated annealing (SA) (Press et al., 1986) algorithm. Each algorithm provides a unique approach to optimization with varying degrees of convergence, reliability, and robustness at different stages during the iterative optimization process. A set of rules and heuristics were coded into the program to switch back and forth among the different optimization algorithms as the process proceeded. These rules will be discussed in this text.

The evolutionary hybrid optimizer handled the existence of equality and inequality constraint functions in three ways: Rosen's projection method, feasible searching, and random design generation. Rosen's projection method (Rao, 1996) provided search directions that guided the descent direction tangent to active constraint boundaries. In the feasible search (Foster and Dulikravich, 1997), designs that violated constraints were automatically restored to feasibility via the minimization of the active global constraint functions. If at any time this constraint minimization failed, random designs were generated about the current design until a new feasible design was reached.

Gradients of the objective and constraint functions with respect to the design variables, also called design sensitivities, were calculated using either finite differencing formulas, or by the much more efficient method of implicit differentiation of the governing equations

(Hafka and Malkus, 1991). The population matrix was updated every iteration with new designs and ranked according to the value of the objective function. During the optimization process, local minimums can occur and halt the process before achieving an optimal solution. In order to overcome such a situation, a simple technique has been devised (Dulikravich and Martin, 1994; 1996). Whenever the optimization stalls, the formulation of the objective function is automatically switched between two or more functions that can have a similar purpose. The new objective function provides a departure from the local minimums and further convergence towards the global minimum.

As the optimization process converges, the population evolves towards the global minimum. The optimization problem was completed when one of several stopping criterion was met; (1) the maximum number of iterations or objective function evaluations was exceeded, (2) the best design in the population was equivalent to a target design, or (3) the optimization program tried all four algorithms but failed to produce a non-negligible decrease in the objective function. The latter criterion was the primary qualification of g is convergence and it usually indicated that a global minimum had been found.

Following is a brief discussion of the most important features of each optimization module that was used in our hybrid constrained optimizer.

12.2.1 Gradient search algorithm

Optimizers based on a gradient search concept require that the negative gradient of a scalar function in the design variable space be multiplied by the optimum line search step size, α^* before adding it to the vector of design variables, $\vec{V}$. Unfortunately, the search direction is only first-order accurate and it is slow in minimizing the objective function, especially near a local or global minimum. Also, this method alone does not support logic to ensure that the constraints are not violated. The simplest way to introduce this capability would be to ensure that the line search does not extend into an infeasible region. Such a technique is very simple, but it can often stall at a stationary point located on a constraint boundary before reaching the global minimum.

The following several sections describe how one can improve convergence and handle constrained gradient-based optimization problems.

12.2.1.1 Improving convergence of gradient based optimizers The DFP algorithm uses quasi-Newtonian or rank-two updates and yields second-order accuracy without excessive calling of the analysis program. The DFP is the most computationally expensive method in our hybrid optimization algorithm, but its convergence is both maximized and guaranteed. This algorithm is susceptible to local minimums and it can get stuck on a constraint boundary. Depending on how this procedure stalls, the hybrid optimizer switches to another optimization routine.

The optimal line search parameter, α^*, is that value that minimizes the objective function along the line search direction. Many one-dimensional minimization algorithms have been published for the purposes of line searching (Golden Search, Fibonacci, quadratic interpolation,etc.). practically all of these line search techniques have difficulties with the constrained non-linear optimization when the objective function variation along the line search direction can have multiple local minimums.

Therefore, a more effective method has been developed for this sub-optimization procedure (Dulikravich and Martin, 1994). A small number (five to ten) of design points are generated between the current design ($\alpha = 0$) and the end of the line search in the searching direction, $\vec{s}$. The end of the line search is determined by calculating the distance to the closest design variable bound in the parametric space. The objective (or constraint) function is evaluated at each of these points. Then, a spline is passed through these control points and interpolated at a large number (for example, 1000) equidistantly spaced points. By sorting these interpolated values, the minimum of this new curve is found and the corresponding α^* is determined. The minimum of the interpolated spline is, in general, different from the true objective function minimum. Therefore, the newly found point, (α^*, F^*), is added to the original set of design points, a new spline curve is fitted and interpolated through this enlarged set, the minimum of this new curve is found, and the corresponding α^* is determined. The process is repeated several times until the new value of α^* is very close to the previous one indicating that the optimal α^* has been found in the particular search direction.

12.2.1.2 ***Enforcement of constraints*** A common method of dealing with constraints is to use a penalty function. Penalty functions are added to the scalar objective function with appropriate weighting factors for scaling of the constraint functions with respect to the objective function value. The use of penalty functions is highly discouraged. Not only do they waste the computing time on evaluating the objective function for infeasible designs, they also change the nature of the design space, often converging to minimums that are nowhere near the global minimum. Penalty methods were not used in the constrained hybrid optimization system because other, more effective procedures were implemented (Rosen's projection method and feasible search).

Rosen's projection method is a constraining approach that was designed to handle the existence of inequality and equality constraint functions (Rao, 1996). It is based on the idea of projecting the search direction into the subspace tangent to any active constraints. After the new search direction has been determined, any standard line search can be performed to update the design. After a line search has been employed and the design updated, the resulting design may become infeasible. A restoration move is then required from the new design point, back to the constraint boundary at the point. The effect of this restoration is to reduce all active constraint functions to zero.

Rosen's restoration is valid only for simple problems that have a local linear behavior. This formula tends to become unstable when nonlinear optimization problems are attempted. In these cases and for the multidisciplinary optimization problems, the feasible search method is highly recommended to restore an infeasible design back to the boundary of the feasible domain. The simplest way to accomplish this task is to employ a sub-optimization problem that minimizes the sum of active constraint functions. Notice that the equality constraints are always active. The evolutionary hybrid optimization algorithm uses DFP method to minimize this function.

12.2.2 Genetic algorithm

The GA's evolutionary approach utilizes its random nature to escape local minimums. When the average cost function of the new generation is not improved, the GA becomes an inefficient optimizer. This most often occurs when its random nature is prevalent, producing several bad and infeasible designs. The GA will be switched to the NM algorithm because the NM works efficiently upon these worst designs. When the variance in the objective function values of the population are very small, the population begins to contract around a possible global minimum. At this point, the optimization switches to the DFP gradient-based algorithm because DFP has the ability to quickly zoom in on that minimum. The GA develops new population members with each iteration, but only those members whose fitness is higher than that of the worst member will be allowed to enter the population.

The GA can handle constraints on the design variable bounds, but it is inherently unable to handle constraint functions. The new set of designs may not, in general, be feasible. Therefore, the feasibility of each generated design is checked and, if any constraints are violated, a feasible search is performed on the new design. If the feasible search fails, a new design is generated randomly about the best design in the population until a satisfactory one is found. Random designs were generated using a Gaussian shape probability density cloud centered on the current design, $\vec{V}^0$, with a randomly generated number $0 < R < 1$.

$$V_i = V_i^0 \pm \sqrt{-2\sigma^2 \left(V_{i,max} - V_{i,min}\right) \ln R}$$

The non-dimensional variance, σ^2, in this function was determined by the conditions of the optimization process. For example, the user specifies the maximum variance in the input of the optimization algorithm.

12.2.3 Nelder-Mead simplex algorithm

For high-dimensional problems, it is known that the sequential simplex-type algorithms are more efficient and robust than gradient based algorithms in minimizing classical unconstrained test functions. The NM method is a zeroth order method that utilizes a simplex generated by the population of previously generated designs. The NM begins by defining a group of solution sets, which, when mapped, form a geometric figure in the N_{var}-dimensional design space, called a simplex. The simplex does not need to be geometrically regular, so long as the distribution of the vertices remains fairly balanced. The NM then becomes a downhill method by obtaining a search direction which points from the worst design in the population through the centroid of the best designs. This algorithm is very easy to program and it has the least amount of objective function evaluations per optimization cycle. The existing population matrix of previously generated feasible designs makes the NM even cheaper to employ. It improves only the worst design in the population with each iteration.

The population of feasible designs, which has been ranked in ascending order according to its objective function values, is utilized such that the centroid of the best designs (omitting the worst design in the population) is computed

$$\vec{V}_{mean} = \frac{1}{N_{var}} \sum_{j=1}^{N\,var} \vec{V}^j$$

A search direction can now be defined as the vector from the worst design to the centroid of the remaining best designs

$$\vec{S} = \frac{\vec{V} - \vec{V}_{N\,var}}{\left|\vec{V}_{mean} - \vec{V}_{N\,var}\right|}$$

Once this search direction is computed, it is projected into the subspace tangent to the active constraints using Rosen's projection method, and then a line search is employed. The initial guess step size of the line search should be set to the average design variable range in the current population. If the line search succeeds, the new design may not be feasible. Then, a restoration move is employed. Eventually, a new feasible design will be obtained that should improve the worst design in the population.

12.2.4 Simulated annealing algorithm

The SA method is analogous with thermodynamic annealing. As a liquid is slowly cooled, thermal mobility of the molecules is slowly lost so that the atoms are able to line themselves up and form a pure crystal without defects. A pure crystal is the state of minimum energy for this system. SA provides a slow reduction in its random searching capabilities that it uses to produce search directions. The continuous minimization SA algorithm uses a modification to the downhill NM simplex method. This SA wanders freely through the local minimum neighborhood so it is used when the optimization slows down or stalls. Unfortunately, it can worsen the objective function in the later optimization cycles. The SA is suitable for large scale optimization problems, especially when the desired global minimum is hidden among many, poorer local extrema. The nature of the SA lends itself to the early optimization cycles. It is also useful for escaping from a local minimum after the DFP algorithm gets stuck in a local minimum. The basic idea of continuous minimization of an objective function using SA is to find an appropriate design variable change, like that of a steepest descent or downhill simplex method. The NM algorithm is used to obtain a search direction.

Logarithmically distributed random numbers, proportional to the cooling temperature, $\tilde{T}$, are added to the function values at the vertices of the simplex.

$$\tilde{F} = F - \tilde{T}\ln(R)$$

Here, R is a random number between 0.0 and 1.0. Another random variable is subtracted from the objective function value of every new point that is tried as a replacement design. The simplex will expand to a size that can be reached at this temperature and then executes a stochastic, tumbling Brownian motion within that region. The efficiency with which the region is explored is independent of its narrowness or aspect ratio. If the temperature is reduced slowly enough, it is likely that the simplex will contract about the lowest minimum encountered.

The cooling scheme that was employed in the hybrid optimizer reduced the temperature $\tilde{T}$ to $(1\text{-}\gamma)\tilde{T}$ every $\tilde{K}$ optimization cycles, where the optimum γ and $\tilde{K}$ are determined empirically for each problem. When the cooling temperature has reduced to a significantly small value, the program switches to a different optimizer.

12.3 AUTOMATED SWITCHING AMONG THE OPTIMIZERS

A set of rules has been added to the hybrid constrained optimization system in order to make the switching among the algorithms automatic, as well as to utilize some of the heuristic understanding of each algorithm's behavior. The purpose of this switching was to increase the hybrid optimizer's robustness and improve upon its convergence. Each rule was based upon and incorporated with the unique behavior of each numerical optimization algorithm. The timing of the switching among the algorithms was forced to occur during those instances in which the particular algorithm performed badly, stalled, or failed. The first algorithm that the optimization process chose to switch to was determined by reasoning and by trial and error. If the subsequent algorithm also failed in its processes, the opportunity was made for every one of the algorithms in the system to have a try at the problem. When all the available algorithms had been tried on a particular population of feasible designs, and all failed at their given tasks, then the program was terminated. The rules for switching will now be discussed for each algorithm in the hybrid optimization system. Figure 12.1 demonstrates the major characteristics of this switching process in flowchart form.

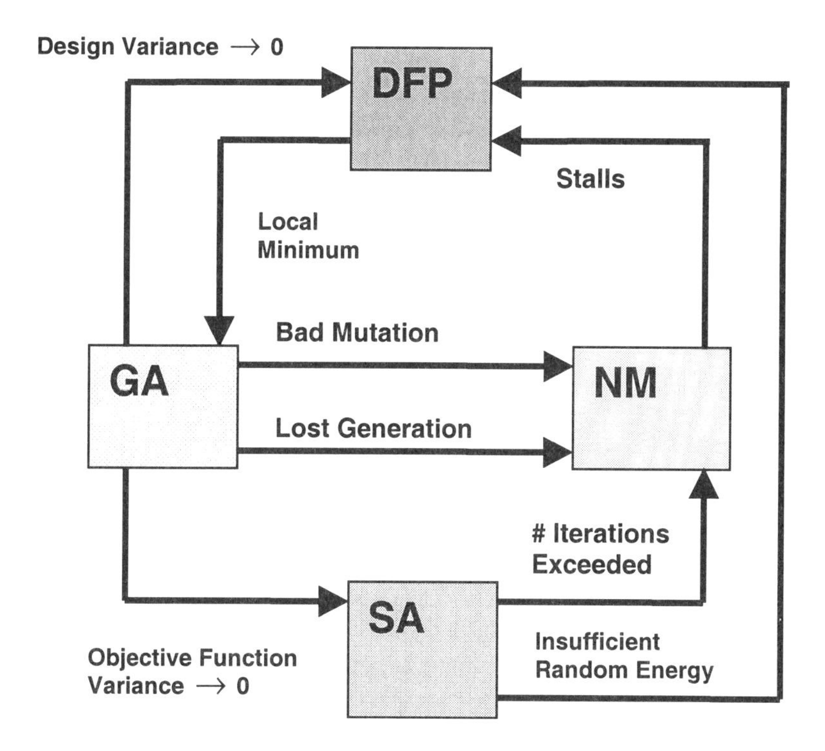

Figure 12.1 Flowchart of automatic switching among modules in a hybrid optimizer.

12.3.1 Rules for switching optimization modules

12.3.1.1 Local minimum rule This rule has been developed for gradient-based methods. The gradient search optimization algorithm is switched whenever the change in the objective is less than a user-specified tolerance, $\overline{\Delta F}$.

$$\frac{|F^* - F_K|}{|F^* - F_0|} < \overline{\Delta F}.$$

Here, F* is the value of the objective function at the new optimum, F_0 is the value of the objective function for the design at the start of the gradient search, and F_K is the value of the objective function from the previous optimization cycle. The GA is the first algorithm chosen, then comes the SA and finally the NM. The genetic algorithm is chosen first because its random searching capabilities are useful in escaping local minimums. This rule is also applicable whenever the program stalls on a constraint boundary.

12.3.1.2 Descent direction rule When the DFP search direction is not a descent direction, the dot product of the search direction, $\vec{S}$, and the gradient, ∇F, is greater than zero. The search direction may be different from the negative gradient direction because of the DFP update formula and because that search direction is projected onto the subspace of active constraints (Rosen's projection method).

$$\vec{S} \bullet \nabla F > 0$$

When this condition is met, the inverse Hessian matrix is re-initialized to the identity matrix. If the inverse Hessian is already equal to the identity due to re-initialization, the program is switched to simulated annealing (SA – NM – GA). The randomness and simplex searching methods of the simulated annealing process provide quick and effective ways of navigating through the irregular design spaces. New designs (optimum along the line search direction) created by the DFP are added to the population matrix and the DFP always works on the best member in the population matrix.

The GA has several criterions that can qualify its performance and so several switching rules have been developed. The most often used switching criterion is based upon the variance in the population. As the GA proceeds, it tends to select members in the population with like qualities to breed more often. The result is that the population tends to acquire a similar set of characteristics and the variation in the population reduces. This can happen too quickly when the specified mutation is too infrequent. In the ideal situation, the design variables of the population will tend to collect around the global minimum, but may have difficulty in finding it.

12.3.1.3 Design variance limit rule This is the first rule for switching from the GA. It is defined as

$$\sigma_V = \frac{1}{N_{VAR} N_{POP}} \sum_{i=1}^{N_{VAR}} \frac{\sqrt{\sum_{j=1}^{N_{POP}} \left(P_{i,j} - \overline{V_i}\right)^2}}{\left(V_{max,i} - V_{min,i}\right)} < \sigma_{V\,min}$$

In this equation, the non-dimensional standard deviation, σ_V, for all design variables in the population is measured with respect to the average design variable in the population.

$$\overline{V}_i = \frac{1}{N_{POP}} \sum_{j=1}^{N_{POP}} P_{i,j} \qquad \text{for } i = 1,\ldots,N_{VAR}$$

When the design variance in the population becomes less than the limit, the GA is switched to the DFP. The reasoning is that the population variance is contracting around a minimum and the DFP can be used to quickly home in on that minimum. The order of the switching is DFP – NM – SA.

12.3.1.4 Objective RMS limit rule This is similar to the aforementioned rule, except that the variance in the objective function is computed rather than the variance in the design variables.

$$\sigma_F = \frac{1}{N_{POP}} \frac{\sqrt{\sum_{j=1}^{N_{POP}} \left(F_j - \overline{F}\right)}}{\left(F^* - F^0\right)}$$

Here, the F_j is the objective function value of the jth population member and the average objective function of the population is computed from them.

$$\overline{F} = \frac{1}{N_{POP}} \sum_{j=1}^{N_{POP}} F_j$$

The difference between this rule and the design variance limit is that the population may be dispersed over a wide design space but each may have very similar objective function values. This can occur is the objective function space is a large flat area with little or no gradient. The program is then switched first to the SA method (SA – NM – DFP).

12.3.1.5 Bad mutation rule The average objective function value of the population is also used as a switching criterion of its own for the GA. The bad mutation rule causes the GA to switch to the NM if the average objective function increases from the previous optimization cycle with the GA. This will most likely occur if the random mutation rate is too large or if it produces one or more really bad designs. Since the NM specializes in bringing the poorest design within the centroid of the best designs, it is the most obvious first choice (NM – SA – DFP).

12.3.1.6 Lost generation rule The GA develops sequential populations of new 'child' designs that are entered into the population only if the population size is allowed to increase, or if the 'child' design is better than the worst member in the population. If no new designs are entered into the population, the GA fails. This lost generation causes the program to switch to the SA algorithm (SA – NM – DFP).

12.3.1.7 Stall rule The NM simplex searching algorithm has only one failure mode, stalled. The NM is said to stall whenever the line search (produced by the direction from the worst design in the population through the centroid of the best designs) fails to improve itself so that it is better than the second worst member in the population.

$$F^*_{N_{POP}} \geq F_{N_{POP}-1}$$

This rule causes the hybrid optimizer to switch to the DFP method (DFP – GA – SA). If the best design in the population was generated by the DFP, then the DFP is passed by and the GA takes over.

12.3.1.8 Insufficient random energy rule The purpose of the SA algorithm is to add energy into the optimization system and allow it to escape the local minimums. The objective functions of the design variables in the population might get worse. Also, the random energy added to the system allows for the modification of any member in the population. Therefore, the capabilities of the SA would be wasted if the optimization process was switched to some other algorithm in case when any one objective function value became worse. It was also found that the SA terminates prematurely when the worse average objective criterion was met. The SA has, therefore, been programmed to end whenever the cooling protocol did not add sufficient energy into the system. The insufficient random energy criterion can be stated as follows.

$$\sqrt{\sum_{j=1}^{Npop} \left(\tilde{F}_j - \bar{F}\right)^2} < \Delta F_{min}$$

Here, the algorithm ends whenever the variance of the objective function values with added random energy, $\tilde{F}$, are less than a user-specified limit. The program is switched from the SA to the DFP method (DFP – GA – NM). After several cycles, the random energy in the SA may have been reduced to a negligible amount, while the insufficient random energy criterion might not be met because of the large variance in the population. Therefore, the SA switches to the NM method (NM – GA – DFP) after $\tilde{K}_{max}$ optimization cycles.

12.4 AERO-THERMAL OPTIMIZATION OF A GAS TURBINE BLADE

Internal cooling schemes of modern turbojet and turbofan engines bleed air from the compressor and pass this air into the serpentine coolant flow passages within the turbine blades. The maximum temperature within a turbine blade must be maintained below a certain value in order to avoid thermal creep, melting, and oxidation problems. Increased coolant heat transfer and increased coolant flow rate directly decrease the amount of air delivered to the combustor and increase specific fuel consumption. Thus, the coolant mass flow rate and the coolant supply pressure should be minimized, while maintaining the inlet temperature of hot gases as high as possible and temperature in the blade material below a specified limit. These objectives can be met (Dulikravich and Martin, 1995; 1996; Martin and Dulikravich, 1997; Dulikravich et al. 1998; Martin et al., 1999) by the constrained optimization of the coolant passage shapes inside the turbine blade.

12.4.1 Geometry model of the turbine blade coating and coolant flow passages

The outer shape of the blade was assumed to be already defined by aerodynamic inverse shape design or optimization. It was kept fixed during the entire thermal optimization procedure. The thermal barrier coating thickness was described by a wall thickness function versus the airfoil contour following coordinate, s. The metal wall thickness variation around the blade was also defined by a piecewise-continuous beta-spline (Barsky, 1988). The number of coolant flow passages in the turbine blade was kept fixed.

The x-coordinates of the intersections of the centerlines of each of the internal struts with the outer turbine airfoil shape were defined as x_{Ssi} and x_{Spi}, for the suction and pressure sides of the blade, respectively. The range over which each strut could vary was specified. In addition to the coordinates of the strut intersections, the strut thickness, t_{Si}, and a filleting exponent on either the trailing or leading edge sides, e_{Sti} and e_{Sli}, respectively, were used to complete the geometric modeling of each strut (Figure 12.2a). The strut fillets were described by a super-elliptic function that varied from a circular fillet ($e_{Si} = 2$) to an almost sharp right angle ($e_{si} \rightarrow \infty$).

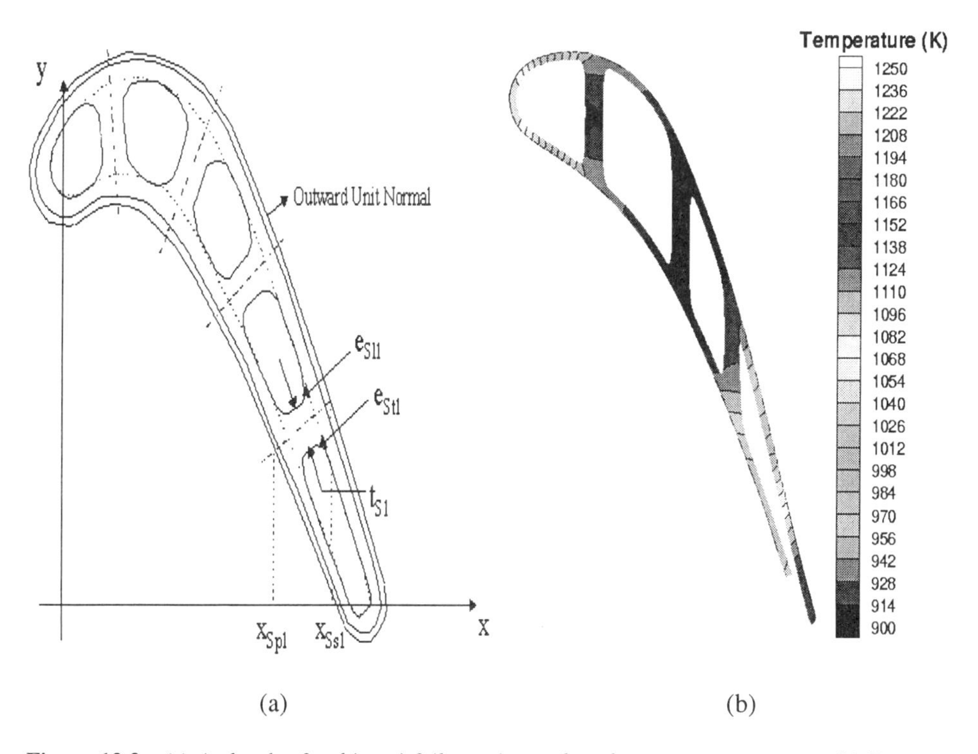

Figure 12.2 (a) A sketch of turbine airfoil, coating and coolant passage geometry. (b) Temperature field computed on the initial guess geometry used for the minimization of coolant temperature at the triling edge ejection location.

The boundary conditions and constraints were: hot gas pressure at the inlet to the blade row (588131 Pa), hot gas pressure at the exit of the blade row (134115 Pa), hot gas inlet Mach number (0.1772), initial guess for the coolant mass flow rate (0.025 kg/s), initial guess for the hot gas inlet temperature (1592.6 K), thermal conductivity of thermal barrier coating (1.0 W/m K), thermal conductivity of the blade metal (30.0 W/m K), thermal barrier coating thickness (100 microns), maximum allowable temperature in the blade material (1250.0 K).

Total number of design variables per section of a three-dimensional blade was 27. These variables were: eight beta-spline control points defining coolant passage wall thickness, six strut end-coordinates (two per strut), three strut thicknesses (one per strut), six strut filleting exponents (two per strut), four relative wall roughnesses (one per each coolant flow passage). Two additional global variables were: one mass flow rate, and one inlet turbine hot gas temperature. The initial guess geometry is depicted in Figure 12.2b.

12.4.2 Turbine cooling scheme optimization for minimum coolant ejection temperature

The uniform temperature and heat flux extrema objectives were not entirely successful (Dulikravich et al. 1998a; Martin et al., 1999) at producing internal blade geometry that would be considered economical or even physically realizable. The minimization of the integrated hot surface heat flux objective tried to eliminate the coolant passages (Martin and Dulikravich, 1997). The maximization of the integrated hot surface heat flux thinned the walls and produced an extreme range of temperatures that questioned the validity of the use of heat transfer coefficients on the outer surface of the blade.

Therefore, another objective was formulated (Martin et al., 1999) that minimizes the coolant temperature at the very end of the coolant flow passage (the ejection slot at the blade trailing edge). This is an indirectly formulated objective since mass flow rate was used as a design variable and could not be simultaneously used as the objective function. The reasoning was that reduced heat transfer coefficients require lower surface roughness on coolant passage walls therefore resulting in lower coolant supply pressure requirements. Thus, compressor bleed air can be extracted from lower compressor stages in which the air is cooler. This in turn should lead to the lower coolant mass flow rate requirement.

First, a turbulent compressible flow Navier-Stokes solver was used to predict the hot gas flow-field outside of the blade subject to specified realistic hot surface temperature distribution. As a byproduct, this analysis provides hot surface normal temperature gradients thus defining the hot surface convection heat transfer coefficient distribution. This and the guessed coolant bulk temperature and the coolant passage wall convection heat transfer coefficients creates boundary conditions for the steady temperature field prediction in the blade and thermal barrier coating materials using fast boundary element technique. The quasi-one-dimensional flow analysis (with heat addition and friction) of the coolant fluid dynamics was coupled to the detailed steady heat conduction analysis in the turbine blade material. By perturbing the design variables (especially the variables defining the internal blade geometry) the predicted thermal boundary conditions on the interior of the blade will be changing together with the coolant flow parameters. As the optimization algorithm ran, it also modified the turbine inlet temperature. Once the turbine inlet temperature changed significantly, the entire iterative procedure between the thermal field analysis in the blade material and the computational fluid dynamic analysis of the external hot gas flow-field was performed again to find a better estimate for thermal boundary conditions on the blade hot surface. This global coupling process was performed only a small number of times during the course of the entire optimization. This semi-conjugate optimization uses sectional two-dimensional blade hot flow-field analysis and a simple quasi one-dimensional coolant flow-field analysis. Consequently, it requires considerably less computing time than would be needed if a full three-dimensional hot gas flow-field and coolant flow-field analysis (Stephens and Shih, 1997) would be used.

Two different minimizations of the coolant ejection temperature were performed; one with the maximum temperature equality constraint, $T_{max} = \overline{T_{max}}$, and the other with the inequality constraint $T_{max} < \overline{T_{max}}$. The design sensitivity gradients were calculated using finite differencing and the optimization program was initiated with the DFP. The optimization with the equality constraint required 15 cycles and 794 objective function evaluations. The large number of function evaluations was needed because forward finite differencing was used to obtain the sensitivity gradients. After 4 optimization cycles, the program switched to the GA, switching finally to the NM in the 14th cycle.

The reduction in coolant bulk temperatures (Figure 12.3a) also significantly reduced the coolant pressure losses and the coolant convection heat transfer coefficients (Figure 12.3b). This suggests that it might be possible to remove the heat transfer enhancements (mechanical wall turbulators) such as trip strips and impingement schemes from the coolant passages thus leading to a substantial reduction in turbine blade manufacturing costs. The ultimate goal (reduction in the coolant mass flow rate) was achieved (Figure 12.4a) by reducing the heat transfer coefficients and by making the passage walls thinner (Figure 12.5a). It should be pointed out that the turbine inlet temperature changed very little when the maximum temperature equality constraint was enforced with this objective.

But, when the maximum temperature inequality constraint was enforced, the coolant mass flow rate was reduced even more dramatically (Figure 12.4b) but the turbine inlet temperature decreased from 1600 K down to 1340 K which is unacceptable. The final optimized configuration had extremely thin walls and struts (Figure 12.5b) which were nearly at the lower limits enforced by the bounds on their design variables. This configuration is clearly unacceptable because of the reasonable doubts that such a thin walled blade could sustain the mechanical stresses. It is interesting to note that the optimization with the maximum temperature inequality constraint ran for more cycles (40), but required fewer objective function evaluations (521). This was because of the fewer number of gradients needed of the inequality constraint functions. That is, the equality constraint was always active, but the inequality constraint was only active when the maximum temperature in the blade was equal to or greater than the target maximum temperature, $\overline{T_{max}}$.

It can be concluded that even a seemingly minor change in the constrained multidisciplinary (aero-thermal) optimization can have a profound influence on the acceptability of the optimized design. It can also be concluded that the hybrid constrained optimizer switching logic proved to be robust and efficient when solving a realistic multidisciplinary problem.

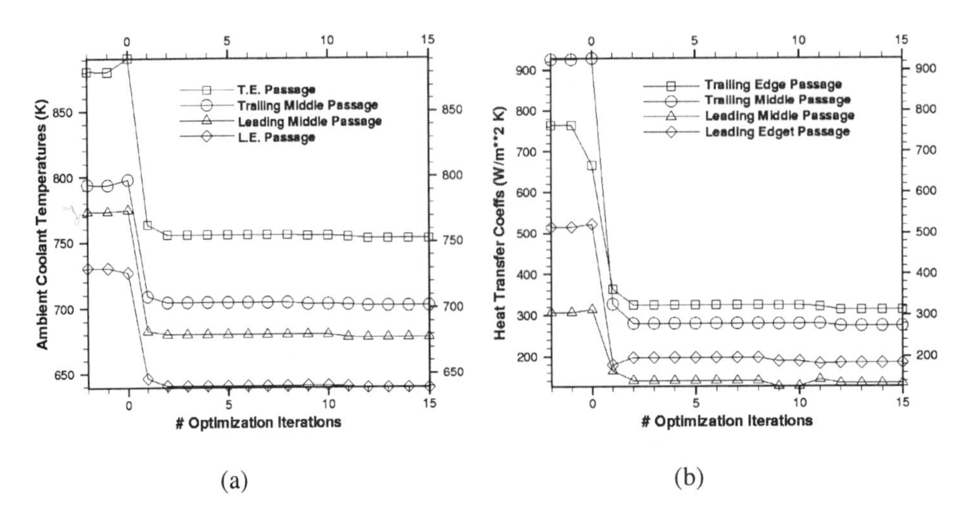

(a) (b)

Figure 12.3 Evolution of optimum bulk coolant temperatures (a) and the coolant passage wall heat transfer coefficients (b) in each of the four coolant flow passages for the minimization of coolant ejection temperature when using the maximum temperature equality constraint.

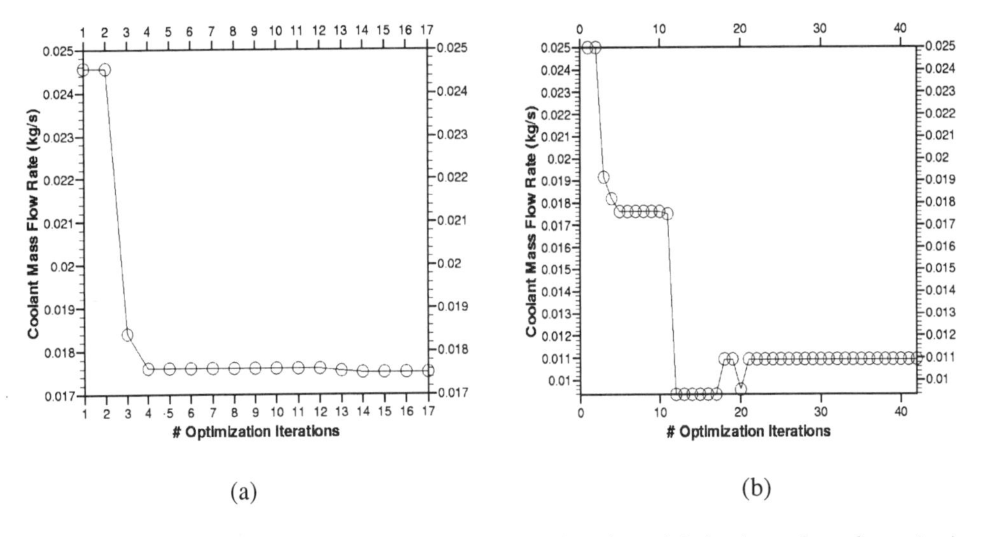

(a) (b)

Figure 12.4 Evolution of coolant mass flow rate for the minimization of coolant ejection temperature using: a) the maximum temperature equality constraint, and b) the maximum temperature inequality constraint.

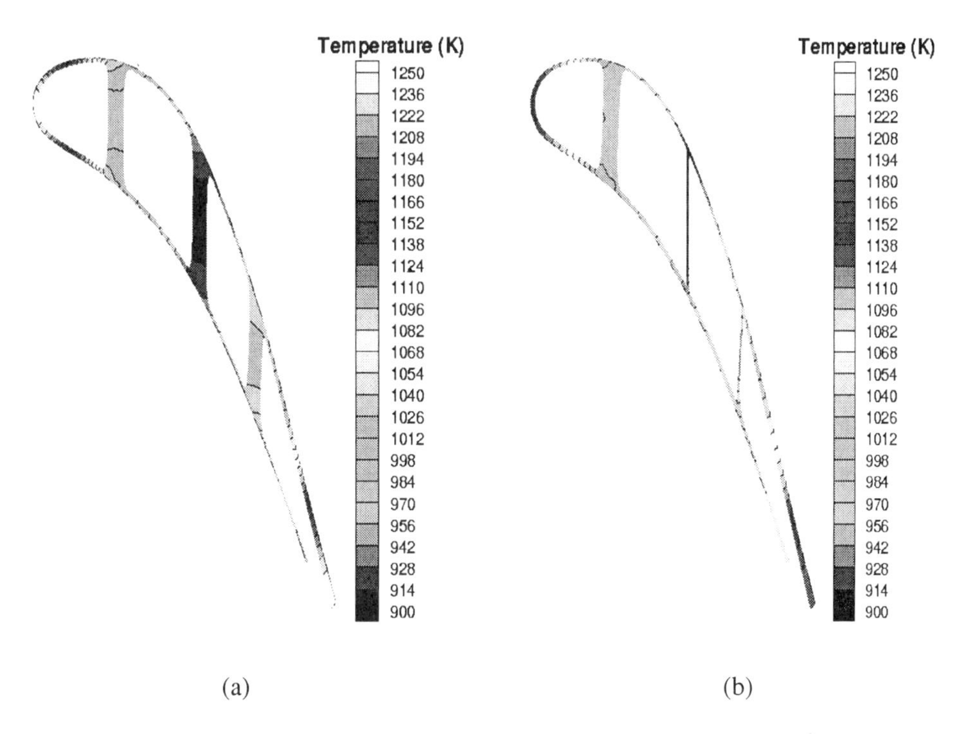

Figure 12.5 Optimized blade interior geometries and temperatures for minimized coolant ejection temperature design of an internally cooled turbine blade using; a) the maximum temperature equality constraint, and b) the maximum temperature inequality constraint.

12.5 HYPERSONIC SHAPE OPTIMIZATION

Two main objectives during shape optimization of hypersonic missile shapes are lowering of the acrodynamic drag and lowering of the aerodynamic surface heating. The aerodynamic analysis was performed using an extremely simple model known as modified Newtonian impact theory (MNIT). This is an algebraic equation which states that the local coefficient of aerodynamic pressure on a body surface is linearly proportional to the square of the sine of the angle between the body tangent at that point and the free stream velocity vector. The MNIT is known to give embarrassingly accurate predictions of integrated aerodynamic lift and drag forces for simple body shapes at hypersonic speeds. It is often used instead of more appropriate (and orders of magnitude more computationally expensive) physical models based on a system of non-linear partial differential equations known as Navier-Stokes equations. In the following examples, the MNIT was used for the flow field analyses. Two optimization techniques were included: the constrained DFP and the constrained GA technique, both using Rosen's projection methodology for improved equality constraint treatment.

12.5.1 Optimum ogive shaped missile

First, results from the program were verified against analytically known solutions. Specifically, the geometry of an axisymmetric body was optimized to reduce compression wave drag at zero angle of attack. Optimal bodies of revolution that minimize drag have previously been analytically determined. Two such solutions are known as the Von-Karman and Sears-Haack bodies (Ashley and Landahl, 1965). These two bodies yield the minimum wave drag under two different sets of constraints. The Von-Karman body assumes that the body terminates with a flat plane, that the base area in this plane is known, and that the total length of the body is specified. The Sears-Haack body assumes that the body is pointed at both ends, and that the total volume and length of the body are given. The constrained DFP optimizer was used to determine computationally the body of revolution that minimizes wave drag at Mach = 10 and an altitude of 18 km. Initially, the body was specified to be a 10-meter-long, 15-degree angle right-circular cone (Figure 12.6a). The design variables for this exercise were specified to be the radii of the body at 10 cross sections. Each design variable (the cross sectional radii) was allowed to vary from 0 to 10 meters. During the optimization process, the length was kept fixed and the total volume of the body was constrained (with an equality constraint) not to change by more than 1.0 cubic meter from its initial value of 75.185 cubic meters. The constrained optimization process converged to the 'bulged' axisymmetric body called an *ogive* (Figure 12.6b). The base area of the optimized body, and the total volume (fixed) were then used to compute Von-Karman and Sears-Haack bodies from analytical expressions (Anderson, 1989). The numerically optimized body was in excellent agreement with the analytically optimized body shapes (Foster and Dulikravich, 1997).

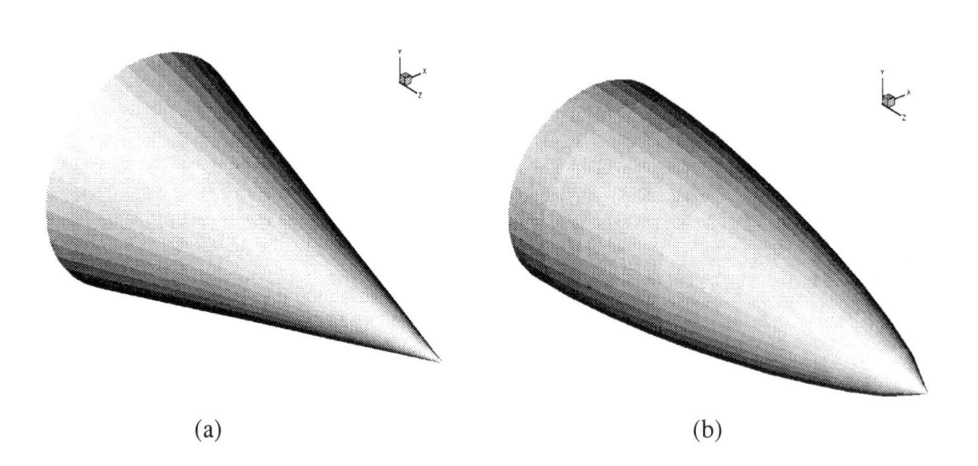

Figure 12.6 Initial cone configuration (a) and optimized ogive shape (b) at the onset of longitudinal 'ridges'. Constrained DFP optimizer was used with MNIT as the analysis code.

12.5.2 Optimum star shaped missile

All of the body surface nodes on the first cross section could move together radially and were controlled by one design variable. On the other five cross section planes, all of the 38 surface nodes had two degrees of freedom except for the two 'seam' points whose x-coordinate is zero (the points on the vertical plane of symmetry). These 'seam' points were allowed to move only vertically (in the y-direction) in their plane. Thus, there were 78 design variables per each of the five cross sections and one design variable (radius) at the sixth cross section giving a total of 391 design variable in this test case.

The constrained DFP algorithm could not converge to anything better than the smooth surface ogive shape which happened around 30th iteration. Therefore, the optimization was switched to a constrained GA algorithm. After the completion of the 45th design cycle which is towards the end of the 'leveled-off' portion of the convergence history (approximately from design cycle 25 to cycle 48), a slightly perturbed ogive geometry was obtained (Figure 12.7a). Due to the use of GA, the convergence after that point again dramatically increased leading to the deepening of the body surface 'ridges', or 'channels', and the narrowing of the spiked nose. It is understandable that these channels appeared because the flow solver that was used for the optimization process was the MNIT. According to the MNIT, the pressure on the surface of the body is solely a function of the local inclination angle with respect to the free stream. Therefore, the optimizer was able to reduce the pressure on all of the body surface panels (after the first cross section) by creating the star-shaped body (Figure 12.7b) which caused the outward normal vectors on each panel to rotate away from the free stream direction.

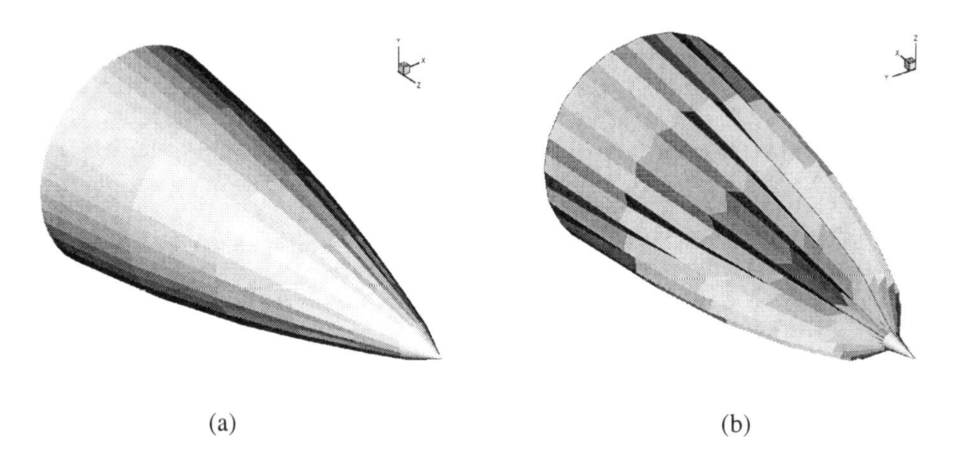

Figure 12.7 The star shaped hypersonic missile obtained when using a constrained GA optimizer and MNIT as a flow-field analysis code: a) front view, and b) axiometric view.

Since the use of MNIT in this shape optimization example causes every second point defining the body cross section to become a 'ridge', the number of optimized 'ridges' will always equal the half of the grid points defining the body cross section. This is strictly the result of the fact that an oversimplifying flow-field analysis model (MNIT) was used. It is interesting to note that star-shaped hypersonic projectile shapes have been studied experimentally for several decades, but the number of 'ridges' or 'fins' on these bodies was considerably smaller. That is, a more physically appropriate non-linear flow-field analysis code would have created a number of optimized 'ridges' that is considerably smaller than half the number of all the grid points defining a body cross section. This illustrates the possible dangers of using overly simplistic surrogate models for the evaluation of the objective function in the latter stages of the optimization process.

The final star cross section missile configuration obtained with the MNIT reduced the aerodynamic drag by 77 percent. When the optimized star missile shape was analyzed using the parabolized Navier-Stokes flow-field analysis code, the total drag of the optimized body was found to be 53 percent lower than the drag of the initial conical geometry. The difference between 77 and 53 percent in the reduction of the missile drag can be represented as a relative error of (77 – 53)/53 = 45.3 percent that is strictly due to the use of an overly simplistic surrogate model (MNIT) for the flow-field analysis. The shape optimization effort took 75 optimization cycles, and called the MNIT flow solver 60001 times. The execution took 4282 seconds on a Cray C-90 computer.

12.5.3 Maximizing aerodynamic lift/drag

Next, an initial 10-meter long 15-degree cone was optimized to maximize the lift-to-drag (L/D) ratio at zero angle of attack using the hybrid gradient optimizer and the MNIT flow solver. The shape was optimized holding its length and volume fixed. Six cross sectional planes with forty surface nodes described the geometry. Every surface node was allowed to vary only radially on its cross sectional plane thus creating a total of 240 design variables. The DFP optimizer was executed until convergence was reached. Execution terminated after 40 design iterations that consumed 1458 CPU seconds on a Cray C-90 and required 19961 objective function analyses. The converged optimal lift-to-drag ratio was L/D = 1.29. Figure 8a shows the final optimized body that is cambered and has ridges that have formed on its upper surface. The optimizer cambered the body so that a greater surface area on the underside faced the free stream so as to increase lift, and formed ridges on top of the body so that downward pressure was minimized. Yet the body still has an ogive appearance, which helps to reduce overall drag. At this point (after the DFP optimizer had converged), the optimization method was changed to the GA optimizer which continued to further reduce the cost function. The execution of the GA was stopped after it had performed an additional 52 design cycles that required only an additional 748 objective function analyses. The converged optimized lift-to-drag ratio was L/D = 1.54. The important fact learned from this case is that the constrained GA optimization technique was able to significantly reduce the cost function after the gradient based constrained DFP had converged and could go no further. Not only did the constrained GA optimizer reduce the cost function, but it did so while performing fewer objective function analyses.

The final optimized geometry is obviously aerodynamically nonsensical (Figure 12.8b). The reason why the hybrid genetic optimizer produced such a nonsensical-looking geometry is a consequence of the flow-field analysis method that was used. The MNIT flow solver is extremely simple and can yield inaccurate results for complex and undulating geometries such as the one in Figure 12.6. Specifically, MNIT cannot account for the viscous forces or the complex shock wave interactions that would occur in the flow-field around such a geometry. Therefore, the geometry depicted in Figure 12.6 is certainly not optimal in light of a correctly modeled flow.

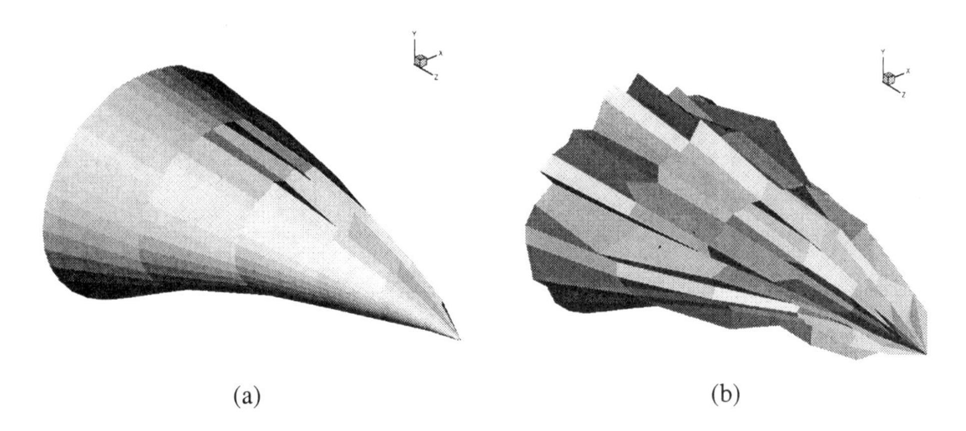

(a) (b)

Figure 12.8 Maximum lift/drag hypersonic body shapes obtained with MINT as the aerodynamic analysis code and: a) constrained DFP, and b) constrained GA.

12.6 AERODYNAMIC SHAPE OPTIMIZATION OF TURBOMACHINERY CASCADES

In the case of a turbomachinery aerodynamics, sources of entropy production other than viscous dissipation and heat transfer could be neglected. For a given set of inlet and exit flow boundary conditions, the amount of entropy generated in the blade row flow-field is determined by the shape of the turbomachinery blade row. Minimization of the entropy generation (flow losses) can therefore be achieved by the proper reshaping of the blade row. Besides the task of designing an entirely new turbomachine, designers frequently face a task of retrofitting an existing compressor or a turbine with a new, more efficient, rotor or a stator. This is a challenging task since it has a number of constraints. Specifically, the axial chord of the new row of blades must be the same or slightly smaller than the axial chord length of the original blade row. Otherwise, the new blade row will not be able to fit in the existing turbomachine. Inlet and exit flow angles must be the same in the redesigned blade row as in the original blade row or the velocity triangles will not match with the neighboring blade rows. Mass flow rate through the new blade row must be the same as through the original blade row or the entire machine will perform at an off-design mass flow rate which can lead to serious unsteady flow problems. Torque created on the new rotor blade row must be the same as on the old rotor blade row or the new rotor will rotate at the wrong angular speed. To make certain that the new blades will be structurally sound, the cross section area of the new blade should be the same or slightly larger than the cross section area of the original rotor blade. In case of a turbine blade, trailing edge thickness should not be smaller than a user specified value or it will overheat and burn.

This amounts to four very stringent equality constraints and one inequality constraint. An additional equality constraint is the gap-to-axial chord ratio that will be kept fixed. In this work we will keep the total number of blades fixed and will optimize only a two-dimensional planar cascade of airfoils subject to a steady transonic turbulent flow.

The method used for calculating the shape of an airfoil in a turbomachinery cascade was chosen to require only nine parameters. These variables include the tangential and axial chord, the inlet and exit half wedge angle, the inlet and outlet blade angle, the throat, unguided turning angle, and the leading and trailing edge radii (Pritchard, 1985; Dulikravich et al., 1998; Dulikravich, Martin, and Han, 1998).

A GA based optimizer was used to design a turbine airfoil cascade shape by varying these nine parameters. The objective of the optimization was to determine the airfoil shape that gives the minimum total pressure loss while conforming to the specified constraints of producing 22,000.0 N of lift, an average exit flow angle of –63.0 degrees, a mass flow rate of 15.50 kg s^{-1}, a cross-sectional area of 2.23E-3 m^2, and an axial chord of 0.1 m. The mathematical form of this objective function is:

$$F = P_0^{outlet} - P_0^{inlet} + c_0(220000 - L)^2$$

$$+ c_1(-63.0 - \theta^t)^2 + c_2(155.0 - \dot{m})^2 + c_3(.223 - A)^2 + c_4(tde)$$

where L is the lift, P_o is the total pressure, θ is the average exit flow angle, $\dot{m}$ is the mass flow rate, and A is the cross-sectional area of the airfoil. The variable tde is the largest relative error in the airfoil thickness distribution compared to an airfoil with a thickness distribution that is considered to be minimum allowable. This constraint prevents airfoil from becoming too thin so that it would not be mechanically or thermally infeasible. The constants c_i are user specified penalty terms. The inlet total pressure, total temperature, and inlet flow angle were set to 440,000 Pa, 1600.0 K, and 30.0°, respectively. The exit static pressure was specified as 101,330 Pa. Adiabatic wall conditions were enforced along the airfoil surface.

The genetic algorithm based optimizer used the micro-GA technique (Krishnakumar, 1989) with no mutation. A binary string that used nine bits for each design variable represented each design in the population. A tournament selection was used to determine the mating pairs (Goldberg, 1992). Each pair produced two children who then replaced the parents in the population. Uniform crossover with a 50% probability of crossover was used to produce the children. Elitism was also implemented in the optimizer; the best individual found from the previous generations was placed in the current generation. Two runs were made, each used a different method for enforcing the constraints.

12.6.1 Penalty method for constraints

This run used penalty terms alone to enforce the constraints. A constant penalty term of 4.0E6 was applied to each normalized constraint. The calculation consumed 70 hours of CPU time on a 350 MHz Pentium II based PC. The genetic optimizer was run for 30 generations with a population of 20. The best airfoil designs from generation 1 and generation 30 are shown in Figure 12.9a. The best design from the GA generation 1 had a total pressure loss of 8200 Pa, which was after 30 generations reduced to 6850 Pa (Figure 12.10). The maximum fitness for this method increased monotonically, although somewhat slowly (Figure 12.11). Also, after 30 generations, a 3.5% violation of the lift constraint, a 1% violation in the exit flow angle constraint, a 3.3% violation of the mass flow rate constraint, a 3.1% violation of the area constraint, and a 4% violation in the thickness distribution constraint were achieved (Figure 12.12).

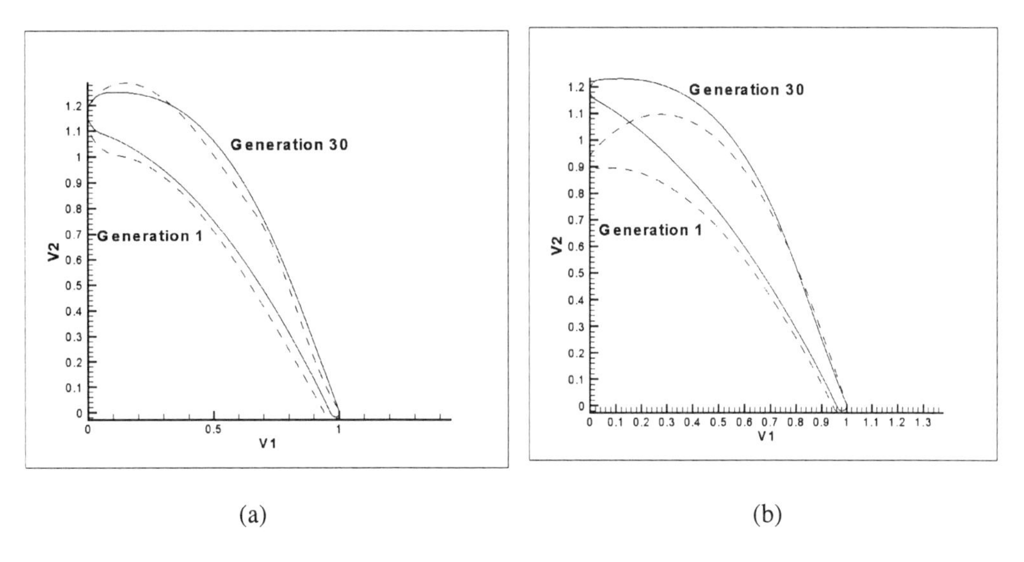

(a) (b)

Figure 12.9 Shapes of the best members of the population after the generations number 1 and 30: a) when constraints were enforced via penalty function only, b) when constraints were enforced via SQP and the penalty function.

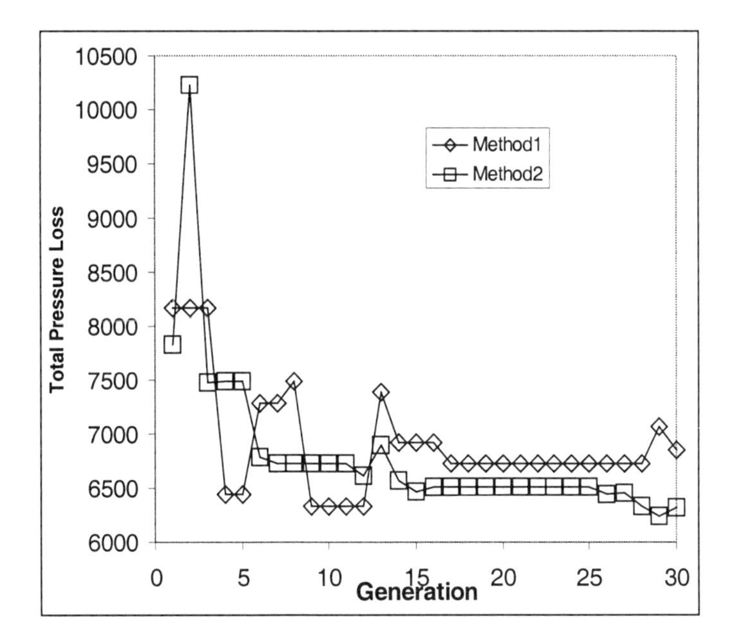

Figure 12.10 Evolution of the total pressure loss for both methods of enforcing constraints.

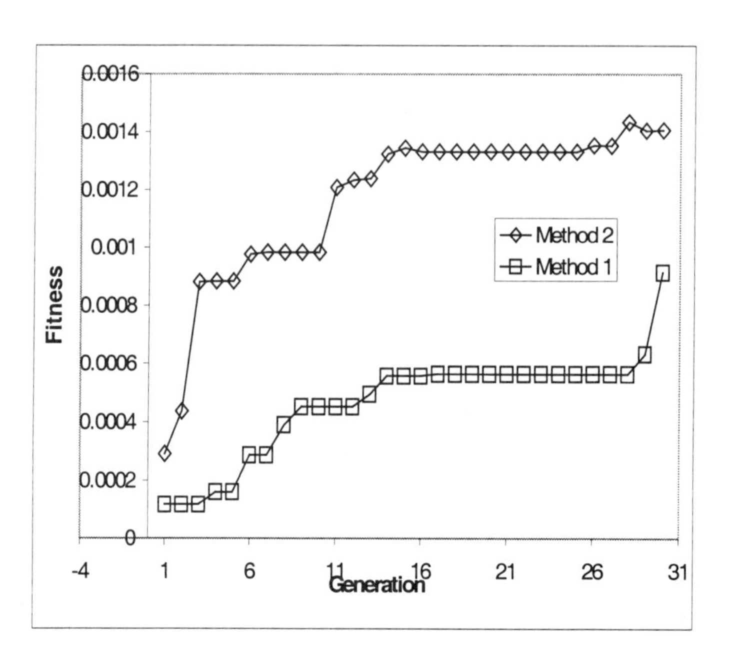

Figure 12.11 Fitness function evolution for both methods of enforcing constraints.

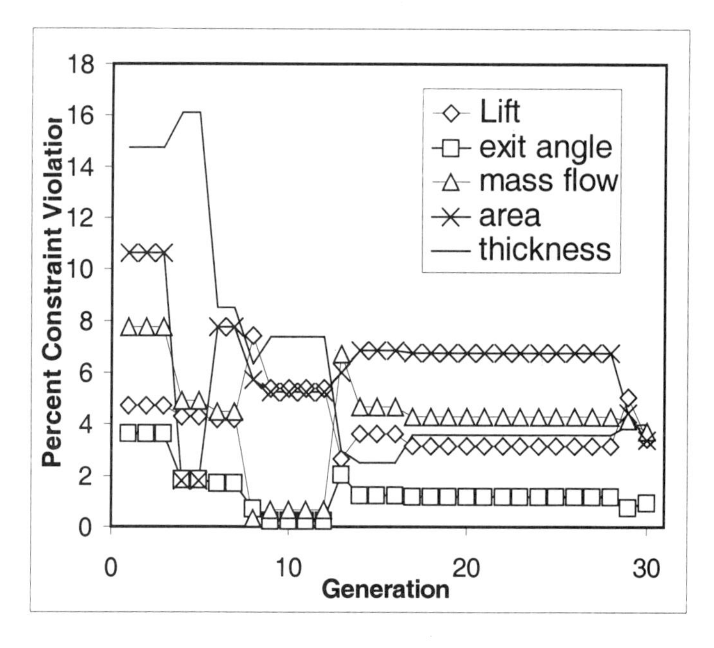

Figure 12.12 Violation of the constraints for the best design of each generation when enforcing constraints via penalty function only.

12.6.2 Gene correction SQP method for constraints

In this run, a gene correction method based on sequential quadratic programming (SQP) (Rao, 1996) was used to enforce the cross sectional area and thickness distribution constraints while penalty terms were used to enforce the lift, mass flow rate, and average exit angle constraints with a penalty constant of 4.0E6. The SQP was used to minimize the thickness distribution error with the specified cross sectional area of the airfoil as an equality constraint. This was done to every individual in the population for each generation.

With this method, essentially all the designs become geometrically feasible before the expensive flow-field analysis is performed. This allows the genetic algorithm to focus on satisfying the lift, mass flow rate, and exit angle constraints only. The combined genetic and SQP optimizer was then run for 30 generations with a population of 15 and consumed 50 hours of computing time on a 550 MHz AlphaPC workstation. The best airfoil designs from the GA generation 1 and generation 30 are shown in Figure 12.9b. The best design from generation 1 had a total pressure loss of 7800 Pa which was reduced after 30 generations to 6546 Pa (Figure 12.10). The maximum fitness for this method increased rapidly during the initial stages of the optimization process (Figure 12.11). Also, after 30 generations, the design had a 5.8% violation of the lift constraint, a 1% violation in the exit flow angle constraint, a 1.9% violation of the mass flow rate constraint, a 0% violation of the area constraint, and a 0% violation in the thickness distribution constraint. Figure 12.13 shows the violation of the constraints for the best design of each generation. This method consistently produced higher fitness designs over the previous method that enforces all constraints via a composite penalty function.

There was an attempt to use SQP to minimize all the constraint violations in lift, mass flow rate, exit angle, area, and thickness for each design in each generation, but this proved to be too computationally expensive. Besides, for every new design generated by the GA, the SQP code failed to find a feasible solution before converging to a local minimum. However, with just the area and thickness distribution constraints, the SQP was able to find a design meeting those constraints most of the time. Also, the computations required to find area and thickness distribution are very inexpensive. This proved to be more economical than trying to use SQP to satisfy all the constraints simultaneously. It was also more effective than treating all the constraints with just a penalty function.

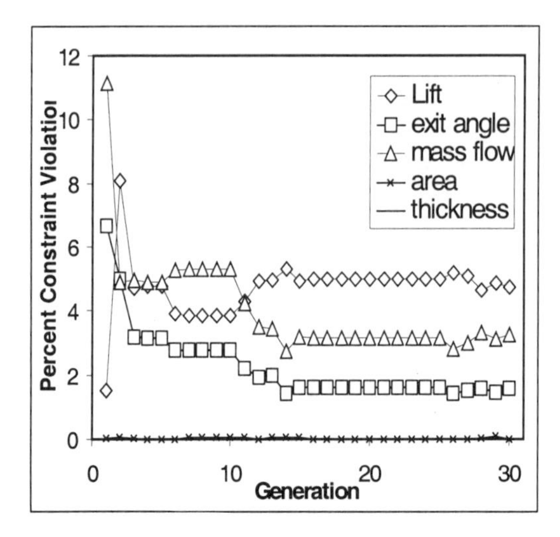

Figure 12.13 Violation of the constraints for the best design of each generation when enforcing constraints via SQP and the penalty function.

12.7 MAXIMIZING MULTISTAGE AXIAL GAS TURBINE EFFICIENCY

Instead of using an extremely computationally demanding three-dimensional multistage viscous transonic flow-field analysis code based on Navier-Stokes equations, we have used a well established fast through-flow analysis code (Petrovic and Dulikravich, 1999). When combined with our hybrid constrained optimization algorithm, it can efficiently optimizes hub and shroud geometry and inlet and exit flow-field parameters for each blade row of a multistage axial flow turbine. The compressible steady state inviscid through-flow code with high fidelity loss and mixing models, based on stream function method and finite element solution procedure, is suitable for fast and accurate flow calculation and performance prediction of multistage axial flow turbines at design and significant off-design conditions.

An analysis of the loss correlations was made to find parameters that have influence on the multistage turbine performance. By varying seventeen variables per each turbine stage it is possible to find an optimal radial distribution of flow parameters at the inlet and outlet of every blade row. Simultaneously, an optimized meridional flow path is found that is defined by the optimized shape of the hub and shroud.

The design system has been demonstrated (Petrovic and Dulikravich, 1999) using an example of a single stage transonic axial gas turbine, although the method is directly applicable to multistage turbine optimization. The comparison of computed performance of initial and optimized design shows significant improvement in the multistage efficiency at design and off-design conditions. The optimization was performed while keeping constant rotational speed, mass flow rate, total enthalpy drop, number of blades, rotor tip clearance, blade chord lengths, and blade trailing edge thicknesses. There was only one geometric constraint in this test case: blade tip radius was allowed to change up to a prescribed value. It was assumed that both blade count and chord lengths have been already determined by some preliminary design procedure. In principle, it is possible to include both of these parameters (and other parameters deemed to be influential) in this optimization method.

To find the turbine configuration that gives the maximum total-to-total efficiency, only 15 minutes on an SGI R10000 workstation were necessary with the constrained hybrid optimizer which mainly utilized DFP algorithm. Figure 12.14 shows a comparison of meridional flow paths for the initial and for the optimized configuration. The geometric changes are relatively small. However, these changes of hub and shroud shapes together with the optimized radial distribution of flow angles at inlet and outlet of the stator and the rotor have as a result a significant decrease in the flow losses and the entropy generation. In the particular single stage transonic turbine the total-to-total efficiency was improved by 2%.

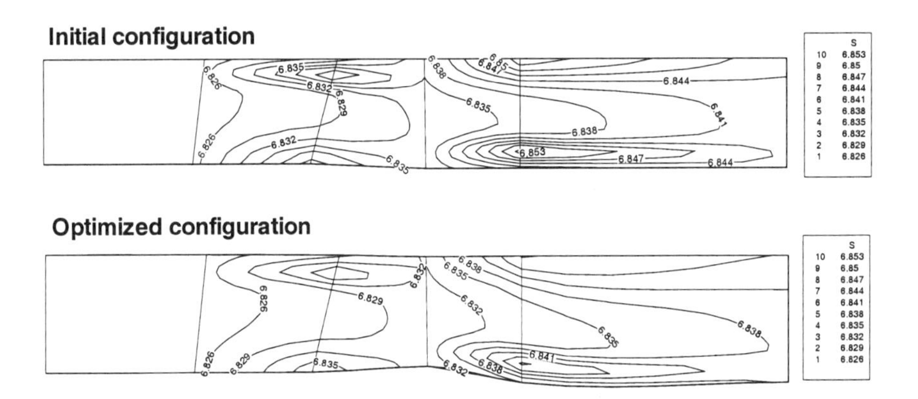

Figure 12.14 Predicted entropy distribution in a meridional plane of the initial and the optimized single stage axial gas turbine configuration.

12.8 SUMMARY

A cluster of standard optimization algorithms was assembled in a hybrid optimization tool where a set of heuristic rules was used to perform automatic switching among the individual optimizers in order to avoid local minimums, escape from the local minimums, converge on a minimum, and reduce the overall computing time. The constraints were enforced either via penalty function or via Rosen's projection method. The hybrid optimizer was applied in aero-thermal optimization of internally cooled blades, aerodynamic shape optimization of hypersonic projectiles, and efficiency optimization of a retrofit turbine airfoil cascade. Lessons learned from these efforts are valuable. Most importantly, hybrid optimization is a very robust and cost-effective optimization concept. Automatic switching among the individual optimizers can be further improved by incorporating certain aspects of neural networks. Use of simplified models (surrogates) for evaluation of the object function is highly cost effective, although progressively more complete physical models should be used as the global optimization process starts converging. Otherwise, ludicrous results are possible where the deficiencies of the surrogate models are fully exploited by the optimizer. Parameterization of the design space plays a crucial role in the hybrid constrained optimization. Coarse parameterization usually, but not always, leads to a converged result at an acceptable cost in computing time. A refined parameterization definitely widens the feasible region in the case of a highly constrained optimization. Finally, a gene correction method based on sequential quadratic programming could be effectively used to enforce certain inexpensive constraints while penalty terms could be used to enforce the remaining constraints. The complex multidisciplinary optimization problems seem to call for a multiobjective automated decision making process which could involve weighting the objective functions or alternating between the objectives and the constraints.

Acknowledgments

The authors are grateful for NASA-Penn State Space Propulsion Engineering Center Graduate Student Fellowship facilitated by Professor Charles Merkle, National Science Foundation Grants DMI-9522854 and DMI-9700040 monitored by Dr. George Hazelrigg, NASA Lewis Research Center Grant NAG3-1995 facilitated by Dr. John Lytle, and supervised by Dr. Kestutis Civinskas, the ALCOA Foundation Faculty Research Fellowship facilitated by Dr. Yimin Ruan, and for Lockheed Martin Skunk Works grant monitored by Mr. Thomas Oatway.

REFERENCES

1. Anderson, J. D. Jr. (1989). *Hypersonic and High Temperature Gas Dynamics*, McGraw-Hill, New York.

2. Ashley, H., and Landahl, M. (1965). *Aerodynamics of Wings and Bodies*, Addison-Wesley Publishing Company, Inc., MA.

3. Barsky, B. A. (1988). *Computer Graphics and Geometric Modeling Using Beta-Splines*, Springer-Verlag, Berlin, Germany.

4. Davidon, W.C. (1959). Variable Metric Method for Minimization, Atomic Energy Commission Research and Development Report, ANL-5990 (Rev.).

5. Dulikravich, G. S. (1997). Design and Optimization Tools Development, Chapters no. 10-15 in *New Design Concepts for High Speed Air Transport,* (editor: H. Sobieczky), Springer, Wien/New York, pp. 159-236.

6. Dulikravich, G. S., and Martin, T. J. (1994). Design of Proper Super Elliptic Coolant Passages in Coated Turbine Blades With Specified Temperatures and Heat Fluxes, *AIAA Journal of Thermophysics and Heat Transfer* **8** (2) 288-294.

7. Dulikravich, G. S. and Martin, T. J. (1996). Inverse Shape and Boundary Condition Problems and Optimization in Heat Conduction, Chapter no. 10 in *Advances in Numerical Heat Transfer - Volume I* (editors: W. J. Minkowycz and E. M. Sparrow), Taylor and Francis, pp. 381-426.

8. Dulikravich, G. S., Martin, T. J., and Han, Z.-X. (1998). Aero-Thermal Optimization of Internally Cooled Turbine Blades, *Proceedings of Fourth ECCOMAS Computational Fluid Dynamics Conference,* (editors: K. Papailiou, D. Tsahalis, J. Periaux, D. Knoerzer) Athens, Greece, September 7-11, 1998, John Wiley & Sons, NY, **2**, pp. 158-161.

9. Dulikravich, G. S., Martin, T. J., Dennis, B. H., Lee, E.-S., and Han, Z.-X. (1998). Aero-Thermo-Structural Design Optimization of Cooled Turbine Blades, *AGARD - AVT Propulsion and Power Systems Symposium on Design Principles and Methods for Aircraft Gas Turbine Engines*, Editor: G. Meauze, Toulouse, France, May 11-15, 1998.

10. Fletcher, R. and Powell, M.J.D. (1963). A Rapidly Convergent Descent Method for Minimization, *Computer Journal* **6**, 163-168.

11. Foster, N. F., and Dulikravich, G. S. (1997). Three-Dimensional Aerodynamic Shape Optimization using Genetic Evolution and Gradient Search Algorithms, *AIAA Journal of Spacecraft and Rockets* **33** (3) 33-33.

12. Goldberg, D. E. (1989). *Genetic Algorithms in Search, Optimization and Machine Learning*, Addison-Wesley.

13. Hafka, R. T., and Malkus, D. S. (1981). Calculation of Sensitivity Derivatives in Thermal Problems by Finite Differences, *International Journal of Numerical Methods in Engineering* **17**, 1811-21.

14. Krishnakumar, K. (1989). Micro-Genetic Algorithms for Stationary and Non-Stationary Function Optimization, *SPIE: Intelligent Control and Adaptive Systems* **1196**, Philadelphia, PA, 1989.

15. Martin, T. J., and Dulikravich, G. S. (1997). Aero-Thermal Analysis and Optimization of Internally Cooled Turbine Blades, *XIII International Symposium on Airbreathing Engines (XIII ISABE),* Editor: F. S. Billig, Chattanooga, TN, Sept. 8-12, 1997, ISABE 97-7165, **2**, 1232-1250.

16. Martin, T. J., Dulikravich, G. S., Han, Z.-X., and Dennis, B. H. (1999). Minimization of Coolant Mass Flow Rate in internally Cooled Gas Turbine Blades, ASME paper GT-99-111, ASME IGTI'99, Indianapolis, IN, June 7-10, 1999.

17. Nelder, J. A. and Mead, R. (1965). A Simplex Method for Function Minimization, *Computer Journal* **7**, 308-313.

18. Petrovic, V. M., and Dulikravich, G. S. (1999). Maximizing Multistage Turbine Efficiency by Optimizing Hub and Shroud Shapes and Inlet and Exit Conditions of Each Blade Row, ASME paper GT-99-112, ASME IGTI'99, Indianapolis, IN, June 7-10, 1999.

19. Press, W. H, Teukolsky, S. A., Vetterling, W. T. and Flannery, B. P. (1986). *Numerical Recipes in FORTRAN, The Art of Scientific Computing*, 2nd Edition, Cambridge University Press, Cambridge.

20. Pritchard L. J. (1985). An Eleven Parameter Axial Turbine Airfoil Geometry Model, ASME 85-GT-219.

21. Rao, S. (1996). *Engineering Optimization: Theory and Practic*e, Third edition, John Wiley Interscience, New York.

22. Stephens, M. A., and Shih, T. I.-P. (1997). Computation of Compressible Flow and Heat Transfer in a Rotating Duct with Inclined Ribs and a 180-Degree Bend, ASME paper 97-GT-192, Orlando, Florida.

13 Genetic Algorithm as a Tool for Solving Electrical Engineering Problems

M. RUDNICKI[1], P. S. SZCZEPANIAK[1] and P. CHOLAJDA[2]

[1]Institute of Computer Science
Technical University of Lodz
Sterlinga 16/18, 90-217 Lodz, Poland
E-mail: rudnicki@ics.p.lodz.pl, piotr@ics.p.lodz.pl

[2]Systems Research Institute
Polish Academy of Sciences
Newelska 6, 01-447 Warsaw, Poland
E-mail: cholajda@ibspan.waw.pl

Abstract. The paper aims at presenting genetic algorithms as an efficient numerical tool in the field of electrical engineering. After an introduction to evolution algorithms methodologies we present two real world applications. The first one enables creation of an expert system which diagnoses technical conditions of an oil transformer basing on results of chromatography of gases dissolved in transformer oil (Dissolved Gas Analysis – DGA). The system is designed to diagnose small and medium size units for which DGA is an essential source of information on their technical conditions. The diagnosis is based on a set of rules which form a database in the expert system. In order to select as small group of rules in the database as possible, a genetic algorithm is applied. The second application is that of parameter identification of an induction motor from available performance data using the equivalent circuit model of an machine. For identification we apply a multilayer feedforward network, genetic algorithm and simulated annealing strategies. Numerical results compare well with the actual parameters of small induction motors.

13.1 EVOLUTIONARY ALGORITHMS

13.1.1 A short overview

Stochastic optimisation methods based on the principles of natural evolution have been receiving more and more attention in research communities due to their robustness and global nature. Generally, we speak of *evolutionary algorithms* (EA) to which the following techniques belong to: *genetic algorithms* (GA: Holland [14]), *evolutionary programming* (EP: Fogel et al. [12]*), evolution strategies* (Rechenberg [16]) *and genetic programming* (Koza [15]). Evolutionary algorithms operate on a population of candidate solutions using the principle of survival of the fittest. At each generation, a new population of individuals is created by selecting best fit chromosomes. The process mimics natural evolution rules by giving best chances for reproduction only those individuals that are better suited to their environment. That way each next population is made of individuals that are better fit than their parents. Individuals (current approximations) are encoded as strings (*chromosomes*) over some alphabet so that the *genotypes* (chromosome values) are uniquely mapped into the decision space (phenotypic domain). Having decoded chromosomes into the decision space one can assign the performance index or fitness of individuals using an objective function of an underlying problem. The objective function is used to provide a measure of performance. The *fitness function* is used to transform the objective function value into a measure of relative fitness. It is worthwhile noting that examining the chromosome string in isolation gives no information about the problem we are solving. Any meaning can be applied to the representation only with the decoding of the chromosome into its phenotypic values. The process of evaluating the fitness of an individual consists of the following steps: (1) converting the chromosome's genotype into its phenotype, (2) evaluation of the objective function, (3) conversion of the objective function value to a *raw fitness*, (4) conversion the raw fitness to a scaled fitness and (5) conversion the scaled fitness to an expected frequency of selection as a parent. Converting the objective function's value to a raw fitness is problem dependent. A function that has been found to be generally suitable is the exponential:

$$f(\mathbf{x}) = e^{K\mathbf{x}}, \quad K \geq 0$$

To ensure that the resulting fitness values are non-negative a linear transformation which offsets the objective function is used. Using the linear scaling the expected number of offspring is approximately proportional to that of individuals performance.

$$F(\mathbf{x}) = g(f(\mathbf{x})) \quad F(\mathbf{x}) = af(\mathbf{x}) + b$$

where f is the objective function and F is relative fitness. If the raw fitness is used to determine parent selection probabilities, two problems might arise. Firstly, in the first few generations, one or a very few extremely superior individuals usually appear. Their fitness values are high enough for them to be selected as parents too many times and their genetic material would quickly dominate the gene pool resulting in early loss of the population diversity which is crucial for genetic optimisation. Secondly, after many generations, inferior individuals will have been died out. The population will consist of highly fit chromosomes and the maximum fitness will be only slightly higher than the average. This may result in missing best fit individuals as parents: they are not selected in the high rate necessary for rapid improvement. Scaling solves both of these problems. *Selection* is the process of finding the number of trials a chromosome is selected for reproduction thus giving the number of offspring the individual can produce. The selection process can be described using three measures of performance: *bias*, *spread* and *efficiency* (Baker [4]). Bias is defined as the absolute difference between the expected and actual selection probability. Spread is the range in the possible number of trials for an individual.

13.1.2 Genetic algorithms

Genetic algorithms are optimisation algorithms based on Darwinian models of natural selection and evolution. The basic idea behind GA is that an initial population of candidate solutions are chosen at random and each solution is evaluated according to the objective function. The original concepts of GA were developed by Holland and have been shown to provide near-optimal performance for a search in complex decision spaces. GA often outperform direct methods on difficult problems, i.e. involving highly non-linear, high-dimensional, discrete, multimodal or noisy functions. However, GA are not guaranteed to find the global function optima because: (1) the precision limits in the encoding process can significantly reduce the solution accuracy, and (2) the search process does not ergodically cover and search the design space.

The above means that using genetic algorithms often does not yield a correct solution because of evolving the process into a "genetic corner" or *niche*, i.e. the system can reach a stable state.

Genetic algorithms traditionally use *binary* or *Gray coding* although other representations are now employed. The Gray coding system was devised to simplify and facilitate the binary representation of numbers. In this system, a unit change in the number causes exactly one bit to change. Using strictly binary approach each gene can have one of two alleles: 0 or 1 and each parameter is encoded as a binary number, and all parameters are linked into one long string. The opposite scheme is direct parameter storage in which there is one gene for each parameter, and the *allele* values for that gene are the possible values of that parameter. Conversion between the genotype and the phenotype is trivial, since they are one and the same.

Population can be initialised randomly or may be seeded with known good individuals. Since the diversity is crucial for to genetic methods it is sometimes recommended to overinitialize an initial population by randomly generating more than the required population size, then discarding the worst. In most GA population size is kept constant. As GA do not require gradient information and because of stochastic nature of the search, GA can search the entire design space being more likely to find global optimum than conventional optimisation algorithms which usually require the objective function to be well behaved. On the contrary, GA can tolerate noisy and discontinuous objective functions. The main selection technique is so called *roulette wheel approach* using which each individual is assigned a slot whose probability is proportional to the fitness of that individual, and the wheel is spin each time a parent is needed. However, it is evident that chance plays too much role in this method: one unlucky generation could wipe out generations of success. Therefore, the basic selection method is *stochastic sampling with replacement*. There are other roulette based selection methods: *stochastic sampling with partial replacement*, *remainder sampling methods* and *stochastic universal sampling*. Details can be found e.g. in Chipperfield et al. [5].

13.1.3 Genetic operators

The basic genetic operators are reproduction, crossover and mutation. Reproduction is a process of selecting individuals based on their fitness relative to that of the population. After reproduction the crossover operation is performed in two steps. In the first step, two chromosomes are selected at random from the mating pool produced by the reproduction operator. Next, a crossover point is selected at random along the chromosome length and the binary digits (alleles) are swapped between the parents. The simplest is a single-point crossover with a single crossing site. The natural extension leads to *multi-point crossover* with a number of crossover points. For example, two-point crossover splits the chromosome into four quarters. The first and third quarters go to one child, while the second and fourth go to the other child. The idea can be extended to *uniform crossover*, *shuffle crossover* and the *reduced surrogate crossover*.

The fitness function which drives the system towards the optimum points may sometimes take the wrong turn. The *mutation* operator can help move the process out of niche. In natural evolution, mutation is a random process where one allele of a gene is replaced by another to produce a new individual. Considering mutation as a background operator, its role is to provide a guarantee that the probability of searching any given string will never be zero. The binary mutation flips the value of the bit at the mutation point. With non-binary coding, mutation is achieved by either perturbing the gene values or random selection of new values within the allowed range. The real goal of mutation is the introduction of new genetic material or recovering of valuable genes that were lost through poor selection of mates. Simultaneously, the valuable gene pool must be protected from wanton destruction. Thus, the probability of mutation should be tiny. If the mutation rate is too high, the system starts mutating backwards because it overwhelms the force of reproduction. Many variations of the mutation operator have been put forward in the literature. They include *trade mutation* and *reorder mutation*, among others.

If one or more the most fit individuals is allowed to propagate through future generations, the GA is said to use an elitist strategy. To keep the size of the original population constant, the new individuals have to be reinserted into the old population. Hence, a reinsertion scheme must be used to decide which chromosomes are to exist in the new population. The least fit members of the old population are usually replaced deterministically.

An important high-level genetic operator is that of multiple *sub-population*. Each sub-population is evolved over generations by a traditional GA and from time to time individuals migrate from one sub-population to another. This is known as the *migration model*.

As the GA is a stochastic search method, the formal specification of convergence criteria is difficult. It may happen that the average fitness of a population remains unchanged for a number of generations before a superior individual is found. It is commonplace to terminate the GA after a specified number of generations. If no acceptable solutions are found, the GA may be restarted or a new search may be initiated.

13.1.4 Data structures in GA

The main data structures in GA are chromosomes, phenotypes, object function values and fitness values. This is particularly easy implemented when using Matlab package as a numerical tool, Chipperfield et al. [5]. An entire chromosome population can be stored in a single array given the number of individuals and the length of their genotypic representation. Similarly, the design variables, or phenotypes that are obtained by applying some mapping from the chromosome representation into the design space can be stored in a single array. The actual mapping depends upon the decoding scheme used. The objective function values can be scalar or vectorial (in the case of multiobjective problems) and are not necessarily the same as the fitness values. Fitness values are derived from the object function using scaling or ranking function and can be stored as vectors.

13.1.5 Why do genetic algorithms work?

The search heuristics of GA are based upon Holland's schema theorem. A *schema* is defined as a template for describing a subset of chromosomes with similar sections. The schema consists of bits 0, 1 and meta-character. The template is a suitable way of describing similarities among patterns in the chromosomes. Holland derived an expression that predicts the number of copies of a particular schema would have in the next generation after undergoing exploitation, recombination and mutation. It follows that particularly good schemata will propagate in future generations. Thus, schemata that are low-order, well-defined and have above average fitness are preferred and are termed *building blocks*. This leads to a building block principle of GA: low order, well-defined, average fitness schemata will combine through recombination to form higher order, above average fitness schemata. Since GAs process many schemata in a given generation they are said to have the property of *implicit parallelism*.

13.1.6 Genetic algorithms and constraint handling

An important problem in genetic optimisation is that of constraint handling. The GENOCOP package of Michalewicz can handle non-linear constraints of different type (Michalewicz [2]).

The GAs basically find a maximum of an unconstrained optimisation problem. To solve a constrained problem we need to transform the primary problem. The first step is to transform the original constrained problem into an unconstrained one using penalty function:

$$\min f(\mathbf{x}) + R\sum_{j=1}^{m} \Phi(g_j(\mathbf{x}))$$

$$g_j(\mathbf{x}) \leq 0$$

$$x_i^{(l)} \leq x_i \leq x_i^{(u)}, \quad i = 1, 2, \ldots, n$$

where the second term to the right-hand side defines the penalty function weighted with the penalty parameter. The second transformation is to minimise the objective function through the maximisation of the fitness function defined as

$$F(\mathbf{x}) = F_{\max} - \left(f(\mathbf{x}) + R\sum_{j=1}^{m} \Phi(g_j(\mathbf{x})) \right)$$

The penalty function approach for inequality constrained problems can be divided into two categories: interior and exterior methods. In the first approach the penalty term is chosen such that it tends to infinity as the constraint boundaries are approached. The interior penalty function methods are also known as barrier methods. In the exterior penalty function method the penalty term generally has the form:

$$\Phi(\mathbf{x}, R) = f(\mathbf{x}) + R\sum_{j=1}^{m} \langle g_j(\mathbf{x}) \rangle^q$$

$$\langle g_j(\mathbf{x}) \rangle = \max\langle g_j(\mathbf{x}), 0 \rangle = \begin{cases} g_j(\mathbf{x}) & g_j(\mathbf{x}) > 0 \\ 0 & g_j(\mathbf{x}) \leq 0 \end{cases}$$

The effect of the second term on the right hand side is to increase the penalty function proportionally to the qth power of the amount of constraint violation. If q>1 then the penalty function will be once continuously differentiable. The value of q=2 is mostly used in practice.

13.2 GENETIC INTERPRETATION OF DGA RESULTS IN POWER TRANSFORMERS

13.2.1 Introduction

Conversion of alternating voltage into higher or lower one is performed by means of transformers, Stein [8]. Therefore it is of great importance to ensure proper operation of transformers in a power network since to a great extent proper economic condition and development of a given country depends on it.

One of the methods applied to diagnose technical condition of transformers is DGA – the analysis of gases dissolved in transformer oil. Basing on their own experience, different countries have worked out different transformer diagnosing methods, Przybylak and Roganowicz [9], based on DGA results. There are systems, Przybylak and Roganowicz [9], which apply many methods from different countries to process DGA results in order to arrive at a transformer diagnosis. The aim of such processing procedures is to present human-expert with different interpretations of DGA results to help take a final decision on the technical condition of the transformer whose oil has been subject to examination. On the other hand, different interpretation techniques applied to DGA results yield different outcomes [11], which may complicate arriving at reliable estimation of the technical condition of a transformer. The present paper proposes a genetic technique, Ishibuchi et al. [10], instead of rigid mathematical dependencies characteristic of the methods used so far for such diagnoses. The algorithms proposed here enable designing of an expert system comprising many different methods. The final diagnosis rests with the human-expert. After a certain statistical analysis is performed, answers obtained from the expert system are close to those given by the human-expert.

13.2.2 Database construction

Data used to design an expert system which is to diagnose technical condition of transformers are defined by three variables (x,y,z). The meaning of these variables, described by (1)–(3), is the same as in IEC code [11]. The code is one of the basic methods recommended for interpretation of DGA results, and for that reason such variables have been applied to describe gas concentration in transformer oil.

$$x = \frac{C_2H_2}{C_2H_4} \tag{1}$$

$$y = \frac{CH_4}{H_2} \tag{2}$$

$$z = \frac{C_2H_4}{C_2H_6} \tag{3}$$

where H_2 denotes the amount of hydrogen in a gas under examination (in ppm units – parts per one million), while CH_4, C_2H_2, C_2H_4, and C_2H_6 – that of methane, acetylene, ethylene, and ethane, respectively. Apart from variables *(x,y,z)* which reflect the DGA results for a transformer subject to examination, each data (placed in database) contains also additional information about the technical condition of the transformer. This information is obtained from a human-expert who has made the diagnosis using available interpretation methods of DGA results or basing on own intuition (Fig. 13.1).

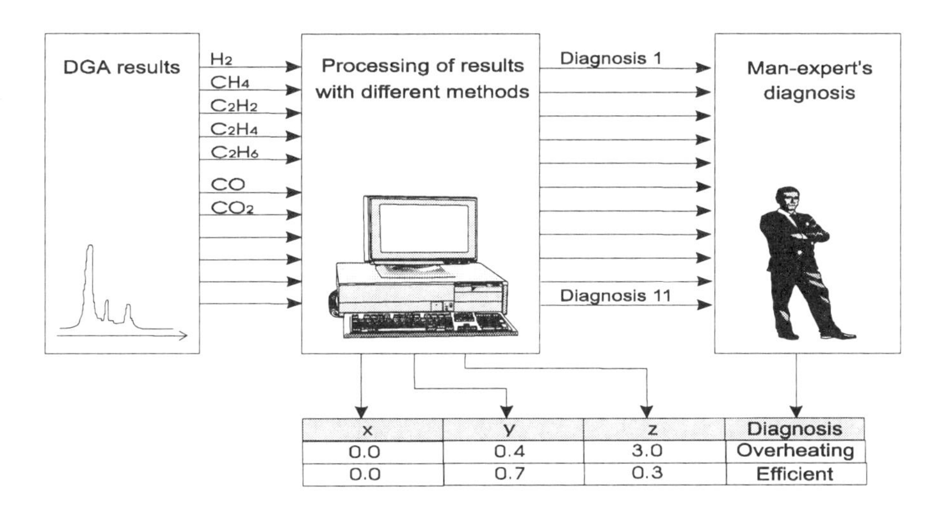

Fig. 13.1 Database structure.

This enables classification of a point (or speaking more generally – object) described by the triple *(x,y,z)*. It has been assumed that there are nine classes (according to IEC code), namely: No fault, Partial discharge of low energy, Partial discharge of high energy, Disruptive discharge of low energy, Disruptive discharge of high energy, Overheating below 150°C, Overheating between 150°C and 300°C, Overheating between 300°C and 700°C, Overheating over 700°C. When the database is ready, co-ordinates *x,y,z* describing each known object are transformed so that all objects in the database belong to a unit cube. To obtain the space $\langle 0,1\rangle\times\langle 0,1\rangle\times\langle 0,1\rangle$ the following formulas are used:

$$\bigwedge_{i \in \langle 1,N \rangle} x_i^{new} = \frac{x_i - \min\{x_i\}}{\max\{x_i\} - \min\{x_i\}}$$

$$\bigwedge_{i \in \langle 1,N \rangle} y_i^{new} = \frac{y_i - \min\{y_i\}}{\max\{y_i\} - \min\{y_i\}}$$

$$\bigwedge_{i \in \langle 1,N \rangle} z_i^{new} = \frac{z_i - \min\{z_i\}}{\max\{z_i\} - \min\{z_i\}}$$

where N is the number of objects in the database, and

$$\begin{aligned}
\min\{x_i\} &= \min\{x_i: i=1,2, \ldots, N\}, \\
\min\{y_i\} &= \min\{y_i: i=1,2, \ldots, N\}, \\
\min\{z_i\} &= \min\{z_i: i=1,2, \ldots, N\}, \\
\max\{x_i\} &= \max\{x_i: i=1,2, \ldots, N\}, \\
\max\{y_i\} &= \max\{y_i: i=1,2, \ldots, N\}, \\
\max\{z_i\} &= \max\{z_i: i=1,2, \ldots, N\}.
\end{aligned}$$

Fig. 13.2 presents objects placed in a unit cube and used for determination of regions in which objects of the same class are expected.

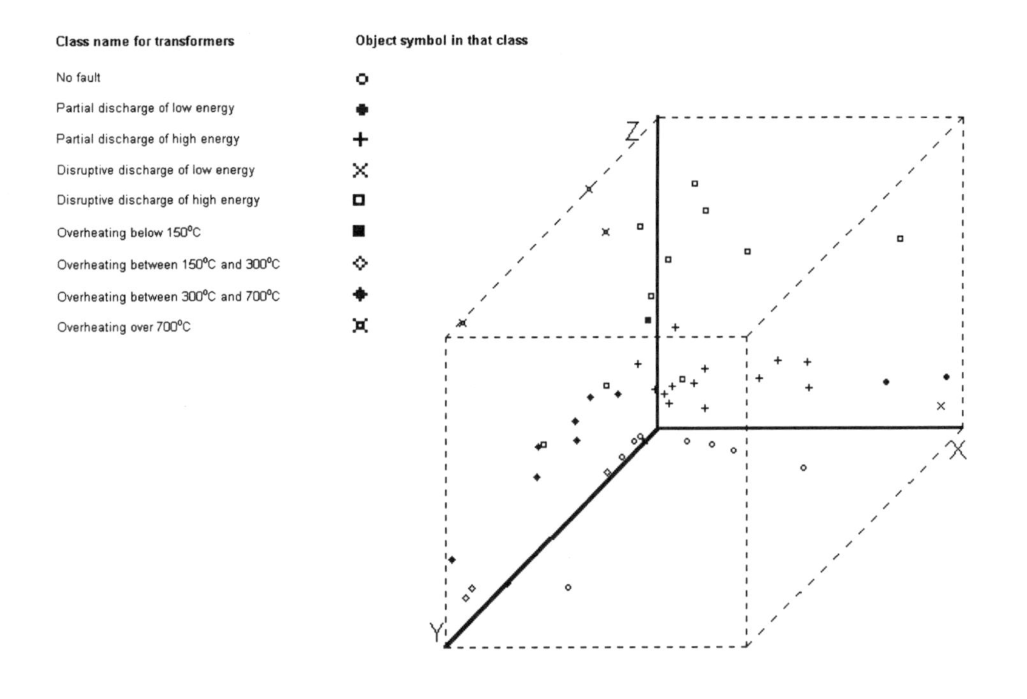

Fig. 13.2 Database objects in a unit cube.

13.2.3 Determination of a set of rules

The 3D space *XYZ* containing data from the database is divided into smaller fuzzy regions defined by triangular membership functions (Figs. 13.3 and 13.4).

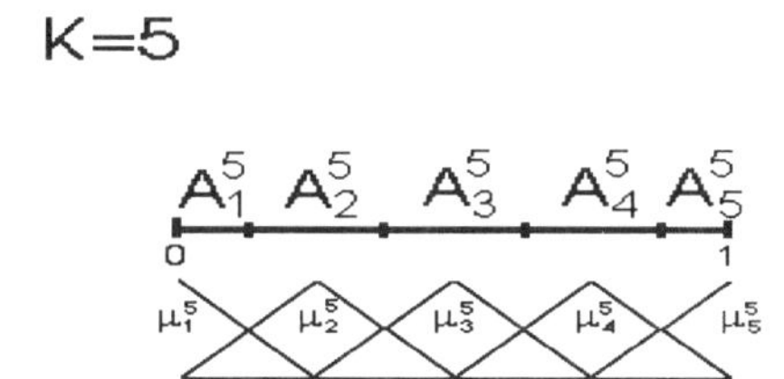

Fig. 13.3 1D space (here a unit segment) division into fuzzy regions $A^K_{i=1,2,...K}$ by K=5 triangular membership functions $\mu^K_{i=1,2,...,K}:\langle 0,1\rangle \to \Re$.

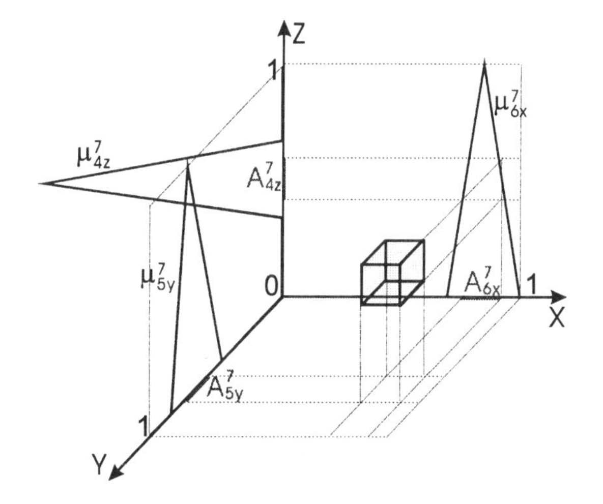

Fig. 13.4 Fuzzy region $A^7_{6X} \times A^7_{5Y} \times A^7_{4Z}$ in a 3D space defined by triangular membership functions $\mu^7_{6X}:\langle 0,1\rangle \to \Re$, $\mu^7_{5Y}:\langle 0,1\rangle \to \Re$, $\mu^7_{4Z}:\langle 0,1\rangle \to \Re$ (the function graphs are presented conventionally here as it is not possible to show successive dimensions in the diagram).

Each region is assigned a fuzzy if-then data-classifying rule that reads:

> "**If** a new object Q = (x,y,z) belongs to a given region $A^K_{iX} \times A^K_{jY} \times A^K_{kZ}$ (where i,j,k∈ ⟨1,2,...,K - the number of divisions⟩), **then** it is of certain class **with certainty** of certain percentage."

To define what the expression "a *certain* class" means the occurrence of objects of different classes from the database in the fuzzy region $A^K_{iX} \times A^K_{jY} \times A^K_{kZ}$ should be examined. The term "certain" defines the name of that class C whose objects are the most numerous in the fuzzy region $A_{iX}^K \times A_{jY}^K \times A_{kZ}^K$. The dependence of the number of objects from the selected class occurring in $A_{iX}^K \times A_{jY}^K \times A_{kZ}^K$ on the number of the remaining objects which are also in $A_{iX}^K \times A_{jY}^K \times A_{kZ}^K$ is given by formula (4) which defines the value of "*certain* percentage"

$$Certainty = Z_{ijk}^K \cdot \mu_{iX}^K(x) \cdot \mu_{jY}^K(y) \cdot \mu_{kZ}^K(z) \tag{4}$$

where

$$Z_{ijk}^K = \frac{C_{ijk}^K - \dfrac{W_{ijk}^K - C_{ijk}^K}{liczba\_klas - 1}}{W_{ijk}^K} \cdot 100\%,$$

C_{ijk}^K - the number of objects of a selected class belonging to the fuzzy region $A_{iX}^K \times A_{jY}^K \times A_{kZ}^K$,

liczba_klas – the number of classes,

W_{ijk}^K - the number of objects of all classes belonging to the fuzzy region $A_{iX}^K \times A_{jY}^K \times A_{kZ}^K$.

It should be noted here that Z_{ijk}^K in (4) takes the value *100%* when all objects in the fuzzy region $A_{iX}^K \times A_{jY}^K \times A_{kZ}^K$ are of one class C. If the number of objects of particular classes is the same, then Z_{ijk}^K = *0%*. Moreover, if in the fuzzy region there are no objects at all, then the value of Z_{ijk}^K is fixed arbitrarily as *0%*, and the name of the class is not defined. The fuzzy if-then rule characterised by the zero value of Z_{ijk}^K is called artificial and does not take part in classification of new objects. Since the value of Z_{ijk}^K depends exclusively on how the known database objects are distributed in $A_{iX}^K \times A_{jY}^K \times A_{kZ}^K$, it is characteristic of that region and is called its confidence.

All triangular functions $\mu_{iX}^K \times \mu_{jY}^K \times \mu_{kZ}^K$ which define fuzzy regions operate over the same 3D space. Consequently, all rules obtained from the division of the space into regions can be used to classify a new object. However, it may turn out that although a new object Q is in a region $A_{iX}^K \times A_{jY}^K \times A_{kZ}^K$, the classification certainty is higher for a rule assigned to some neighbouring region (i.e. $A_{i'X}^K \times A_{j'Y}^K \times A_{k'Z}^K$ $i' = i\text{-}1,\ i,\ i\text{+}1;\ j' = j\text{-}1,\ j,\ j\text{+}1;\ k = k\text{-}1,\ k,\ k\text{+}1$). Such a situation can occur when the confidence value in $A_{iX}^K \times A_{jY}^K \times A_{kZ}^K$ is much lower than in a neighbouring region. Differences in particular confidence values result from different distribution of known database objects in particular regions. It is therefore assumed that a new object Q belongs to class C of that rule for which the object classification confidence value is the highest.

If not all database objects (each treated as a new one) can be correctly classified (i.e. with the same result as the one arrived at by a human-expert) by means of the obtained rules, a successive division into still smaller regions should be performed. In this way, a greater number of rules can be obtained, which, in turn, yields better classification precision. Successive divisions of *XYZ* are carried out until correct classification of all objects by means of the obtained fuzzy rules has been attained.

It should be noted here that the greater the number of rules in successive divisions, the greater also the number of operations needed to classify a new object, and unfortunately the longer the time necessary to perform classification. The most serious drawback of this technique is that in still smaller regions that it eventually creates there are fewer and fewer known objects from the database. Consequently, there appear a few tiny regions with confidence equal to *100%* together with numerous regions with zero confidence which cannot be used for classification purposes. Therefore more frequently classification of new objects is not performed, which is illustrated in Fig. 13.5.

To prevent this, it is advised to use for classification not only the rules from the last division of a given space into regions but all the rules obtained through all successive divisions. This procedure allows a considerable reduction in the number of new objects that cannot be subject to classification. However, this technique is not devoid of drawbacks, either, since it leads to a significant increase of the number of classification rules (Fig. 13.6).

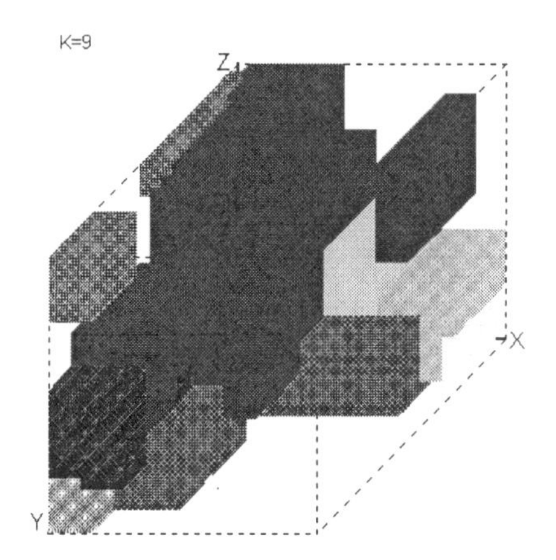

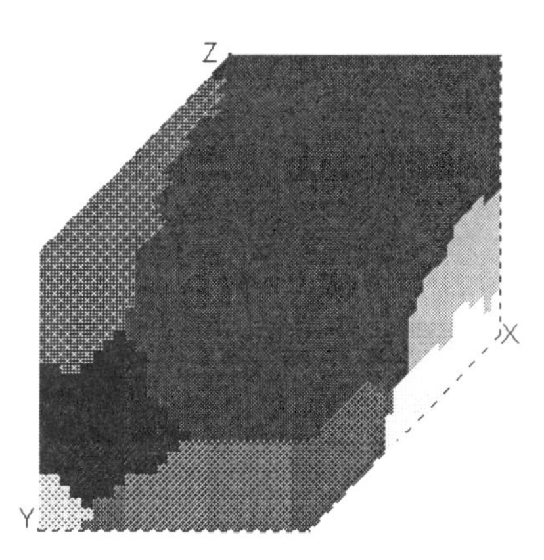

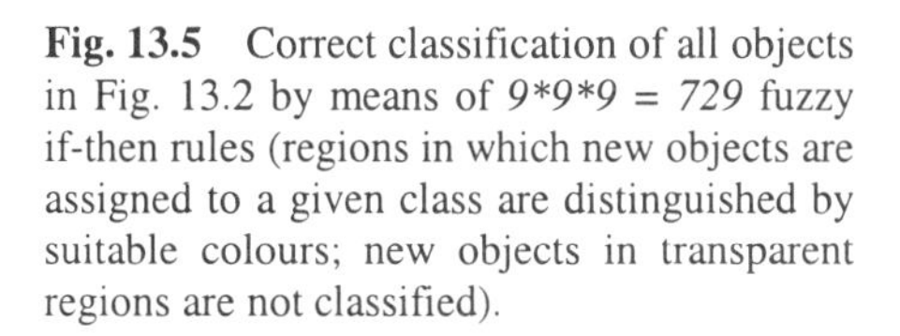
Fig. 13.5 Correct classification of all objects in Fig. 13.2 by means of $9*9*9 = 729$ fuzzy if-then rules (regions in which new objects are assigned to a given class are distinguished by suitable colours; new objects in transparent regions are not classified).

Fig. 13.6 Correct classification of all objects in Fig. 13.2 by means of all $2^3+3^3+4^3+5^3+6^3+7^3+8^3+9^3 = 2024$ fuzzy rules.

It should be possible to attain a significant reduction in calculations needed to classify a new object if, under the condition that classification correctness is kept for the greatest possible number of known database objects, the number of rules used for classification could be reduced.

Let *S* denote the set of all rules obtained from all performed divisions of the space and let *B* stand for the set of rules selected for classification. In order to carry out the selection of rules from S to B, with the condition that objects from the database are as correctly classified as possible, the genetic algorithm can be applied.

13.2.4 Selection of significant rules

The genetic algorithm aims at minimisation of the number of fuzzy classification rules so that correctness of the performed classification be unaffected. To this purpose, the genetic algorithm finds the minimal value of a function $f(B) \rightarrow \Re$ given by (5)

$$f(B) = a \cdot k(B) - b \cdot |B| \tag{5}$$

where B is a set of fuzzy if-then rules selected for classification, $|B|$ is the power of B, $k(B)$ denotes the number of correctly classified data describing the technical condition of a given oil transformer, a and b are real positive numbers and $a \geq b > 0$.

To make the genetic algorithm work, set B must be encoded in a so-called chromosome of an individual. Each gene in the chromosome is assigned one rule from S. If the rule is used for classification, the gene takes the value 1, if it is not – the gene has the value 0. Rules whose confidence value is 0% are always represented in the chromosome by a gene of value 0. In this way, set B becomes characterised by the distribution of 1's in the chromosome. The genetic algorithm tends to minimisation of the number of 1's in chromosomes of a given population so that the final distribution of 1's can represent the set of rules that classify correctly the greatest possible number of the known objects from the database. Adaptation function in the genetic algorithm is defined by dependence (5).

13.2.5 Results

Since the genetic algorithm does not guarantee that the global maximum of function f is obtained, its m best results are stored in a knowledge base (in the form of a base of rules). The fuzzy rules in the base are used for classification of new DGA results. A new object is located in a class assigned to it by that particular result B_m of the genetic algorithm for which the multiplication of the performed classification confidence and result adaptation function $f(B_m)$ takes the highest value. Using this procedure one eventually arrives at an expert system which operates like a team of experts.

Analysing Figs 13.7 and 13.8 it is not difficult to notice that despite considerable differences there are also certain similarities in the classification methods applied by different experts. The differences follow from expert's arbitrariness in interpretation of regions in which no known objects have occurred, which is also characteristic of human behaviour. Similarities result from the necessity of correct classification of objects which are in the database.

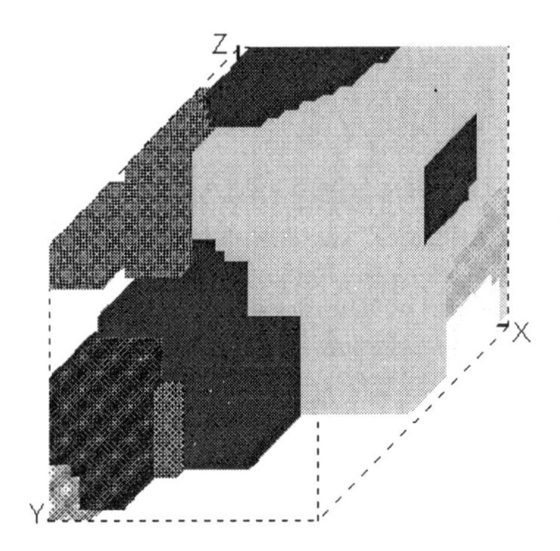

Fig. 13.7 Correct classification of all objects in Fig. 13.2 by means of 30 fuzzy if-then rules.

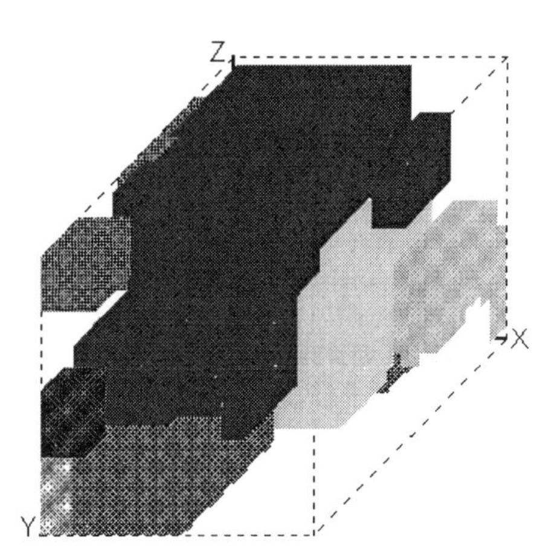

Fig. 13.8 Correct classification of all objects in Fig. 13.2 by means of 35 fuzzy if-then rules.

Figs 13.7 and 13.8 show that the regions in which objects are classified as overheating look alike. That confirms views of human-experts in whose opinion the diagnosis of overheating is relatively easy since overheating results in quite specific gas concentrations in transformer oil.

The best result of the genetic algorithm, a set of 30 fuzzy if-then rules, has been reached for $a=26$ and $b=1$ in (4.1) (the classification performed with these rules is shown in Fig. 13.7). For these values of a and b all results obtained with the use of the genetic algorithm and stored in the base of rules were able to classify all objects from the database.

Thus, an expert system constructed in the way described above comprises all techniques that prove useful for a person diagnosing the technical condition of a transformer on the basis of DGA results. This is because the base of rules is created from the data in the database, which in turn are obtained on the basis of a final diagnosis provided by a man-expert. Consequently, in the way it operates the expert system reflects preferences of the human-expert in the choice of suitable techniques of interpretation of DGA results.

13.3 NEURAL NETWORKS AND STOCHASTIC OPTIMISATION FOR INDUCTION MOTOR PARAMETER IDENTIFICATION

13.3.1 Introduction

In order to predict the performance of a motor one should know its electrical and mechanical parameters, which are sometimes hardly available. On the choice of such parameters is crucial in order to achieve the optimal performance of the motor. Among the most important performance characteristics there are: starting torque, breakdown torque, full-load torque, starting current, efficiency and so on. The present paper aims at identifying the motor parameters from available performance data using the equivalent circuit model of an induction machine. The success of the previously used methods depended highly on the selection of good initial estimates, Nolan et al. [18].

As known, artificial neural networks and stochastic algorithms work well even for discontinuous functions, given any initial guess. Therefore, we apply a multilayer feedforward neural network with the Levenberg-Marquardt learning rule which is much faster and reliable than the traditional steepest descent rule. For the same purpose we use a genetic algorithm with elitist strategy. To assess and verify the results a simulated annealing algorithm is finally used. We minimise three torque error functions with the following design variables: stator resistance, rotor resistance and leakage reactance which is composed of the stator and rotor leakage reactances. The numerical results compare well with the actual parameters of small induction motors.

13.3.2 Neural networks and function approximation

Neural networks are intrinsically function approximators and can be classified as functions since they produce a unique output for a given input. Multilayer feedforward networks can easily handle complex non-linear relations, provided there is enough hidden layers in the network. As proved by Kolmogorov, Hornik et al. [20] any continuous function of N variables can be computed by a three-layer perceptron-like network with N(2N+1) neurons using a continuously increasing activation function, Neittaanmäki et al. [3]. Such a function performs a sort of smooth threshold on the signal of a neuron. The sigmoid activation (or transfer function) is mostly used, e.g. logistic or hyperbolic tangent function. A three-layer feedforward network can generate arbitrarily complex decision regions in the space spanned by the inputs. The capabilities of multilayer feedforward networks for learning a function approximation can be summarised as follows:

A network with one hidden layer can learn the function consisting of a finite set of points and the continuous function defined on a compact domain. Under some conditions discontinuous functions or functions without a compact support can also be learned. All other functions can be learnt using a network with two hidden layers.

13.3.3 Artificial neural networks for inverse electromagnetic modelling

Explicit inversion of complex physical models requires, in general, a numerical approach. Artificial neural network can help in solving this problem. If the cause (represented by a real valued vector **X**) produces a result **Y**, then a model *m* by which we can predict **Y** for any appropriate **X** can be represented by the relation

$$\mathbf{Y} = m(\mathbf{X})$$

A representative set of **Y** is collected in order to apply the model to each **X**. Multiple samples of **Y** must be generated if the model is stochastic. The **X** and **Y** are used as training sets. Each **Y** is presented to the network inputs and the network is trained to produce the corresponding **X** on its output according to the model *m*. Once appropriately trained, the network is capable of representing an unknown relation: in particular that of representing the performance criteria for an induction motor in question. For that purpose we apply a three-layer feedforward neural network with the Levenberg-Marquardt non-linear least-squares learning rule. This algorithm is capable of recovering from numerous local minima of the network error function, thus enabling us to find a global minimum of the error function. The used network has the hyperbolic tangent hidden layer followed by an output layer of linear neurons. As stated above, such a network can approximate any function with a finite number of discontinuities arbitrarily well, given sufficient neurons in the hidden layer.

13.3.4 Simulated annealing

This approach is nowadays well known and recognised as a remedy for the optimisation of very difficult functions, Neittaanmäki et al. [3]. The basic idea is to introduce a thermal noise into the optimisation problem, thus considering it as an artificial physical system in search of thermal equilibrium (metastable configurations of lowest energy). Starting at high enough initial temperature the system is gradually cooled until the convergence is achieved. It is important to keep the system long enough at each temperature. Moreover, the cooling rate should be slow enough to ensure the exhaustive search of the design space. This may be achieved by using the so called Cauchy machine which is much faster than classical Boltzmann scheme, Szu and Hartley [23].

13.3.5 The problem

An induction machine can be represented by the equivalent circuit shown below (Fig. 13.9).

We wish to identify the equivalent circuit parameters of the induction motor. The three torque functions are as follows, Nolan et al. [18]:

$$F_1(R_1, R_2, X_L) = \frac{U^2 R_2}{s\omega_s\left[\left(R_1 + \frac{R_2}{s}\right)^2 + X_L^2\right]} - T_{fl}$$

$$F_2(R_1, R_2, X_L) = \frac{U^2 R_2}{\omega_s\left[(R_1 + R_2)^2 + X_L^2\right]} - T_{lr}$$

$$F_3(R_1, R_2, X_L) = \frac{U^2}{2\omega_s\left[R_1 + \sqrt{R_1^2 + X_L^2}\right]} - T_{bd}$$

where

R_1, R_2 - stator and rotor resistances, respectively,

$X_L = X_1 + X_2$ - leakage reactance,

X_1, X_2 - stator and rotor leakage reactances, respectively,

U - applied voltage,

ω_s - synchronous speed,

s - slip,

T_{fl}, T_{lr}, T_{bd} - full load torque, locked rotor torque and breakdown torque, respectively.

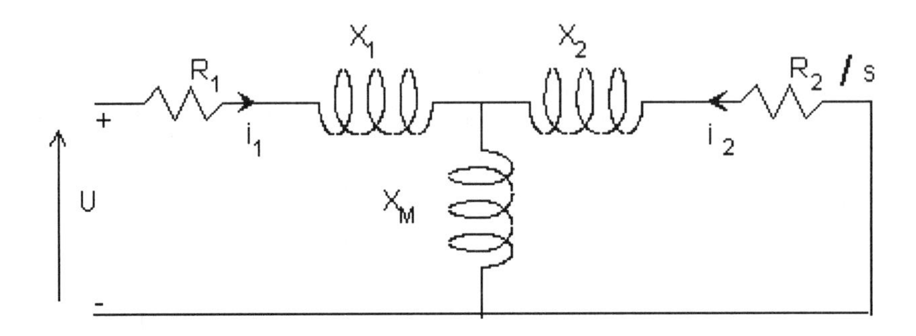

Fig. 13.9 Induction motor equivalent circuit.

The objective function to be maximised is the fitness defined as

$$\theta = \frac{1}{F_1^2 + F_2^2 + F_3^2}$$

The design parameters are the stator and rotor resistances and leakage reactance which is a combination of the stator and rotor reactances. The magnetising reactance can be calculated after the design parameters have been calculated using the full load power factor equation.

In the following we examine three induction motors with known parameters (Table 13.1) to test our three methods: genetic algorithm, three layer feedforward neural network and fast simulated annealing.

Table 13.1 Actual parameters of three induction motors.

Horsepower	Speed	Voltage	R_1	R_2	X_L
5	188,5	230	0,434	0,303	2,511
50	178,55	460	0,087	0,228	0,604
500	185,67	2,30	0,262	0,187	2,412

In Figs 13.10 and 13.11 we present the results for the first motor. In Fig. 13.10 we demonstrate the results of our neural network testing. The original surface plot of the objective function is shown along with the learned mapping by the neural network (circles).

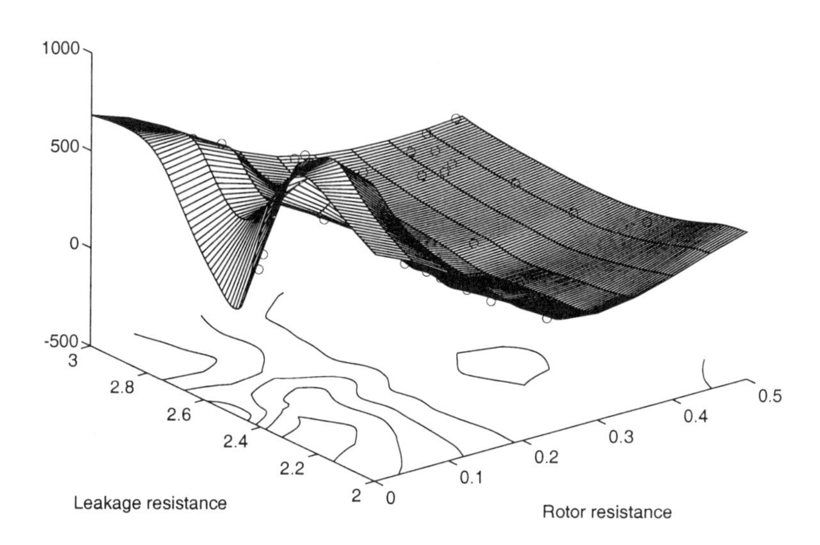

Fig. 13.10 Objective function for the first motor.

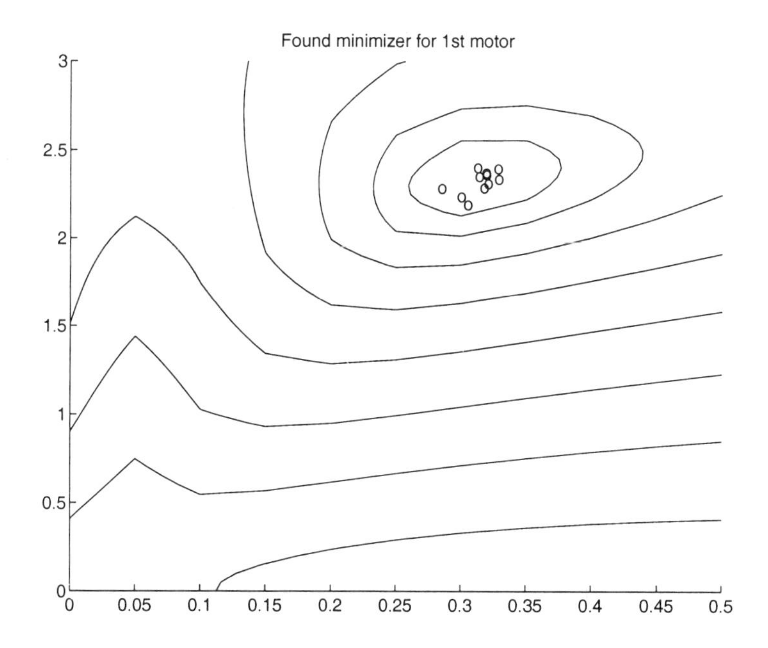

Fig. 13.11 The results found by the genetic algorithm.

The results of numerical simulation for the first motor are summarised in Table 13.2.

Table 13.2 Comparison of numerical results for the first motor.

Actual values		Fast simulated annealing	Genetic algorithm	Newton-Raphson method
R_1	0,434	0,39	0,367	0,392
R_2	0,303	0,313	0,314	0,313
X_M *)	2,511	2,377	2,386	2,375

*) magnetising (mutual) reactance

In Figs 13.12 and 13.13 the results of genetic optimisation for the second motor are shown.

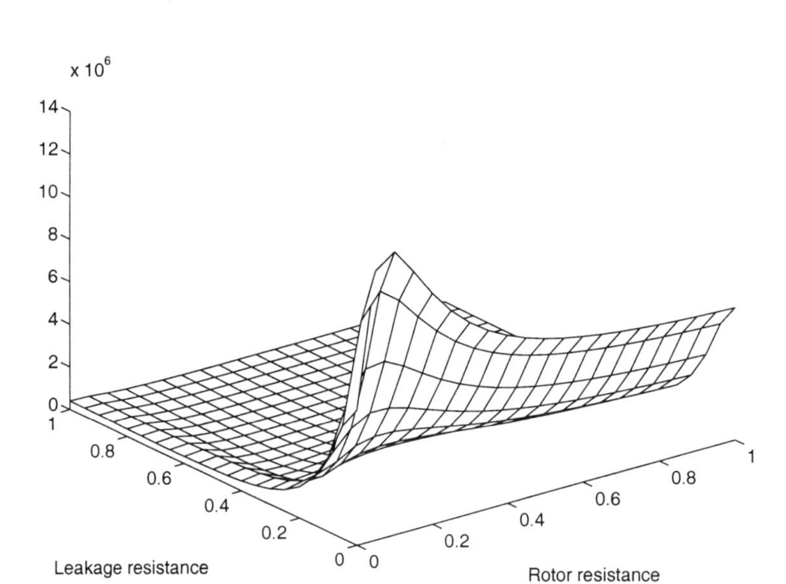

Fig. 13.12 Objective function for the second motor.

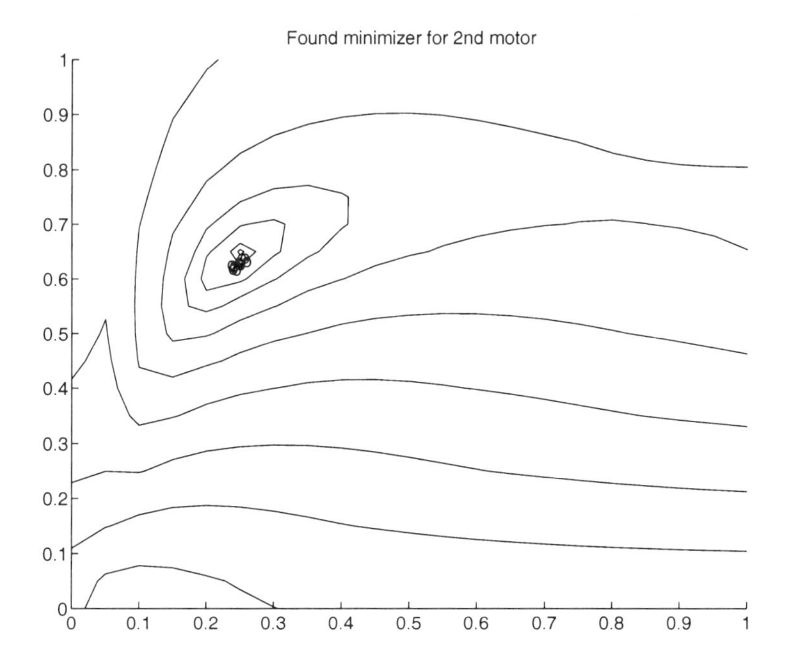

Fig. 13.13 Found minimizer for the second motor.

13.4 CONCLUSION

As demonstrated above, genetic algorithms and artificial neural networks are reliable tools for identification of induction motor parameters. As opposed to the Newton-Raphson method which depends heavily on the starting point, genetic algorithm always converges to a sub-optimal solution from any initial guess. A trained feedforward network is capable of approximating an unknown mapping, the objective function. The same conclusions hold true for other two induction motors.

13.5 SUMMARY

In the paper we demonstrated that genetic algorithms can be effectively used for the solution of electrical engineering problems. Two applications are given. In each case final conclusions are drawn.

REFERENCES

1. D.E. Goldberg, Genetic Algorithms in Search, Optimisation and Machine Learning, Addison-Wesley, 1989.

2. Z. Michalewicz, Genetic Algorithms + Data Structures = Evolution Programs, Springer-Verlag, 1992.
3. P. Neittaanmäki, M. Rudnicki and A. Savini, Inverse Problems and Optimal Design in Electricity and Magnetism, Clarendon Press, Oxford, 1996.
4. J. E. Baker, "Reducing bias and inefficiency in the selection algorithm", Proc. ICGA, pp. 14-21, 1987.
5. A. Chipperfield, et al., Genetic Algorithm Toolbox for Use with Matlab: User's Guide, University of Sheffield, U.K., 1997.
6. M. Rudnicki and A. Savini, "Application of neural networks and genetic algorithms to an induction motor parameter identification", Proc. X Int. Conference on Servomotors and Microdrives, Rydzyna, Poland, September 1996, pp. 79-83.
7. P. Szczepaniak and M. Cholajda, "Genetic interpretation of DGA results in power transformers", Proc. Int. Conf. SSCC'98: Advances in Systems, Signals, Control and Computers, Durban, South Africa, 1998, vol. II, pp. 216-220.
8. Z. Stein, Machines and Electric Drive, Warsaw, WSP, 1995 (in Polish).
9. D. Przybylak, J. Roganowicz, An Expert System to Monitor Technical Condition of Transformers, Technical Report, Power Engineering Institute, Dept. of Transformers, Lodz, Poland, 1995 (in Polish).
10. H. Ishibuchi, et al., Selecting Fuzzy If-Then Rules for Classification Problems using Genetic Algorithms, IEEE Trans. on Fuzzy Systems, Vol.3, No.3, 1995, pp. 260-270.
11. International Electrotechnical Commission, Interpretation of the Analysis of Gases in Transformers and other Oil-Filled Electrical Equipment in Service, Geneva, 1979.
12. L. J. Fogel, A. J. Owens and M. J. Walsh, Artificial Intelligence through Simulated Evolution, Wiley , New York, 1966.
13. L. Davis (Ed.), Handbook of Genetic Algorithms, Van Nostrand Reinhold, New York, 1991.
14. J. H. Holland, Adaptation in Natural and Artificial Systems, Ann Arbor, MI: University of Michigan Press, 1975.
15. J. R. Koza, Genetic Programming, MIT Press, Cambridge, MA, 1992.
16. I. Rechenberg, Evolutionsstrategie: Optimierung technischer Systeme nach Prinzipen der biologischen Evolution, Fromman-Holzboog Verlag, Stuttgart, 1973.
17. H. P. Schwefel, Numerical Optimisation for Computer Models, Wiley, Chichester, 1981.
18. R. Nolan, R. Pilkey and T. Haque: "Application of Genetic Algorithms in Motor Parameter Determination", Conference Records of IAS Annual Meeting, October 2-5, 1994, Denver, CO, USA.
19. Neural Network Toolbox User's Guide for Matlab, The MathWorks, 1994.
20. K. Hornik, M. Stinchcombe and H. White: Multilayer feedforward networks are universal approximators, Neural Networks, 4(2), 1991.

21. S.R.H. Hoole: Artificial neural networks in the solution of inverse electromagnetic problems, IEEE Trans. on Magnetics, 29(2), 1993.

22. L. Ingber and B. Rosen: Genetic algorithms and very fast simulated reannealing: a comparison, Math. Comput. Modelling, 16, 1992.

23. H. Szu and R. Hartley: Fast simulated annealing, Physics Letters A, 122, 1987.

14 Genetic Algorithms in Shape Optimization: Finite and Boundary Element Applications

M. CERROLAZA and W. ANNICCHIARICO

Institute for Materials and Structural Models, Faculty of Engineering
Central University of Venezuela
PO Box 50.361, Caracas 1050-A, Venezuela
Email: mcerrola@reacciun.ve, annwill@sagi.ucv.edu.ve

Abstract. This work deals with the continuous and discrete optimization of 3D truss structures as well as with the shape optimization of bidimensional Finite Element (FE) and Boundary Element (BE) models, by using Genetic Algorithms and β-splines boundary modelling. Several successful attempts have been made in order to optimize shape of continuous models to reduce their area and to reduce the internal stresses. Here, the authors present and discuss a new methodology to shape optimization based on the use of GAs and β-spline modelling. Internal variables to be controlled are discussed, then introducing some concepts of β-splines formulation. A software based on Genetic Algorithms developed for FE models is also presented. Finally, some engineering examples are included in order to show that the technique proposed is able to deal with FE and BE models.

14.1 INTRODUCTION

Shape optimization of continuum models is a subject of the most concern to design engineers. For a long time, this field of research has deserved great attention from the numerical analysis community, and many available optimization techniques have been developed and successfully used in engineering analysis and design. In this class of problems, aspects such as geometric definitions, mesh generation, analysis and displaying of results are usually involved. Furthermore, other elements play a decisive role in the optimization process, such as sensitivity analysis and numerical optimization programming.

The problem can be divided into three main tasks. The first step is to define the geometric and the analytical models. The geometric model is where the design variables are easily imposed and it allows an explicit integration with other design tools, such as CAD or CAM systems. On the other hand, the analytical model is used to obtain the structural response of the part, subjected to external actions. Then, a sensitivity analysis must be done to obtain a solution of the problem; and finally, an appropriate optimization algorithm must be selected to solve the optimization problem in an effective and reliable way.

Since Zienkiewicz and Campbell (1973) initiated the numerical treament of shape optimization problems, a large number of publications have appeared in the field of shape sensitivity analysis. Haug et al. (1986) presented a unified theory of continuum shape design sensitivity using a variational formulation for the governing equations of structural mechanics and material derivatives concepts. Several researches, see for instance Yang and Botkin (1986), Yao and Choi (1989), have investigated the accuracy of the shape design sensitivity theory utilizing the finite element method because of its variational formulation. Generally, the shape design sensitivity formula requires the evaluation of accurate stress quantities on the boundary, which is usually difficult to obtain by the FEM. The Boundary Element Method has become a popular alternative in shape design sensitivity analysis due to its accuracy of the boundary displacement and stress solutions as well as the fact that remeshing is easier for the BEM than for the FEM. The evaluation of design sensitivities are the adjoint method and direct differentiation method. The reader can find many works in the technical literature on this subject. Choi and Kwak (1990), Baron and Yang (1988), Kane and Saigal (1988), Mellings and Aliabadi (1993) and Mellings (1994), have presented numerical results in optimization of potential and two-dimensional elasticity problems. Also, Meric (1995) presented a sentivity analysis of 2D shape optimization with BEM, comparing integral and differential formulations in heat conduction problems. Another work by Parvizian and Fenner (1997) optimizes 2D boundary element models using mathematical programming and normal movement techniques.

The recent paper of Kita and Tanie (1997) proposed an approach based on GAs and BEM, where the initial mesh is setted up with a rectangular mesh of nodes. These nodes move their position, following genetic optimization, until the optimal shape is reached. The results are encouraging although the user can't define an initial real mesh. Also, Periaux et al (1997, 1998) have presented interesting applications of shape optimization of aircraft-wings by using GAs and FE models.

Initially, some authors such as Zienkiewicz and Campbell (1973), Ramakrishnan and Francavilla (1975) among others, did not use geometric modeling in the shape optimization problems addressed by them. Instead, they defined the nodal coordinates of the discrete finite element model as design variables. This approach has limitations such as:

1. It requires a large number of design variables and tends to produce jagged-edges shapes.

2. A large number of constraints must be added in order to generate smooth boundaries, which complicates the design task.

3. The lack of an associated geometric model does not allow the integration with powerful design tools like CAD or CAM systems.

The success of any optimization methodology hangs on its ability to deal with complex problems as is the case of shape optimization design. Solving non-linear optimization problems efficiently is challenging. Furthermore, it is quite common in practice that methods are modified, combined and extended in order to construct an algorithm that matches best the features of the particular problem at hand. For these reasons, emphasis have to be made on the basis of various methods in order to choose the most appropriate.

The search for a robust optimization algorithm, with good balance between efficiency and efficacy, necessary to survive in many different environments, has led to the use of Genetic Algorithms (GAs). Genetic Algorithms have many advantages over traditional searching methods. Among other considerations, they do not need further additional information than objective function values or fitness information. This information is used to determine the success or failure of a design in a particular environment. Moreover, its ability to find out the optimum or quasi-optimum solution gives it a privileged position as a powerful tool in non-conventional optimization problems. All of these advantages have led to propose the combination of GAs with geometric modeling, by using β-splines curves, in order to solve non-conventional problems like shape optimization problems.

In CAD systems, the geometric modeling is carried out by using Lagrange polynomials, Bezier curves, B-splines and β-splines curves and surfaces and coons patches (see, for instance, Schramm and Pilkey (1993), Othmer (1990) and Olhoff et al. (1991)). Among these approaches, the β-spline curve properties can be easily controlled by using its parameters. Other advantages that encourage the use of this technique are:

- The curve lies in a convex hull of the vertices.
- The curve does not depend on an affine-transformation.
- The tangents of the initial and final points are defined repectively by the first and the last edge of the polygon. The above characteristic allows the imposition of C^1 continuity (Schramm and Pilkey 1993).

In previous works (Annicchiarico and Cerrolaza 1998a, 1999a) it has been shown that GAs provide a powerful tool to optimize bidimensional finite-element (FE) models. However, the use of nodal coordinates in the FE model has proven to be somewhat difficult to control, which suggested the use of a class of spline-curves to model the boundary variation.

Thus, recent developments by the authors in shape optimization using the Finite Element Method and β-spline curves (Annicchiarico and Cerrolaza 1998b, 1999b), suggested the extension of some of the ideas previously presented to their application to the genetic optimization of bidimensional engineering problems discretized by the Boundary Element Method. This paper collects some of the results presented in the previous cited works.

14.2 A SHORT REVIEW ON GENETIC ALGORITHMS

Nowadays, engineers are faced up to optimize more and more their designs, due to the high costs involved in their production. The first phase in optimal design consists in the definition of the basic characteristics of the final product that will be optimized. These features could be the size, thickness, shape or topological configuration of their members. In genetic algorithms, a particular design (a) is represented by a set of its characteristics a_i, called phenotype and defined as real numbers

$$A = a_1 * a_2 * a_3 * \cdots * a_c = \prod_{i=1}^{c} a_i \qquad a_i \in \Re . \tag{1}$$

The phenotype of a structure is formed by the interaction of the total package with its environment and his **genotype** is obtained by encoding each a_i into a particular code (binary in this case). The transformation of the phenotype structure into a string of bits leads to the so-called *chromosomes*, and it represents, like in natural systems, the total genetic prescription for the construction and operation of some organism

$$A = \prod_{i=1}^{c} a_i = \prod_{i=1}^{c} \left(e\colon a_i \rightarrow \{0,1\}\right) \tag{2}$$

Genetic Algorithms operate on populations of strings (structures) and progressively (t= 0,1,2...) modifies their genotypes to get the better performance of their phenotype environment E.

The adaptation process is based on the mechanics of natural selection and natural genetics. They combine the survival of the fittest together with string structures, with a structured yet randomized information, which is exchanged to form a search algorithm. In each generation, a new set of artificial creatures (strings) is generated by using bits and pieces of the fittest of the previous generation. They efficiently exploit historical information to speculate on new search points with expected improved performance.

In order to use GAs we have to define an objective function or fitness function, that measures the behavior of each individual into its environment. This function provides a direct indication of the performance of each individual to solve the optimization problem subjected to the imposed constraints of the environment. With this population ranked according to fitness, a group of chromosomes are selected from the population. There exist several methods to select parents. In this work, the following methods have been used (Brindle, 1981):

- Stochastic sampling with replacement.
- Remainder stochastic sampling without replacement.

The selected chromosomes are then reproduced through the crossover and mutation operators. The crossover operator consist in taking two selected chromosomes as parents. Then, they are either crossed by using a certain probability value in order to create two new chromosomes (children) or they are directly included into the new population. It can be found many procedures to carry out this task in the technical literature (Davis, 1991). The present work used both one break-point (simple crossover) and two break-points (double crossover), since these methods lead to a simple and reliable solutions.

In simple crossover, an integer position n_1 along the string is randomly selected between 1 and the string length less one [1,1,-1]. Two new strings are then created by swapping all the characters between positions $n_1 + 1$ and 1 (see Figure 14.1).

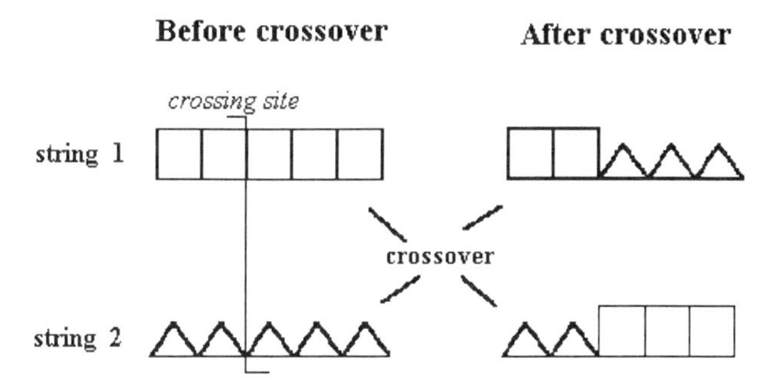

Figure 14.1 Graphic scheme of simple crossover procedure.

The two break-points method is based in the interchange of the sub-string placed between bits n_1 and n_2 of the parents. Both numbers are randomly chosen, in the same way as simple crossover.

The mutation operator gives each bit in a chromosome the opportunity of changing (if it is 1 to 0 or vice versa). The selection according to the fitness, combined with the crossover, provides genetic algorithms the bulk of their processing power. Mutation plays a secondary role in the operation of the genetic algorithms and it is needed because, even though selection and crossover effectively search and recombine extant notions, occasionally they may become overzealous and they can lose some potentially useful genetic material (1's or 0's at particular locations). Thus, the mutation operator protects against such irrecoverable premature loss of important notions. Due to the secondary importance of this operator, a low mutation probability value is usually considered.

What is the power involved behind these simple transformation over a random population of n strings, that allow genetics algorithms to find the optimum point (or nearly the optimum point) in complex and non-linear situations? The answer of this question was found by John Holland and it is exposed in the 'Schema Theorem' or 'The Fundamental Theorem of Genetic Algorithms' (Holland, 1971, 1973).

A schema is a similarity template describing a subset of string displaying similarities at certain string positions. It is formed by the ternary alphabet {0.1,*}, where * is simply a notational symbol, that allows the description of all possible similarities among strings of a particular length and alphabet. In general, there are 2^l different strings or chromosomes of length l, but schemata display an order of 3^l. A particular string of length l inside a population of n individuals into one of the 2^l schemata can be obtained from this string. Thus, in the entire population the number of schematas present in each generation is somewhere between 2^l and $n.2^l$, depending upon the population diversity. But, how many of they are processed in a useful way? Clearly, not all of them have the same possibility to carry on its genetic information through each generation, since genetic operators could destroy some of them. J. Holland (1975) estimated that in a population of n chromosomes, the GAs process $O(n^3)$ schematas into each generation. He gives this a special name: 'Implicit Parallel Process', which can be observed in Figure 14.2.

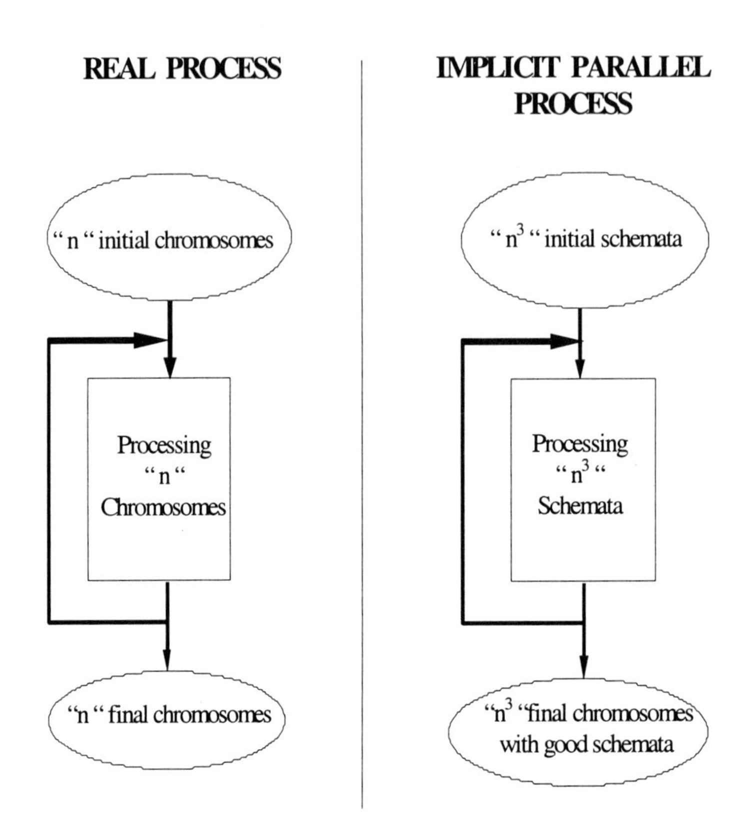

Figure 14.2 Implicit parallel process.

Even though at each generation we perform a proportional computation to the size of the population n, we get useful processing of n^3 schematas in parallel with memory other than the population itself. Holland has shown that 'good schemata', mainly related to high fitness values, are propagated generation to generation by giving exponentially increasing samples to the observed best; and bad schemata disappear. At the end of the generation process, the chromosomes in the population are only formed from good schemata, which he called 'Building Blocks'.

14.2.1 Some refined operators

There exists several new operators (see for instance Davis, 1991) that can be used together with the basic reproduction, crossover and mutation operators, explained before.

In the present work, it has been used the following refined operators: rebirth, elithism and sharing functions to induce niche exploitation and speciation throughout the population, when dealing with multicriteria optimization.

14.2.1.1 Rebirth Rebirth isn't really a genetic operator, because it doesn't work directly over the chromosome. This operator works over the procedure to obtain the best chromosomes. The basic idea of this operator is to obtain a good fit, in order to get the desired optimum.

Genetic algorithm is a stochastic and dynamic search process, which achieves a near optimal solution of the objective scalar function in every evaluation. This near optimal solution means:

$$\frac{\left|O_i - O_{bj}\right|}{O_i} \leq \varepsilon_k \;;\; 0 \leq \varepsilon_k << 1 \quad , k = 1,2,\cdots,nc \tag{3}$$

where

O_I : the ideal objective scalar function;

O_{bj}: is the best objective scalar function in run j;

ne : is the number of runs.

Also in the phenotype structure

$$\frac{\left|I_k - b_{jk}\right|}{I_k} \leq \varepsilon_k \;;\; 0 \leq \varepsilon_k << 1 \quad , \; k = 1,2,\cdots,nc \tag{4}$$

where

I_k : is the phenotype k of the ideal chromosome I.

b_{jk}: Is the phenotype k of the best chromosome in the process j.

nc : is the number of chromosomes.

When the genetic algorithm converges to the near optimal solution, the process is stopped and it is made the 'rebirth' of a new population, thus creating random chromosomes in a subspace of the initial phenotype space. This subspace is defined by taking a range of variation (r) for each phenotype, which is less than the one used during the initial step, and by taking the best last value obtained before rebirth as the starting of the interval for each new chromosome. So, if P is the best last chromosome, then the new phenotype structures are bound in the following way

$$P_i - r \leq a_i \leq P_i + r \qquad i = 1,2,\cdots,nc \tag{5}$$

In this way, it is possible to fit the searching of the optimum, because a new process is started with a new random population created into a reduced phenotypical space of the best chromosomes belonging to the initial step.

14.2.1.2 Elitism The operations of crossover and mutation can affect the best member of the population, which doesn't produce offspring in the next generation. The elitist strategy fixes this potential source of losing by copying the best member of each generation into the succeeding generation. The elitist strategy may increase the domination speed exerted by a super-individual on the population. However, it seems to improve the genetic algorithm performance.

14.2.1.3 Niche and speciation Sometimes, when several criteria are used simultaneously and there is no possibility to combine by using a single number. When dealing with this situation, the problem is said to be a multiobjective or multicriteria optimization problem. This sort of problems do not have a unique optimal solution in the strict sense of the word, since the algorithm provides the better way to combine the criteria in order to obtain a plausible solution, but it isn't the unique. The final selection depends on the designer's criterion.

When using genetic algorithms, the way to deal with multicriteria problems is by employing nichelike and specieslike behavior in the members of the population (Goldberg, 1989; Horn and Nafpliotis, 1993). In order to induce stable sub-population of strings (species) serving different sub-domains of a function (niches), it is then necessary to define a sharing function to determine the neighborhood and degree of sharing for each string in the population (Goldberg and Richardson, 1987).

First of all, it is needed to count the number of individuals belonging to the same niche. This is carried out by accumulating the total number of shares, which is determined by summing the sharing function values provided by all other individuals in the population, in a factor called *niche count* (m_i)

$$m_i = \sum_{j \in Pop} Sh[d(i,j)] \tag{6}$$

where d(i,j) is the distance between individuals i and j and Sh[d] is the sharing function. Sh[d] is a decreasing function of d(i , j), such that Sh[0] =1 and $Sh[d \geq \sigma_{share}] = 0$. Usually a triangular sharing function is used

$$\begin{aligned} Sh[d] &= 1 - \frac{d}{\sigma_{share}} && \text{for } d \leq \sigma_{share} \\ Sh[d] &= 0 && \text{for } d > \sigma_{share} \end{aligned} \tag{7}$$

here σ_{share} is the niche radius, defined by the user at some estimates of the minimum separation desired or expected between the desired species.

Then, the selection of the best no-dominate individuals to go on reproduction will be made over the one that has the less niche count m_i. In this way, the pareto frontier will be formed by a wide variety of individuals.

14.3 BOUNDARY MODELLING USING β-SPLINES CURVES

The β-spline curves are geometric design elements, which stem from the well-known β-spline curves. This formulation (Barsky, 1988; Bartels et al, 1987; Zumwalt and El-Sayed, 1993) defines additional parameters that control the bias (β_1) and tension (β_2) of each curve segment. The effect of these parameters is to change the parametric continuity between the curve segments while maintaining the geometric continuity.

The parametric equations of β-Splines can be visualized as the path of a particle moving through space. Increasing β_1 above unity, the velocity of the particle immediately after a knot point increases. This serves to push it further in the direction of travel before it turns as influenced by the next control point. This is said to bias the curve to the right. Decreasing β_1 below unity, the particle velocity decreases and thus, it biases the path towards the left. The parameter β_2 controls the tension in the curve. As β_2 is increased above zero, the knot points are pulled towards their respective control points. For negative β_2, the knot points are pushed away. There are two types of formulation:

- Uniformly shaped β-splines, where the parameters β_1 and β_2 have the same value along the entire curve.

- Continuously shaped β-splines, where β_1 and β_2 can have different values along the different curve segments.

The β-spline curves are defined with a set of points called control vertices (V_m). Although these points are not part of the generated curve, they completely define its shape. The vertices of a curve are sequentially ordered and connected in succession to form a control polygon (Figure 14.3). In this way, the β-spline curve approximates a series of m+1 control vertices with m parametric curve segments, $Q_i(u)$. The coordinates of a point $Q_i(u)$ on the ith curve segment are then given by

$$Q_i(u) = \sum_{r=-2}^{1} b_r(\beta_1, \beta_2; u)V_{i+r} \qquad 0 \le u < 1, \quad i = 0 \cdots m-1 \tag{8}$$

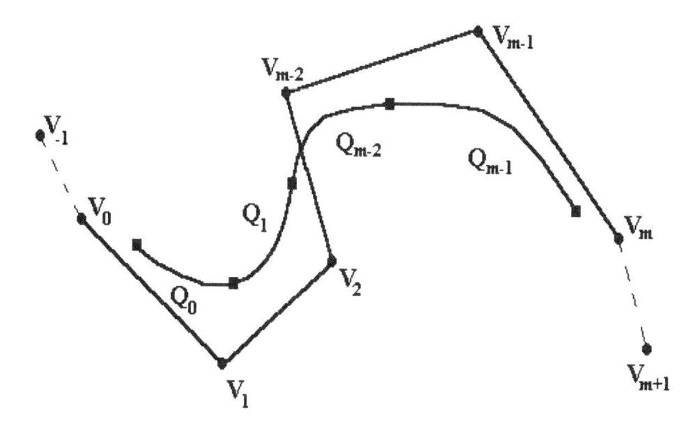

Figure 14.3 β-splines control polygon with end-vertex interpolation.

One of the most important advantages of this formulation is the local control. This control is obtained by the exploitation of the piecewise representation of the β-splines formulation, which is based in a local basis; that is, each β-spline basis function has local support (nonzero over a minimum number of spans). Since each control vertex is associated with a basis function, it only influences a local portion of the curve and it has no effect on the remaining part of the curve. Then, when moving a single control vertex, a localized effect on a predetermined portion of the curve or surface is obtained. In order to get local control, a β -spline curve segment is completely controlled by only four of the control vertices; therefore, a point in this curve segment can be regarded as a weighted average of these four control vertices. Associated with each control vertex is a weighting factor. The weighted factors b_r($\beta_1,\beta_2;u$) are the scalar-valued basis function, evaluated at some value of the domain parameter u, and of each shape parameter β_1 and β_2. If $\beta_1>0$ and $\beta_2 \geq 0$ then they form a basis; that is, they are linearly independent, and any possible β-spline curve segment can be expressed as a linear combination of them. Each basis function is a function of β_1 and β_2 and the u such that it is a cubic polynomial in u whose polynomial coefficients are themselves functions of β_1 and β_2:

$$b_r(\beta_1, \beta_2; u) = \sum_{g=0}^{3} C_{gr}(\beta_1, \beta_2) u^g \qquad (9)$$

$$\text{for } 0 \leq u < 1 \text{ and } \quad r = -2,-1,0,1.$$

Now, by applying the geometric continuity constraints to the joint of the ith and $(i+1)^{st}$ curve segments, the following conditions yield:

$$\begin{aligned} Q_{i+1}(0) &= Q_i(1) \\ Q_{i+1}^{(1)}(0) &= \beta_1 Q_i^{(1)}(1) \\ Q_{i+1}^{(2)}(0) &= \beta_1^{\,2} Q_i^{(2)}(1) + \beta_2 Q_i^{(1)}(1) \end{aligned} \qquad (10)$$

One more constraint is required in order to uniquely determine the coefficients functions C_{gr} . A useful constraint for axis independence and convex hull properties is to normalize the basis functions (that is, their sum is the unity), at u=0

$$C_{0,-2}(\beta_1,\beta_2) + C_{0,-1}(\beta_1,\beta_2) + C_{0,0}(\beta_1,\beta_2) + C_{0,1}(\beta_1,\beta_2) = 1 \qquad (11)$$

The first and second derivative vectors can be easily obtained from eq. (8). By substituting these expressions in eqs. (10) and (11), a system of linear equations follows. The solution of this system gives the Cgr coefficients. A detailed discussion of β-spline curves formulation can be found in Barsky (1988) and Bartels et al (1987).

The complete definition of a curve by an open β-splines formulation requires the specification of an end condition. Different techniques can be used to solve this problem. In this paper, the technique of phantom vertices was used. In this technique, an auxiliary vertex is created at each end of the control polygon. The auxiliary vertices are created for the sole purpose of defining the additional curve segments, which satisfy some end conditions. As these vertices are unaccessible to the user they are not displayed; thus, they will be referred to as phantom vertices.

It is frequently desirable and convenient to constrain the initial and terminal positions of the curve to coincide with the initial and terminal vertex, respectively; that is, the curve starts at V_o and ends at V_m. The phantom vertices can be obtained by using the following expressions:

$$V_{-1} = \frac{1}{\beta_1^3}\left[V_0 - V_1\right] + V_0$$

$$V_{m+1} = \beta_1^3\left[V_m - V_{m-1}\right] + V_m$$

14.4 A GAs SOFTWARE FOR FE MODELS OPTIMIZATION

This section describes a GAs software produced by the authors for the 2D optimization of FE models. The details of the software can be found in Annicchiarico and Cerrolaza (1999a). The **GENOSOFT** system is a powerful computational tool designed for the optimization of some structural engineering problem such as:

1. Sizing optimization of 2D and 3D truss structures and frame structures.
2. Geometric optimization of 2D and 3D truss structures and framed structures.
3. Shape optimization of 2D finite element models.

This software deals with many real engineering constraints and restraint conditions imposed on the analytical model and in the optimization process. This section describes the main features of the GENOSOFT system, which is basically organized as follows

GENOSOFT: Main module to drive the use of the four program modules.

GENOPRE: Interactive preprocessor to generate and edit the data of the analytical and optimization model.

GENOPT: Genetic optimization module that carry out all the tasks related with the optimization process.

GENOPRO: analysis module. It is made up by the following analysis modules:

- 2D and 3D truss structures,
- 2D and 3D frame structures,
- 2D finite element models.

GENOPOST: Graphic interactive post-processor for the graphic displaying and interpretation of the processor results.

The software is written in TurboPascal for personal computers and in Pascal for Unix workstations. The main flow chart of GENOSOFT system is depicted in Figure 14.4. It displays the relation among the different modules. Also, two inter-phases which allow the easy interchange of data between GENOPT and GENOPRO modules were developed.

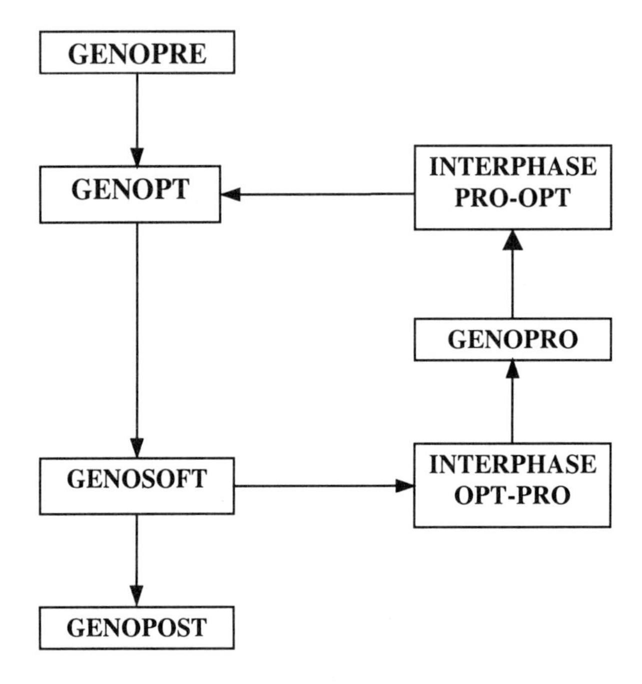

Figure 14.4 Main flow chart of GENOSOFT system.

The first inter-phase module is called **OPT-PRO** (optimization module-analysis module). Its main function is the identification of the design variables and their relation with Genopro. This inter-phase carries out the transformation of all the variables used in the optimization process to those parameters which modify the analytical model. A large variety of design variables can be defined, which can be classified into two types:

a) Variables that are defined over element properties such as:

- areas,
- thicknesses,
- dimension of cross sections: base and height.

b) Variables that are defined over geometry such as:

- nodal coordinates,
- restraints positions.

Once the mechanical behavior of the analytical model is obtained, it is needed to transform the different results in words that can be understood by the optimization module (Genopt). This task is carried out by the **PRO-OPT** interphase (analysis module-optimization module). This module calculates the restrictions imposed over the optimization model and the objective function values. In what follows, we will describe the main features of the GENOSOFT system programs.

14.4.1 Genopre: Data Preprocessor

The Data Preprocessor is divided in two main modules carrying out the following tasks: the preprocessing of the structural model and the preprocessing of the optimization model. The user can define the following types of elements:

- 2D and 3D truss structures,
- 2D and 3D frame structures,
- 2D finite element models.

In order to define the analytical model and in agreement with the above types of elements, the user has to supply the following information:

a) general: number of nodes, elements, restricted nodes, number of nodal and element loads, and the type of elements to be used in the structural model;

b) nodal coordinates;

c) element connectivities and attributes;

d) material properties information such as Young's module, Poisson's ratio, yield stress and ultimate stress of the concrete or steel;

e) geometric characteristics of the elements as, for example, cross sectional areas, inertia moment and gyration ratio;

f) boundary conditions, and

g) loading cases in nodes and elements.

Once the structural model is defined, the user can inspect it through an interactive graphic routine. This routine allows the checking of all the information that has been previously defined.

The optimization model is defined through an interactive module. In this module the user is asked for the following information:

a) Genetic information parameters, such as:

- number of generations,
- number of individuals of the population,
- selection scheme,
- crossover scheme and crossover probability,
- mutation probability,
- number of objective functions.

b) Number and types of design variables. In the problems addressed by GENOSOFT it is possible to define the following design variables:

- for 2D and 3D truss structures: the cross section type (t_i) of each group of bars and the increment value (Δ_k) of each joint coordinate defined to be a moving coordinate;
- for 2D and 3D frame structures, the height and the width of the cross section of the member to be optimized;
- for finite element models, the increment value (Δ_k) of each joint coordinate in finite elements along the moving boundary.

c) The objective functions and constraints: In the case of truss optimization, the primary objective function is to find the minimum weight:

$$\text{Min}\,(W) = \sum_{j=1}^{\text{Min}\,N} A_j \sum_{i=1}^{nj} \rho_i L_i$$

where

A_j the area of the section type j,

ρ_i , L_i : density and length of the member i,

subjected to

- Tensile and compression stress constraints

$$\sigma_{il} - \sigma^{up} \leq 0 \quad i = 1,2,\cdots,nbar \quad l = 1,\cdots,ncases$$
$$\sigma^{up} = \sigma_i^t \text{ or } \sigma_i^c$$

being

σ_{il} : the stress of the member i under the load case l,

σ^t_i , σ^c_i : the upper bounds of tensile and compression stress of member i.

- Slenderness constraints:

$$S_i - S^{up} \leq 0 \qquad S^{up} = S^t \text{ or } S^c$$

where

S_i : the slenderness ratio of bar i.

S^t , S^c : the upper slenderness bounds for tensile and compression bars.

- Displacement constraints:

$$u_{kl} - u_k^{up} \leq 0 \qquad k = 1,2,\cdots,n_r$$

being

U_{kl} : the elastic displacement of the k degree of freedom (d.o.f.) under load case l.

$u_k{}^{up}$: the upper o lower limit of the elastic displacement of k d.o.f.

n_r : number of restrained displacements.

The shape optimization problem consists in to find the shape of the model which has a minimum area and the minimum stress concentration zones. So that, the objective function is stated by

$$\text{Min (W)} = \sum_{1}^{Nel} V_i \rho_i$$

If it is a multi-objective optimization problem, the above objective function is used together with the minimization of the stress concentration factor (k)

$$\text{Min (K)} = \frac{\sigma_{max}}{\sigma_{med}}$$

where

V_i , ρ_i : Volume and specific weight of element i.

Nel : Number of finite elements of the model.

σ_{max} , σ_{med} : maximum and mean Von-Mises stresses of the model.

subjected to

a) Stress restrictions: the Von-Mises stresses, calculated at Gauss points of the finite element, must not exceed the limit value σadm

$$\sigma_{von_i} - \sigma_{adm} \leq 0$$

The Von-Mises stress is calculated as

$$\sigma_{von} = \left[\left(\sigma_x - \sigma_y \right)^2 + 3\tau_{xy}^2 \right]^{\frac{1}{2}}$$

b) Nodal coordinates restrictions: nodal coordinates of some nodes of the mesh should not move beyond certain limit values in X and Y directions in order to maintain the mesh topology.

c) Restrictions on the shape of the elements of the moving boundary, in order to avoid singular or negative Jacobians.

14.4.2 Genopro: Analysis module

The GENOPRO module carries out the calculation of the stresses and deformations of the different kind of elements. Figure 14.5 shows how this process is done. Notice that elementary stiffness matrices are computed in different procedures.

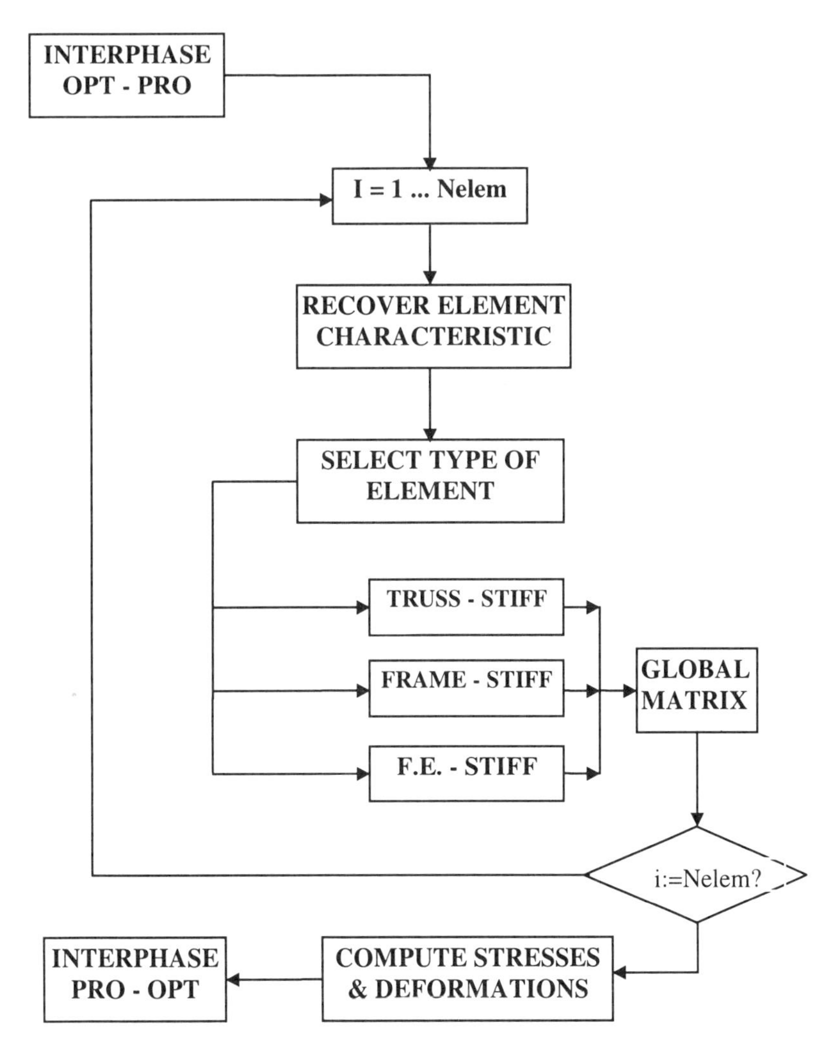

Figure 14.5 Flow chart of processor module GENOPRO.

14.4.3 Genopt: Genetic optimization module

In this module resides all the power of Genetic Algorithms used as the optimization method. This module was developed from those basic routines from Goldberg (1989) and others specially written to work with structural optimization problems. Figure 14.6 displays a simple flow chart of GENOPT module.

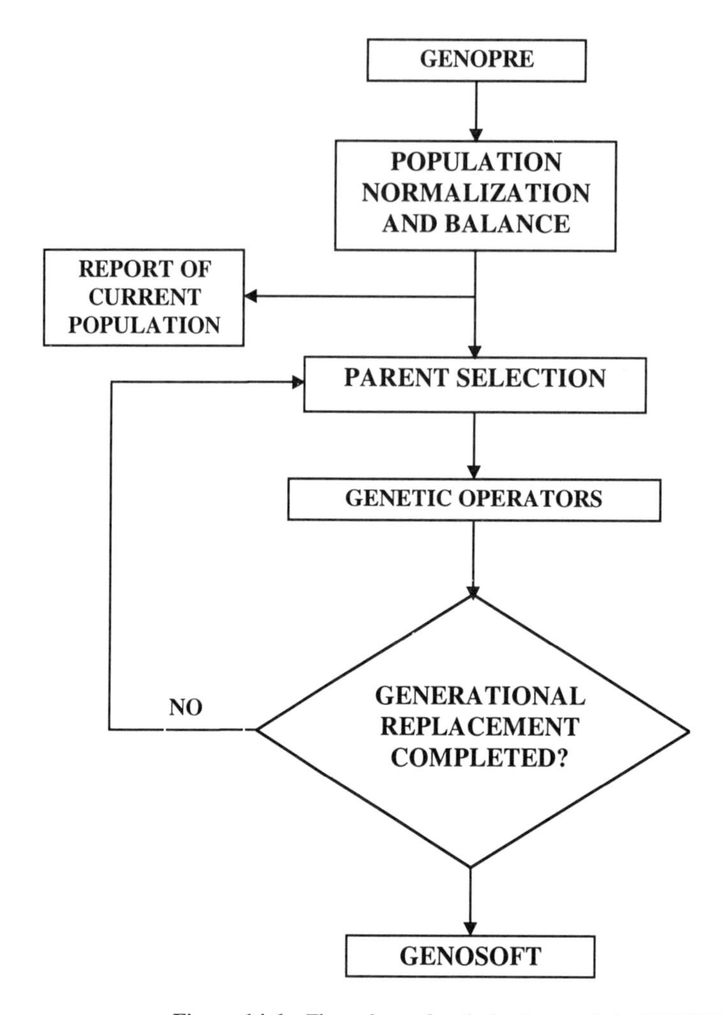

Figure 14.6 Flow chart of optimization module GENOPT.

14.4.4 Genopost: Graphic interactive post-processor

The described algorithm allows the optimization of complex problems, which usually produce a large amount of numerical results rather difficult to inspect and to analyze. Thus, the post-processing or output presentation of the results must be a simple process. The graphic capabilities should be powerful to allow the analyst to inspect the results in a wide variety of manners. The results consist mainly in stresses at elements, displacements of nodes and the final geometric configuration of the model. In this sense, the user can compare the structural initial state with the optimization/final state of the model in terms of stresses, displacements and configurations. All this can be possible due to GENOPOST, a completely user friendly module, where all the graphic options are displayed on the screen and its multi-window environment capabilities allows the definition up to four windows simultaneously.

14.5 SOME ENGINEERING NUMERICAL EXAMPLES

This section is devoted to present and to discuss some engineering examples optimized by using GAs and B-splines modelling, as well as nodal-position monitoring. Truss structures and bidimensional FE and BE models are included to show the ability and power of the proposed approach.

14.5.1 Three-dimensional steel tower: continuous and discrete optimization

The topology and dimensions of a three dimensional 25 -bar-truss (Marti et al, 1987; Rajeev and Krishnamoorthy, 1992) are shown in Figure 14.7. The goal is to optimize the weight of the structure by considering the following cases:

a) Continuum optimization of the design variables such as cross section types and geometry configuration (joint coordinates).

 a1. Optimization of cross-sectional types with fixed geometry configuration,

 a2. Optimization of cross-sectional types and variable geometry configuration.

b) Discrete optimization of the cross-sectional areas of the members. Table 14.5 displays the sections used to optimize the structure (Rajeev and Krishnamoorthy, 1992)

Tables 14.1, 14.2, 14.3 and 14.4 contain the design properties, loading cases and the design variables chose for optimization with the genetic algorithm.

Table 14.1 Design properties (three-dimensional 25-bar truss tower).

Young modulus	68950 Mpa
Density	27.13 KN/m^3
Maximum tensile or compression stress	275.8 Mpa
Maximum displacement (nodes 3 and 4)	$\pm$ 0.89 cm
Maximum buckling stress	Following the AISC methodology

Table 14.2 Loading cases (three-dimensional 25-bar truss tower).

Hypothesis	Node	Fx (KN)	Fy (KN)	Fz (KN)
1	1	2.224	0.0	0.0
	2	2.224	0.0	0.0
	3	4.448	44.48	-22.24
	4	4.448	44.48	-22.24
2	3	0.0	88.96	-22.24
	4	0.0	88.96	-22.24

Table 14.3 Physical design variables (three-dimensional 25-bar truss tower).

Variable	Element	Minimum Value (cm^2)
1	1	0.645
2	2 3 4 5	0.645
3	6 7 8 9	0.645
4	10 11	0.645
5	12 13	0.645
6	14 15 16 17	0.645
7	18 19 20 21	0.645
8	22 23 24 25	0.645

Table 14.4 Geometric design variables (three-dimensional 25-bar truss tower).

Variable	Coordinate (nodes)	Vmax (± cm)	Vmin (± cm)
9	Z(1,2,5,6)	254	25.4
10	X(1,2,5,6)	95.25	25.4
11	Y(1,2,5,6)	95.25	25.4
12	X(7,8,9,10)	254	25.4
13	Y(7,8,9,10)	254	25.4

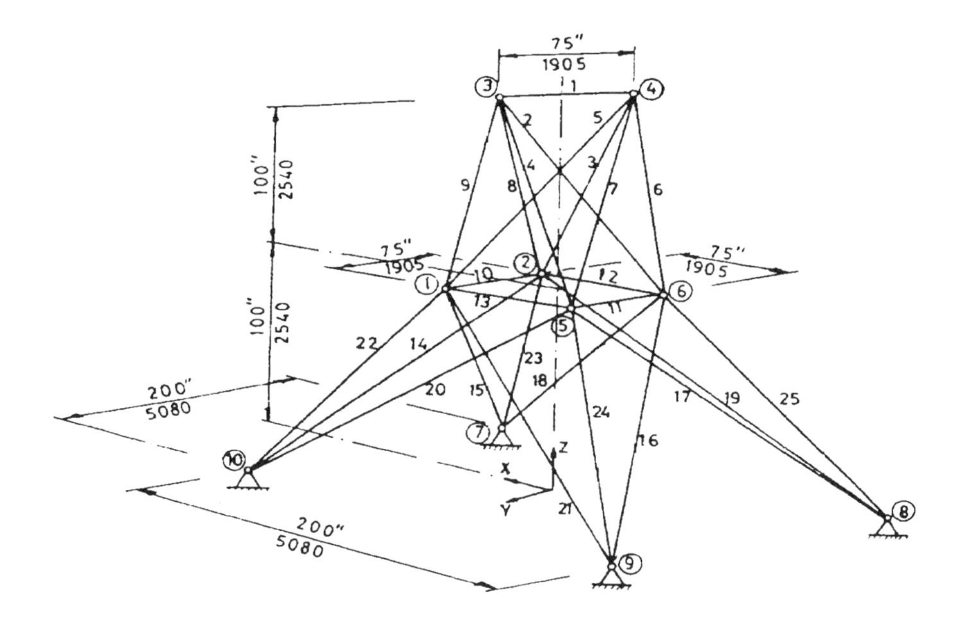

Figure 14.7 Three-dimensional 25-bar transmission tower. Node and member numeration (dimensions in mm).

Table 14.5 Member types used for discrete optimization (three-dimensional 25-bar truss tower).

Type	Area (cm^2)	Type	Area (cm^2)	Type	Area (cm^2)
1	0.645	11	7.096	21	13.548
2	1.290	12	7.741	22	14.193
3	1.935	13	8.387	23	14.838
4	2.580	14	9.032	24	15.483
5	3.225	15	9.677	25	16.129
6	3.870	16	10.322	26	16.774
7	4.516	17	10.967	27	18.064
8	5.161	18	11.612	28	19.356
9	5.806	19	12.258	29	20.645
10	6.451	20	12.903	30	21.935

Figure 14.8 illustrates the best three runs of the weight evolution of the truss versus the number of generations. Those analyses have considered stress, displacement and buckling restrictions, by assuming a fixed geometry. Note that the initial weight was about 525 Lbs and the final one was 381.63 Lbs.

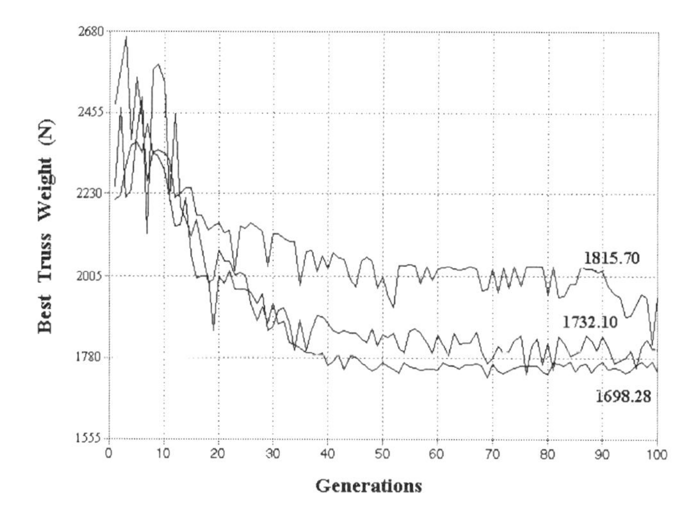

Figure 14.8 Genetic history with fixed geometry (three-dimensional 25-bar truss tower).

Figure 14.9 shows also the weight evolution of the truss versus the number of generations, but in this case the variation of the geometry of the truss was included into the optimization criteria. It can be observed that the final weight was less than 100 Lbs. in the above optimization.

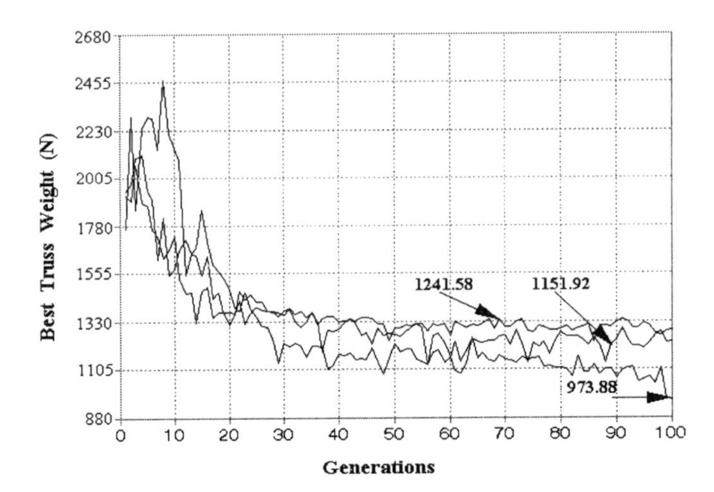

Figure 14.9 Genetic history with variable geometry (three-dimensional 25-bar truss tower).

The results of the three best runs when dealing with discrete optimization are shown in Figure 14.10. In this case, the optimization was carried out including stress and displacement restrictions.

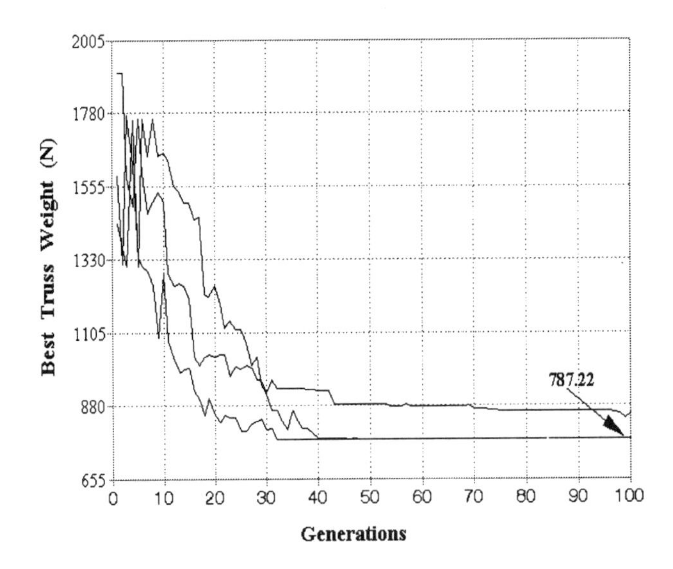

Figure 14.10 Genetic history with discrete optimization (three-dimensional 25-bar truss tower).

In Figure 14.11, the buckling stress of the bars are compared against the maximum value allowed for them. As it can be noted, all stresses in the bars are less than this limit.

Table 14.6 summarized the results obtained in the different studied cases.

Table 14.6 Final results of the different runs (three-dimensional 25-bar truss tower).

Variable active restriction	Fixed Geometry (stress/displ/buckling)/(cm^2)	Variable Geometry (stress/displ/buckling)/(cm^2)	Discrete Optimization (stress/displacement)/ (cm^2)
1	3.193	2.380	0.645
2	8.909	14.006	10.967
3	15.987	8.335	5.806
4	1.742	1.0064	0.645
5	0.935	1.735	1.290
6	6.561	2.632	0.645
7	4.219	4.206	0.645
8	4.361	3.703	0.645
Final Weight (N)	1698.18	973.88	787.22

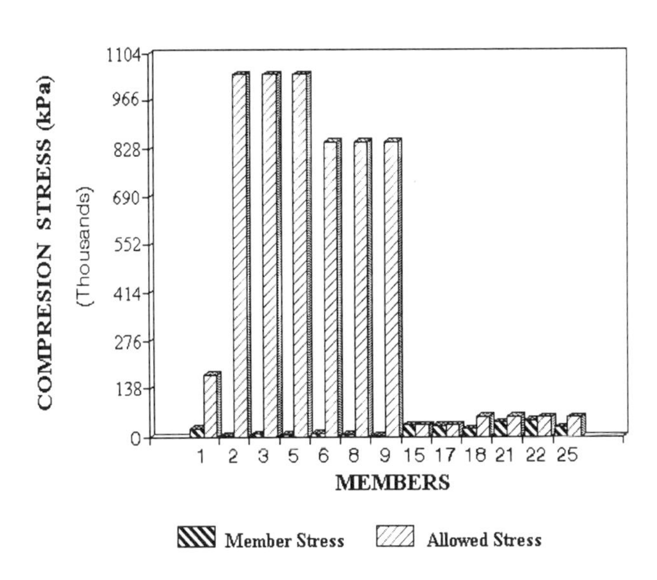

Figure 14.11 Comparison between buckling stress in members and the maximum allowed stress (three-dimensional 25-bar truss tower).

Finally, in Figure 14.12 we can appreciate the initial and final geometry configuration of the tower.

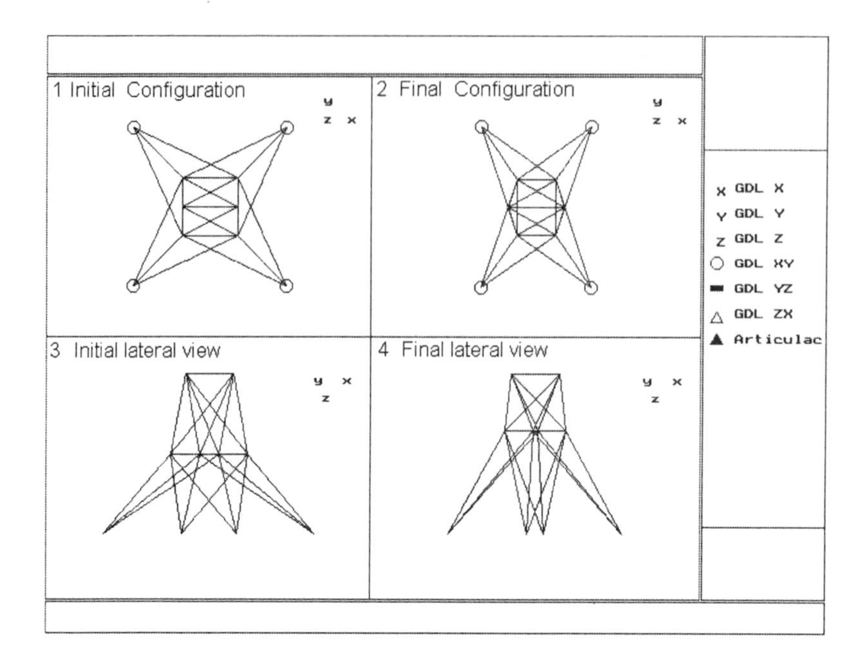

Figure 14.12 Final three-dimensional 25-bar truss tower configuration.

14.5.2 2D finite element models: monitoring nodal coordinates

The following examples illustrate the ability of the algorithm to optimize finite element model shapes, in order to get the minimum area and an uniform stress distribution. The nodal coordinates are used as genetic variables to obtain the best shape.

14.5.2.1 Plate with moving boundary As a demonstration of effectiveness of shape optimization using GAs, in this example it is considered the optimization of the surface area of a plate.

Figure 14.13 presents the geometry, dimensions and the moving boundary AB of the plate. The loading and boundary conditions are depicted in Figure 14.14.

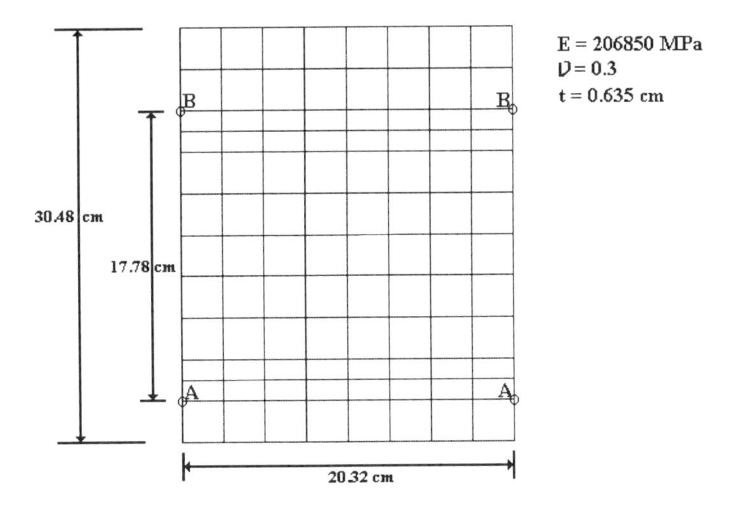

Figure 14.13 Plate with moving boundary (A-B), geometry and dimensions.

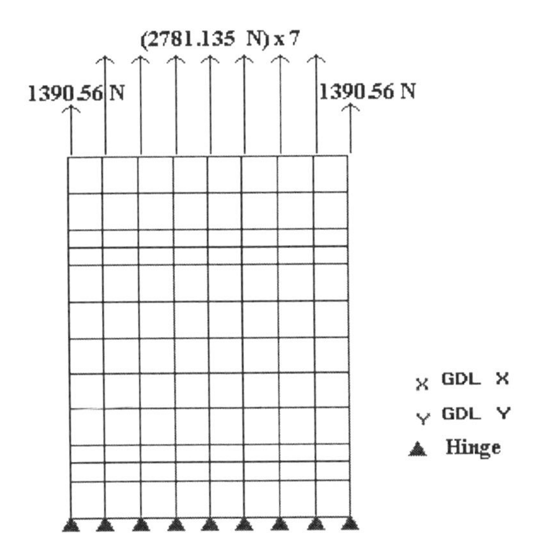

Figure 14.14 Loads and boundary conditions.

The objective function of the problem is the optimization of the surface area of the plate. The constraints are the Von-Mises stresses, which must not exceed 5000 psi. The initial area of the shape is shown in Figure 14.13, which is 96.0 in^2.

The optimized shape is shown in Figure 14.15. It can be noted how the area of the plate was reduced significantly. The final area obtained was 70.78 in^2, which is a 26.27% reduction.

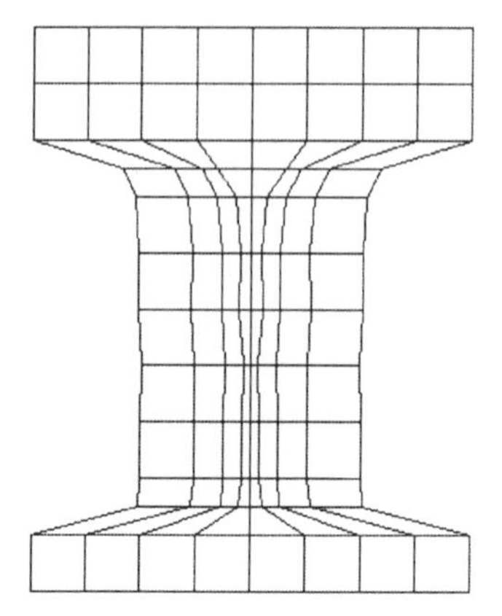

Figure 14.15 Final shape of the optimized piece.

The comparison between the initial and final stress is shown in Figures 14.16 and 14.17. Note how the final shape is the result of the interaction between allowed stress and the element stress in the narrow zone of the final shape.

Finally, in Figure 14.18 the evolution curves through generations are presented.

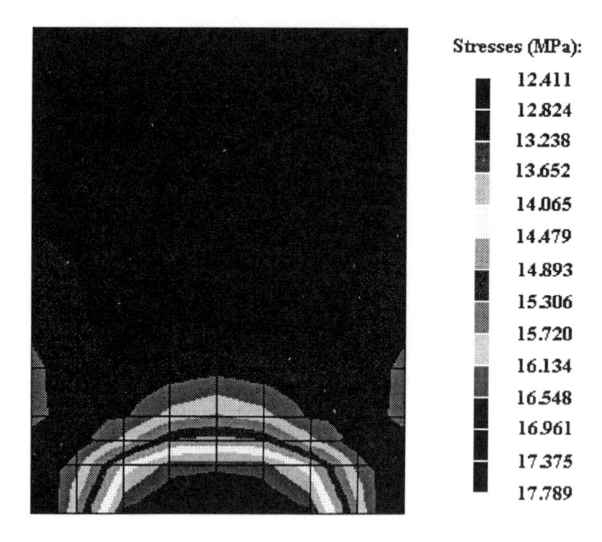

Figure 14.16 Initial stress distribution (plate subjected to traction).

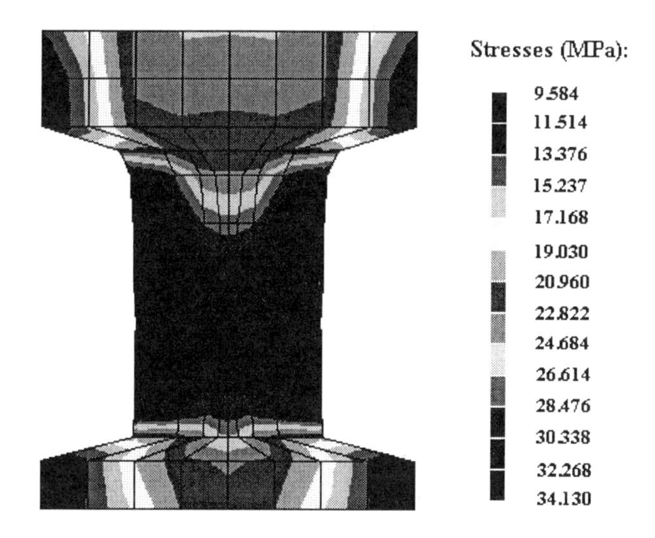

Figure 14.17 Final stress distribution (plate subjected to traction).

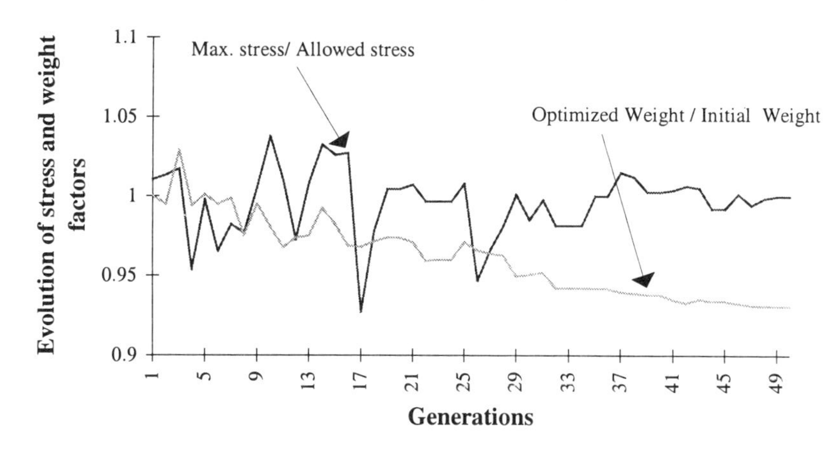

Figure 14.18 Genetic evolution of stresses and weight (plate subjected to traction).

14.5.2.2 ***Cross section of a pipe*** Figure 14.19 shows a quarter part of a cross section of a pipe, which is loading with an internal constant pressure of $P_1 = 100$ Kg/cm^2. The goal is to find the best cross section with the less weight, by providing that the following constraints are not violated:

1. Inner radius.
2. Thickness of the pipe's wall.
3. Maximum allowed Von-Mises stress.

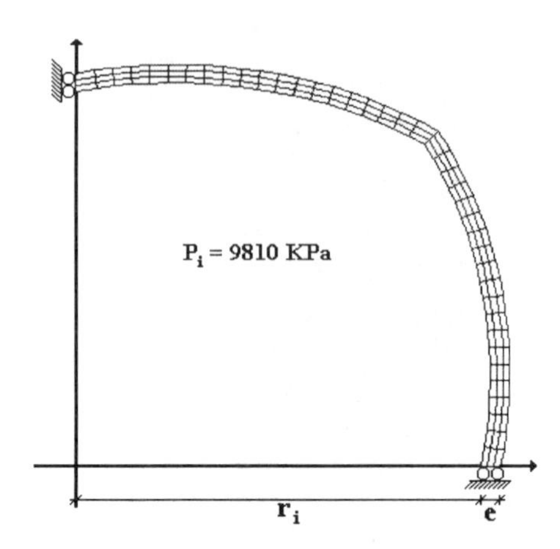

Figure 14.19 Initial geometry, loading and boundary conditions (cross sectional pipe problem).

In Figure 14.20, the final mesh of the model is displayed. It corresponds to the best design obtained by the algorithm. It can be appreciated how the optimization process smoothed the piece's contour, thus getting a circular shape, as expected.

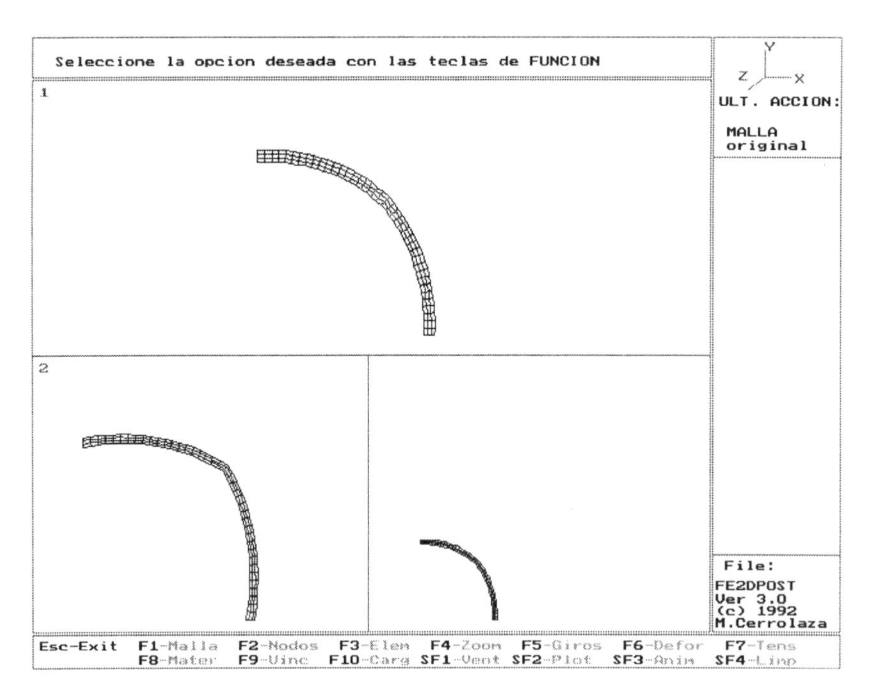

Figure 14.20 Top window: Zoom on the final geometry reached by the algorithm; bottom left window: Initial geometry of the model; bottom right window: final geometry of the model.

Figures 14.21 and 14.22 display a comparison between the initial stress and the final one. It can be noted an important reduction in the stress values. The maximum stress moved from 24175,90 Kg/cm^2 to 13387,17 Kg/cm^2, which represents a nearly decrease of 50% in the maximum stress.

Figure 14.23 shows the weight evolution of the model, versus the number of generations.

The evolution of the maximum stress (σ_{max}) divided by the allowed stress (σ_c) is depicted in Figure 14.24. Note how this factor is lower than one.

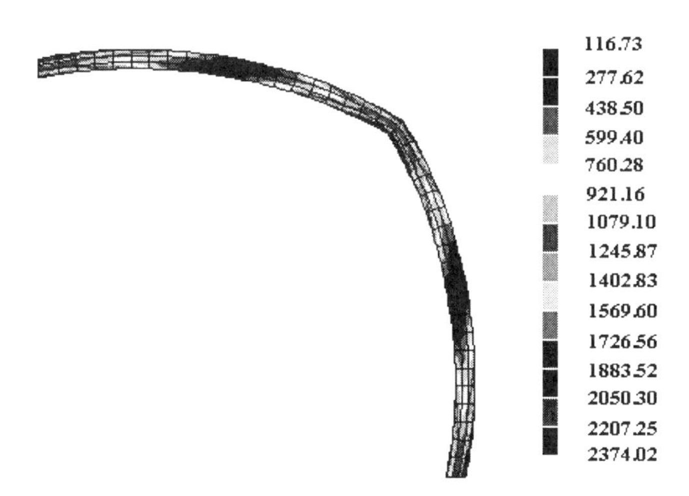

Figure 14.21 Initial Von Mises stress distribution (stresses in MPa).

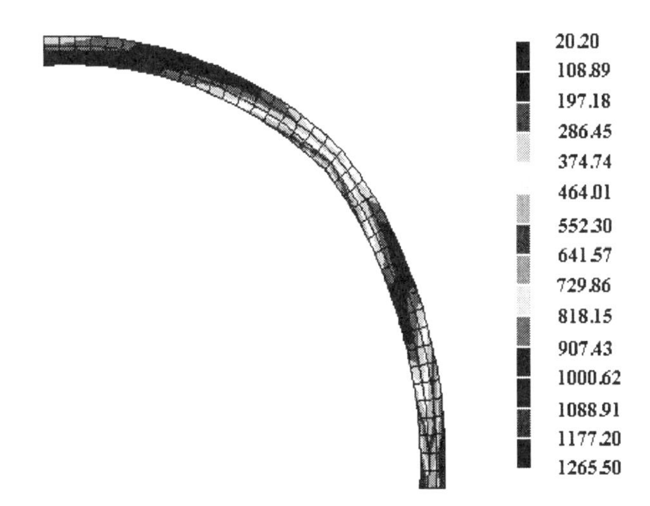

Figure 14.22 Final Von Mises stress distribution (stresses in MPa).

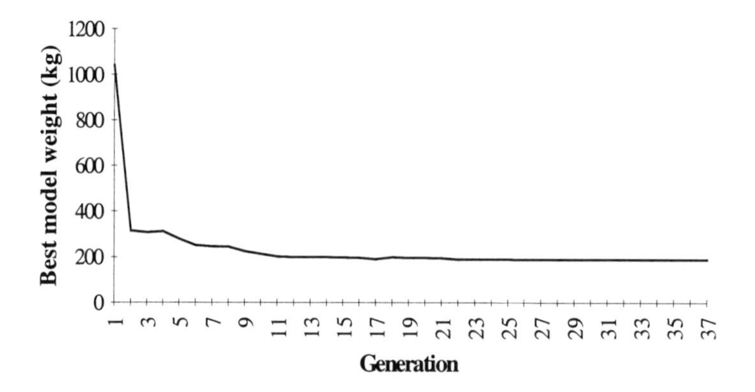

Figure 14.23 Genetic History of cross sectional pipe problem.

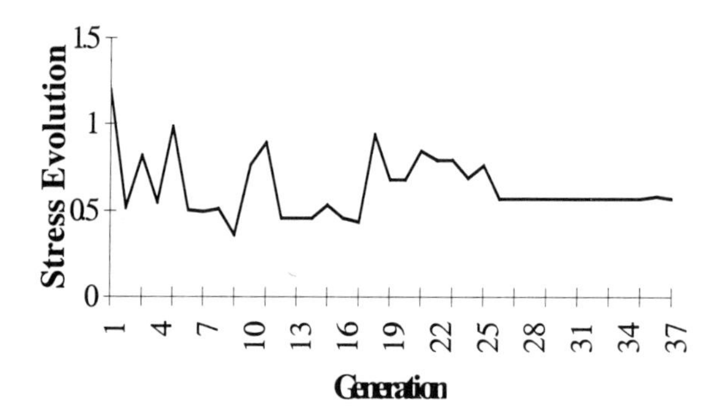

Figure 14.24 Evolution of the tension factor (σ_{max}/σ_c) versus number of generations (cross sectional pipe problem).

14.5.3 2D Finite element models: β-splines approach

Here, some examples of FE models optimized using β-splines will be discussed.

14.5.3.1 Plate subjected to uniform traction The geometry and dimensions of a plate subjected to traction is depicted in Figure 14.25 The objective of this problem is to find out the optimal surface area, providing that the Von-Mises stresses must not exceed 34,50 Mpa. Also, Figure 14.25 shows the boundary AB where the β-spline curve is defined. The control polygon is defined with 5 vertices. Three of these vertices can move in X direction.

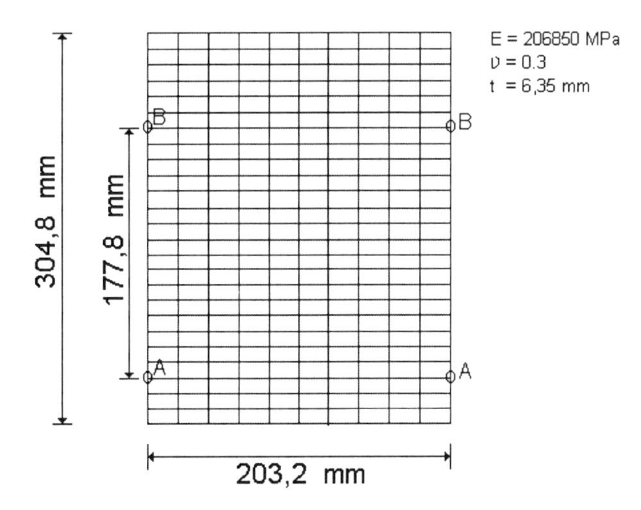

Figure 14.25 Plate with β-splines boundary (A-B), geometry and dimensions.

As shown in Figure 14.25 the initial area of the plate is 61935,00 mm^2. Table 14.7 contains the genetic parameters and operators used to optimize the plate.

Table 14.7 Genetic parameters for plate under traction.

Parameter	Value
Population size	100
Number of generations	50
Selection scheme	RSSWR
Crossover scheme	Double crossover
Crossover probability	0.8
Mutation probability	0.005

The loading and boundary conditions are depicted in Figure 14.26.

The optimized shape of the plate is shown in Figure 14.27. It can be noted how the plate area was reduced according to the stress constraints. The final area is 45632,00 mm^2, which means a 26,32% reduction.

The same problem was previously solved by the authors, by using some mesh nodal coordinates of the analytical model as design variables, as described in Annicchiarico and Cerrolaza (1998a). In that case, the number of moving joints was 64 and it required 8 design groups. On the contrary, in the case shown in Figure 14.27 the number of moving joints was 3 and 3 design groups were used. This group reduction was mainly due to the use of a geometric model, and the way of defining the moving boundary AB. In both cases similar results were obtained, as it is shown in Table 14.8.

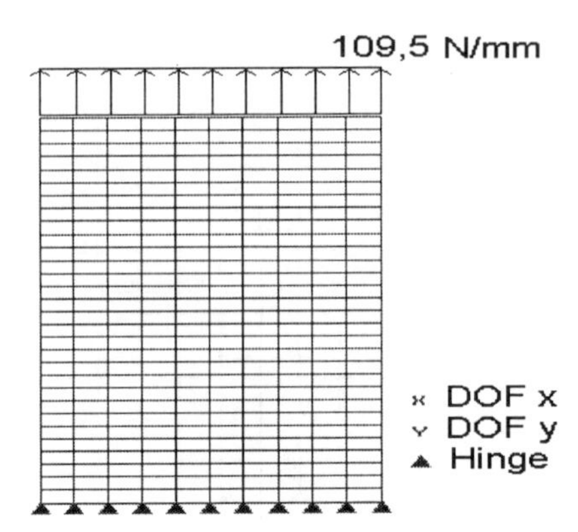

Figure 14.26 Loads and boundary conditions of the plate.

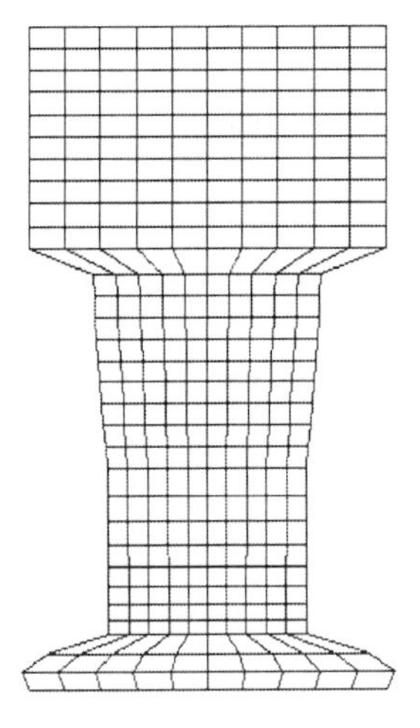

Figure 14.27 Final shape of the optimized plate.

Table 14.8 Comparison between using joint nodes or β-splines as optimization technique.

Optimization technique	Final area (mm^2)	Final maximum stress (Mpa)
Joint nodes	45665,00	34,13
β-splines	45632,00	34,20

The initial and final stresses are shown in Figures 14.28 and 14.29, respectively. Note how the final shape is smoothed by the splines curves and its final configuration is the result of the interaction between the allowed stresses and the element stresses in the moving zone of the model.

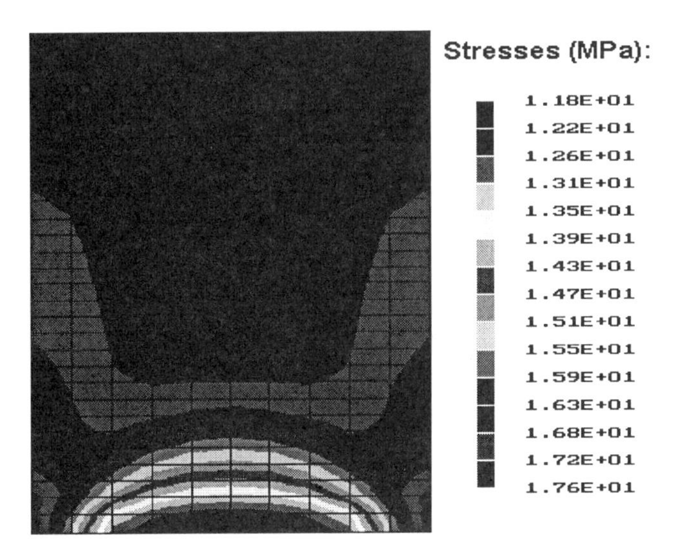

Figure 14.28 Initial Von-Mises stresses for the plate subjected to traction.

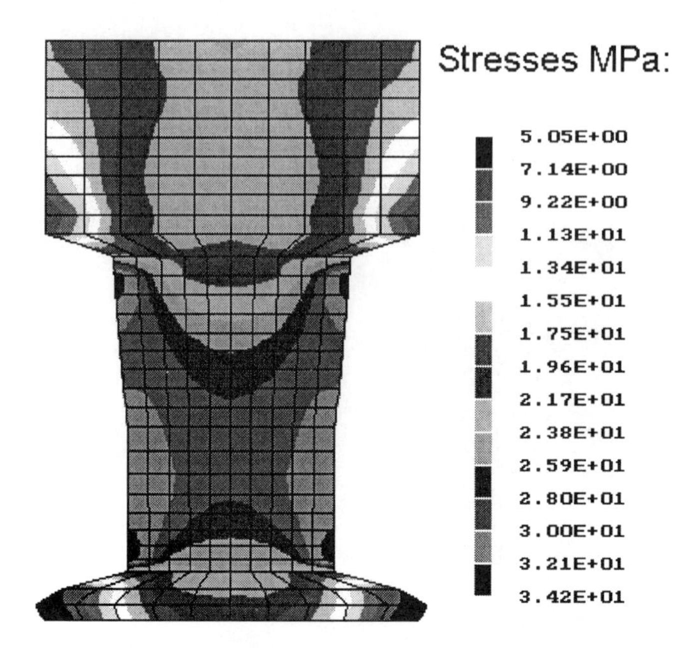

Figure 14.29 Final Von-Mises stresses for the plate.

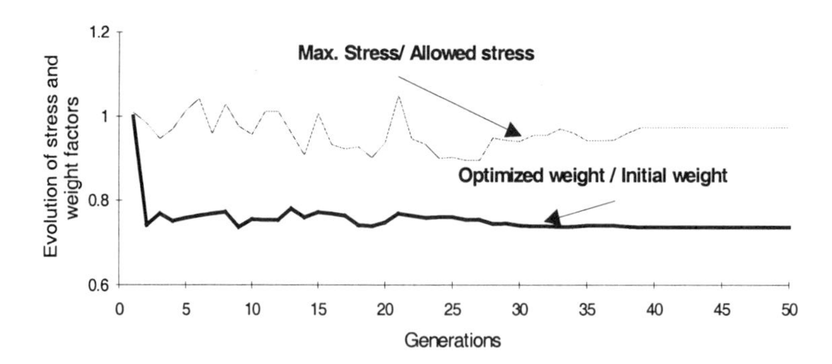

Figure 14.30 Genetic evolution of stresses and weight (plate subjected to traction and design variables defined with β-splines).

Figure 14.30 displays the evolution curves of the stresses and weight of the plate. It can be noted that the non-dimensional weight factor is always below one, which clearly shows the stability of the proposed optimization method.

14.5.3.2 Connecting rod The goal of this problem is to show the ability of the algorithm to handle real-life problems. For this purpose the well-known application of a connecting rod was chosen, which is designed to minimize the surface area and to maintain the structural integrity. The finite element model of the rod is shown in Figure 14.31. This model was defined with 232 finite elements and 295 nodes, having an initial area of 3622,37 mm^2. The different design variables used to optimize the piece are defined as: h_1, h_2, h_3, h_4, r_1, r_2, r_3 and the external boundary AB was modeled using β-splines.

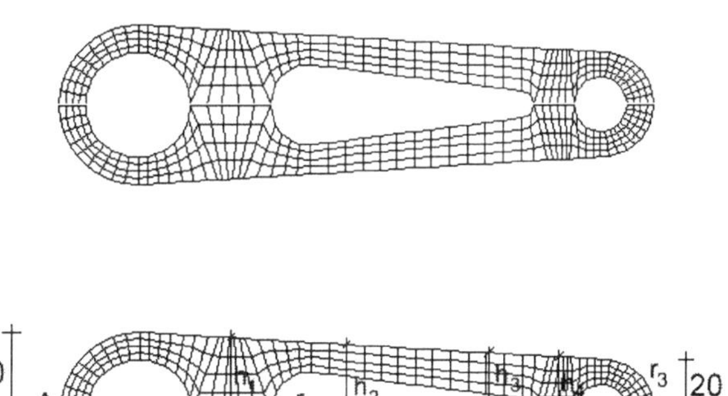

Figure 14.31 Connecting rod geometry (dimensions in mm). Design variables and moving external boundary A-B.

The boundary conditions are a distributed force at the pin-end and fixed conditions at the crankshaft end. The total distributed forces applied on the pin-end are 12000 N in the x-direction and 10000 N in the y-direction. So, the half symmetry model was used (6000 N, 5000 N). The boundary conditions, including those necessary for symmetry, and loading are presented in Figure 14.32.

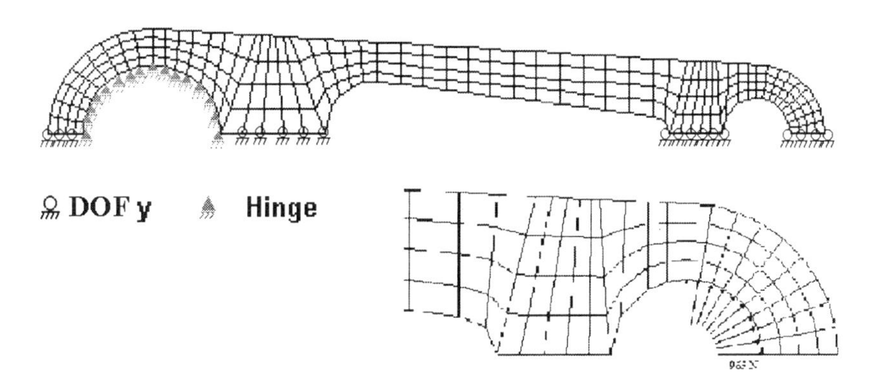

Figure 14.32 Boundary conditions and loading for connecting rod.

The material properties were Young's modulus = 2,07 x 10^5 Mpa and Poisson´s ratio = 0,30. The cost function is specified as the area and the constraints are prescribed to limit the Von Mises stresses throughout the connecting rod, which must be less than 720 Mpa.

The side constraints used in order to maintain structural integrity are collected in Table 14.9. These constraints prevent against designs that will produce no realistic geometric models.

Table 14.9 Side constraints used in the optimization of a connecting rod.

Constraint number	Side Constraints
1	$1 - h_1/20 \leq 0$
2	$1 - 3h_2/(r_2 + 2r_1)$
3	$1 - 3h_3/(r_1 + 2r_2)$
4	$1 - r_3/12 \leq 0$
5	$1 - h_4/(r_2 + 1) \leq 0$

The genetic operators and parameters used to optimize the structure are displayed in Table 14.10.

Table 14.10 Genetic operators and parameters used in the optimization of a connecting rod.

Population size	150
Number of generations	55
Selection scheme	RSSWR
Crossover scheme	Uniform crossover
Crossover probability	0,8
Mutation probability	0,005

The optimal design, shown in Figure 14.33, yields a 51,75% reduction of the surface area.

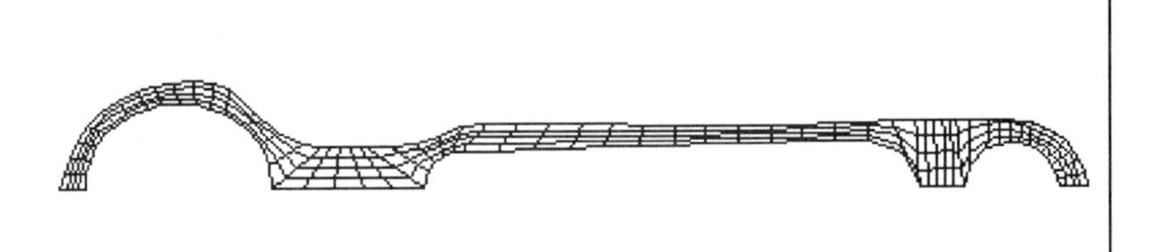

Figure 14.33 Final shape of the optimized connecting rod.

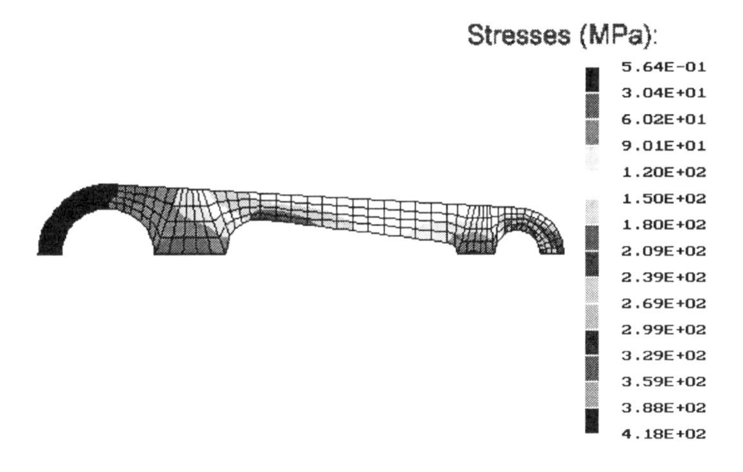

Figure 14.34 Initial Von-Mises stresses distribution (connecting rod).

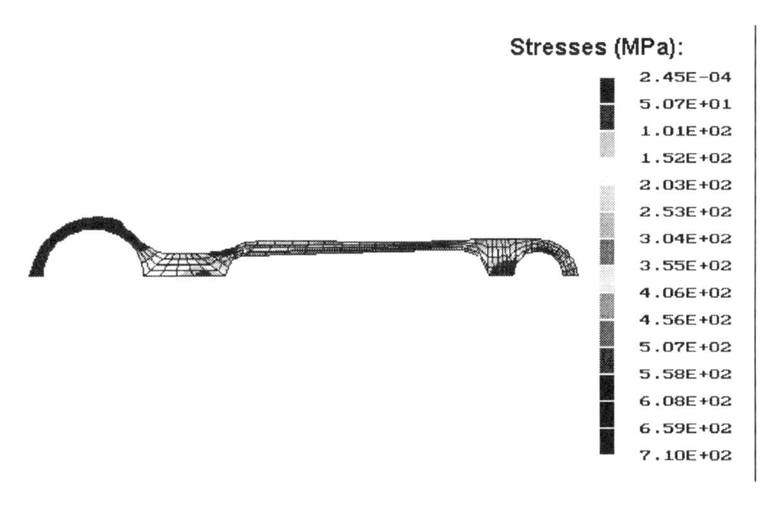

Figure 14.35 Final Von-Mises stresses distribution (connecting rod).

Both initial and final stresses are displayed in Figures 14.34 and 14.35. It can be observed how the maximum Von-Mises stresses move from 418 MPa in the initial shape to 710 MPa in the optimal shape, which is below the maximum allowed value 720 MPa.

The optimization history, illustrated in Figure 14.36 for the non-dimensional weight factor (Best $Area_i$/Initial Area) and the non-dimensional stress factor (Maximum $stress_i$/allowable stress), shows convergence in 35 generations. Moreover, the weight factor is less than one and it behaves in a very stable manner around the optimum value. This is a clear indication that the algorithm is stable and is able to deal with complex shape optimization problems.

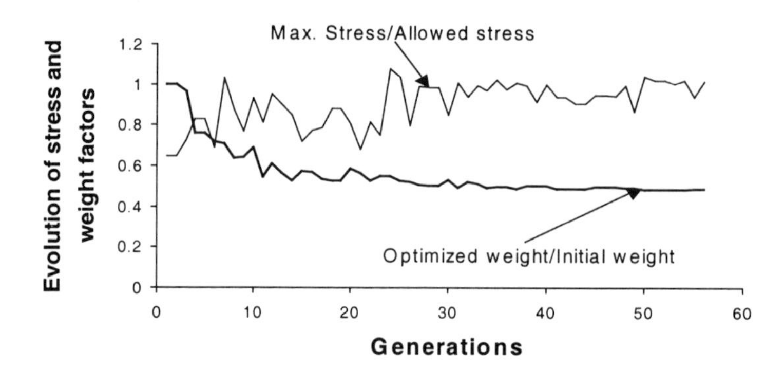

Figure 14.36 Genetic evolution of stresses and weight (connecting rod).

14.5.4 2D Boundary element models: β-splines approach

Some research is currently being done in order to optimize boundary element models by using GAs. This section reports some preliminary results obtained in this direction.

14.5.4.1 L-shaped plate The geometry, boundary conditions and loading of an L-Shaped plate are displayed in Figures 14.37a and 14.37b. In order to define the boundary of the model, eight cubic super-elements were used and β-splines curves were asociated to super-elements 6 and 7. Only these splines are allowed to move, since the analyst is trying to reduce the stress concentration arising near the entrant corner of the piece, which in this case is around 25 MPa (Cerrolaza et al, 1999). The shear module is G=500 Mpa where the Poisson's ratio is ν=0.25. These academic values are reported here only for illustrative purposes.

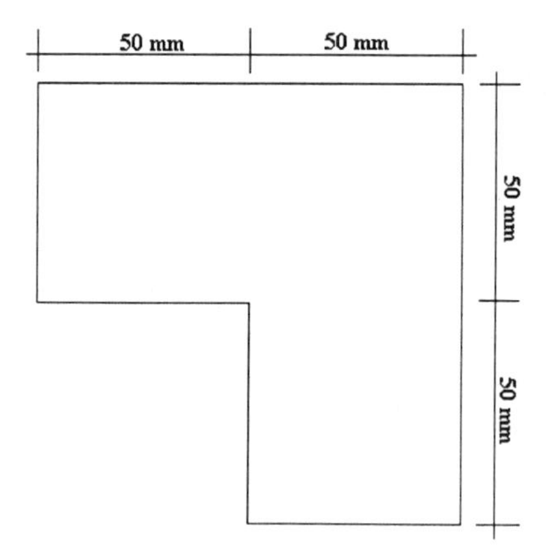

Figure 14.37a L-shaped plate geometry defined by using boundary elements.

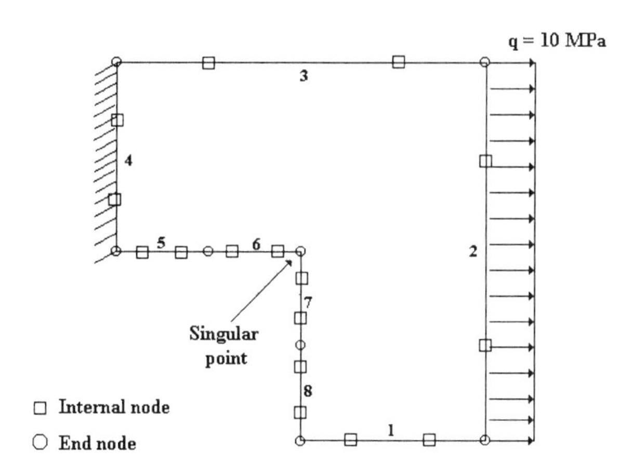

Figure 14.37b Superelements, loads and boundary conditions of L-shaped plate. Cubic β-splines number 6 and 7 are the only allowed to move.

Table 14.11 contains the genetic parameters used to drive the genetic optimization process.

Table 14.11 Genetic parameters used to optimize the L-shaped plate.

Population size	100
Number of generations	80
Selection scheme	RSSWR
Crossover scheme	uniform
Mutation scheme	normal
Crossover probability	0.8
Mutation probability	0.02

The processing module in this case is based on the Boundary Element Method. The reader interested in the details of the BEM formulation is referred to Kane (1994) and Brebbia et al (1984). The analysis module generates the boundary element model according to the shape of the superelements defined at the boundary. In this analysis, 10 bilinear boundary elements are generated in splines 6 and 7, where 5 boundary elements are generated at the other superelements (1,2,3,4,5 and 8), thus involving a 50-element model.

The initial area of the plate is 7500 mm^2 and the allowed Von-Mises stress is 15 Mpa. Figure 14.38 shows the evolution of the shape of the plate, according to the generations evolution in GAs process. Note that the entrant corner is smoothed by the algorithm, in order to reduce the stress concentration which normally arise at this region of the model.

The Figure 14.39 displays the final geometry of the plate, obtained at the generation number 80. Note that the area of the plate is now 7652 mm^2, which is somewhat larger than the original area, but now the stress concentration was reduced up to 12 MPa.

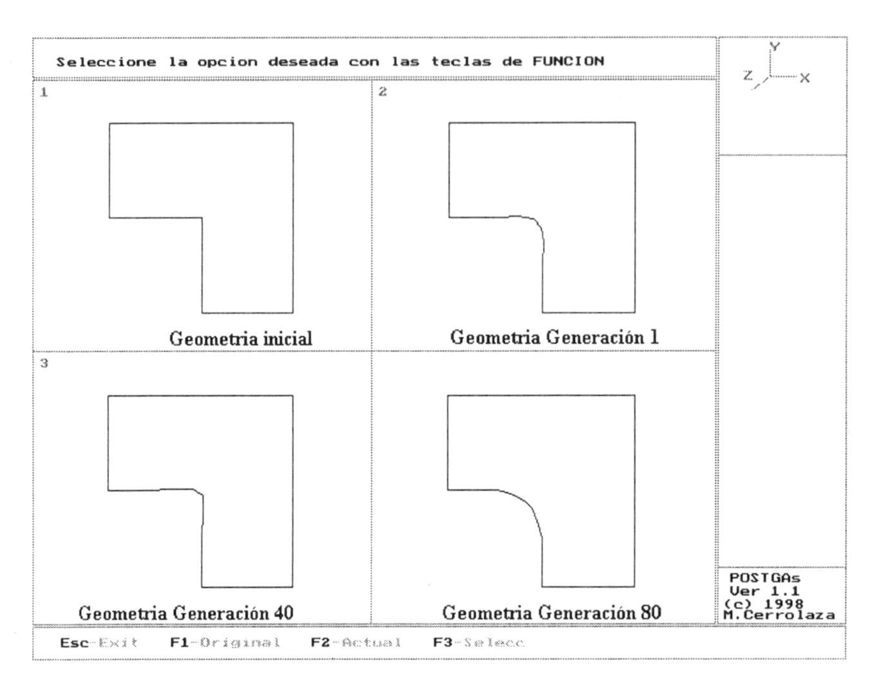

Figure 14.38 Geometry evolution of L-shaped domain. Upper left window (1): initial geometry. Upper right window (2): best geometry of generation number 1. Lower left window (3): best geometry of generation number 40. Lower right window (4): best geometry of generation number 80.

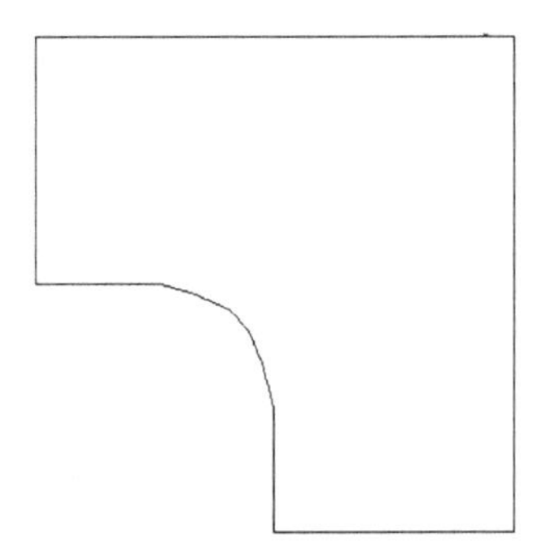

Figure 14.39 Final shape of L-shaped plate: note that the entrant corner (singular point) is smoothed to reduce the stress concentration close to the corner.

Once again, the proposed technique, coupled with a boundary element model, was successful in optimizing the domain.

14.6 CONCLUDING REMARKS

The combination of four powerful tools in shape optimization, finite elements, boundary elements, geometric modelling and genetic algorithms, is presented and discussed in this work. As shown in the numerical examples, the use of this combined approach improves the final results and saves computer time, since few design variables are required. Also, in order to get these improvements, automatic FEM and BEM mesh generators were used to generate the analytical FE and BE models from the geometric models, through a set of geometric parameters.

The use of genetic algorithms as an optimization technique improves the performance of the approach, due to its great advantages as compared with traditional optimization techniques. One of the most attractive points of this technique is that it requires no calculation of sensitivities and it reaches an optimum or quasi-optimum solution.

The BE technique has shown that 2D models can be optimized by using a few number of boundary superelements, defined with β-spline curves, which reduce the human and computational effort in data preprocessing. Also, when dealing with FE 2D models, the technique proposed herein displayed a good performance. Optimum results were obtained with a relatively few number of generations.

The use of β-splines curves as geometric elements has shown very promising results, when modelling complex boundaries. The analyst is required to define only those boundaries (in FE and BE approaches) which are allowed to move, thus reducing the number of genetic and geometric variables involved in the optimization process.

Further efforts are being done to apply this approach for the optimization of 3D FEM and BEM complex models, also using β-splines-based surfaces.

Acknowledgments

The authors wish to acknowledge the grant by the Consejo de Desarrollo Científico y Humanístico (CDCH) from the Central University of Venezuela to this research, as well as the support provided by the Institute for Materials and Structural Models from the Central University of Venezuela.

REFERENCES

Annicchiarico, W. and Cerrolaza, M. (1998a), Optimization of Finite Element Bidimensional Models: An Approach Based on Genetic Algorithms, *Fin. Elem. Anal. Des.,* **29(3-4)**:231-257

Annicchiarico, W. and Cerrolaza, M. (1998b), Structural shape optimization of FE models using B-splines and Genetic Algorithms, *Proc. IV World Cong. on Comp. Mechanics*, Buenos Aires, Argentina

Annicchiarico, W. and Cerrolaza, M. (1999a), A Structural Optimization Approach and Software Based on Genetic Algorithms and Finite Elements , *J. Eng. Opt., (to appear)*

Annicchiarico W. and Cerrolaza M. (1999b), Finite Elements, Genetic Algorithms And β-Splines: A Powerful Combined Technique For Shape Optimization, *Fin. Elem. Anal. Design,* (to appear)

Baron M.R. and Yang R.J. (1988), Boundary integral equations for recovery of design sensitivities in shape optimization, *J. of AIAA,* **26/5)**:589-594

Barsky B.A.(1988), *Computer Graphics and Geometric Modeling Using β-Splines*, Springer-Verlag, New York

Bartels, R.H., Beatty, J. C. and Barsky B. A (1987), *An Introduction to Splines for Use in Computer Graphics & Geometry Modeling,* Morgan Kaufmann publishers Inc, Los Altos, CA

Brebbia C.A., Telles J.F. and Wrobel L. (1984), *Boundary Element Techniques,* Springer Verlag, Berlin

Brindle, A. (1981), Genetic Algorithms for Function Optimization, *Doctoral dissertation*, University of Alberta, Edmonton

Cerrolaza M., Annicchiarico W. and Martinez M. (1999), Optimization of 2D boundary element models using B-splines and genetic algorithms, *Eng. Anal. Bound. Elem.,* (submitted)

Choi J.H. and Kwak B.M. (1990), A unified approach for adjoint and direct method in shape design sensitivity analysis using boundary integral formulation, *Eng. Anal. Boundary Elements*, **9(1):**39-45

Davis, L. (1991), *Handbook of Genetic Algorithms,* Van Nostrand Reinhold, New York

Goldberg, D. E. (1989), *Genetic Algorithms in Search Optimization & Machine Learning*, Addison-Wesley, Reading, MA, Chaps. 3 and 4, p.89.

Goldberg, D. E. and Richardson, J. (1987), Genetic Algorithms with Sharing for Multimodal Function Optimization, *Proceeding of the Second International Conference on Genetic Algorithms*, USA

Haug E.J., Choi K.K. and Komkov V. (1986), *Design Sensitivity Analysis of Structural Systems*, Academic Press, Inc. Florida

Holland, J.H. (1971), Processing and Processors for Schemata, In E. L. Jacks (Eds), *Associative information Processing*, Elsevier, New York, pp. 127-146.

Holland, J.H. (1973), Schemata and Intrinsically Parallel Adaptation, *Proc. NSF Workshop of Learning System Theory and its Applications*, Gainesville, University of Florida, pp. 43-46.

Holland, J.H. (1975), *Adaptation in Natural and Artificial Systems*, University of Michigan Press

Horn, J. and Nafpliotis, N. (1993), Multiobjective Optimization Using the Niched Pareto Genetic Algorithms, *Illinois Genetic Laboratory*, report N° 93005

Kane J.H. (1994), *Boundary Element Analysis in Engineering Continuum Mechanics,* Prentice Hall, Englewood Cliffs, NJ

Kane J.H. and Saigal S. (1988), Design-sensitivity of solids using BEM, *Eng. Mechanics*, **114(10):**1703-1722

Kita E. and Tanie H. (1997), Shape optimization of continuum structures by genetic algorithm and boundary element method, *Eng. Anal. Bound. Elem.,* **19:**129-136

Marti, P., Company, P. and Sanchis, M. (1987), Acoplamiento Elementos Finitos-Técnicas de Optimización en el Sistema Disseny, *Proc. Of the II Cong. Met. Num. Ing.*, Spain

Mellings S.C. (1994), Flaw Identification using the Inverse Dual Boundary Element Method, *PhD Thesis*, University of Porstmouth

Mellings S.C. and Aliabadi M.H. (1995), Flaw identification using the boundary element method, *Int. J. Num. Meth. Eng.* **38**:399-419

Meric R.A. (1995), Differential and integral sensitivity formulations and shape optimization by BEM, *Eng. Anal. Bound. Elem.,* **15:**181-188

Olhoff, N., Bendsoe, M.P. and Rasmussen, J. (1991), On CAD-Integrated Structural Topology and Design Optimization. *Comp. Mech. In Appl. Mech. And Engng.*, **89**, 259-279

Othmer, O. (1990), Marriage of CAD, FE and BE, in *Advances in Boundary Element Methods in Japan & USA.*, M. Tanaka, C.A., Brebbia and R. Shaw, eds., Comp. Mech. Publications, Boston, MA

Parvizian J. and Fenner R.T. (1997), Shape optimization by the boundary element method: a comparison between mathematical programming and normal movement approaches, *Eng. Anal. Bound. Elem.,* **19:**137-145

Periaux J., Mantel B., Sefrioui M., Stoufflet B., Desideri J-A, Lanteri S. and Marco N. (1997), Evolutionary computational methods for complex design in aerodynamics, *Proc. 36th American Inst. Of Aeronautics and Astronautics Conference,* **AIAA-98-0222,** Reno, USA

Periaux J., Sefrioui M., Mantel B. and Marco N. (1998), Evolutionary CFD design optimization problems using Genetic Algorithms, *in Simulación con Métodos Numéricos (Prado, Rao and Cerrolaza, Eds),* Ed. SVMNI, Caracas, pp. CI89-CI105

Rajeev, S. and Krishnamoorthy, C. S. (1992), Discrete Optimization of Structures Using Genetic Algorithms, *J. of Structural Engineering*, . **118**, N° 5

Ramakrishnan, C.V. and Francavilla, A. (1975), Structural Shape Optimization Using Penalty functions, *Journal of Structural Mechanics*, **3(4),** 403-422

Schramm, U. and Pilkey, W.D. (1993), Parameterization of Structural Shape Usign Computer Aided Geometric Design Elements, *Structural Optimization 93*, Proceedings of the World Congress on optimal design of Structural Systems, Río de Janeiro (Brazil), Vol 1, 75-82

Yang R.J. and Botkin M.E. (1986), Comparison between the variational and implicit differentiation approaches to shape design sensitivities, *AIAA Journal*, **24(6)**

Yao T.M. and Choi K.K. (1989), 3-D shape optimal design and automatic finite element regridding, *Int. J. Num. Meth. Eng,* **28**:369-384

Zienkiewicz, O. C. and Campbell, J. C. (1973), Shape Optimization and Sequential linear Programming, in *Optimum Structural Design*, John Wiley and Sons, New York, 101-126

Zumwalt, K. W. and El-Sayed, M. E. M. (1993), Structural Shape Optimization Using Cubic β-Splines, *Proceedings of Structural Optimization 93*, The World Congress on Optimal Design of Structural Systems, Río de Janeiro (Brazil), Vol. 1, pp 83-90

15 Genetic Algorithms and Fractals

E. LUTTON

INRIA Rocquencourt, B.P. 105, 78153 LE CHESNAY Cedex, France
Tel. +33 1 39 63 55 23 Fax +33 1 39 63 59 95
E-mail: Evelyne.Lutton@inria.fr http://www-rocq.inria.fr/fractales/

Abstract. Fractals are largely known as a theory that is used to generate nice (and complex) images, fractals are less known as efficient tools in the framework of signal analysis. However, some fractal notions developed for signal analysis are very convenient in the analysis of GA behaviour. The aim of this paper is to show that there are deep justifications to consider together Genetic Algorithms (GA) and fractals, both from applied and theoretical viewpoint. We present below a synthesis of research that have been done on this topic at INRIA in the FRACTALES group.

15.1 INTRODUCTION

Genetic Algorithms (GA) and more generally evolutionary algorithms (EA) are currently known as efficient stochastic optimization tools, and are widely used in various application domains. These techniques are based on the evolution of a population of solutions to the problem, the evolution being driven by a "fitness" function that is maximized during the process. Successive populations of solutions are thus built, that fit increasingly well (the values of the fitness function increase). Their evolution is based on stochastic operators: selection (the fitness function is used as a sort of "bias" for a random selection in the population), and the "genetic" operators, mainly crossover (combination of two solutions) and mutation (perturbation of a solution). This technique is based on the assumption that well fitted solutions (also called individuals) can provide better solutions with help of the genetic operators. This assumption can be proven to be connected to some notions of "GA-difficulty" of the function to be optimized: one usually talks about "deception". Notions such as "epistasis" [11] or "fitness landscape"[43] may also be related to "GA-difficulty".

Theoretical investigations on GA and EA generally concern convergence analysis (and convergence speed analysis on a locally convex optimum for Evolution Strategies), influence of the parameters, GA-easy or GA-difficulty analysis. For GA, our main concern here, these analyses are based on different approaches:

- proof of convergence based on Markov chain modeling [12], [8], [1], [45]
- deceptive functions analysis, based on Schema analysis and Holland's original theory [23], [18], [19], [21], which characterizes the efficiency of a GA, and sheds light on "GA-difficult" functions.
- some rather new approaches are based on an explicit modeling of a GA as a dynamic system [28], [32], [57].

It has to be noted first that in the modeling of GA as dynamic systems, some fractal features have been exhibited [28]. This approach has mainly led to the generation of fractal images, and may be considered as anecdotal. The aim of this paper is to offer evidence that there are stronger justifications, both theoretical and applied, for considering Genetic Algorithms and Fractals together.

Fractals are largely known as a way to generate "nice" images (Julia sets, Mandelbrot sets, Von Koch curves, Sierpinski gasket), that present the characteristic of having an infinity of details and that show a sort of "self-similarity". We are dealing here with other aspects of fractals, that is the use of fractal tools in order to perfom analyses of complex signals. The use of fractal or multifractal theoretical tools in order to perform analyses of signals that are not necessarily "fractal" is now an important trend in this area, and has been proven successful in various domains of applications, such as image analysis, finance, or network traffic analysis, see [56].

In this framework, evolutionary algorithms have been proven to be efficient tools for several applications:

- resolution of the inverse problem for IFS [59], [58], [42], [40], [53], [34] with some applications to speech signal modeling [54],
- image compression [55], [17],
- fractal antennas optimization [9].

This success is mainly due to the fact that when we are dealing with fractal analysis of signals, we often encounter complex optimization problems, with very irregular functions (it is difficult to define local derivatives on such functions), having many local optima, and on large search spaces. Stochastic optimization methods and Evolutionary algorithms in particular are well suited to this type of optimization problems.

From the theoretical viewpoint some tools, developed in the framework of fractal theory, can be used in order to perform a finer analysis of Genetic Algorithms behaviour (mainly based on the schema theory). As we will see in the following, an analysis of how GA optimize some "fractal" functions makes it possible to model the influence of some parameters of the GA. Such an analysis can then be generalized and gives some clues about how to tune some of the GA parameters. Finally, a further analysis on the same theoretical basis allows the influence of the coding in a GA to be analyzed.

This theoretical analysis is presented in section 15.2. In section 15.3 applications that have been developed in the FRACTALES group are described (details can be found at `http://www-rocq.inria.fr/fractales/`).

15.2 THEORETICAL ANALYSIS

The analysis developed below is based on schema theory and deception analysis. The GA modeled in this framework is the so-called canonical GA, i.e. with proportionate selection (roulette wheel selection), one point crossover and mutation, at fixed rates p_c and p_m throughout the GA run.

If we suppose that the fitness function f, defined on $\{0,1\}^l$ is the sampling with precision $\varepsilon = \frac{1}{2^l}$ of a Hölder function F defined on the interval $[0,1]$ (this hypothesis is always valid, even if the function F does not reflect in a simple way the behaviour of f):

$$\forall x \in \{0,1\}^l, \quad f(x) = F\left(\frac{I(x)}{2^l}\right)$$

$I(x) \in [0, 2^l - 1]$ is the integer whose binary expansion is x, $I(x) = \sum_{t=0}^{l-1} x_t 2^t$.

A first result relates l, associated to the sampling precision of the underlying Hölder function F, to the precision of the maximum location that can be obtained, see [35]. This result is valid for any optimization method, including Genetic Algorithms.

A deception analysis on Hölder function has then been derived (see section 15.2.2), that provides a relation between a measure of "deceptiveness" of f, the Hölder exponent of F, and some parameters of the GA (l, p_m, and p_c). This relation suggests that an adjustment of these parameters may tend to improve the performances of a GA. One can also feature an *a posteriori* validation method of the results provided by a GA, see [38]. Anyway, experiments have proven that even if the behaviour of the GA corresponds to the theoretical relation, one need a finer theoretical analysis in order to be able to build a practical and robust *a posteriori* validation method. The main practical implication of this study is to indicate in which way the parameters must be tuned in order to reduce deceptiveness.

The previous analysis is based on an underlying distance that is the euclidean distance on $[0, 2^l - 1]$. A more "natural" distance on the search space $\{0,1\}^l$ is the Hamming distance. A similar analysis is described in section 15.2.6, which provides a more precise relation between deceptiveness and some irregularity measures on the function. This second approach has two main advantages: the irregularity measures (a set of l coefficients, called bitwise regularity coefficients) are easier to estimate than the two coefficients of an underlying Hölder function, and the deceptiveness bound is more precise. A direct practical application is then the estimation of the influence of the bits order in the chromosome coding. Of course this estimation has some limits (mainly due to the theoretical limitations of the static deception model) and does not reflect the behaviour of the simple GA on "very epistatic" functions.

Another important point concerning practical implications of such models is the simple GA model, that is usually not used on real world applications. This is one reason why we have extended the previous model to a GA with uniform crossover in section 15.2.8. The same relations as for one point crossover can be established in both cases: underlying Hölder functions and general case. Experiments on the deceptiveness and convergence behaviour of a GA with and without Gray encoding of the chromosomes are reported in [30].

15.2.1 Hölder functions

Definition 1 (Hölder function of exponent h) *Let (X, d_X) and (Y, d_Y) be two metric spaces. A function $F : X \rightarrow Y$ is called a* Hölder function of exponent $h > 0$, if for each $x, y \in X$ such that $d_X(x, y) < 1$, we have:

$$d_Y(F(x), F(y)) \leq k \cdot d_X(x, y)^h \qquad (x, y \in X) \tag{15.1}$$

for some constant $k > 0$.

Although a Hölder function is always continuous, it need not be differentiable (see the example of Weierstrass functions below).

Intuitively a Hölder function with a low value of h looks much more irregular than a Hölder function with a high value of h (in fact, this statement only makes sense if we consider the highest value of h for which (15.1) holds).

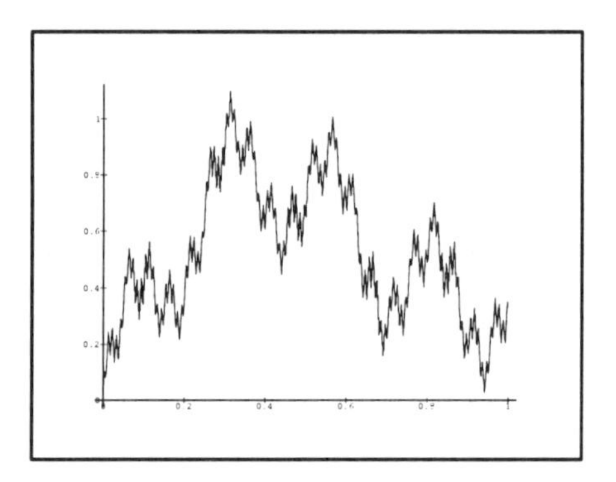

Fig. 15.1 Weierstrass function of dimension 1.5.

The frame of Hölder functions, while imposing a condition that will prove useful for tuning the parameters of the GA, makes it possible to consider very irregular functions, such as the Weierstrass function displayed in Fig. 15.1 and defined by:

$$W_{b,s}(x) = \sum_{i=1}^{\infty} b^{i(s-2)} \sin(b^i x) \qquad \text{with } b > 1 \text{ and } 1 < s < 2 \tag{15.2}$$

This function is nowhere differentiable, possesses infinitely many local optima, and may be shown to satisfy a Hölder condition with $h = s$ [15]. For such "monofractal" functions (i.e. functions having the same irregularity at each point), it is often convenient to talk in terms of box dimension d (sometimes referred to as "fractal" dimension), which, in this simple case, is $2 - h$.

Hölder functions appear naturally in some practical situations where no smoothness can be assumed and/or where a fractal behaviour arises (for example, to solve the inverse problem for IFS [42], in constrained material optimization [52], or in image analysis tasks [37], [3]). It is thus important to obtain even very preliminary clues that allow the parameters of a stochastic optimization algorithm like GA to be tuned, in order to perform an efficient optimization on such functions.

15.2.2 GA-Deception analysis of Hölder functions

This analysis is based on Goldberg's deception analysis [18], [19], which uses a decomposition of the function to be optimized, f, on Walsh polynomials. This decomposition allows the definition of a new function f', which reflects the behaviour of the GA, and which represents the expected fitness value that can be reached from the point x:

$$f'(x) = E(f(x'))$$

where x' is a random variable that represents the individuals that can be reached by mutation and crossover from x.

The GA is said to be deceptive when the global maxima of f and f' do not correspond to the same points of the search space.

15.2.3 Schema theory

A schema represents an affine variety of the search space: for example the schema **01$\star\star$11$\star$0** is a sub-space of the space of codes of 8 bits length ($\star$ represents a "wild-card", which can be 0 or 1).

The GA modelled in schema theory is a canonical GA which acts on binary strings, and for which the creation of a new generation is based on three operators:

- a proportionate *selection*, where the fitness function steps in: the probability that a solution of the current population is selected is proportional to its fitness,
- the *genetic operators*: one point crossover and bit-flip mutation, randomly applied, with probabilities p_c and p_m.

Schemata represent global information about the fitness function, but it has to be understood that schemata are just tools which help to understand the codes structure: A GA works on a population of N codes, and implicitly uses information on a certain number of schemata.

We recall below the so called "schema theorem" which is based on the observation that the evaluation of a single code makes it possible to deduce some (partial) knowledge about the schemata to which that code belongs.

Theorem 1 (Schema theorem (Holland)) *For a given schema H, let:*

- *$m(H,t)$ be the relative frequency of the schema H in the population of the tth generation,*
- *$f(H)$ be the mean fitness of the elements of H,*
- *$\mathcal{O}(H)$ be the number of fixed bits in the schema H, called the* order *of the schema,*

- $\delta(H)$ *be the distance between the first and the last fixed bit of the schema, called the* definition length *of the schema.*

- p_c *be the crossover probability,*

- p_m *be the mutation probability of a gene of the code,*

- $\bar{f}$ *be the mean fitness of the current population.*

Then:

$$E[m(H,t+1)] \geq m(H,t)\frac{f(H)}{\bar{f}}\left[1 - p_c\frac{\delta(H)}{l-1} - \mathcal{O}(H)p_m\right]$$

From a qualitative view point, this formula means that the "good" schemata, having a short definition length and a low order, tend to grow very rapidly in the population. These particular schemata are called *building blocks*.

The usefulness of the schema theory is twofold: first, it supplies some tools to check whether a given representation is well-suited to a GA. Second, the analysis of the nature of the "good" schemata, using for instance Walsh functions [20], [24], can give some ideas on the GA efficiency [11], via the notion of deception that we describe below.

15.2.4 Walsh polynomials and deception characterization

Goldberg has suggested using a method based on a decomposition of f on the orthogonal basis of Walsh functions on $[0..2^l - 1]$, where $[0..2^l - 1]$ denotes the set of integers of the interval $[0, 2^l - 1]$.

On the search space $[0..2^l - 1]$, we can define 2^l Walsh polynomials as:

$$\Psi_j(x) = \prod_{t=0}^{l-1}(-1)^{x_t j_t} = (-1)^{\sum_{t=0}^{l-1} x_t j_t} \qquad \forall x, j \in [0..2^l - 1]$$

x_t and j_t are the values of the tth bit of the binary decomposition of x and j.

They form an orthogonal basis of the set of functions defined on $[0..2^l - 1]$, and we let $f(x) = \sum_{j=0}^{2^l-1} w_j \Psi_j(x)$ be the decomposition of the function f.

The function f' [18], [19] can thus be written as follows:

$$f'(x) = \sum_{j=0}^{2^l-1} w'_j \Psi_j(x) \qquad \text{with} \qquad w'_j = w_j\left(1 - p_c\frac{\delta(j)}{l-1} - 2p_m\mathcal{O}(j)\right) \qquad (15.3)$$

The quantities δ and $\mathcal{O}$ are defined for every j in a similar way as for the schemata: $\delta(j)$ is the distance between the first and the last non-zero bits of the binary decomposition of j, and $\mathcal{O}(j)$ is the number of non-zero bits of j.

15.2.5 Haar polynomials for the deception analysis of Hölder functions

If we consider the fitness function f as the sampling of some Hölder function defined on the interval $[0, 1]$, it is intuitively obvious that the more irregular the function is (i.e. the lower the Hölder exponent is), the more deceptive it is likely to be. The intuition can be reinforced by theoretical arguments, as we will see in the following.

Another decomposition than the previous one is then more suited to a deception analysis of Hölder functions. There exist simple bases which permit, in a certain sense, the irregularity of a function to be characterized in terms of its decomposition coefficients. Wavelet bases possess such a property. The simplest wavelets, i.e. Haar wavelets, are defined on the discrete space $[0..2^l - 1]$ as:

$$H_0(x) = 1 \quad \text{for all } x \text{ in } [0..2^l - 1]$$

$$[2pt]H_{2^q+m}(x) = \begin{cases} 1 & \text{for } (2m)2^{l-q-1} \leq x < (2m+1)2^{l-q-1} \\ -1 & \text{for } (2m+1)2^{l-q-1} \leq x < (2m+2)2^{l-q-1} \\ 0 & \text{otherwise in } [0..2^l - 1] \end{cases}$$

with $q = 0, 1, \ldots, l - 1$ and $m = 0, 1, \ldots, 2^q - 1$: q is the degree of the Haar function.

These functions form an orthogonal basis of the set of functions defined on $[0..2^l - 1]$. Any function f of $[0..2^l - 1]$ can be decomposed as:

$$f(x) = \sum_{j=0}^{2^l-1} h_j H_j(x), \qquad h_j = \frac{1}{2^{l-q}} \sum_{x=0}^{2^l-1} f(x) H_j(x).$$

The following result can be easily proven (see [38]):

$$\forall j, \quad |h_j| \leq \frac{k}{2} 2^{-h(q+1)}.$$

An upper bound for the quantity $|f(x) - f'(x)|$ can be computed (see [38]):

Theorem 2 *Let f be the sampling on l bits of a Hölder function of exponent h and constant k, defined on $[0, 1]$, and let f' be defined as in (15.3). Then:*

$$\forall x \in [0..2^l - 1], \qquad |f'(x) - f(x)| \leq k * B(p_m, p_c, l, h) \tag{15.4}$$

with

$$B(p_m, p_c, l, h) =$$
$$\frac{p_c}{l-1} 2^{-h} \left[\frac{2^{-l(h+1)} - 1}{2^{-(h+1)} - 1} + \frac{(1 - 2^{1-h})(2^{-hl} - 1) - l2^{-hl}(1 - 2^{-h})}{(2^{-h} - 1)^2} \right]$$
$$+ \quad p_m \frac{2^{-h}}{(2^{-h} - 1)^2} \left[1 + 2^{-hl}(l2^{-h} - l - 1) \right].$$

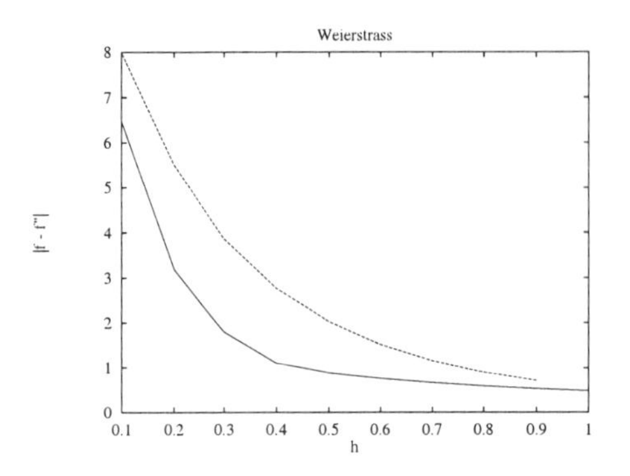

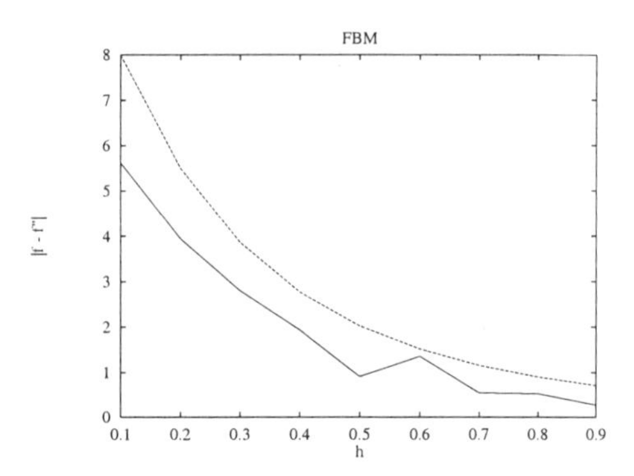

Fig. 15.2 $B(p_m, p_c, l, h)$ (dotted) and computed maximum differences between f and f' (continuous) as a function of h, for Weierstrass functions (left), and FBM's (right), l = 8 bits, $p_c = 0.9, p_m = 0.25$.

Since for all admissible values of l, p_m, p_c, B is an increasing function of h, this relation implies that the smaller h is (i.e., the more irregular the function is), the greater the difference between functions f and f' may be, thus the more deceptive f is likely to be. This first fact is confirmed by numerical simulations (for Weierstrass functions and FBM's[1]) displayed in Fig. 15.2.

A fine analysis of the function $B(p_m, p_c, l, h)$ is rather difficult, because B defines a hyper-surface of $\mathbf{R}^5$, but the following results may be stated (see [38]).

- $B(p_m, p_c, l, h)$ increases with l for small values of l, reaches a maximum l_{max} and then decreases for larger values of l. It has the following asymptotic behaviour when $l \to \infty$:

$$\lim_{l \to \infty} B(p_m, p_c, l, h) = p_m \frac{2^{-h}}{(2^{-h} - 1)^2}$$

- $B(p_m, p_c, l, h)$ decreases with respect to p_c and p_m, and deception is less influenced by p_c than by p_m.

15.2.6 Bitwise regularity analysis

The previous analysis is based on an irregularity characterization with respect to an underlying distance that is the Euclidian distance on $[0, 1]$. This approach is straightforward for fitness functions defined on $\mathbb{R}$, and in the general case it is always possible to consider the fitness

[1]FBM stands for Fractional Brownian Motion. For definition and properties of the Fractional Brownian Motion (FBM) see for instance [39]. As Weierstrass functions, paths of FBM (almost certainly) verify a Hölder property, the irregularity being the same at each point. Thus an FBM with Hölder exponent h has box dimension equal to $2 - h$.

function as the sampling of an underlying one-dimensional Hölder function. It is however less obvious in this latter case that the Hölder exponent reflects the irregularity of the fitness function in a simple way (it may appear for example more irregular than it is in a multidimensional space). This is why we have performed a similar irregularity analysis but with respect to the Hamming distance on the set of binary strings. Another justification is that the use of Hamming distance is more closely related to the action of genetic operators.

Due to the above considerations, we introduce the following coefficients, that are derived from Hölder grained exponents with respect to a distance proportional to the Hamming distance (see [30] for details):

Definition 2 (Bitwise regularity coefficients) *Let f be a function defined on $\{0,1\}^l$:*

$$\forall q \in \{0, \ldots, l-1\},\ C_q = \sup_{x \in \{0,1\}^l} \{|f(x) - f(x'_{l-q-1})|\}$$

with x'_{l-q-1} and x differing only with respect to one bit at the position $(l-q-1)$.[2]

In other terms, the C_q coefficient represents the maximum fitness variation due to a bit flip at the position $(l-q-1)$. Therefore, we can show that:

$$\forall j = 2^q + m,\ |h_j| \leq \frac{C_q}{2}$$

In the same way as in [38], with the help of the Haar basis, the following theorem has been established (see [30] for a proof):

Theorem 3 *Let f be a function defined on $\{0,1\}^l$ with bitwise regularity coefficients $(C_q)_{q \in \{0,\ldots,l-1\}}$, and let f' be defined as in (15.3). Then $\forall x \in \{0,1\}^l$:*

$$|f(x) - f'(x)| \leq \frac{p_c}{l-1} * \sum_{q=0}^{l-1} C_q * \left(\frac{1 + 2^q(q-1)}{2^q}\right) + p_m * \sum_{q=0}^{l-1} C_q * (q+1)$$

Furthermore, this result still holds when the order of the C_q is reversed, so the final bound is the one minimizing the preceding expression.

We also have to note that the bits do not have the same role in this bound expression. In fact their relative weight strictly increases with respect to the index q. Sorting (either in increasing or decreasing order) would then minimize this bound, suggesting that the simple change of coding consisting of a permutation on the bits would make the function easier. This feature can be explained by the fact that the one point crossover more easily disrupts a combination of a few genes spread at each extremity of the chromosome than if these genes were grouped at one extremity. Reordering the bits in order to sort the bitwise regularity coefficients is

[2]The less significant bit being at position 0.

then equivalent to grouping the most "sensitive" genes at one extremity of the chromosome. Some experiments presented in [30] partially support this hypothesis, but also reveal that other phenomena (such as epistasis [11]) have to be taken into account in order to predict the sensitivity of GA to such encoding changes.

15.2.7 Bitwise regularity coefficients compared to Hölder exponent

If we suppose that the fitness function f is the sampling on l bits of a Hölder function of exponent h and constant k, defined on $[0, 1]$, the bound of *Theorem 3* is lower than the bound of *Theorem 2*.

One can easily show, (see [30]), that:

$$C_q \leq \ k * 2^{-(q+1)h} \tag{15.5}$$

as we have:

$$|h_j| \leq \frac{C_q}{2} \quad \text{and} \quad |h_j| \leq \frac{k}{2} * 2^{-(q+1)h}$$

and as the bound on $|f - f'|$ is a linear function of the bounds on the $|h_j|$, it immediately follows that the bound of *Theorem 3* is the lowest. Moreover, the estimation of the bitwise regularity coefficients is computationally cheaper than the estimation of the Hölder exponent and its associated constant k.

15.2.8 Deception analysis of a GA with uniform crossover

As we have seen, the bound on $|f - f'|$ derived from the bitwise regularity coefficients C_q depends on their relative order, due to the use of one point crossover. The aim of this section is to present analogous results that have been established for the uniform crossover ([51], for which the positional bias no longer exists).

The only change is to replace the schema disruption probability p_d for this version of crossover:

$$p_d \leq \left(1 - \left(\frac{1}{2}\right)^{O(H)-1}\right) \tag{15.6}$$

This upper bound is obtained by observing that once the first fixed bit of the schema is allocated to one of the offsprings, it will always survive if all other fixed bits are allocated to the same offspring.

Theorem 4 (Schema theorem with uniform crossover) *For a given schema H, let:*

- *$m(H,t)$ be the relative frequency of the schema H in the population of the tth generation,*
- *$f(H)$ be the mean fitness of the elements of H,*
- *$O(H)$ be the number of fixed bits in the schema H, called the* order *of the schema,*

- *p_c be the crossover probability,*
- *p_m be the mutation probability of a gene of the code,*
- *$\bar{f}$ be the mean fitness of the current population.*

Then:

$$m(H,t+1) \geq m(H,t)\frac{f(H)}{\bar{f}}[1 - p_c\left(1 - \left(\frac{1}{2}\right)^{O(j)-1}\right) - O(H)p_m].$$

Then the new adjusted Walsh coefficients are:

$$w'_j = w_j\left[1 - p_c\left(1 - \left(\frac{1}{2}\right)^{O(j)-1}\right) - 2p_m O(j)\right].$$

Notice that $O(j)$ no longer depends on the defining length of the schema. Furthermore as the order of a schema is invariant with respect to a permutation on the bits, the following theorem has been proven (see [30] for a demonstration):

Theorem 5 *Let f be a function defined on $\{0,1\}^l$ with bitwise regularity coefficients $(C_q)_{q\in\{0,\ldots,l-1\}}$, and let f' be defined as in (15.3). Then for all permutation σ defined on the set $\{0,\ldots,l-1\}$, $\forall x \in \{0,1\}^l$:*

$$|f(x) - f'(x)| \leq p_c * \sum_{q=0}^{l-1} C_{\sigma^{-1}(q)} + p_m * \sum_{q=0}^{l-1} C_{\sigma^{-1}(q)} * (q+1). \tag{15.7}$$

We immediately see that this upper bound is minimal when the $C_{\sigma^{-1}(q)}$ are ordered in decreasing order.

Practically, if it is possible to get the C_q values (or good estimations), it is hard to draw conclusions from the value of the bound (15.7). But if we consider the effect of an encoding change on it, it is interesting to see if its variation is experimentally correlated to the performances of the GA. Intuitively, the hypothesis is formulated as follows: if an encoding change (such as Gray code) induces a decrease of the bound (15.7), the GA should perform better with this new encoding, and conversely. We present experiments with the Gray code in [30].

15.2.9 Conclusions and further work

Besides the intuitive fact that it relates the irregularity of the fitness function to its "difficulty", and provides information about how to modify some of the GA parameters in order to improve its performances, one important application of this theoretical analysis is that it provides a means of measuring (of course to a certain extent, due to the intrinsic limitations of deception theory) the influence of the chromosome encoding. We present in [30] some experimentations with the Gray encoding that prove the interest of such an approach: these tests show that the bound derived from the bitwise regularity coefficients are relatively reliable as long as the variations in the bound between two different encodings is large enough.

It has to be noted finally that Schema theory has some known limitations, and thus that the practical implications of the previous analysis must be considered with care. See [30] [3] for a more detailed criticism. In spite of these limitations, Schema theory has the great advantage of providing a simple model of the GA behaviour, and allows to make some computations, that are much more complicated or even infeasible for other models (other theoretical analyses of GA provide different models, that are also simplified in a different way). Conclusions built from this model, as it is a rough simplification of the behaviour of a real GA, must thus be mainly considered as qualitative analyses.

Further work can concern the analysis of Hölder exponents and bitwise regularity coefficients in the framework of dynamical systems [2] and Markov-based modeling [8]. A critical analysis of epistasis measures can be found in [44], that leads to the conclusion that epistasis is not a reliable performance indicator, but can be considered as a performance measure of the crossover operator for the choice of different encodings. This result is similar to those presented before. Moreover, a look at the expression of the epistatis in the Walsh basis compared to that of f' shows the resemblance and differences between the two measures: the function f' takes into account the influence of the genetic operators while epistasis represents an intrinsic non-linearity measure of the function. These two quantities computed on the fitness function represent two different viewpoints of the fitness difficulty. A first idea would be to analyse, in the same way as for the epistasis, a correlation between f and f'. This may provide a more reliable indicator that the estimation of the bound $|f - f'|$. A second idea would be to use a combination of these two measures in order to better capture the notion of GA-difficulty.

15.3 RESOLUTION OF A "FRACTAL" INVERSE PROBLEM

Most of the applications described below are related to the inverse problems that arise in signal analysis applications. A standard inverse problem can be stated as follows: from a given set of data, it is possible to compute the output of a system, but from a given output (the "target") it is impossible to calculate the input data set.

A classical strategy, a "black box" technique, is to turn this inverse problem into an optimization problem: search for the input data set such that the output of the system resemble the target. Usually, evolutionary algorithms are adapted to the resolution of difficult inverse problems for which there is little *a priori* information (the functions to be optimized are not explicitly known, and *a fortiori* their derivatives). In the framework of fractal data analysis, some difficult inverse problems have been successfully solved using EA, for example:

- inverse problems for IFS [59], [58], [42], [40]. We present some experiments with GA for affine IFS [53], and with GP in the general case of mixed IFS, that is more complex [34]. An application to speech signals modeling has also been developed [54].

- inverse problems for finite automata [31].

[3] An interesting viewpoint on schema theory can also be found in [44] which proves results similar to the schema theorem with the help of a Markov chain model. This work clearly presents the limitation of schema theory model.

The resolution of these "academic" inverse problems has led to experimenting GA on applied problems such as image compression [55], [17], or fractal antennas optimization [9].

The main difficulties of such applications are:

- How to find an adequate coding of the problem, which often makes it possible to efficiently exploit some *a priori* knowledge about the system. In the example of the inverse problem for fixed point finite automata, a coding of the lengths of the words of an automaton yields a much more efficient GA than a direct encoding of the words, see [31].
- How to conveniently handle the contraints of the problem, that also allows some shortcuts in the computation of the fitness function. A good example is the inverse problem for IFS, where contractivity contraints have been exploited in order to prune a lot of extra computations, see [53], [34].

In a general manner, an efficient GA for a difficult problem can be built using as much *a priori* knowledge as possible. This is particularly true when constraints can be identified and handled with the function to be optimized.

Constraints handling is a crucial problem for EA in general and several methods have been proposed for handling nonlinear constraints in a EA [48], [47], [46], [41]. For the inverse problems presented below, all the constraints of the problems were integrated as penalty functions with variable weights within the fitness function (this was the simplest way to properly handle the constraints for feasible spaces that can be very sparse). These weights were designed in order to favour the penalty term at the beginning of the evolution of the GA, in order to first draw the populations near feasible regions, and then to progressively favour the term to be optimized. This structure allows us to easily make computation cut-off when some constraints are not fulfilled. Another justification for using such a strategy is that weighted penalty functions allow "smoother" fitness functions, which clearly tend to make the job easier for the GA: this intuitive fact is also enforced by the conclusions of the theoretical analysis presented above.

These experiments have enforced once more, if needed, the commonly-held opinion known among EA-people, that careful parameter setting, efficient encoding and an "economic" fitness function computation can make a huge difference to the efficiency and accuracy of the GA, and even turn an "infeasible" problem into a "feasible"one.

15.3.1 Inverse problems for IFS

Iterated functions system (IFS) theory is an important topic in fractals, introduced by J. Hutchinson [25]. These studies have provided powerful tools for the investigation of fractal sets, and the action of systems of contractive maps to produce fractal sets has been considered by numerous authors (see, for example, [4], [6], [14], [22]).

A major challenge of both theoretical and practical interest is the resolution of the so-called inverse problem [5], [40], [58], [59]. Except for some particular cases, no exact solution is known.

A lot of work has been done in this framework, and some solutions exist based on deterministic or stochastic optimization methods. As the function to be optimized is extremely complex, most of them make some *a priori* restrictive hypotheses: use of an affine IFS, with a fixed number of functions [6], [16], [27], [26]. Solutions based on GA or EA have been presented for affine IFS [17], [42], [55], [58].

We present below a simple implementation of a GA for the resolution of the inverse problem for affine IFS (Section 15.3.3). If we let the contractions be non affine (as in numerous applications the IFS are implicitely supposed to be affine, we use the term "mixed" IFS, in order to emphazise the fact that the functions are not constrained to be affine, see Section 15.3.4), the inverse problem cannot be addressed using "classical" techniques. A solution based on GP is presented in Section 15.3.5.

15.3.2 IFS theory

An IFS $\mathcal{I} = \{F, (w_n)_{n=1,..,N}\}$ is a collection of N functions defined on a complete metric space (F, d).

Let W be the operator defined on H, the set of non-empty compact subsets of F:

$$\forall K \subset F, \quad W(K) = \bigcup_{n\in[0,N]} w_n(K).$$

Then, if the w_n functions are contractive (the IFS is then called a *hyperbolic* IFS), there is a unique set A, the **attractor** of the IFS, such that:

$$W(A) = A.$$

Recall: A mapping $w : F \to F$, from a metric space (F, d) onto itself, is called **contractive** if there exists a positive real number $\lambda < 1$ such that $d(w(x), w(y)) \leq \lambda \cdot d(x, y)\ \forall x, y \in F$.

The uniqueness of a hyperbolic attractor is a result of the contractive mapping fixed-point theorem for W, which is contractive according to the Hausdorff distance:

$$d_H(A, B) = \max[\max_{x\in A}(\min_{y\in B} d(x, y)), \max_{y\in B}(\min_{x\in A} d(x, y))].$$

From a computational viewpoint, an attractor can be generated according to two techniques:

- **Stochastic method (toss-coin)**
  Let x_0 be the fixed point of one of the w_i functions. We build the point sequence x_n as follows: $x_{n+1} = w_i(x_n)$, i being randomly chosen in $\{1..N\}$.

  Then $\bigcup_n x_n$ is an approximation of the real attractor of $\mathcal{I}$. The larger n is, the more precise the approximation is.

- **Deterministic method**
  From any kernel S_0, we build the set sequence $\{S_n\}$,

$$S_{n+1} = W(S_n) = \bigcup_n w_n(S_n).$$

  When n is large, S_n is an approximation of the real attractor of $\mathcal{I}$.

The computation for any hyperbolic IFS of its attractor and its invariant measure is an easy task, but the inverse problem is a complex one:

Given a subset A of H, find an IFS whose attractor is A.

This problem has been considered by a number of authors. An exact solution can be found in the particular case where each $w_n(A) \bigcap w_m(A)$ is of zero measure (see [5]). In the general case, no exact solution is known[4].

An essential tool for solving this problem is the **collage theorem** :

Theorem 6 (Collage theorem) *Let $E \subset F$ be such that:*

$$d_H(E, \bigcup_n w_n(E)) < \varepsilon \quad \textit{then} \quad d_H(A, E) < \frac{\varepsilon}{1-\lambda}$$

where $\lambda = \sup_n\{\lambda_n\}$ is the largest contractive factor, and A is the attractor of $(F, (w_n))$.

Based on this, several optimization methods have been proposed, see [5], [49], [13, 40].

In [49], the authors propose a GA for solving the 1D inverse problem. Their algorithm proves to be efficient in the examples presented. Note that the 2D problem is much harder than the 1D one, since it involves many more unknowns. Typically, 6 unknowns are present in [49], and for example 24 in the 2D problem for affine IFS with 4 functions.

In the general case, a first problem is the estimation of the number n of contractions w_i, another unsolved problem is the choice of the basis of functions. Most people use affine transforms for simplicity, but the use of other types of functions, such as sine polynomials (see Fig. 15.3) can lead to a much richer class of shapes.

15.3.3 The case of affine IFS: use of a GA

In the case of an affine IFS, each contractive map w_i of $\mathcal{I}$ is represented as

$$w_i(x, y) = \begin{bmatrix} a_i & b_i \\ c_i & d_i \end{bmatrix} \cdot \begin{bmatrix} x \\ y \end{bmatrix} + \begin{bmatrix} e_i \\ f_i \end{bmatrix}$$

[4]The image compression method using IFS does not solve this inverse problem, but a simplified one, based on the separation of the image into small blocks and the finding of optimal mappings between blocks of different sizes. A simple attempt at using a GA for fractal image compression can be found in [55].

Fig. 15.3 The logo of the FRACTALES group: a sine polynomials IFS.

The inverse problem corresponds to the optimization of the values $(a_i, b_i, c_i, d_i, e_i, f_i)$ to get the attractor that most resembles the target.

The error function based on the collage theorem, i.e. the Hausdorff distance between the attractor of (w_n) and the attractor of $(w_n + \varepsilon_n)$ where ε_n is a small perturbation is extremely irregular.

Results were obtained with a simple implementation of a GA on two attractors : the well-known fern [6], and a perturbated Sierpinski triangle. See [33] for details of the experiment.

Improvements of this technique see [35] are based on:

- the use of the sharing method developed in [36], that improves the exploration capability of the GA, and allows the simultaneous detection of several optima (if any),
- the use of a multi-resolution and iterative scheme: a first very rough and then increasingly precise approximation of the fitness function is used in successive runs of the GA (the result of the previous run being used as an inital solution in the next one, at a finer resolution). A reduced number of generations has been necessary in order to obtain an approximation,
- a more precise computation of distances between target images and attractors, with the help of distance images (a well-known tool of mathematical morphology [7]), which seems to be more stable than the Hausdorf distance, and less irregular than L^1 or L^2 distances.

An experimental comparison of two fitness functions: the one based on the classical toss coin algorithm, the other using the collage theorem, yields as a first conclusion that the toss coin fitness provides more reliable results.

15.3.4 Mixed IFS

When the w_i are no longer restricted to being affine functions, the corresponding inverse problem cannot be addessed in a simple way, unless some *a priori* hypotheses on the structure of the IFS (number and type of functions) are made. In the following, we have chosen to call these IFS **mixed IFS** in order to emphasize the fact that they are no longer restricted to being affine. When dealing with mixed IFS, the first point to be addressed is to find an adequate representation. A natural one is to code the mixed IFS as trees.

The attractors of Fig. 15.4 is a random mixed IFS: the w_i functions have been recursively built with the help of random shots in a set of basic functions, a set of terminals (x and y), and a set of constants. In our examples, the constants belong to $[0, 1]$, and the basic functions set is $\{+, -, \times, \text{div}(x, y) = \frac{x}{0.0001+|y|}, \cos, \sin, \text{root}(x) = \sqrt{|x|}, \text{loga}(x) = \log(1 + |x|)\}$

$$w_1(x, y) = \begin{pmatrix} \sqrt{|\sin(\cos 0.90856 - \log(1 + |x|))|} \\ \sin y \end{pmatrix}$$
$$w_2(x, y) = \begin{pmatrix} \cos(\cos(\sqrt{|x|})) \\ \cos(\log(1 + |y|)) \end{pmatrix}$$
$$w_3(x, y) = \begin{pmatrix} \log(1 + |\cos(\log(1 + |y + x|))|) \\ \sqrt{|\sin 0.084698|} \end{pmatrix}$$
$$w_4(x, y) = \begin{pmatrix} \log(1 + |\sin(\sqrt{|0.565372|})|) \\ \sqrt{|0.81366 - ((\log(1 + |0.814259|)) * \cos y)|} \end{pmatrix}$$
$$w_5(x, y) = \begin{pmatrix} \log(1 + |\sqrt{|0.747399 + \cos y|}|) \\ \sin(\frac{0.73624}{0.0001+|0.264553*y+0.581647+x|}) \end{pmatrix}$$

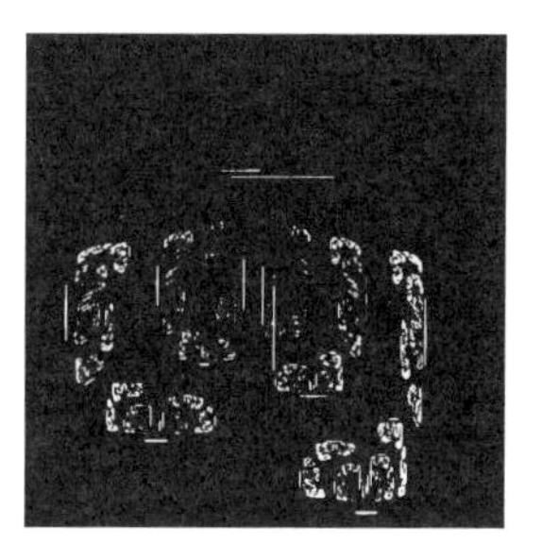

Fig. 15.4 A mixed IFS and its attractor.

Each w_i can thus be represented as a tree. The trees of the w_i are then gathered to build the main tree representing the IFS $\mathcal{I}$. This very simple structure allows an IFS to be coded with different numbers and different types of functions.

The set of possible IFSs depends on the choice of the basic function set and constant set. A difficult problem for a mixed IFS is to verify that the w_i are contractive, in order to select a *hyperbolic* IFS. Unlike the case of an affine IFS, this verification is not straightforward for a mixed IFS and is in fact computationally intractable. We have proposed using some heuristics that reject strongly noncontractive functions (see [34] for details).

15.3.5 Inverse problem for mixed IFS: a solution based on GP

Genetic programming can be efficiently used in order to solve a more "general" inverse problem for IFS: a simultaneous numeric and a symbolic optimization can be efficiently performed with GP. The interest for the inverse problem with "mixed IFS" is important and may enlarge the scope of applications, for example image compression, because it allows a wider range of shapes to be coded.

A GP algorithm has been tested on shapes that were actual attractors of IFSs, some of which being constituted with randomly chosen contractive maps. The choice of basic functions for the GP is the one presented in Section 15.3.4. Initial populations were randomly chosen. See [34] for a detailed description of the algorithm and experiments.

The inverse problem for mixed IFS has been solved within a reasonable computation time (a few hours on Sparc 10 and Dec 5000 stations). This computation time is similar to computation times of GA applied to the inverse problem for the affine IFS [33], although in the case of the mixed IFS the size of the search space is much larger. This fact may be explained by the use of variable-sized structures in the GP algorithm, which seems to perform a more efficient search in a large space. Parameter adjustment remains a challenging task, but we empirically noticed the following facts:

- The distance images are very efficient, and especially their associated multiresoltion scheme: increasingly precise evaluation of the difference between the current IFS and the target are used throughout the evolution of the GP algorithm.

- The mutation operator is important: a specialised mutation on constants has been used in order to make the constants evolve.

Finally, the target images that yield good results are rather compact; the convergence to line-shaped targets is more difficult.

15.3.6 Interactive GP for the generation of mixed IFS attractors

A genetic Progaming algorithm may also be used in an interactive way in order to investigate the space of mixed IFS attractors. The GP algorithm is used as an exploration tool in an image space: the implicitly optimised fitness function is the "users satisfaction". This interactive approach is not new: Karl Sims [50] has extensively shown the power of the method in the framework of computer graphics.

An application has been developed (in JAVA language interfaced with a C GA-toolbox) in order to provide an artistic exploration tool. A beta version of this algorithm is available on the FRACTALES WEB pages: `http://www-rocq.inria.fr/fractales/`.

The originality of our implementation is based on the use of mixed IFS, which are fractal structures represented as variable size trees, and on the use of genetic programing.

Fig. 15.5 Some Mixed IFS attractors generated using interactive GP

15.4 CONCLUDING REMARKS AND FUTURE DIRECTIONS

This paper is an overview of the work done on Genetic Algorithms since 1993 in the FRACTALES Group. The main theoretical contribution concerns a new look at deception theory: the analysis of the adjusted fitness function f' has helped to shed new light on the notion of deception, that can be related – to a certain extent – to some irregularity measures on the function. This adjusted fitness function can be considered as an interesting performance indicator of the GA, as it involves some of the GA parameters[5]. However, a lot of work remains to be done, especially concerning the exact signification of this function f' and its expression in the framework of other models than deception theory.

Another contribution of this work to be emphasized concerns the application of GA and GP to the inverse problems for IFS. The resolution of the inverse problem for IFS is an important topic for fractal analysis of data. We intend to continue on this topic in the following way:

- It is well known that fractal geometrical structures (as they are extremely irregular, but with some scaling homogeneity) have interesting physical properties, see [56]. An application related to structural mechanics has been initialized in collaboration with the CMAPX laboratory, based on the inverse problem for IFS.
- We are currently starting the experimentation of a different approach with GA/GP to the inverse problem for IFS, using a sort of "individual" approach. An IFS is represented in the "classical" approach as an individual of a population: in the "individual" approach, an IFS is represented with a whole population, an individual being a function. This tends to explore the seach space in a more economical way (population sizes are smaller, too), at the expense of a more complex fitness computation, of course. This approach

[5]Of course, deceptivity cannot be related to GA performance in a simple way, and we by no means claim that the bound on $|f - f'|$ is **the** performance indicator. The experimentations presented in [30] tend to prove that this bound is reliable in a relative way to estimate encoding efficiency, but of course extensive experiments must be carried out before claiming such a fact.

has been first implemented with GA for affine IFS, and preliminary experiments tend to prove that the GA converge very rapidly onto a rough approximation of the target shape, precise converge being a more difficult task (where a careful design of the fitness function is extremely important).

- Finally, an efficient resolution of the inverse problem for IFS may influence the classical techniques in the field of fractal compression of data (signal or images). Some studies in the FRACTALES group concerning data compression [54], [10] will be pursued based on more complex transformation and on the use of GA/GP optimization techniques.

REFERENCES

1. A. Agapie. Genetic algorithms: Minimal conditions for convergence. In *Artificial Evolution, European Conference, AE 97, Nimes, France, October 1997,*. Springer Verlag, 1997.

2. Lee Altenberg. The Schema Theorem and Price's Theorem . In *Foundation of Genetic Algorithms 3*, 1995. ed. Darrell Whitley and Michael Vose. pp. 23-49. Morgan Kaufmann, San Francisco.

3. P. Andrey and P Tarroux. Unsupervised image segmentation using a distributed genetic algorithm. *Pattern Recognition, 27*, pages 659–673, 1993.

4. M. Barnsley and S. Demko. Iterated function system and the global construction of fractals. *Proceedings of the Royal Society*, A 399:243–245, 1985.

5. M. Barnsley, V. Ervin, D. Hardin, and J. Lancaster. Solution of an inverse problem for fractals and other sets. *Proc. Natl. Acad. Sci. USA*, 83, 1986.

6. M. F. Barnsley. *Fractals Everywhere.* Academic Press,N Y, 1988.

7. G. Borgefors. Distance transformation in arbitrary dimension. *Computer Vision, Graphics, and Image Processing*, 27, 1984.

8. R. Cerf. *Artificial Evolution, European Conference, AE 95, Brest, France, September 1995, Selected papers*, volume Lecture Notes in Computer Science 1063, chapter Asymptotic convergence of genetic algorithms, pages 37–54. Springer Verlag, 1995.

9. Nathan Cohen. Antennas in chaos: Fractal-element antennas. In *Fractals in Engineering 97*. INRIA, 1997. Hot Topic Session, Arcachon, France, June 25-27.

10. Guillaume Cretin and Evelyne Lutton. Fractal image compression: Experiments on hv partitioning and linear combination of domains. In *Fractals in Engineering 97*. INRIA, 1997. Poster Session, Arcachon, France, June 25-27.

11. Y. Davidor. *Genetic Algorithms and Robotics. A heuristic Strategy for Optimization.* World Scientific Series in Robotics and Automated Systems - vol 1. World Scientific, Teaneck, NJ, 1991.

12. Thomas E. Davis and Jose C. Principe. A Simulated Annealing Like Convergence Theory for the Simple Genetic Algorithm. In *Proceedings of the Fourth International Conference on Genetic Algorithm*, pages 174–182, 1991. 13-16 July.

13. S. Demko, L. Hodges, and B. Naylor. Construction of fractal objects with ifs. *Siggraph*, 19(D 3), 1985.

14. J. H. Elton. An ergodic theorem for iterated maps. In *Georgia Tech. preprint*, 1986.

15. K. J. Falconer. *Fractal Geometry: Mathematical Foundation and Applications*. John Wiley & Sons, 1990.

16. Y. Fisher. Fractal image compression. In *Siggraph 92 course notes*, 1992.

17. B. Goertzel. Fractal image compression with the genetic algorithm. *Complexity International*, 1, 1994.

18. D. E. Goldberg. Genetic Algorithms and Walsh functions: Part I, a gentle introduction. *TCGA Report No. 88006*, University of Alabama, Tuscaloosa, US, 1988.

19. D. E. Goldberg. Genetic Algorithms and Walsh functions: Part II, deception and its analysis. *TCGA Report No. 89001*, University of Alabama, Tuscaloosa, US, 1989.

20. David A. Goldberg. *Genetic Algorithms in Search, Optimization, and Machine Learning*. Addison-Wesley Publishing Company, inc., Reading, MA, January 1989.

21. David E. Goldberg. Construction of high–order deceptive functions using low–order walsh coefficients. IlliGAL Report 90002, University of Illinois at Urbana–Champaign, Urbana, IL 61801, December 1990.

22. D. P. Hardin. *Hyperbolic Iterated Function Systems and Applications*. PhD thesis, Georgia Institute of Technology, 1985.

23. J. H. Holland. *Adaptation in Natural and Artificial System* . Ann Arbor, University of Michigan Press, 1975.

24. A. Homaifar and S. Guan. Training weights of neural networks by genetic algorithms and messy genetic algorithms. In M. H. Hamza, editor, *Proceedings of the Second IASTED International Symposium. Expert Systems and Neural Networks*, pages 74–77, Hawaii, 17-17 Aug 1990. Acta Press, Anaheim, CA.

25. J. Hutchinson. Fractals and self-similarity. *Indiana University Journal of Mathematics*, 30:713–747, 1981.

26. E. W. Jacobs, Y. Fisher, and R. D. Boss. Fractal image compression using iterated transforms. In *Data Compression*, 1992.

27. A.E. Jacquin. Fractal image coding: A review. *Proc. of the IEEE*, 81(10), 1993.

28. J. Juliany and M. D. Vose. The genetic algorithm fractal. *Evolutionary Computation*, 2(2):165–180, 1994.

29. J. R. Koza. *Genetic Programming* . MIT Press, 1992.

30. Benoit Leblanc and Evelyne Lutton. Bitwise regularity coefficients as a tool for deception analysis of a genetic algorithm. Technical Report RR-3274, INRIA research report, 1997. October.

31. Benoit Leblanc, Evelyne Lutton, and Jean-Paul Allouche. *Artificial Evolution, European Conference, AE 97, Nimes, France, October 1997, Selected papers*, volume Lecture Notes in Computer Science, chapter Inverse problems for finite automata: a solution based on Genetic Algorithms. Springer Verlag, 1997.

32. P. Liardet. Algorithmes genetiques d'un point de vue ergodique. In *Mini-colloque "Algorithmes Evolutifs: de la theorie aux applications"*, 1997. Centre de Mathematiques et Informatique, Marseille, 19-20 juin 1997.

33. E. Lutton and J. Levy-Vehel. Optimization of fractal functions using genetic algorihms. Technical Report 1941, INRIA research report, June 1993.

34. E. Lutton, J. Lévy Véhel, G. Cretin, P. Glevarec, and C. Roll. Mixed ifs: resolution of the inverse problem using genetic programming. *Complex Systems*, 9:375–398, 1995. (see also Inria Research Report No 2631).

35. Evelyne Lutton. *Genetic Algorithms and Fractals - Algorithmes Génétiques et Fractales.* Dossier d'habilitation à diriger des recherches, Université Paris XI Orsay, 11 Février 1999. Spécialité Informatique.

36. Evelyne Lutton and Patrice Martinez. A genetic algorithm for the detection of 2d geometric primitives in images. Research Report 2110, INRIA, November 1993.

37. Evelyne Lutton and Patrice Martinez. A Genetic Algorithm for the Detection of 2D Geometric Primitives in Images. In *12-ICPR*, 1994. Jerusalem, Israel, 9-13 October.

38. Evelyne Lutton and Jacques Lévy Véhel. Some remarks on the optimization of holder functions with genetic algorithms. Research Report 2627, INRIA, July 1995.

39. B.B. Mandelbrot and J.W. Van Ness. Fractional Brownian Motion, fractional Gaussian noises and applications. *SIAM Review 10*, 4:422–437, 1968.

40. G. Mantica and A. Sloan. Chaotic optimization and the construction of fractals: solution of an inverse problem. *Complex Systems*, 3:37–62, 1989.

41. Zbigniew Michalewicz and Marc Schoenauer. Evolutionary algorithms for constrained parameter optimization problems. *Evolutionary Computation*, 4(1):1–32, 1997.

42. D. J. Nettleton and R. Garigliano. Evolutionary algorithms and a fractal inverse problem. *Biosystems*, 33:221–231, 1994. Technical note.

43. S. Rochet, G.Venturini, M. Slimane, and E.M. El Kharoubi. *Artificial Evolution, European Conference, AE 97, Nimes, France, October 1997, Selected papers*, volume Lecture Notes in Computer Science, chapter A critical and empirical study of epistasis measures for predicting GA performances: a summary. Springer Verlag, 1997.

44. Sophie Rochet. *Convergence des Algorithmes Génétiques: modŁles stochastiques et epistasie*. PhD thesis, Université de Provence, Janvier 1998. Spécialité Mathématiques et Informatique.

45. G. Rudolph. Asymptotical convergence rates of simple evolutionary algorithms under factorizing mutation distributions. In *Artificial Evolution, European Conference, AE 97, Nimes, France, October 1997,*. Springer Verlag, 1997.

46. Marc Schoenauer. *Evolutionary Computation and Numerical Optimization*. Dossier d'habilitation à diriger des recherches, Université Paris XI Orsay, January 1997. Spécialité Informatique.

47. Marc Schoenauer and Zbigniew Michalewicz. Boundary operators for constrained parameter optimization problems. In *ICGA97*.

48. Marc Schoenauer and Zbigniew Michalewicz. Evolutionary computation at the edge of feasability. In *PPSN 96, Parallel Problem Solving from Nature*.

49. R. Shonkwiler, F. Mendivil, and A. Deliu. Genetic algorithms for the 1-d fractal inverse problem. In *Proceedings of the Fourth International Conference on Genetic Algorithms*, pages 495–502, 1991. 13-16 July.

50. K. Sims. Interactive evolution of dynamical systems. In *First European Conference on Artificial Life*, pages 171–178, 1991. Paris, December.

51. Gilbert Syswerda. Uniform crossover in genetic algorithms. In J.D. Schaffer, editor, *Procedings of the Third International Conference on Genetic Algorithms*. Morgan Kaufmannn publishers, San Mateo, 1989.

52. P. Trompette, J. L. Marcelin, and C. Schmelding. Optimal damping of viscoelastic constrained beams or plates by use of a genetic algotithm. In *IUTAM*, 1993. Zakopane Pologne.

53. J. Lévy Véhel and E. Lutton. Optimization of fractal functions using genetic algorithms. In *Fractal 93*, 1993. London.

54. Jacques Lévy Véhel, Khalid Daoudi, and Evelyne Lutton. Fractal modeling of speech signals. *Fractals*, 2(3):379–382, September 1994.

55. L. Vences and I. Rudomin. Fractal compression of single images and image sequences using genetic algorithms., 1994. The Eurographics Association.

56. Jacques Lévy Véhel, Evelyne Lutton, and Claude Tricot (Eds). *Fractals in Engineering: From Theory to Industrial Applications*. Springer Verlag, 1997. ISBN 3-540-76182-9.

57. M.D. Vose. Formalizing genetic algorithms. In *Genetic Algorithms, Neural Networks and Simulated Annealing Applied to Problems in Signal and Image processing*. The institute of Electrical and Electronics Engineers Inc., 8th-9th May 1990. Kelvin Conference Centre, University of Glasgow.

58. E. R. Vrscay. *Fractal Geometry and Analysis*, chapter Iterated function Systems: theory, applications and the inverse problem, pages 405–468. 1991.

59. R. Vrscay. Moment and collage methods for the inverse problem of fractal construction with iterated function systems. In *Fractal 90 Conference*, 1990. Lisbonne, June 6-8.

16 Three Evolutionary Approaches to Clustering

H. LUCHIAN

"Al.I.Cuza" University of Iasi, Romania
E-mail: hluchian@infoiasi.ro

Abstract. This paper is an overview of our work for solving clustering problems in an evolutionary setting. In approaching these problems, we faced many of the general questions one has to answer when solving real-world problems under such a weak[1] approach as Genetic Algorithms; we use here some of these questions as sub-titles for the paragraphs of the paper. However, it has not been our intention to undertake a study of such questions or to provide general answers to them. Simply, a particular experience is presented: a sequence of four *case studies* – two problems in their general versions and two real-world applications. The structure of this paper does not follow closely the usual one: the emphasis is on the design process rather than, for example, on experiments and their interpretation. Our hope is that those wishing to have a practical insight into either Genetic Algorithms from the viewpoint of Clustering, or into some versions of Clustering from the viewpoint of Genetic Algorithms, will find this paper interesting.

16.1 INTRODUCTION

Clustering and its follow-up, classification, are ubiquitous problems, versions and instances of which are likely to be met, either explicitly or not, in many real-world problems more abstractly labeled as "pattern recognition" problems; clustering and classification processes may have to be applied to data from all kinds of sources and of all types – be it acoustical or visual, temporal or logical, numerical or categorial. Clustering and classification have been performed, in manual, "rule-of-the-thumb"-like ways, in more "classical" times: Karl Linee and Lev Mendeleev are among the most celebrated "classifier scientists" of all times; they found clustering criteria in Botany and Chemistry more as a result of their genius than that of a systematic, reproducible approach. Sir W.M.F.Petrie, one of the greatest Egyptologists, used, by the beginning of this century, another form of clustering: he performed beforehand-unpredictably long sequences of manual column and row swaps in huge binary matrices, in

[1] I.e., widely applicable.

his (successful) attempt to define a temporal order on various objects and facts related to ancient Egyptian history.

A general way of stating and approaching such various, specific problems has not been, however, in order until relatively recently; one can consider clustering as a "modern" research problem. It is therefore not intriguing that nearly all major "modern" techniques have been tried for solving it: Artificial Neural Networks, Fuzzy Sets, Evolutionary Computing (that is, the *Soft Computing* Club members) – see [CAG87], [CAR91a], [CAR91b], [CAR92], [BRE91], [PAB93], [THA95], [RIP96], [AND96]) –, Simulated Annealing ([FLA89], [ZEG92]), and others [KOO75]. Of course, more "classical" analytical and statistical approaches are also being used ([DHA73], [FUK90], [RIP96]).

In the sequel, we present briefly our evolutionary approaches to some clustering problems. The presentation follows the chronology of our study, which started at the general clustering problem level and then focused on real world problems. For excellent monographs on various aspects of clustering and classification, we refer to [HAR75], [DID82], [JAD88], [CEL89], [KAR90], and [RIP96]; we will not go here into details on previously used heuristics. More complete presentations of our approach can be found in [LUC94], [LUC95], [LUC98a]; we have reported more recent results in [LUC99b-d].

Extended as it may seem, the general presentation of the problem at hand – clustering – is necessary in order to illustrate properly the process of designing our genetic algorithms.

For GA notions and techniques we refer to three excellent monographs: [MIT97], [KOZ93], [MIC92]; many of the papers mentioned in this paper can be found in Proceedings of the ICGA 1989-1997.

16.2 CLUSTERING PROBLEM NO. 1: AUTOMATED UNSUPERVISED CLUSTERING – THE QUANTITATIVE CASE

The first version of the clustering problem we are interested in is unsupervised automated clustering – the quantitative case. The problem can be stated as follows:

> Given **n** objects, each one characterized by specific values of **a** numerical attributes, find the best partition of the **n** objects.

Quite often, this is the more formal statement of a real life (sub)problem; note that the number of clusters (classes of the partition) is considered unknown; that is, we are concerned with *unsupervised clustering*.

16.2.1 Are Genetic Algorithms suitable for this problem?

Stated as above, the clustering problem is imprecisely defined (*what is a "good" partition?*) and far from encouraging a GA approach (for selection purposes, *what is a "better" partition?*). Nevertheless, clustering being a well-developed research field, one can soon come across optimization versions of it, which answer the two questions in Italics. Various optimization criteria have been proposed for tackling the problem; starting from these criteria, various *heuristics* have been developed (there is no deterministic algorithm which can find the optimal partition for all instances of the problem); see [BAH65], [FOR65], [MAQ67], [HAK77], [HAW79], [MOO89], [HUA95], [RIP96], for heuristics such as k-means, k-medoids, adaptive resonance etc. The space of all possible partitions of the **n** objects, which is the search space to be scanned by these heuristics, is huge[2]. Thus, we have three indications that an approach by means of Genetic Algorithms may be worth looking at: a) the problem at hand can be stated as an optimization one, b) it has a very large search space[3] and c) no exact algorithm is available.

16.2.2 Trying to assess the level of difficulty of the problem

What would be interesting to assess at this stage is the level of difficulty of the problem, in terms of the fitness function, under an evolutionary approach. A tool for that has been proposed by Michalewicz: a statistic s, which estimates the dispersion of the fitness function, values around its mean value. As pointed out in [MIC93], a value of s close to s_0=0.1 suggests that the problem (actually, the fitness function) may be "easy-to-solve" by means of GAs; the further the value of s from s_0, the more difficulties may appear (very likely, the selection pressure should be either increased or decreased in the attempt to obtain a good convergence).

As will be seen in Section 16.2.6, we actually used several candidate fitness functions and more test cases for this general case of the problem. For two of the six candidate fitness functions, the estimations for s were close to 0.01 (for the first one, the values were between 0.087 and 0.279[4]); the other four candidate functions led to "worse" estimated values for s (typical estimated values for s were grouped in two sets: one set around 0.01 and the other one around 2.0). Hence, this part of the designing process suggested that, from the viewpoint of the likely convergence of the algorithm under design, the first candidate function is the favorite one. Another empirical conclusion has been that though all the candidate fitness functions provide the same linear ordering over the partitions of the set of given points, they do not lead to equivalent scanning processes of the search space, performed by the GA (which was to be expected).

16.2.3 Looking for a proper representation

The first idea comes from the observation that a candidate solution consists of a set of classes defined over the **n** given objects; we have to represent a set of sets obeying the two restrictions of a partition. Trying to match the fact that on one hand the population is a set of individuals and on the other hand the candidate solution is a set of classes, one idea would be to represent a solution by the entire population; this obviously leads to important changes in the general Genetic Algorithm; see [AND96] for such an approach.

Aiming to stay as close as possible to both the problem characteristics and the GA general scheme, we ended up with a representation inspired from the existing clustering theory: the "mobile centers" version of the "dynamic clouds" method [CEL89]. Each chromosome represents a clustering (not a cluster, as in [AND96]), that is, a partition of the set of given objects. Chromosomes have different and variable lengths: the chromosomes in the initial populations are generated with different lengths and at each generation, genetic operators may alter these lengths. A chromosome has **k** genes, where **k** is the number of

---

[2] Two examples: for **n**=50, the order of magnitude is 10^{47}; and **n**=100 objects can be partitioned into **k**=5 classes only, in approx. 10^{68} different ways.
[3] See [JOB91] for a study on partitioning problems in an evolutionary setting.
[4] The s values differ because of the sampling errors and also because of the different input data.

classes in the partition represented by that individual. A gene is an **a**-uple of numerical[5] values; these values represent the coordinates (attribute values) of a point $C \in \mathbf{R}^a$, where $\mathbf{R}^a$ is the corresponding Euclidean space. Each gene **g** representing the point C, encodes a cluster; it consists of those points Q, among the **n** given ones, which C is closest to, of all **k** points represented in the chromosome:

$$d(Q,C) = \min_{j=1}^{k} d(Q,C_j), \tag{1}$$

where C_j, j=1,...,**k**, are the points represented by the **k** genes. Let us call C_j *the center of cluster* j [6] and let Cl_j denote the corresponding class[7]. Any **k** different points define a unique partition of the set of given points into **k** classes; the reverse is not true, since there may exist different **k**-uples of points, which define the same partition of the **n** given points. For technical reasons, the genes of a chromosome are sorted according to the values $d(C_j,O)$, j=1..k, where O is the point in $\mathbf{R}^a$ having coordinates (0,...,0); the order of genes is not relevant.

16.2.4 Is the chosen representation appropriate?

Assessing the appropriateness of a specific representation for a given problem is a desirable step to take – either for choosing among different candidate representations or / and for predicting how easy it will be for the algorithm to find good solutions under that representation. Jones and Forrest proposed a measure of appropriateness – the fitness-distance correlation [JOF95]. It is a statistic intended to detect the level of correlation between the accuracy of a candidate solution and the distance – in the genotype space – to the (closest) optimal solution[8]; Schoenauer e.a. [SCK96] discussed the way FDC could be applied to the case of variable length chromosomes and also to problems with unknown global optimum. We used this technique for testing whether to proceed at the coordinate-genes level or to lower the level of representation (actually, of operators) to bitstring chromosomes. For the minimization approach in Section 16.2.9, the differences were very clear (typical FDC values for coordinate genes were around +0.65, while those for bitstring were around +0.07) and consequently we decided to drop out the bit-level mutation and crossover, which were used in a first version of the algorithm. See Sections 16.4.1 and 16.5.1 for more examples on the use of the FDC.

16.2.5 The criterion to be optimized

We chose to use a popular optimization criterion for clustering, derived from the theorem of Huygens. The frame is an Euclidean space[9], which all the attribute values belong to; in terms of its distance, three "inertia", inspired from solid mechanics theory, are defined: the total inertia **T**, the within-cluster inertia **W** and the between-clusters inertia **B**. The values of **B** and **W** depend on the number **k** of classes and on the particular partition into **k** classes. The Huygens theorem states that, for each set of **n** points to be clustered, the respective inertia satisfy the relation $\mathbf{T}=\mathbf{B}_k+\mathbf{W}_k$. The derived optimization criterion is "for fixed **k**,

---

[5] In our implementations: integers.

[6] C_j needs not be the gravity center of cluster j.

[7] The chromosome is decoded by a *minimum-distance classifier* procedure [RDU97].

[8] This distance expresses the *precision* of a solution.

[9] Our implementation actually uses coordinates on the screen, expressed in number of pixels.

maximize $\mathbf{B}_k$" or, equivalently, "minimize $\mathbf{W}_k$". Obviously, this criterion defines a local optimum (i.e., for a fixed **k**) and not the global one (for any value of **k**). In other words, if we know the value of **k** beforehand (supervised clustering), then this optimization criterion can lead to the optimal clustering. However, since the criterion above does not define a linear ordering on the set of all partitions, in cases where **k** is unknown (unsupervised clustering), one cannot decide, for two partitions of different cardinalities, which one is better; this raises a serious problem for the selection process of the Genetic Algorithm we are to devise. Therefore, the best we could hope at this point is to develop a GA for supervised clustering.

16.2.6 Selecting among candidate models

The critical issue in trying to apply evolutionary techniques to unsupervised clustering is to decide which of two partitions of different cardinalities (i.e., having different numbers of classes) is better; in other words, what we need is to define a fitness function which, on one hand, is a correct model for the problem at hand and, on the other hand, interprets efficiently and properly the chromosomes of various lengths defined above. Such a function would define in a rigorous manner the optimum, which was imprecisely required above.

One can think of various functions to be used for comparing partitions of different cardinalities; selecting the most appropriate such function is a difficult task, if no formal proof of its optimality exists.

In search of a good fitness function for our Genetic Algorithm, we have concentrated on (minimizing) the sum of the squares of the **n** distances from the given points to the respective centers of classes in a chromosome *Chr*:

$$f(Chr) = \sum_{j=1}^{k} \sum_{Q \in Cl_j} \sum_{i=1}^{a} ((x_Q^i - x_{C_j})^2) = W_{\mathbf{k}}(P), \tag{2}$$

where x_Q^i, i=1..**a**, are the **a** coordinates (attribute values) of point Q in $\mathbf{R}^{\mathbf{a}}$ and P is the partition encoded by *Chr*. Minimizing f should lead to "homogenous" clusters, which would follow the "density" of the given points: the within-cluster inertia $\mathrm{W}_{\mathbf{k}}$ is being minimized. However, the best partition under this minimization task is the one, which has exactly **n** single-point clusters (since the minimal value of W above is 0, obtained for such clusters). One way to control the growth of the number of clusters is to include **k** as a weight that increases the values of f given by (2):

$$f_1(Chr) = \mathbf{k} \cdot \sum_{j=1}^{k} \sum_{Q \in Cl_j} \sum_{i=1}^{a} ((x_Q^i - x_{C_{ji}})^2) = \mathbf{k} \cdot W_{\mathbf{k}}(P). \tag{3}$$

We have used function f_1 in early experiments. For this function, neither the computed s values (which lay in the interval [+1.51;+1.95]) nor the FDC values for cases with known optimum (numbers in the interval [-0.78; -0.47]) were encouraging: a high selection pressure and a possibly deceptive Genetic Algorithm, respectively, were suggested by these values. The experiments confirmed these expectations – over 85% of the runs led to premature convergence (local optima); not even the optimal value of **k** has been found. Other candidate fitness functions had to be tried.

For a fixed value of **k**, if one considers several different clusterings with **k** clusters each, then the one that produces the highest value of the ratio B/T is the "best" one, the next highest B/T value comes from the second-best one etc. If unsupervised clustering is in order, then not only the optimal clustering, but also the optimal value of **k** has to be found. However, if clusterings with various numbers of clusters are considered, then the orderings defined by B/T values obtained for different **k** values are interlaced; Fig. 16.1 illustrates the relation between values for (B/T) and values for **k**, for various clusterings obtained as best_so_far chromosomes in a typical run of the Genetic Algorithm **A,** described below.

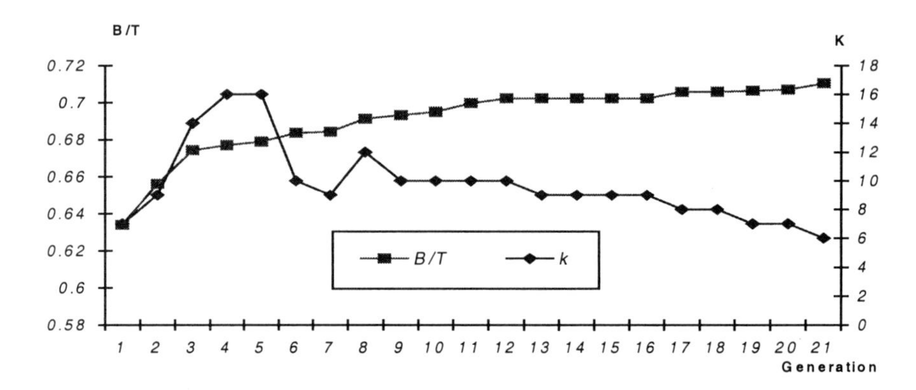

Fig. 16.1 Typical evolution of B/T as a fitness function and of the length for the best_so_far chromosome, in a typical run.

Our idea has been to enforce a more convenient grouping of the B/T values by considering the candidate fitness function *F*:

$$F(Chr) = (B/T)^{\mathbf{k}}, \tag{4}$$

where *Chr* is a chromosome of length **k** which encodes a clustering characterized by the total inertia T and the between-clusters inertia B[10]; note that $B/T \in (0,1)$ and hence **k** decreases the fitness value, as needed for control under maximization. A side effect of using the power **k** in (4) is, as in (3), the control on the number of clusters in chromosomes in future generations.

Function *F* was the starting point for a study on several similar candidate fitness functions:

$$F_1 = \left(\frac{B}{T}\right)^{\ln \mathbf{k}};\ F_2 = \left(\frac{B}{T}\right)^{\sqrt{\mathbf{k}}};\ F_3 = \frac{B}{\mathbf{k} \cdot T};$$

$$F_4 = \frac{B}{T \cdot \ln \mathbf{k}};\ F_5 = \frac{B}{T \cdot \sqrt{\mathbf{k}}}. \tag{5}$$

The monotonicity of each of the six functions suggests identical orderings on the set of clusterings. However, the search process for two otherwise identical Genetic Algorithms,

---

[10] We dropped here the subscript **k** in order to suggest that *any* partitions are to be compared.

one using F as fitness function, and the other one using, say, F_5, may differ dramatically: for chromosomes of different lengths **k**, the values $|F(Chr_i)-F(Chr_j)|$ and $|F_5(Chr_i)-F_5(Chr_j)|$, $i,j \in 1...pop\_size$, may lead to different structures of the"roulette-wheel" for the selection process and thus produce different behaviors of the algorithm[11].

Suppose we have several candidate functions to model a given problem and no theoretical results to help choosing the best one. If "best" meant "most appropriate for optimization using a GA", then one way to select a model is to repeatedly run a properly designed Genetic Algorithm for that problem, keeping all the rest unchanged and activating as fitness function each time another candidate function (of course, each version of the GA has to be run more than once). Some functions will lead to very good solutions for the problem, some others will find less satisfactory solutions; the function producing the best results[12] will be the guess for the proper model. We do not discuss here the fact that for each fitness function, the best setting of the GA may differ dramatically (e.g., in each case a different scaling strategies could improve the performance)[13].

A possible implementation of the strategy outlined above is to run a Genetic Algorithm many times changing the fitness function every 20 runs; the best obtained results[14] in one series of 20 runs should indicate the most appropriate fitness function.

For selecting the fitness function, we have tested each of the six functions against eight sets of randomly generated input data (**n** values between 1000 and 10000); the optimal clusterings were known beforehand in each case. The Genetic Algorithm **A** has been kept unchanged, except for the procedure calculating the fitness value; let $\mathbf{A}_F$, $\mathbf{A}_{F1}$, $\mathbf{A}_{F2}$, $\mathbf{A}_{F3}$, $\mathbf{A}_{F4}$, $\mathbf{A}_{F5}$ denote the six algorithms. Each algorithm has been run 50 times for each data set; the percentages of successful runs (optimal clustering found) are given in Table 16.1.

Table 16.1 Percentages of successful runs out of 50. Each row corresponds to a candidate fitness function and each column, to a test case.

	T1	T2	T3	T4	T5	T6	T7	T8
A	100	100	100	92	94	86	100	60
A_1	100	96	84	90	78	76	86	52
A_2	90	88	72	54	24	38	58	10
A_3	58	46	48	32	14	8	22	6
A_4	50	38	36	42	22	14	30	14
A_5	62	34	20	36	28	12	24	16

The decision has been to use function F in all experiments.

16.2.7 From function-to-optimize to fitness function

The main concern of the previous paragraph has been a modeling problem: select among several candidate fitness functions. A related topic is how to obtain, from a function-to-

[11] Rank-based selection would not produce this effect.
[12] Most often, or on the average etc.
[13] One solution to such an objection is to design a more complicated experiment by using a supervisor GA with a *fitness gene* included in the genotype.
[14] Under some statistics (average, maximum, standard deviation etc.).

optimize, a fitness function. In our case, there has been no need for translating the fitness function values. We introduced an exponent **k**, which alters the theoretically founded B/T. The exponent is meant for a twofold goal: regroup the B/T values for different **k** values and control the growth of **k** as the evolution process proceeds. We can also interpret the power **k** as a *sui generis* exponential scaling of the B/T ratio, which lowers the selection pressure.

16.2.8 Choosing / adapting operators (I)

As it usually is the case, we had to adapt operators both to the problem at hand and to the chosen representation.

Mutation We used three mutation operators; their rates have been independent and denoted the expected number of chromosomes to be affected by mutation.

- *basic mutation*: a randomly chosen gene is randomly altered (a point C_j is replaced by randomly changing the values of its coordinates);

- *length mutation*: a randomly generated bit decides whether this operator inserts a random gene in the chromosome or, on the contrary, deletes a randomly chosen gene from the chromosome. This operator allowed for very quick convergence to the optimum even when the initial population contained only chromosomes of length 1;

- *binary mutation*: this is the standard mutation operator, when the chromosome is seen as a bitstring.

The three mutation operators yield valid chromosomes, since the centers of clusters need not be among the **n** given points.

Crossover Three crossover operators are used; they share a global crossover rate (for example, for $\mathbf{p}_c$=0.8, the three operator rates could be $\mathbf{p}_{c1}$=0.3, $\mathbf{p}_{c2}$=0.3 and $\mathbf{p}_{c3}$=0.2):

- *same-position one-point crossover*: for two parents of lengths $\mathbf{k}_1$ and $\mathbf{k}_2$, respectively, a cutting point between 1 and $min(k_1,k_2)$ is chosen at random; standard crossover is then performed at the gene level (not the bit level). Each descendant inherits the length of a parent;

- *Two-position one-point crossover*: for each parent, a cutting point between 1 and **k** (the number of genes) is randomly chosen. The lengths of the two descendants may differ from the lengths of the parents.

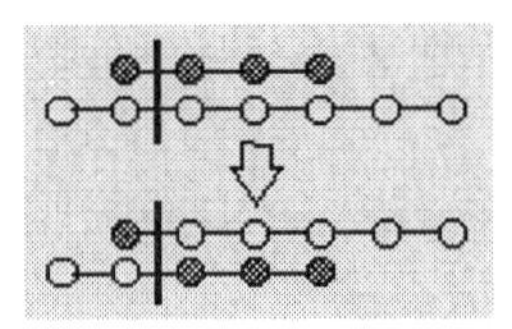

Fig. 16.2 Two-position one-point crossover: parents of lengths 4 and 7 yield descendants of lengths 6 and 5.

- *binary crossover*: same as the first crossover operator above, but cutting is performed at the bit level, (not the gene level).

Other features Besides these six genetic operators, a couple of technical improvements have been made by means of other operators, which are applied *during evaluation*:

- *write-back*: replaces, upon evaluation, the center C of a cluster by the gravity center of that cluster[15]. There is empirical evidence that a 100% rate of the write-back operator slows down the convergence; this may be because after write-back, the chromosomes represent more stable clusterings. We used a write-back rate of 50 %;

- *delete*: there may exist genes of a chromosome, which represent clusters containing at most one point. Such genes are always deleted from the chromosome, since void or single clusters are irrelevant for non-trivial clusterings.

Adjusting the rates We used a supervisor Genetic Algorithm [GRE86] for setting the parameters of our basic Genetic Algorithm, **A;** the elements of this supervisor GA are called "meta-elements" (from "meta-GA").

A meta-chromosome has nine meta-genes, encoding respectively: the population size, the three mutation rates, the write-back operator rate, the three crossover operators rates. The *alleles* for each meta-gene correspond to the encoded element; for example, we used 101 possible values for the mutation rates, in the range [0.0;1.0], with the step 0.01.

The evaluation process for a meta-chromosome followed the ideas from [GRE86]:

- set the parameters of the algorithm **A** to the values given by the meta-genes;

- run **A** 10 times;

- the arithmetic mean of the ten best chromosomes in the ten runs is the fitness value of the meta-chromosome under evaluation.

Note that the fitness function is not deterministic.

The other meta-elements are similar to those of the standard GA. For efficiency reasons, the maximum number of generations for the meta-GA has been set to 50 and the meta-population size to 10.

The supervisor GA indicated the following parameter values for **A**: pop-size=20; the three mutation rates 0.03, 0.04 and 0.04, respectively (a global rate of 0.11); write-back rate 0.50; the three crossover rates 0.30, 0.24 and 0.06, respectively (a global rate of 0.60). In each case, larger or smaller values led to either slow convergence, or premature convergence.

For input data that was generated to favor the Euclidean distance (see next paragraph), the optimal clustering has generally been found relatively quickly – 20-100 generations –, even for **n**=10,000. Our implementation of the existing heuristic which inspired us ("mobile centers") converged to the optimum in half the time needed by the Genetic Algorithm, provided that the correct number of classes was used as an input; however, existing

[15] A similar idea can be found in [FOR65]. As for the GA, such an operator illustrates "Lamarckianism" – the idea that parents can pass on to their descendants non-inherited characters.

heuristics became much slower if several or all possible values for **k** had to be tested successively: as a typical example, for an input with optimal **k**=9, the "mobile centers" heuristic needed more than 90 seconds to converge if **k** was not known beforehand, while the GA found the optimum in less than 25 seconds on the same computer.

However, for input data which does not favor the use of the Euclidean distance (optimal clustering with clusters of various shapes and orientations), our GA most often failed to find the optimum; premature convergence occurred, for some input data at a 100% rate. Usually, an optimal cluster was divided, in the found solution, in several, more "Euclidean-like", clusters.

16.2.9 Further hybridization

The Genetic Algorithm developed so far can benefit further from problem-specific knowledge. The next step we undertook was to let metrics adapt themselves to the input data; this idea is inspired from the so-called "adaptive metrics" approach [CEL89], [RIP96]. A *local* metric is considered for each cluster. As before, a chromosome of length **k** represents a clustering of **n** **a**-dimensional points (1≤**a**≤20), with **k** clusters. When the chromosome is decoded for evaluation, a local metric is used for assigning points to centers; after obtaining the partition, the local metrics are updated for fitness calculations and for decoding in the next generation. The between-clusters inertia B is computed in a global metric[16]; the Euclidean distance is used in the first generation for decoding all chromosomes. This is a deterministic way of handling local metrics; a possible improvement for further development of this algorithm could be to consider the *evolution* of local metrics – a metric gene could be attached to each ordinary gene, with proper *allele* and operators. We give below the general formulae describing the way distances are obtained; see [DID82], [CEL89], [RIP96] for complete presentations and [LUC95], [LUC98a] for more details on our approach.

Generalized Euclidean distances are given by:

$$d(\mathbf{x}_i, \mathbf{x}_j) = (\mathbf{x}_i - \mathbf{x}_j) \cdot \mathbf{M} \cdot^t (\mathbf{x}_i - \mathbf{x}_j), \tag{6}$$

where **M** is a symmetrical, positively defined matrix. For each cluster $\mathbf{Cl_j}$ with $\mathbf{n_j}$ objects, *the Mahalanobis distance* ([MAH36]), which we use in the current implementation, instantiates **M** by the covariance matrix $\mathbf{V}_i$, i=1..**k**:

$$\mathbf{V}_i = (\mathrm{cov}(\mathbf{x}^i, \mathbf{x}^j))_{i=1,\mathbf{a}}^{j=1,\mathbf{a}} = (\sum_{m=1}^{\mathbf{n}_i} (\mathbf{x}^i_m - \overline{\mathbf{x}}^i) \cdot (\mathbf{x}^j_m - \overline{\mathbf{x}}^j))_{i=1,\mathbf{a}}^{j=1,\mathbf{a}}, \tag{7}$$

where subscripts refer to objects and superscripts refer to attributes. The local distance for cluster $\mathbf{Cl}_i$ is then:

$$d_i^2(\mathbf{x}_m, \mathbf{x}_j) = (\mathbf{x}_m - \mathbf{x}_j) \cdot \mathbf{V}_i^{-1} \cdot^t (\mathbf{x}_m - \mathbf{x}_j), \ \mathbf{x}_m, \mathbf{x}_j \in \mathbf{Cl}_i. \tag{8}$$

Analogous relations are used for the global metric needed to compute the between-clusters inertia B.

It is well known that Euclidean metric is proper for data displaying clusters of quasi-equal volumes and quasi-spherical shapes. The Mahalanobis distance allows for more

[16] The global metric is given by the covariance matrix of the **n** points.

general shapes and structures of the clusters to be detected; automated scaling on the coordinate axes, proper adjustment of the correlation between attributes and the possibility to find more general *decision boundaries* are the advantages of using this kind of distance (see, e.g., [RDU97]).

Since we are considering here the problem of *automated decision* as to which metric is the proper one for a given input, we have changed slightly the formula for the fitness function, for allowing the local metrics to be used more conveniently:

$$F(Chr)=(1/(1+W/B))^{\mathbf{k}}. \quad (9)$$

16.2.10 Choosing/adapting operators (II)

At this stage of developing the algorithm, we changed the set of operators: the *join/split* mutation replaced the binary mutation and the *phenotype* crossover became the only crossover operator. The goal has been to favor the otherwise deterministic dynamic of local metrics, by using cluster-oriented operators rather than center-oriented ones.

A random bit is used for deciding whether the join/split mutation will join together two clusters in the chromosome or split an existing cluster into two ones (see details in [LUC98a]). The only crossover operator uses the phenotype as a template: for example, one cutting point splits the **a**-dimensional hypercube which contains the **n** given points into $2^{\mathbf{a}}$ regions; each descendant inherits the centers of clusters in some regions from one parent and the centers in the other regions, from the other parent, depending on the elements of a binary matrix. See [LUC98a] for details on operators, parameters setting and experiments.

Some of the input data sets for which the first version of our Genetic Algorithm (§8) could not find the optimal clusterings were properly interpreted by the new version of the algorithm. We show here two such cases. Each of Figs 16.3 and 16.4 below illustrate, for two different 2D inputs (**n**=500 and **n**=1000, respectively), two stages in the evolution of the best chromosome – an intermediary one and that in the last generation; actually, the corresponding phenotype is displayed. The two figures show the gravity centers of the classes and, for each cluster, the shape of the ellipsis in the Mahalanobis distance which, under the standard Euclidean distance, would be a circle; these ellipsis are drawn only for illustrating the shapes – they do not necessarily cover all the points of the respective cluster.

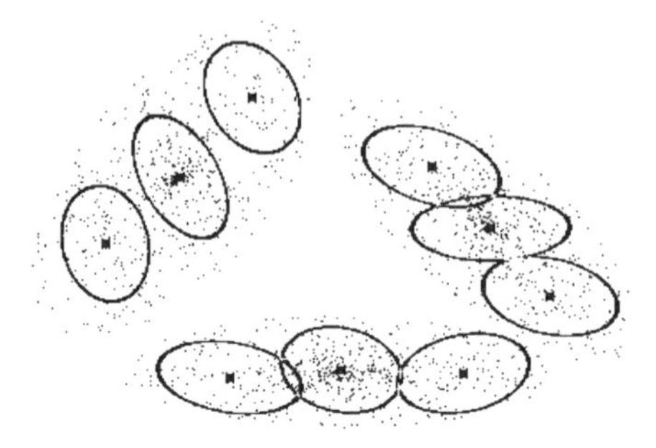

a) The 9 clusters in the 10th generation.

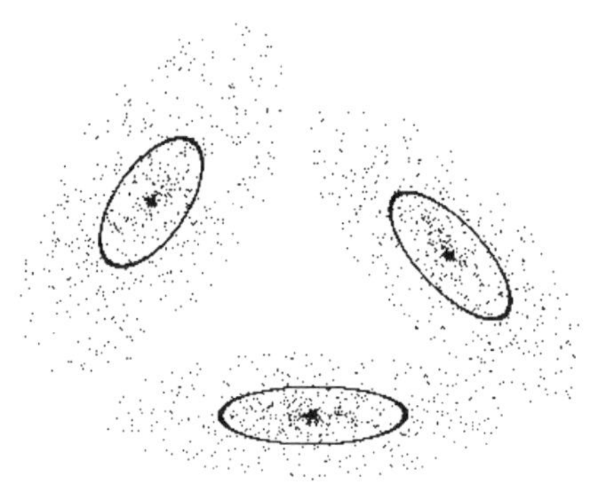

b) **k**=3 clusters in generation 50.

Fig. 16.3 a), b): The gravity centers of all clusters and the shapes of the Euclidean circles under the local metrics for a 2D input with **n**=500 points and optimal **k**=3; best solutions in the 10th and 50th generation.

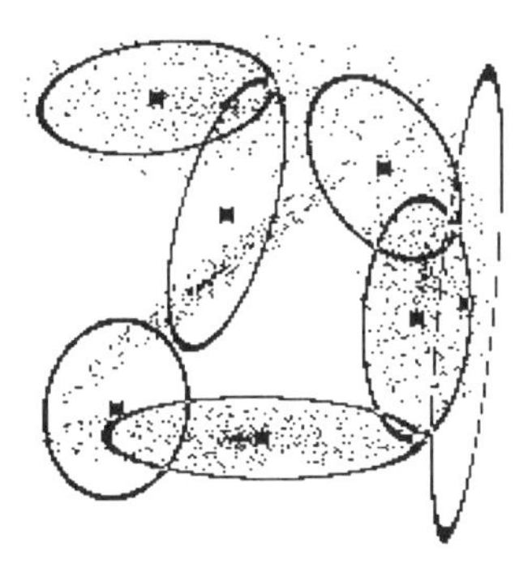

a) The 7 clusters in the 10th generation.

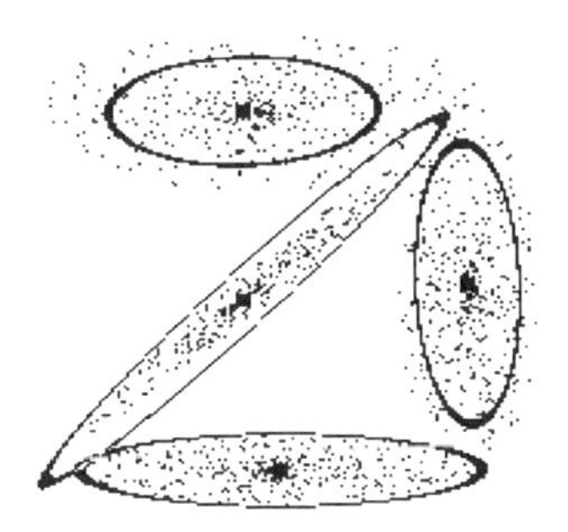

b) Optimal solution, found in generation 27.

Fig. 16.4 a), b): The gravity centers of clusters and the shapes of the Euclidean circles under the local Mahalanobis metrics for a 2D input with **n**=1000 points and optimal **k**=4; best solutions in the 10th and 50th generation.

16.3 CLUSTERING PROBLEM NO. 2: A MOBILE-PHONE PROBLEM

A real world problem provided us the opportunity for testing the algorithm described above; real data was used for tests. For details on this application, see [LUC98b].

A mobile-phone company faces the challenge of unfair calls – so-called "clone calls": a snooper unfairly uses communication resources of a client, thus "cloning" the latter's calls. In order to avoid situations when the client refuses to pay such calls, the company strives to detect, in less than one second from the beginning of a call, whether or not it is a clone one; if the decision is "clone", then the call has to be automatically cut. The admissible error threshold is set by the company at 5% erroneous "fair" decisions and 2% wrong "clone" decisions.

For implementing such a strategy, the company has created a training database, consisting of numerical descriptions for 20,000 calls – half of them fair calls and half clone calls. One record contains 13 attributes; twelve values of numerical parameters[17] characterizing the call, and a binary attribute, which indicates the nature of the call: "1" for "fair" and "0" for "clone". The data recorded in the database is to be used for finding a way to decide in real time, on the basis of the twelve values of any new call, whether or not it is a clone. We present below questions raised by two approaches we undertook.

16.3.1 Adapting the general algorithm

A convenient approach, after devising the algorithm above, would be to use it for the problem at hand. The critical question is whether or not there is a natural interpretation of the given problem in terms of the clustering problem above. Fortunately, the answer is positive; a solution to the clone calls problem could be designed in the following three steps:

1. find an optimal clustering of the 20,000 given points;
2. find a rigorous description of the found clusters;
3. for each new call, apply the rules to its 12 parameter values and decide whether or not it is a clone.

We are only concerned here with step 1. As described above, this step seems to require solving a "supervised pattern recognition" problem [RIP96] – the number of classes **k** seems to be 2. Note, however, that the granularity of the found clustering need not be at this level. More than two clusters may be detected, meaning that each of the two main classes, "0" and "1", can be actually split into several clusters. On one hand, the set of measured parameters, the corresponding set of units of measure and, on the other hand, the algorithm used to solve the problem, the metrics involved, decide the level of granularity for the optimal clustering. It is desirable – and it eventually proved to be true – that, at the given level of granularity, each found cluster be completely included[18] in either class "1" or class "0". Hence, the

---

[17] There are two relevant attributes for each of six characteristics of a call; these characteristics are scalar quantities calculated from the radio frequency measurements of the phone's transmission.

[18] If this were not the case, then the set of parameters and/or the set of units of measure and or the set of metrics had to be improved.

problem at hand is actually an unsupervised automated clustering one and the Genetic Algorithm presented in previous paragraphs can be applied to solve it.

The only important issue in moving to 12 dimensions for the algorithm above, is time: calculations, especially matrix inversions, are very time consuming. The results were, however, acceptable: the required precision has been nearly reached after running the algorithm for nearly 30 hours on a Pentium computer; the accuracy achieved by this "training" procedure has been within a total of 8% errors (instead of the required 7%). The rules describing the found clusters were similar to those presented in [MIT97] (pp. 56-60, where a different problem is discussed) and hence easy to implement for the third step above. The next section discusses an alternative approach, which led to an algorithm with improved efficiency and accuracy.

16.3.2 A Genetic Programming approach

Under the Genetic Programming approach, one implicitly uses Koza's [KOZ93] answer to the fundamental question "What is a solution to a problem?". Indeed, why look for the clustering and not for a function which can perform it accurately and efficiently? For a detailed presentation of our algorithm, see [LUC98b].

Under this approach, one can think of a single solution for steps 1 and 2 above: finding a function of twelve variables (the parameter values of a call), which takes, say, strictly positive values for records corresponding to fair calls and strictly negative values for records corresponding to clone calls. However, the feasible approach is to seek for a function which minimizes the number of records in the database representing "fair calls" for which the function takes negative values and the number of "clone calls" which lead, via the function, to positive values. Since a) this is an optimization task and b) the sought solution is a function, which can be represented by a tree, a Genetic Programming approach can be considered.

The usual representation is used: a candidate function is encoded by a binary tree, whose nodes are the *genes*; the *allele* represent elementary functions of arity 1 or 2, for the interior nodes, and constants or variables (among the 12 ones) for the leaf nodes. The 9 elementary functions were: +, –, *, /, $c*x^i$, e^{-t*x}, log|x|, cos(x), sin(x), where c, i, t are parameters of the algorithm.

16.3.3 Initialization issues

In order to control the depth and the number of nodes in the trees of the initial population, we considered the following procedure:

- at each step, keep a counter **L** equal to the number of leaf nodes needed to complete the tree;
- generate interior nodes until **L**=m*12, where m is a parameter of the algorithm ;
- generate the necessary leaf nodes, labeled with the 12 variable names or with randomly chosen constants in given intervals, with uniform probability.

An alternative, more complex, initialization procedure can be found in [LUC98b]; the idea is to generate a single random value for taking several initialization decisions.

16.3.4 The threat of evaluation complexity and a well-known solution: sampling

Evaluating such a chromosome is straightforward: count the correctly interpreted records out of the 20,000 in the database. On a Pentium 200/64Mo RAM[19], the equivalent of 1,000 generations for a population of 100 individuals lasted for about 100 hours. Since little can be down to accelerate the decoding of a tree and the calculations for each record, the obvious idea is to use only a part of the database for evaluating each chromosome. Random sampling for each chromosome evaluation may affect seriously the accuracy of the solution. A proper sampling of the database would allow for more efficient processing, while preserving as much as possible the accuracy.

One solution is to define a hierarchy on the set of given records (points) and use only the first **j** (**j**<<20,000)[20] of them for evaluating the chromosomes; the hierarchy should be given by the "relevance" of each point for finding a good solution. A procedure that gave good results for this problem is given in [LUC98b]; the basic idea is to mark each point according to the "density" of the region it belongs to and then sort the points using these marks.

Another solution for sampling is *co-evolution* [HIL90], [KOZ93]: the Genetic Algorithm GA1 running on the population of trees is accompanied by another Genetic Algorithm, GA2, which has as population *pop_size_2* chromosomes encoding subsets of the initial set of 20,000 records. Since the sought function is supposed to make an approximate classification, both the evaluation of trees and the evaluation of subsets are based on an acceptance threshold, which is used for deciding whether or not the points in a subset, as a whole, are correctly classified by a tree (see [LUC98b] for details).

We also mention the fact that we used nine mutation operators, but with very small rates. There are various reasons for using mutation operators; for example, allowing the parameters of the functions to evaluate has been the reason for introducing one of them.

With the sampling techniques described above, our algorithm had better accuracy than the one commercially used: 1,45% wrong "1" decisions and 0.58% wrong "0" decisions, in only 15 hours – much less than the time taken by the other algorithm. A typical solution is given in Fig. 16.5; the use of editing operators would have led to a more elegant expression.

---

[19] This allowed for the whole database to be kept in memory.
[20] **j** may be increased during the run.

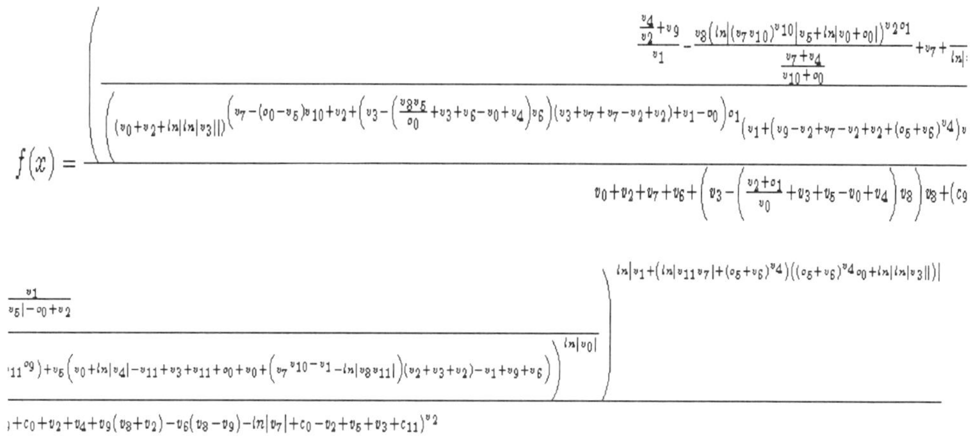

$$\frac{\frac{v_4}{v_2}+v_9}{v_1} - \frac{v_8\left(\ln\left|(v_7 v_{10})^{v_{10}}\right|v_5+\ln|v_0+c_0|\right)^{v_2 c_1}}{\frac{v_7+v_4}{v_{10}+c_0}} + v_7 + \frac{}{\ln|}$$

$$f(x) = \frac{\left(\left((v_0+v_2+\ln|\ln|v_3||)^{\left(v_7-(c_0-v_5)v_{10}+v_2+\left(v_3-\left(\frac{v_8 v_5}{c_0}+v_3+v_6-v_0+v_4\right)v_6\right)(v_3+v_7+v_7-v_2+v_2)+v_1-c_0\right)c_1}\left(v_1+\left(v_9-v_2+v_7-v_2+v_2+(c_5+v_6)^{v_4}\right)v\right.\right.\right.}{v_0+v_2+v_7+v_6+\left(v_3-\left(\frac{v_2+c_1}{v_0}+v_3+v_5-v_0+v_4\right)v_8\right)v_8+(c_9}$$

$$\ln|v_1+\left(\ln|v_{11}v_7|+(c_5+v_6)^{v_4}\right)\left((c_5+v_6)^{v_4}c_0+\ln|\ln|v_3||\right)|$$

$$\frac{\frac{v_1}{v_5|-c_0+v_2}}{{}_{11}{}^{c_9})+v_5\left(v_0+\ln|v_4|-v_{11}+v_3+v_{11}+c_0+v_0+\left(v_7{}^{v_{10}-v_1}-\ln|v_8 v_{11}|\right)(v_2+v_3+v_2)-v_1+v_9+v_6\right)\Bigg)^{\ln|v_0|}\Bigg)}$$

$$)+c_0+v_2+v_4+v_9(v_8+v_2)-v_6(v_8-v_9)-\ln|v_7|+c_0-v_2+v_5+v_3+c_{11})^{v_2}$$

Fig. 16.5 A solution found by the algorithm, interpreted in Latex®.

16.4 CLUSTERING PROBLEM NO. 3: CROSS-CLUSTERING

A special case of clustering is the so-called "cross-clustering" [DID82]. Intuitively, the problem asks for a Boolean (usually, sparse) matrix to be brought to a quasi-canonical form. Such a form can be achieved by gathering the 1's in as compact as possible blocks, only by means of elementary operations; such an operation consists of swapping either two rows or two columns. Since quasi-canonical forms are vaguely defined and also because of the intrinsic complexity, the problem is a difficult one; indeed, there is no exact deterministic algorithm to solve it. The existing heuristics rely on topological sorting or on a random iterative procedure (ibid.). The problem is important, since it has numerous applications in image analysis, ecology, archaeology – data sequencing – etc.

An evolutionary approach is possible if the task is formulated in terms of optimization – e.g., minimizing the number of 0's included in blocks of 1's. For details concerning our evolutionary solutions to this problem, see [LUC99a].

16.4.1 The "two-permutations" approach: facing epistasis

The first evolutionary solution we tried has been an "optimal permutation" one: given a Boolean matrix **M**, find permutations[21] of rows and columns that bring **M** to the "best" quasi-normal form as described above. Under this approach, it is straightforward to represent a solution by a pair of permutations – one for the rows and the other one for the columns[22]. As expected, the experiments showed that there is a high level of epistasis

[21] The identity permutation is given by the initial order of rows and columns.

[22] One can choose to represent only one of these permutations and leave the search of the other one to another heuristic.

between the two permutations (let alone the epistasis inside any permutation representation). In this context, the simultaneous evolution of the two permutations seldom leads to convergence to an optimum, or even to an acceptable solution. One way to avoid this is to let the two candidate permutations evoluate by turns rather than simultaneously: the fitness function and operators switch every g generations – g being a parameter of the algorithm – between optimizing the row-permutation and optimizing the column-permutation. The experiments showed that this switching strategy avoids to a certain extent the effect of epistasis.

The optimization criterion used in this version of the algorithm is related to the number of positions between the first and the last 1 in either each row or each column; the number of 1's in the respective row/column is also taken into consideration.

The ten operators[23] are typical for permutations: swap mutation and three generalized swap mutations (more than two genes are interchanged in some specific way); cycle crossover [OLI87].

The available software product, currently used for archaeological applications, compared favorably to our algorithm for matrices under 50*50, but less and less favorable as the dimensions of the matrices increased. We looked for a better evolutionary solution.

The weakest point of our "permutation optimization" algorithm is that, no matter what fitness function one would define for it, the blocks of 1's given by a candidate solution are not properly processed (see [LUC99a] for details). This is clearly shown by the Fitness-Distance Correlation (Fig. 16.6): neither at the permutation level, nor at the bit level, the FDC values (between 0.088 and 0.17) and plots do not illustrate a good correlation between genotype and phenotype distances under this representation. We give the corresponding plots for two genotypes (permutation and bitstring) and two different fitness functions, both of them based on the distances between 1 values in all rows/columns. Following these observations, one initial decision has been that no operator will be defined at the bit level; later on, the permutation approach has been abandoned altogether.

[23] Actually, only 5 operators, each one with row and column versions.

a), b): The permutation level

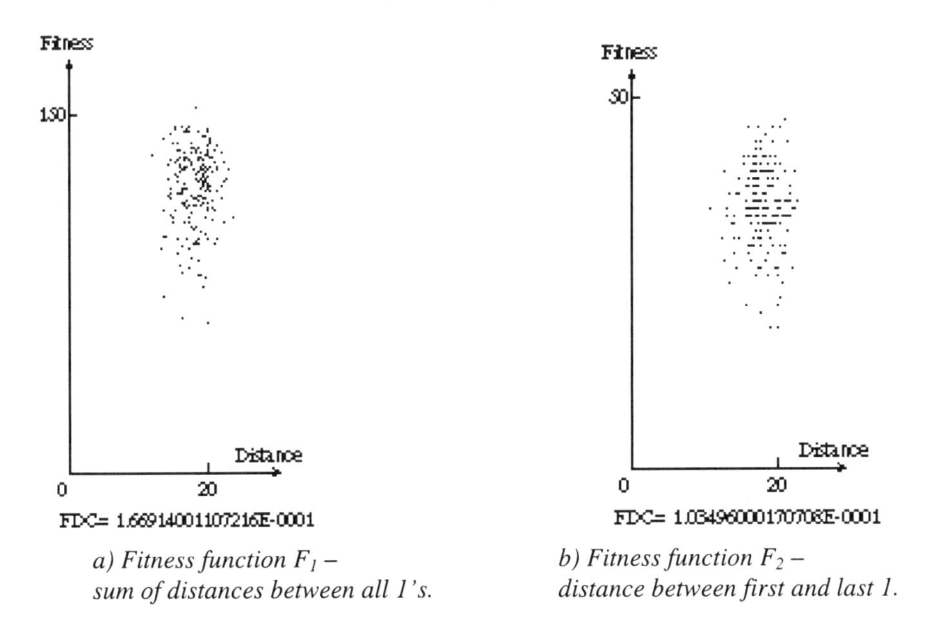

a) Fitness function F_1 – sum of distances between all 1's.

b) Fitness function F_2 – distance between first and last 1.

c), d): Bitstring level (Hamming distance)

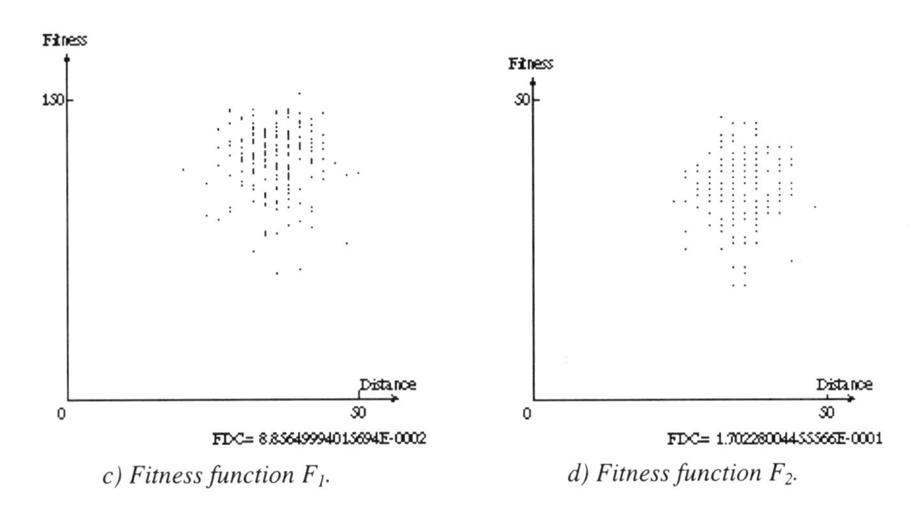

c) Fitness function F_1.

d) Fitness function F_2.

Fig. 16.6 a)-d): Typical fitness-distance correlation values and plots for two genotypes (permutation and bitstring) and two fitness functions.

16.4.2 More hybridization: the "blocks" approach

The name "cross-clustering" indicates a relation with the general clustering problem. Using this relation, another evolutionary approach could be devised. However, the general clustering algorithm presented in the first part of this paper cannot be applied here directly: the notions of distance, gravity center cannot be used as such.

Our second algorithm, called BLOCKS, is based on hybridization with existing heuristics; defining a fitness function that would properly interpret the quality of a candidate solution has been our main goal at this stage. We used the cross-clustering heuristic presented in [DID82]: cluster the rows and columns of the initial matrix and treat the elements at the intersections of each row-cluster with each column-cluster as a block; optimize the clusterings by improving the blocks w.r.t. the quasi-normal form asked for and by reducing the number of clusters.

Each gene in a chromosome represents a subset of either the set of rows ("row-gene") or the set of columns ("column-gene"). The chromosome as a whole represents a partition of the set of rows and a partition of the set of columns. Obviously, two chromosomes need not have the same length. In the evaluation procedure, each sub-matrix defined by a pair (row-gene; column-gene) receives either the label "0" or the label "1", depending on which binary value the majority of the elements in the block are equal to. A penalty is computed for each sub-matrix block; the fitness function to be minimized is built starting from the sum of all penalties; in a way similar to (4) and (9), an exponent controls the growth of the number of clusters; other technical details on the fitness function and on the algorithm BLOCKS in general can be found in [LUC99a].

The operators are typed: row-operators and column-operators. The crossover is a kind of uniform crossover, adapted for the restrictions of a partition. The three mutation operators for each dimension perform:

- an exchange of rows/columns between genes of the respective type;
- deletion of a gene with random re-allocation of its rows/columns (thus decreasing the length of the chromosome);
- creation of a new gene by randomly putting together elements from other genes,

respectively. Fig. 16.7 illustrates the behavior of the algorithm under various values for the total mutation rate and for the crossover rate. The respective rates are: a) p_m=0.05; p_c=0.60; b) p_m=0.03; p_c=0.60; a) p_m=0.05; p_c=0.50; a) p_m=0.05; p_c=0.40. We used the values in c) for further experiments.

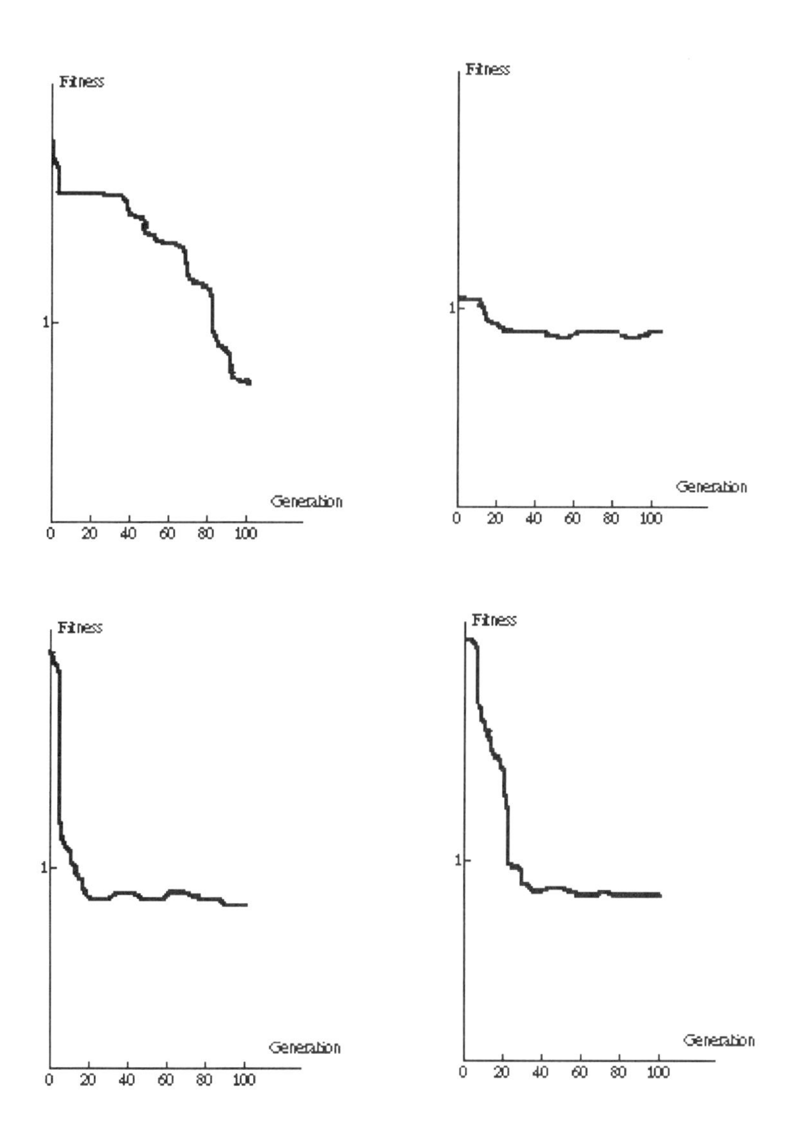

Fig 16.7 Typical evolutions of fitness values for the Best_so_far chromosome, with various settings of mutation and crossover rates.

A critical issue for this algorithm proved to be the replacement strategy; we had to perform a supervisor GA, the chromosomes of which contained a meta-gene for the replacement strategy, encoding various choices (generational fitness-based; generational rank-based; tournament; random; a combination of the first and last one). Probably due to the difficulties in scanning the huge search space (for a 100*100 matrix, it has $(100!)^2$ points), the last strategy has been the winner. The experimental results indicated a clear improvement over our first solution in terms of both accuracy and efficiency.

16.4.3 A two-step approach

We are currently experimenting a further improvement of our algorithm. The idea is somewhat similar to "Δ-coding" ([SCB90]) or to "Dynamic Parameter Encoding" ([WMF91]): the level which the optimization is performed at, varies during the run. The first step of the improved algorithm is identical to the BLOCKS algorithm above: the input matrix M_1 is brought to a "good" quasi-normal form – the matrix BEST. In the second phase, BEST is used for building a new Boolean matrix M_2; each element of M_2 is the label of a block from BEST. Finally, BLOCKS is run again, this time with M_2 as the input matrix. For matrices of large dimensions, this new algorithm seems to improve dramatically the quality of the solutions.

16.5 (CLUSTERING) PROBLEM NO. 4: ELECTRIC POWER DISTRIBUTION

An electric power distributor[24] wishes to use its database of previous consumption levels ("loading curves") for the analysis and for a real-time prediction of future consumption levels; see [CAN79], [CLI96], [FON94], [ILC97] for a few examples of previous approaches to this problem.

The *analysis* is expected to detect seasonalities and/or types of consumption profiles and the moments when the consumption level switches from one profile to another. A typical *prediction* task is: "estimate the level of consumption one hour in advance, based on the last day's hourly-measured levels of consumption". See, e.g., [CHA89] for standard approaches to time series analysis and prediction. Lately, such problems are quite often undertook in an Artificial Neural Networks setting; evolutionary approaches are also being used for this kind of task (see, for example, [MIT97], pp.56-61). We designed two evolutionary solutions – one mostly for the analysis of the time series, the other one for prediction. We give in the §18 a short presentation only for the second one, despite the fact that only our solution for analyzing the time series has to do with clustering. The reason is that the implementation of the "analysis" solution is ongoing, while for "prediction" we already have working implementations. Here is, however, an outline of our design for the analysis approach:

- represent the information in the consumption database in a multidimensional setting[25];
- perform unsupervised clustering on the multidimensional data, for explicitly separating the consumption patterns;
- obtain matrices containing estimated probabilities of switching from any consumption pattern to another one.

Under this approach, prediction is performed along the following lines:

- use the ARIMA approach for finding prediction formulae (one for each consumption pattern implicitly detected here, but explicitly found by clustering);

---

[24] Conel Bacău, Romania.

[25] The basic idea is to consider **m** successive measured values as a vector, then study the new, vector-valued time series.

- for any real-time prediction:
  - identify the cluster the given point leads to;
  - compute, using the appropriate prediction formula, the most likely next level of consumption.

This is a "modular hybridization" approach: problem-specific knowledge is used not necessarily in the elements of the Genetic Algorithm, but rather in separate, non-evolutionary steps.

16.5.1 Finally, no clustering – just predicting

The use of Genetic Algorithms for regression is not new (see, for example, [KOZ93], pp. 245-255). We describe here the two algorithms we devised, LIN and TREE, for the problem at hand.

First, we tried to find a linear formula for predicting the level of consumption for the next hour, S(t), based on the 24 previous hourly measurements. In an initial version of LIN, a chromosome consisted of 25 genes, each one being a candidate coefficient c_i, i=0,..,24, for the formula below:

$$S(t) = c_0+c_1 \cdot S(t-1) + c_2 \cdot S(t-2) + \ldots + c_{24} \cdot S(t-24); \qquad (10)$$

a gene encoded a real number with a precision of 10^{-6} (see [MIC93], p.19).

The three mutation operators performed the following elementary operations:

- setting a non-zero coefficient to 0;
- setting a nil coefficient to a random value;
- randomly changing the value of a non-zero coefficient into another value.

Uniform crossover has been used.

The evaluation of a chromosome starts by calculating the estimation for every measured value which the 24 previous hourly measurements exist for, in the input data. The fitness value of the chromosome is then the arithmetic mean of all differences between estimated and measured values.

Since the algorithm constantly provided good solutions with just three significant coefficients – c_1, c_{23}, and c_{24} – and all the others very close to 0, we decided to keep only four genes in the chromosomes. This representation and the evaluation above led to the FDC values and plots given in Fig. 16.8, for consumption during one month in two towns. The relatively high values of the FDC and the orientation of the sets of points predict, as Jones and Smith suggest, a good behavior of the algorithm.

MĂRGINENI

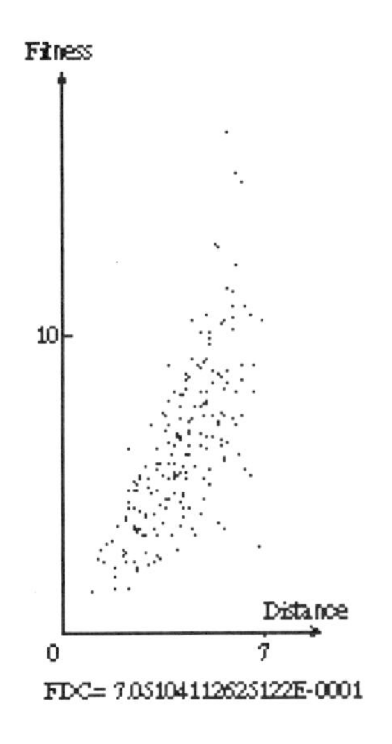

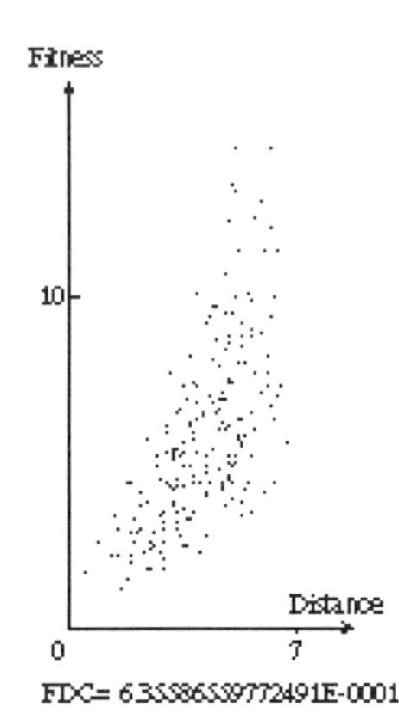

Fig. 16.8 The fitness-distance correlation maps given by the electricity consumption predictions (LIN), based on input data for two towns.

A typical solution found in 500 generations by this version of LIN had an average error of 1.72% of the observed values, and the four coefficients were: c_0=0.000486, c_1=0.911694 , c_{23}=0.157452 , c_{24}=-0.077180.

In a second phase, we used a standard Genetic Programming approach. The algorithm TREE only had the four arithmetic operations as elementary functions (division in protected form). The mutation rate was 0 and the usual crossover had a rate of 0.9. The variables set consisted only of the four significant ones in (12), detected by LIN. TREE converged in 100 generations to good solutions: for 700 estimated values, a typical value for the total error of the best solution has been 150, with an average relative error of 0.5% (the measured values were in the interval [25; 95]).

16.6 CONCLUSIONS

We presented several evolutionary solutions to a few clustering problems.

We started with the general automated unsupervised clustering by partition problem – the quantitative case; the Genetic Algorithm for this problem is given in some detail. Current and future directions include the expansion of the set of metrics considered – Tchebyshev, City-block etc. – and the actual evolution (not deterministic update) of metrics. Another general problem considered here has been cross-clustering. An evolutionary solution based on permutations and another one, inspired from an existing heuristic, were presented; a further improvement, by means of a two-step run of the Genetic Algorithm, each time at a different level, has been suggested.

We also discussed two real world applications where clustering has been applied. We have presented, in each case, two different evolutionary approaches: for the "clone-calls" problem, a clustering solution based on the initial algorithm and a Genetic Programming one; for the power consumption example, we sketched a "modular hybridizing" approach and then two algorithms for symbolic regression were given. Although all the algorithms presented here find accurate solutions and have a reasonable time complexity, part of our work on these problems is still ongoing – hence some of the elements presented here may be subject to future reconsideration.

Acknowledgments

I gratefully acknowledge the sponsorships, at various stages of this research, from Lockheed Martin International, Microsoft – Romania and The Department of Computer Studies at Napier University of Edinburgh. I also wish to acknowledge the contributions of M. Petriuc (Microsoft, Redmond, U.S.A.), B. Paechter (Napier University, U.K.), S. Luchian, P. Cotofrei and V. Radulescu ("Al.I.Cuza" University, Romania); they carried out much of the computer work and were supportive and creative during numerous discussions on evolutionary clustering.

REFERENCES

[AND96] Andrey, P.: "Selectionist Relaxation: Genetic Algorithms Applied to Image Segmentation", personal communication, Ecole Normale Superieure, Paris, 1996.

[BAH65] Ball, G.B., Hall, D.J.: "ISODATA: a Novel Method of Data Analysis and Pattern Classification", Technical Report, Stanford Research Institute, Menlo Park, California, 1965.

[BRE91] Bhuyan, J., e.a.: "Genetic Algorithm for Clustering with an Ordered Representation", in Proc. of the 4th ICGA, Morgan Kaufman, 1991, pp.408-415.

[CAG87] Carpenter, G.A., Grossberg, S.: "A Massively Parallel Architecture for a Self-Organizing Neural Pattern Recognition Machine", in *Computer VisionGraphics and Image Processing*, vol.37, 1987, pp.54-115.

[CAN79] Canal, M.: "Analyse et classification de la clientele EDF en fonction de la forme de la courbe de charge", Bull. De la DER, No.2, 1979;

[CAR91a] Carpenter, G.A., Grossberg, S., Reynolds, J.H.: "ARTMAP: Supervised Real-Time Learning and Classification of Nonstationary Data by a Self-Organizing Neural Network", in *Neural Networks Journal*, vol. 4, nr.3, 1990, pp.565-588.

[CAR91b] Carpenter, G.A., Grossberg, S., Rosen, D.B.: "Fuzzy ART: Fast Stable Learning and Categorization of Analog Patterns by an Adaptive Resonance System", in *Neural Networks Journal*, vol. 4, no. 4, 1990, pp.759-771.

[CAR92] Carpenter, G.A., Grossberg, S., Markuzon, N., Reynolds, J.H., Rosen, D.B.: "Fuzzy ARTMAP: a Neural Network Architecture for Incremental Supervised

Learning of Analog Multidimensional Maps", in *IEEE Transactions on Neural Networks*, vol. 3, 1992, pp.698-713.

[CEL89] Celeux, G., Diday, E., Govaert, G., Lechevallier, Y., Ralambondrainy, H.: "Classification automatique de donnees", Dunod, Paris, 1989.

[CHA89] Chatfield, C.: "The Analysis of Time Series", fourth edition, Chapman and Hall, London, New York, 1989.

[CLI96] Clocotici, V., Ilinca, M.: "Analiza consumului de energie electrică prin metode clasice", research report, Renel, Iasi, 1996.

[DHA73] Duda, R.O., Hart, P.E.: "Pattern Classification and Scene Analysis", Wiley, New York, 1973.

[DID82] Diday, E., Lemaire, J., Pouget, J., Testu, F.: "Elements d'analyse de donnees", Dunod, Paris, 1982.

[FLA89] Flanagan, J.A., e.a.: "Vector Quantization Codebook Generation Using Simulated Annealing", in *Proc. of International Conf. on Acoustics, Speech and Signal Processing*, Glasgow, May 1989, pp.1759-1762.

[FON94] Fon, Y.J., e.a.: "A Real-Time Implementation of Short-Term Load Forecasting for Distribution Power Systems", IEEE Trans. On Power Syst., No.2, May 1994;

[FOR65] Forgy, E.W.: "Cluster Analysis of Multivariate Data: Efficiency vs. Interpretability of Classifications", in *Biometrics Journal*, vol. 21, 1965, pp.768-769.

[FUK90] Fukunaga, K.: "An Introduction to Statistical Pattern Recognition", Academic Press, New York, 1990.

[GRE86] Grefenstette, J.J.: "Optimization of Control Parameters for Genetic Algorithms", in IEEE Transactions on Systems, Man and Cybernetics, vol. 16, no.1, 1986, pp.122-128.

[GRE87] Grefenstette, J.J.: "Incorporating Problem Specific Knowledge into Genetic Algorithms", in Genetic Algorithms and Simulated Annealing (L. Davis - editor), Morgan Kaufman Publishers, Los Altos, CA, 1987, pp. 42-60.

[HAK77] Hall, D.J., Khanna, D.: "The ISODATA Method of Computation for Relative Perception of Similarities and Differences in Complex and Real Computers", in *Statistical Methods for Digital Computers*, vol. 3, (ed.: K.Enslein, A.Ralston, H.S.Wilf), Wiley, New York, 1977, pp.340-373.

[HAR75] Hartigan, J.A.: "Clustering Algorithms", Wiley, New York, 1975.

[HAW79] Hartigan, J.A., Wong, M.A.: "A **k**-means Clustering Algorithm", in *Applied Statistics*, vol. 28, 1979, pp.100-108.

[HIL90] Hillis, W.D.: "Co-evolving Parasites Improve Simulated Evolution as an Optimization Procedure", *Physica D*, vol.42, 1990, pp.228-234.

[HUA95] Huang, J., Georgiopoulos, M., Heileman, G.L.,: "Fuzzy ART Properties", in *Neural Networks Journal*, vol. 8, 1995, pp.203-213.

[ILC97] Ilinca, M., Clocotici, V.:"On the Relation between Climatic Factors and the Loads of the Electrical Power Distribution Networks", in *Acta Universitatis Cibiniensis*, vol. XXVII, Sibiu, 1997, pp.202-209.

[IOS78] Iosifescu, S.: "Lanţuri Markov finite şi aplicaţii", Editura Tehnicã, Bucharest, 1978.

[JAD88] Jain, A.K., Dubes, R.C.: "Algorithms for Clustering Data", Prentice Hall, Englewood Cliffs, New Jersey, 1988.

[JOB91] Jones, D.R., Beltramo, M.A.: "Solving Partitioning Problems with genetic Algorithms", in Proc. of the 4th ICGA, Morgan Kaufman, 1991, pp.442-449.

[JOF95] Jones, T., Forrest, St.: "Fitness Distance Correlation as a Measure of Problem Difficulty for Genetic Algorithms", in Proc. of the 6th ICGA, 1995, Morgan Kaufman, pp.184-192.

[KAR90] Kaufman, L., Rousseeuw, P.J.: "Finding Groups in Data. An Introduction to Cluster Analysis", Wiley, New York, 1990.

[KOO75] Koontz, W.L.G., Narendra, P.M., Fukunaga, K.: "A Branch and Bound Clustering Algorithm", in *IEEE Transactions on Computers*, vol. 24, 1975, pp.908-915.

[KOZ93] Koza, J.: "Genetic Programming – on the Programming of Computers by Means of Natural Selection", The MIT Press, Cambridge, Massachusetts, London, England, 2nd edition, 1993.

[LUC94] Luchian, H., Petriuc, M., Luchian, S.: "Evolutionary Automated Classification", in Proceedings of the 1st IEEE World Congress on Evolutionary Computing, Orlando, Florida, 1994, pp.585-598.

[LUC95] Luchian, S., Petriuc, M., Luchian, H.: "An Evolutionary Approach to Unsupervised Automated Classification", in Proceedings of the European Forum for Intelligent Techniques, Aachen, 1995, pp. 377-392.

[LUC98a] Luchian, H., Luchian, S., Cotofrei, P., Radulescu, V.: "Seven Evolutionary Clustering Algorithms", Research report, "Al.I.Cuza" University of Iasi, 1998.

[LUC98b] Luchian, H., Cotofrei, P., Radulescu, V., Paechter, B., Luchian, S.: "Clone Calls Detection in an Evolutionary Setting", research report.

[LUC99a] Luchian, H., Radulescu, V., Paechter, B., Luchian, S.: "Evolutionary Cross-clustering with Applications in Archaeology and Ecology", research report, "Al.Cuza" University of Iasi, 1999.

[LUC99b] Luchian, H, Ilinca, M., Cotofrei, P., Radulescu, V.: "Two Evolutionary Approaches to Power Consumption Prediction", presented at 10th ECMI Congress, Goteborg, Sweden, 1998; research report, "Al.Cuza" University of Iasi, 1999.

[MAH36] Mahalanobis, P.C.: "On Generalized Distance in Statistics", in *Proceedings of the National Indian Institute of Science*, vol. 12, 1936, pp.49-55.

[MAQ67] MacQueen, J.: "Some Methods of Classification and Analysis of Multivariate Observations", in *Proceedings of the 5th Berkeley Symposium on Mathematical Statistics and Probability*, (editors L.M.LeCam, J.Neyman), University of California Press, vol. I, Berkeley, California, 1967, pp.281-297.

[MIC92] Michalewicz,Z.: "Genetic Algorithms + Data Structures = Evolution Programs", 1st edition Springer Verlag, Berlin, 1992.

[MIT97] Mitchell, M.: "An Introduction to Genetic Algorithms", MIT Press, Cambridge, Massachusetts, 1997.

[MOO89] Moore, B.: "ART1 and Pattern Clustering", in *Proceedings of the 1988 Connectionist Models Summer School* (editors D. Touretzky, G. Hinton, T. Sejnowski), Morgan&Kaufman, San Mateo, California, 1988, pp.174-185.

[MUL94] Muller, C., e.a.:"SYNAPSE - Systeme Neuronal d'Anticipation de la Puissance; guide d'utilisation", Bull. De la DER, Aout 1994.

[OLI87] Oliver, I.M., Smith, D.J., Holland, J.R.C.: "A Study of Permutation Crossover Operators on the Traveling Salesman Problem", în *Proceedings of the 2nd ICGA* (editor J.J.Grefenstette), Lawrence Erlbaum Associates, Hillsdale, New Jersey, 1987, pp.224-230.

[OTE96] Oterson, S.: personal communication, Mangalia, 1996.

[PAB93] Parodi, A., Bonelli, P.: "A New Approach to Fuzzy Classifier Systems", Proc. of the 5th ICGA, Morgan Kaufman, 1993, pp.223-230.

[RDU97] Duda, R.O.: "Pattern Recognition for HCI", course notes, San Jose State University, 1997.

[REW95] Reeves, C.R., Wright, C.C.: "Epystasis in Genetic Algorithms: An Experimental Design Perspective", in Proceedings of the 6th International Conference on Genetic Algorithms, editor L.Eshelman, Morgan Kaufman Publishers, San Francisco, 1995, pp.217-224.

[RIP96] Ripley, B.: "Pattern Recognition and Neural Networks", Cambridge University Press, England, 1996.

[SCB90] Schraudolf, N., Belew, R.: "Dynamic Parameter Encoding for Genetic Algorithms", CSE Technical Report #CS90-175, University of San Diego, La Jolla, 1990.

[SCK96] Schoenauer, M., Kallel, L.: "Fitness-Distance Correlation for Variable Length Representations", Raport de recherche, Ecole Polytechnique de Palaiseau, CNRS URA756, 1996.

[THA95] Thangiah, S.R.: "An Adaptive Clustering Method Using a Geometric Shape for vehicle routing problems with Time Windows", in Proc. of the 6th ICGA, Morgan Kaufman, 1995, pp. 536-544.

[WMF91] Whitley, D., Mathias, K., Fitzhorn, P.: "Delta-coding – An Iterative Search Strategy for Genetic Algorithms", in Proc. of the 4th Int. Conf. on Genetic Algorithms, (Editori Belew şi Booker) Morgan Kaufman Publishers, Los Altos, CA, 1991, pp.77-84.

[ZEG92] Zeger, K., Vaisey, J., Gersho, A.: "Globally Optimal Vector Quantization Design by Stochastic Relaxation", in *IEEE Transactions on Signal Processing*, vol. 40, 1992, pp.310-322.

Part III

Industrial applications

17 Evolutionary Algorithms Applied to Academic and Industrial Test Cases

T. BÄCK[1], W. HAASE[2], B. NAUJOKS[1], L. ONESTI[3] and A. TURCHET[4]

[1] Informatik Centrum Dortmund, Joseph–von–Fraunhofer–Str. 20, D–44227 Dortmund
[2] DaimlerChrysler Aerospace, Department MT63, P.O. Box 80 11 60, D–81663 München
[3] Engine Soft Tecnologie per l'Ottimizzazione, via Malfatti 1, Trento
[4] Sincrotrone Trieste S.C.P.A., Strada Statale 14, Localita' Basovizza, 34012 Trieste

Abstract. Within the framework of the European "Thematic Network" project INGENET, evolutionary algorithms such as genetic algorithms (see e.g. [5, 8]), evolutionary programming (e.g., [4]), and evolution strategies (e.g., [1, 13]) are applied to challenging academic and industrial test cases. The major goal is to evaluate the performance of evolutionary algorithms for practical applications.

This paper reports on results which are obtained by evolution strategies on two well–known mathematical test cases (Rastrigin's function and Keane's function) and outlines the application of an evolution strategy for a multi–point 2D–airfoil design. Results for the latter case, utilizing a genetic algorithm, are briefly discussed in the present paper.

17.1 INTRODUCTION TO EVOLUTIONARY STRATEGIES

In general, evolutionary algorithms mimic the process of natural evolution, the driving process for the emergence of complex and well adapted organic structures, by applying variation and selection operators to a set of candidate solutions for a given optimization problem. The following structure of a general evolutionary algorithm, algorithm 1, reflects all essential components of an evolution strategy [2]. For a formulation of genetic algorithms and evolutionary programming in the framework of algorithm 1, the reader is referred to [2].

In case of a (μ,λ)–evolution strategy, the following statements regarding the components of algorithm 1 can be made:

- $P(t)$ denotes a population (multiset) of μ individuals (candidate solutions to the given problem) at generation (iteration) t of the algorithm.

- The initialization at $t = 0$ can be done randomly, or with known starting points obtained by any method.
- The evaluation of a population involves calculation of its members quality according to the given objective function (quality criterion).
- The variation operators include the exchange of partial information between solutions (recombination) and its subsequent modification by adding normally distributed variations (mutation) of adaptable step sizes. These step sizes are themselves optimized during the search according to a process called self–adaptation.
- By means of recombination and mutation, an offspring population $P'(t)$ of $\lambda >> \mu$ candidate solutions is generated.
- The selection operator chooses the μ best solutions from $P'(t)$ (i.e. Q=0) as starting points for the next iteration of the loop. Alternatively, a $(\mu+\lambda)$–evolution strategy would select the μ best solutions from the union of $P'(t)$ and $P(t)$ (i.e. $Q=P(t)$).
- The algorithm terminates if no more improvements are achieved over a number of subsequent iterations or if a given amount of time is exceeded.
- The algorithm returns the best candidate solution ever found during its execution.

Algorithm 1

```
t := 0;
initialize P (t);
evaluate P (t);
while not terminate do
   P'(t) := variation(P (t));
   evaluate (P'(t));
   P (t + 1) := select(P'(t) ∪ Q);
   t := t + 1;
od
```

For a more detailled explanation concerning the basic components of an evolution strategy, the reader is referred to [1, 9, 13].

For a given optimization problem

$$f : M \subseteq \mathbb{R}^n \rightarrow \mathbb{R}, \quad f(\vec{x}) \rightarrow \min$$

an individual of the evolution strategy contains the candidate solution $\vec{x} \in \mathbb{R}^n$ as one part of its representation. Furthermore, there exist a variable amount (depending on the type of strategy used) of additional information, so–called strategy parameters, in the representation of individuals. These strategy parameters essentially encode the n–dimensional normal distribution which is to be used for the mutational variation of the solution.

The mutation in evolution strategies works by adding a normally distributed random vector $\vec{z} \sim N(\vec{0}, C)$ with expectation vector $\vec{0}$ and covariance matrix C^{-1}, where the covariance matrix is described by the mutated strategy parameters of the individual.

In evolution strategies, recombination is incorporated into the main loop of the algorithm as the first variation operator and generates a new intermediate population of λ individuals by λ–fold application to the parent population, creating one individual per application from ϱ $(1 \leq \varrho \leq \mu)$ individuals. Normally, $\varrho=2$ or $\varrho=\mu$ (so–called global recombination) are chosen. The types of recombination for object variables and strategy parameters in evolution strategies often differ from each other, and typical examples are discrete recombination (random choices of single variables from parents, comparable to uniform crossover in genetic algorithms) and intermediary recombination (arithmetic averaging). A typical setting of the recombination consists in using discrete recombination for object variables and global intermediary recombination for strategy parameters. For further details on these operators, see [1].

Essentially, the evolution strategy offers two different variants for selecting candidate solutions for the next iteration of the main loop of the algorithm, (μ,λ)–selection and $(\mu+\lambda)$–selection. The notation (μ,λ) indicates that μ parents create $\lambda>\mu$ offspring by means of recombination and mutation, and the best μ offspring individuals are deterministically selected to replace the parents (in this case, $Q = 0$ in algorithm 1). Notice that this mechanism allows that the best member of the population at generation $t+1$ might perform worse than the best individual at generation t, i.e. the method is not elitist, thus allowing the strategy to accept temporary deteriorations that might help to leave the region of attraction of a local optimum and reach a better optimum.

There are several options for the choice of the termination criterion, including the measurement of some absolute or relative measure of the population diversity (see e.g. [1], pp. 80–81), a predefined number of iterations of the main loop of the algorithm, or a predefined amount of CPU time or real time for execution of the algorithm.

17.2 MATHEMATICAL TEST CASES

In the present work, we focus on the Rastrigin function and on Keane's function as representatives of single–objective, multimodal, and constrained (Keane's function) objective functions. The multi–criteria optimization problems proposed by Poloni and Quagliarella are not reported here, however, the interested reader is referred to the INGENET report [3] for detailed information.

17.2.1 Rastrigin's function

The analytical expression of the Rastrigin function was cited in [16] in its 2–dimensional original form and has been generalized by Rudolph [10] for scalability to arbitrary dimensionality. Although it is a separable multimodal test function, it has been widely used in the evolutionary computation community after its first explicit publications in [6, 11] in the following form:

$$f(\vec{x}) = nA + \sum_{i=1}^{n}(x_i^2 - A\cos(2\pi x_i)) \rightarrow \min \tag{1}$$

with the additional box constraints $\forall i \in [1..n],\ x_i \in [-5.12, 5.12]$ and $A=10$.

A plot of the function surface for n=2 is shown in Fig. 17.1, inverting the original minimization problem into a maximization problem to illustrate the region of the global optimum.

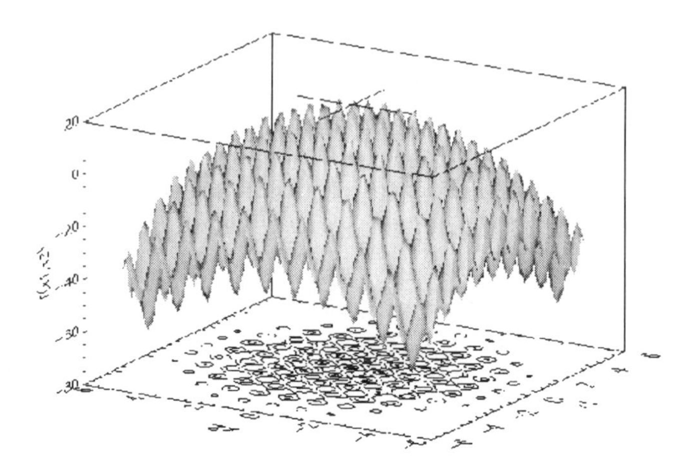

Figure 17.1 Graph of Rastrigin's function for n = 2, shown as a maximization task to visualize the region around the global optimum.

For such a highly multimodal objective function, a standard (15,100)–evolution strategy imposes a too strong selective pressure, causing convergence of the algorithm in local optima without reaching the global optimum. Therefore, a (98+100)–ES with a single self–adaptive σ, standard operators and the parameter settings as given in Table 17.1 was used for optimizing this objective function.

Table 17.1 Parameter settings of the evolution strategy.

Parameter	Setting
Number of parents	98
Number of offspring	100
Selection method	(98+100)
Initial σ	4.0
Initial x_i	U[–5:12 ... 5::12]
Learning rate τ_o	0.18
Recombination	global discrete

This evolution strategy finds globally optimal solutions with an error of less than 10^{-10} after less than 30,000 objective function evaluations, i.e. 300 generations. The evolution process was stopped when the difference of the fitnesses of the best and the worst individual was less than 10^{-10}.

For ten runs, the corresponding final best objective function values, the number of generations and the number of function evaluations are summarized in Table 17.2. These results demonstrate that the current evolution strategy variant is a reliable global optimization algorithm for this objective function.

Table 17.2 Results of the evolution strategy on Rastrigin's function.

Run Number	Objective Value	Generations	Evaluations
1	5.99911231802253e–11	280	28,000
2	5.88684656577243e–11	245	25,400
3	7.37756522539712e–11	272	27,200
4	4.21351842305739e–11	244	24,400
5	8.97415475265007e–11	253	25,300
6	7.46283035368833e–11	262	26,200
7	7.27737869965495e–11	286	26,200
8	6.92921275913250e–11	273	28,600
9	8.29558643999917e–11	250	27,300
10	5.48610046280373e–11	257	25,700
Average	6.79023e–11	263.1	26,310

17.2.2 Keane's function

As a second example, a constrained multimodal optimization problem is considered to investigate the capabilities of an evolution strategy to deal with constraints. The analytical expression of Keane's function is given as follows (refer to [7] for a more detailed discussion of the problem and solutions found so far):

$$f(\vec{x}) = \left| \left[\sum_{i=1}^{n} \cos^4(x_i) - 2\prod_{i=1}^{n} \cos^2(x_i) \right] \Big/ \sqrt{\sum_{i=1}^{n} i x_i^2} \right| \rightarrow \max \tag{2}$$

where:

$$\prod_{i=1}^{n} x_i \geq 0.75 \tag{3}$$

$$\sum_{i=1}^{n} x_i \leq 0.75n \tag{4}$$

and $0 \leq x_i \leq 10$. The constraints make the problem particularly difficult, because the constraint (3) is active at the global optimum. For n=2, Fig. 17.2 provides a hidden–line plot of Keane's function.

Following the approach outlined in [7] for the initialization method, the mutation operator of the evolution strategy is modified to include a repair mechanism that guarantees feasibility. The mutation operator works as a two–step process by first mutating all object variables according to the standard mutation operator, and then eventually performing a repair as outlined in the following algorithm:

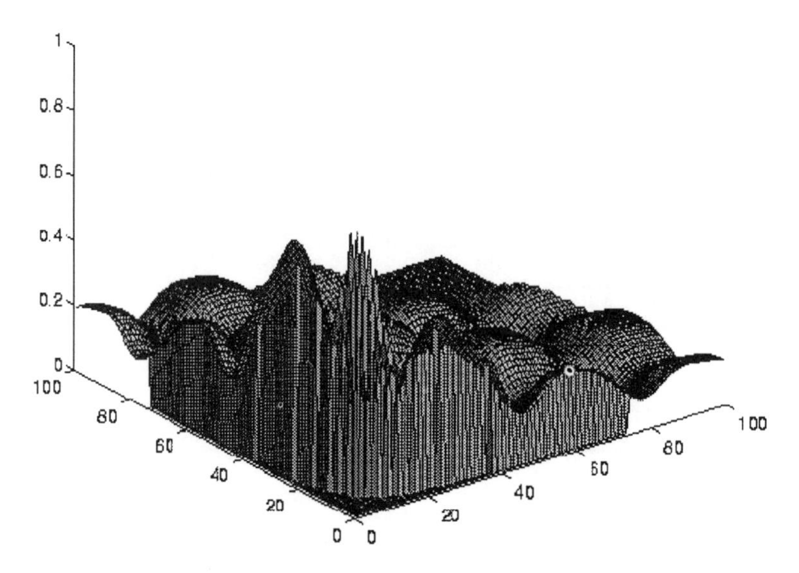

Figure 17.2 Graph of Keane's function for n = 2.

Algorithm 2

1: **for** $i := 1$ **to** n **do**
2: $x_i :=$ mutate(x_i) ;
3: **if** $x_i < 0$
then $x_i := -x_i$;
4: **if** $x_i > 10$
then $x_i := 10 - (x_i \bmod 10)$;
5: $k \sim U(\{1, \ldots ,n\})$;
6: $x_k := 0.75 \,/\, \left(\prod_{i=1}^{k-1} x_i \cdot \prod_{i=k+1}^{n} x_i \right)$;
od

In the agorithm–2 repair procedure outline above, line 2 indicates the usual normally distributed mutation of x_i, and lines 3 and 4 summarize the handling of the box constraints $0 \le x_i \le 10$. In line 5, a random index k is sampled and line 6 presents the repair method for x_k which guarantees constraint (3) to be taken into account. For this objective function, runs were performed with a standard evolution strategy with parameter settings shown in Table 17.3.

For this function, several combinations of recombination operators were tested for object variables and standard deviations, and ten runs were performed for every pair of operators from the combinations (discrete, global discrete), (discrete, intermediate), (global discrete, global discrete), and (global discrete, intermediate). For n=20, all runs are terminated after 20000 function evaluations, and the results of these runs are given in Table 17.4.

Table 17.3 Parameter Settings of the Evolution Strategy for Keane's Function.

Parameter	**Setting**
Number of parents	15
Number of offspring	100
Selection method	(15,100)
Number of σ_i	n
Initial σ_i	0.01
Initial x_i	U[0 ... 10]
Global learning rate τ'	$\sqrt{(2n)}^{-1}$
Local learning rate τ'	$\sqrt{(2n)}^{-1}$
Recombination	Different operators tested

As it becomes clear from Table 17.4, the algorithm repeatedly finds a feasible solution of quality **0.803619**, which is an improvement over the so far best known solution of quality 0.803553 reported in [7].

Table 17.4 Results of the evolution strategy on Keane's function, n = 20.

Trials	**Discrete global Discrete**	**Discrete Intermediate**	**global Discrete global Discrete**	**global Discrete Intermediate**
1	0.798182	0.803613	0.803619	0.803274
2	0.803619	0.803607	0.803619	0.803520
3	0.803619	0.803509	0.803619	0.803536
4	0.803619	0.803588	0.803619	0.802955
5	0.803619	0.803577	0.792608	0.803590
6	0.794662	0.803376	0.803619	0.803594
7	0.803619	0.803519	0.803619	0.803387
8	0.803619	0.803558	0.792608	0.803387
9	0.803619	0.794350	0.803619	0.785699
10	0.803619	0.803435	0.803619	0.803611
best	0.803619	0.803613	0.803619	0.803611
average	0.80218	0.802613	0.801417	0.800623

17.3 MULTI–POINT AIRFOIL DESIGN

17.3.1 Description of test case

The two–point airfoil design problem – originally proposed by Theo Labrujere from NLR – is defined by the minimization of an objective function being the difference between the computed/optimized pressure (more precisely: pressure coefficient) at two different design points and pre–defined target pressures.

The objective function reads:

$$F(\alpha_1, \alpha_2, x(s), y(s)) = \sum_{n=1}^{2} \left[W_n \int_0^1 \left(C_p^n(s) - C_{p,target}^n(s) \right)^2 ds \right] \qquad (5)$$

with s being the airfoil arc–length measured around the airfoil and W_n are the weighting factors. $C_p{}^n$ is the pressure coefficient distribution of the current and $C_{p,target}{}^n$ the pressure coefficient distribution of the target airfoil, respectively.

In the following, Table 17.5 provides a summary of aerodynamic (flow) data chosen for the multi–point design.

Table 17.5 Summarized design conditions (c=chord length).

Property	Case	high lift	low drag
$M\infty$	[–]	0.20	0.77
Re_c	[–]	$5 \cdot 10^6$	10^7
$X_{transition}$ upper/lower	[c]	3% / 3%	3% / 3%
α	[°]	10.8	1.0

It becomes obvious from Table 17.5 that this test case is complicated with respect to the objective to receive a profile shape which holds for both transonic and subsonic (high angle of attack) flow conditions. The airfoil incidences are fixed in this investigation in order to support a more comprehensive validation, although it is possible to use α_1 and α_2 as additional design parameters. When taking the incidences as additional design parameters the desired optimum is altered because the targets (with non–fixed α_n) are changed.

In order to check carefully the accuracy of the optimization process in general, one or two initial test cases can be run prior to the multi–point case, e.g. the pressure re–design case of either the subsonic or transonic flow.

In the European Commission's CFD project, ECARP [17], a hill–climber was used to compute the described multi–point test case with weighting factors of W_1=W_2=0.5 on the basis of an viscous–inviscid–interaction approach. When a multi–objective genetic algorithms is applied, two objective functions can be used, thus splitting eqn. (5), in order to obtain a complete Pareto curve and not a single–point optimum.

17.3.2 Test–case data

Surface data for both airfoils, the transonic low–drag and subsonic high–lift airfoil, are prescribed. All data sets needed for validation purposes can be provided by the second author or directly from the INGENET Web site (www.inria.fr/sinus/ingenet). These data sets include pressure data that have been obtained from a fine (512x128) underlying mesh, thus providing a better surface accuracy due to the use of a Spline (Component–Spline) interpolation. These data are accompanied by target pressure data computed using a Navier–Stokes approach.

In general, all relevant data presented in this paper, accompanied by additional results on finer meshes and different parametrization, can be taken from the above mentioned INGENET Web site.

As a starting airfoil for the optimization process, the NACA four–digit airfoil NACA4412 should be used.

17.3.3 Test–case environment

It was mentioned already that the multi–point design case uses a sophisticated CFD method, i.e. 2D Navier–Stokes approach. However, no particular attention is payed to the problem of flow–physics modelling. The turbulence models applied for the present cases are the Johnson–King model for subsonic flow (including possible separation) and the Johnson Coakley model for transonic flow, respectively. These models have been chosen because of their predictive capabilities to accurately describe flow physics phenomena in the desired flow domains, compare [17].

One of the major results of the European project FRONTIER [18] is an optimization software which can be applied to multi–disciplinary, multi–objective optimization. It employs a genetic algorithm, a hill climber and a tool called MCDM (Multi–Criteria–Decision–Making). A graphical user interface is guiding the (non–experienced) user. All results obtained for the multi–point airfoil optimization are utilizing this FRONTIER [18] software technology.

The parametrization [19] used for the airfoil investigations is based on Bezier splines for the upper and lower airfoil surface, respectively. In total, 6 or 12 Bezier "weighting points" have been chosen, their influence on the optimization results for the airfoil pressure–reconstruction case are going to be discussed below. The parametrization used by C. Poloni [20] has not been adopted for this work, but will be likely considered for future comparisons. Instead it was suggested to use the same parametrization as it was used in the ECARP [21] project for better comparison.

All computation are carried out using a structured grid with 128x32 mesh volumes which denotes the lowest value that can be taken for airfoil flow investigations. Further results are currently carried out with 256x64 mesh volumes which will provide a first tendency concerning mesh dependence aspects and how this relates to the optimization process.

17.3.4 Pressure re–design

In particular for multi–point optimization it turns out to be valuable to first carry out a pressure re–design case in order to test the accuracy of both the parametrization and the analysis tool (Navier–Stokes method). Following [19], a NACA 0012 airfoil is used as a starting airfoil and a NACA 4412 as the target airfoil.

The objective is now to minimize the difference in pressure between current and target airfoil. 64x36 (=2304!) individuals per generations have been chosen for 12 design variables and 64x16 individuals per generations for 6 design variables. Of course, this large total number of individuals is somewhat academic, however, they have been chosen to drive the solution to the absolute minimum. It was not taken into account to constrain the range of the design parameters after some generations have been computed in order to receive – in a continuation run – an objective function which gets nearer to the absolute minimum.

The flow conditions chosen are: Ma = 0.77, Re = 10 million, angle of attack, $\alpha = 1^{o}$. Transition from laminar to turbulent flow is set to 3% chord length on lower and upper surface.

Results can be taken from Fig. 17.3 and 17.4, respectively. While Fig. 17.3 exhibits the pressure distributions for the starting airfoil, target airfoil and the "best individual", Fig. 17.4 presents the corresponding airfoil shapes (starting, target and current) together with the Bezier weighting points. The y–axis is zoomed by a factor of 10 to better show the shape differences.

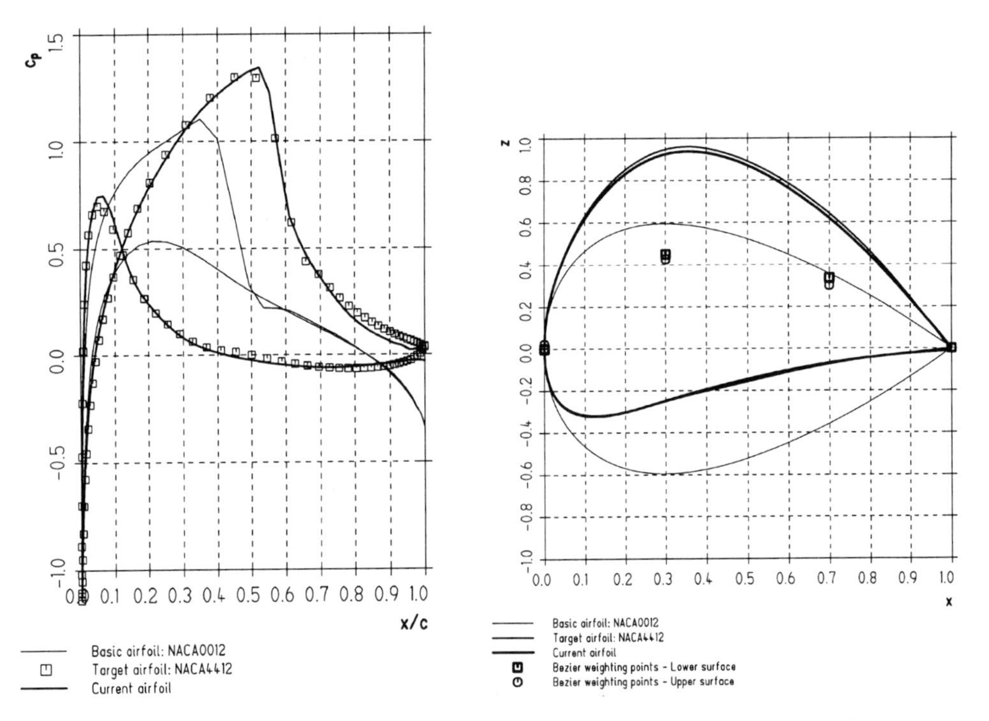

Figure 17.3 Pressure distributions for airfoil re–design case.

Figure 17.4 Airfoil shapes and Bezier parameters for airfoil re–design case.

Fig. 17.5 exhibits the total number of individuals needed for the two different numbers of design variables (= Bezier points). One can easilt observe that the 6–parameter case is

much faster. The reason is that the target airfoil geometry (NACA 4412) is a rather smooth profile shape – in particular in the nose area – and three design parameters on both upper and lower surface are sufficient enough to describe the (smooth) target airfoil. Moreover, oscillations in the geometry, as "usually" predicted by the genetic algorithm as intermediate individuals, are reduced, thus leading to a much smaller number of individuals needed for the reaching the optimum.
It should be pointed out that the current parametrization with 6 Bezier points at fixed x/c locations, does not really hold for the pressure re–design of the transonic target airfoil specified for the multi–point optimization. The reason is that this particular airfoil shape exhibits a drastic curvature variation near the trailing edge. Hence, in this very case, 12 design parameters are clearly superior.

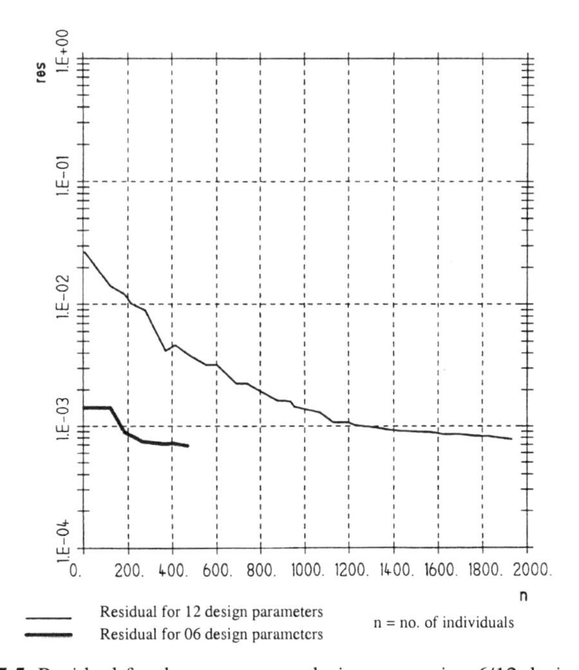

Figure 17.5 Residual for the pressure re–design case using 6/12 design parameters.

17.3.5 Multi–point optimization

As mentioned above, the multi–point airfoil optimization denotes a rather complex test case as it does not aim at one specific point–design but is aiming at a compromise between the two completely different flow situations.

On the nasis of the multi–point test case, it becomes obvious that evolution strategies provide efficient means to receive a complete set of results (Pareto frontier) for airfoil shapes ranging from purely transonic airfoil shapes up to purely high–lift airfoils. An experienced engineer may now choose particular solutions even for an improved and better understanding of flight physics in general. I.e., he may easily choose one solution which is near to the transonic flow – as this can be the most important flight regime in a flight envelope – and try to get airfoil shapes which also provide better performance in the high–lift domain. Particular constraints (e.g. moment forces) can be treated additionally, however, have not been treated in the context of this paper.

Fig. 17.6 contains all shape information necessary to understand the complete trial. At first, the two target airfoil shape are given, exhibiting the different needs (for geometry) for transonic and subsonic flow about an airfoil. As mentioned before, the use of only three Bezier points on lower and upper surface might not be the ultimate choice, however, it more fastly provides "non–oscillating" shapes. It is of importance that CPU–time–intensive sophisticated CFD methods should provide results with good accuracy and the genetic algorithm becomes somewhat "guided" in order not to produce physically irrelevant airfoil shapes.

In addition, it can be taken from Fig. 17.6 that the starting airfoil shape is quite far away from the "best solution". Thus a continued optimization run with reduced design parameter ranges might be a possibility for even improved results.

It becomes very obvious, Fig. 17.6, that the "optimal" shape combines the geometric (and physical) properties of both the subsonic and the transonic target airfoils. Additionally, the Bezier spline weighting points indicate the importance of proper, i.e. "trial–aligned" parametrization. Hence, it might be more adequate to choose a starting airfoil shape which is closer to the suggested solution, although this is not a trivial and easy undertaking.

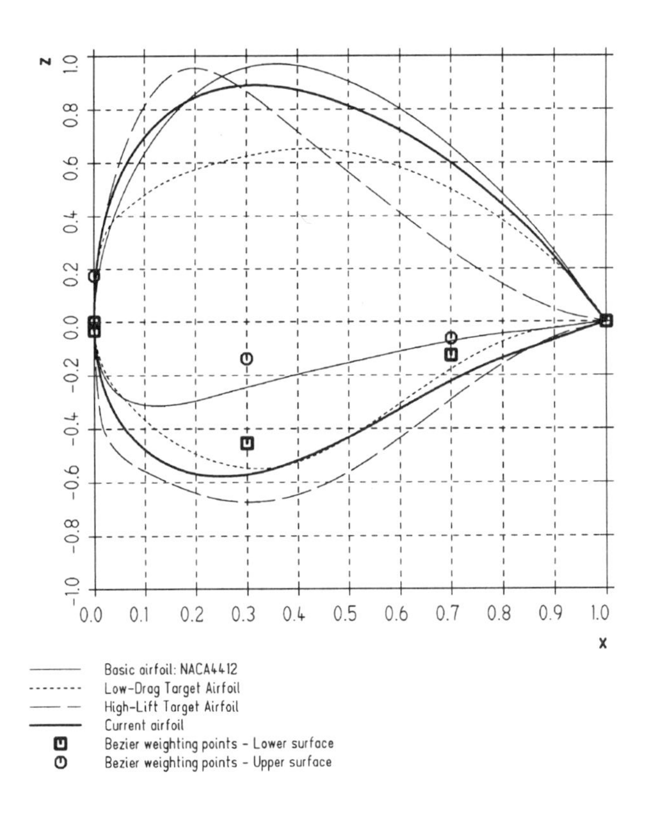

Figure 17.6 Airfoil shapes and Bezier weighting points for multi–point airfoil optimization.

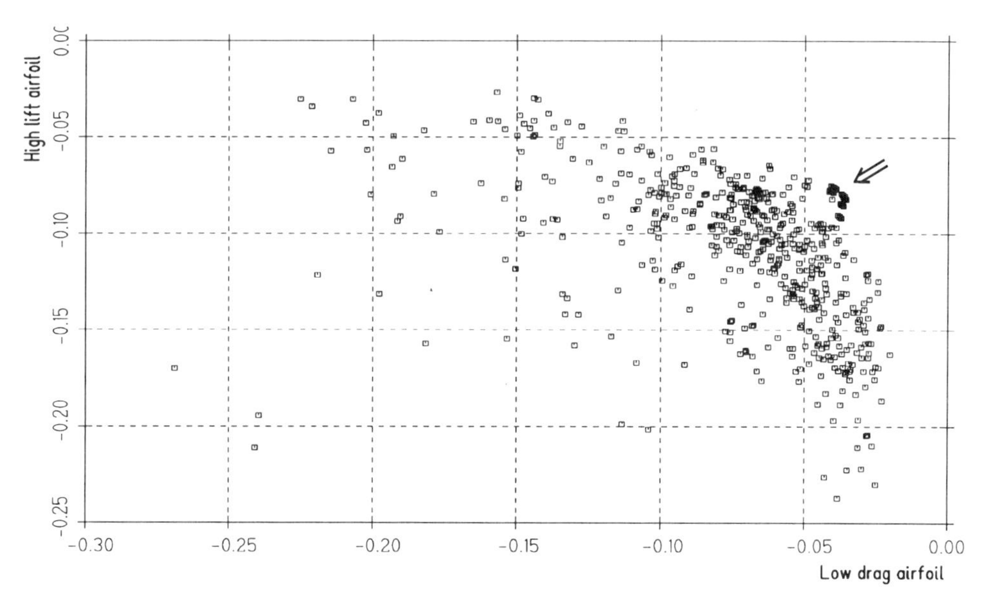

Figure 17.7 Design space for multi–point airfoil optimization.

The computed (complete) Pareto curve can be detected from Fig. 17.7, the arrow is pointing at the "best solution", provided already in Fig. 17.6. It has to be clear that the objectives (transonic and the subsonic pressure difference), eqn. (5) has been split into two parts, do not really converge, reaching a discrete mimimum. Instead, a compromise is found between the two objectives. Again, it is now the engineer who should consider all cases which are located on the Pareto frontier in order to get as close as possible to the intended goal.

17.4 CONCLUSIONS AND OUTLOOK

The current paper describes the successful application of evolutionary algorithms on academic/mathematical and industrial test cases. These cases are part of the work carried out in the INGENET project. New and improved results are currently carried out and will be placed on the mentioned data base.

Although it is sometimes mentioned that evolutionary algorithms do not really hold for industrial applications, but provide very reasonable and accurate results for "non–complex" test cases, the present paper very clearly demonstrates the capabilities of these methods in an illustrative way. Cooperative actions – academic–industrial – should be carried out in order to learn via cross–fertilization and to get close(r) to industry requirements.

For the current case of multi–point airfoil design, it even seems as if an evolutionary algorithm is clearly ahead of the use of a (single–objective) hill climber.

In future investigations on airfoil optimization, mesh dependence and flow–physics–modelling studies as well as different parametrization techniques should be treated in order to gain an enhanced knowledge about complex optimization processes.

The FRONTIER technology will be one of Dasa–M's technologies for future work. A fruitful cooperation with fundamental research will improve this tool and broaden industry's optimization capabilities.

Acknowledgements

The research of the first and third author is supported by grant 01IB802B1 of the German Federal Ministry of Education, Science, Research and Technology (BMBF) and the European Commission's INGENET project, BRRT–CT97–5034. The second and the last two authors worked together in the framework of the INGENET project.

REFERENCES

1. Th. Bäck. Evolutionary Algorithms in Theory and Practice. Oxford University Press, New York, 1996.
2. Th. Bäck, U. Hammel, and H.–P. Schwefel. Evolutionary computation: History and current state. IEEE Transactions on Evolutionary Computation, 1(1):3–17, 1997.
3. Th. Bäck, M. Laumanns, B. Naujoks, and L. Willmes. Test case computation results. INGENET Project Report D 5.1, Center for Applied Systems Analysis, Informatik Centrum Dortmund, December 1998.
4. D. B. Fogel. Evolutionary Computation: Toward a New Philosophy of Machine Intelligence. IEEE Press, Piscataway, NJ, 1995.
5. D. E. Goldberg. Genetic Algorithms in Search, Optimization and Machine Learning. Addison Wesley, Reading, MA, 1989.
6. F. Hoffmeister and Th. Bäck. Genetic algorithms and evolution strategies: Similarities and differences. In H.–P. Schwefel and R. Männer, editors, Parallel Problem Solving from Nature Proceedings 1st Workshop PPSN I, volume 496 of Lecture Notes in Computer Science, pages 447–461. Springer, Berlin, 1991.
7. Z. Michalewicz, G. Nazhiyath, and M. Michalewicz. A note on usefulness of geometrical crossover for numerical optimization problems. In L.J. Fogel, P.J. Angeline, and T. Bäck, editors, Proceedings of the Fifth Annual Conference on Evolutionary Programming, pages 305–312. The MIT Press, Cambridge, MA, 1996.
8. M. Mitchell. An Introduction to Genetic Algorithms. The MIT Press, Cambridge, MA, 1996.
9. I. Rechenberg. Evolutionsstrategie '94, volume 1 of Werkstatt Bionik und Evolutionstechnik. Frommann–Holzboog, Stuttgart, 1994.
10. G. Rudolph. Globale Optimierung mit parallelen Evolutionsstrategien. Diplomarbeit, Universität Dortmund, Fachbereich Informatik, 1990.
11. G. Rudolph. Global optimization by means of distributed evolution strategies. In H.–P. Schwefel and R. Männer, editors, Parallel Problem Solving from Nature. Proceedings 1st Workshop PPSN I, Volume 496 of Lecture Notes in Computer Science, pages 209–213. Springer, Berlin, 1991.
12. G. Rudolph. On correlated mutations in evolution strategies. In: R. Männer and B. Manderick, editors, Parallel Problem Solving from Nature 2, pages 105–114. Elsevier, Amsterdam, 1992.
13. H.–P. Schwefel. Evolution and Optimum Seeking. Sixth–Generation Computer Technology Series. Wiley, New York, 1995.

14. H.–P. Schwefel and Th. Bäck. Artificial evolution: how and why? In: D. Quagliarella, J. Periaux, C. Poloni, and G. Winter, editors, Genetic Algorithms in Engineering and Computer Science, chapter 1, pages 1–19. Wiley, Chichester, 1997.
15. H.–P. Schwefel and G. Rudolph. Contemporary evolution strategies. In: F. Moran, A. Moreno, J. J. Merelo, and P. Chacon, editors, Advances in Artificial Life. Third International Conference on Artificial Life, volume 929 of Lecture Notes in Artificial Intelligence, pages 893–907. Springer, Berlin, 1995.
16. A. Törn and A. Zilinskas. Global Optimization, volume 350 of Lecture Notes in Computer Science. Springer, Berlin, 1989.
17. W. Haase, E. Chaput, E. Elsholz, M. Leschziner, D. Schwamborn (Eds.). ECARP: European Computational Aerodynamics Research Project – Validation and Assessment of Turbulence Models. Notes on Numerical Fluid Dynamics, Vol. 58, Vieweg Verlag (1997).
18. D. Spicer, J. Cook, C. Poloni, P. Sen. EP 20082 Frontier: Industrial multiobjective design optimisation. In: D. Papailiou et al (Eds.), Computational Fluid Dynamics '98, John Wiley & Sons Ltd., 1998.
19. A. Turchet. Utilizzo di codici fluidodinamici di ottimizzazione per il progetto di profili aereonautici. Diploma work University of Trieste, 1998.
20. C. Poloni, V. Pediroda. GA coupled with computationally expensive simulations: tools to improve efficiency. In: D. Quagliarella, J. Periaux, C. Poloni, and G. Winter, editors, Genetic Algorithms in Engineering and Computer Science, chapter 1, pages 1–19. Wiley, Chichester, 1997.
21. J. Periaux, G. Bugeda, P.K. Chaviaropoulos, K. Giannakoglou, St. Lanteri, B. Mantel (Eds.). Optimum Aerodynamic Design & Parallel Navier–Stokes Computations, ECARP – European Computational Aerodynamics Research Project. Notes on Numerical Fluid Dynamics, Vol. 61, Vieweg Verlag (1998).

18 Optimization of an Active Noise Control System inside an Aircraft, Based on the Simultaneous Optimal Positioning of Microphones and Speakers, with the Use of a Genetic Algorithm

Z. G. DIAMANTIS[1], D. T. TSAHALIS[1] and I. BORCHERS[2]

[1] LFME: Laboratory of Fluid Mechanics & Energy, Chemical Engineering Department, University of Patras, P.O. Box 1400, 26500 Patras, Greece.

[2] Dornier F1M/GV, Daimler-Benz-Aerospace, 88039 Friedrichschafen, Germany

Abstract. In the last decade, Active Noise Control (ANC) has become a very popular technique for controlling low-frequency noise. The increase in its popularity is a consequence of the rapid development in the fields of computers in general, and more specifically in digital signal processing boards. ANC systems are application specific and therefore they should be optimally designed for each application. Even though the physical background of the ANC systems is well known and understood, efficient tools for the optimization of the sensor and actuator configurations of the ANC system, based on classical optimization methods, do not exist. This is due to the nature of the problem that allows the calculation of the effect of the ANC system only when the sensor and actuator configurations are specified. An additional difficulty in this problem is that the sensor and the actuator configurations cannot be optimized independently, since the effect of the ANC system directly depends on the combined sensor and actuator configuration. For the solution of this problem several *intelligent* techniques were applied. In this paper the successful application of a Genetic Algorithm, an optimization technique that belongs to the broad class of evolutionary algorithms, is presented.

18.1 INTRODUCTION

Nowadays, in general, noise control is achieved mainly through the application of i) passive, and/or ii) active measures. The application of the passive measures (consisting basically on the application of insulation materials, sound barriers, etc.), has achieved up to now good results in reducing the high-frequency noise. However, in the low frequency, where most of the machinery noise exists, the latter measures do not perform very well and active measures should be applied. The technological advances of the last decade, concerning the introduction of more powerful computer processors at a rapidly decreasing price, led to a massive growth of digital signal processing. Consequently, the active measures were also developed and applied successfully to more difficult noise control problems.

The main application of active measures is the Active Noise Control (ANC). This relies in i) the measurement of the initial noise (primary field) in a control area and ii) the introduction of a proper artificially-created "anti-noise" field that, when superimposed to the primary field, results in a combined noise field of lower intensity than the primary one. An Active Noise Control System (ANCS) consists of a) a set of sensors, b) a control unit and c) a set of actuators.

Each ANCS is application specific, i.e., it is designed for a specific case (i.e., for the same type of car, or the same type of aircraft). During the development of an ANCS for an application, all the three basic components of the ANCS should be designed optimally. However, the optimality of (b) is limited by the existing digital signal processing technology (hardware and software). Therefore, the optimal selection of (a) and (c) has to be performed for an ANC application. This is a difficult problem, which has been solved up to now, not always successfully, with the use of classic optimization techniques. Also, this problem is a difficult one since both sensor and the actuator configurations should be optimized simultaneously. Moreover, the results of the ANCS can be calculated only after the sensor and actuator configurations are specified.

In this paper [1], a new approach for the simultaneous selection of the optimal set of sensors and actuators for an ANCS, based on Genetic Algorithms (GAs), is presented in more detail. The results point out the efficiency of the GA-based optimization technique for this type of problem. A description of the test case article and the optimization problem are given in Section 18.2. A general description of the GAs is given and the developed GA is described in detail in Section 18.3. Finally, in Section 18.4, the results of the GA application on the problem are summarized.

18.2 DESCRIPTION OF THE TEST CASE ARTICLE AND THE OPTIMIZATION PROBLEM

For the application of an ANCS in a control area, initially the positions where sensors and actuators may be installed are identified. A general overview of how the sensors and actuators are placed can be seen in Fig. 18.1. The selection of these positions is usually made after taking into account engineering experience and application constraints that may be summarized as a) considerations of where the sensors will be able to receive a signal relevant to the global noise condition in the aircraft, b) considerations of where the actuators will be able to influence the global noise field with their action (positions where the signal is not blocked by other installations, like seats), c) wiring considerations, etc..

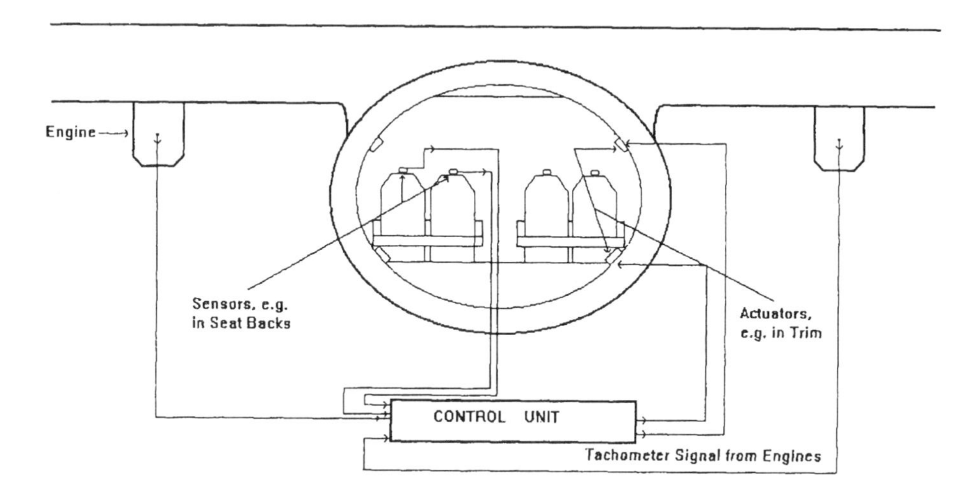

Fig. 18.1 General overview of the plane.

In general, it is obvious that there will be several available positions. The ideal case would be to use all these positions for the installation of sensors and actuators. Nevertheless, both technical and economical constraints limit the number of sensors and actuators that may be applied. The number of the sensors and actuators that will be used is usually dictated by the number of available input/output channels of the ANCS control unit. Moreover, certain specific constraints may arise during the application. For instance, in the case of an ANCS applied in an aircraft, the final number of the actuators that will be used depends gravely on their weight which is a significant limiting factor.

In the case at hand the demonstrator of the ANCS was selected to be a propeller aircraft. The total number of available sensor positions $\mathbf{S_T}$, which were identified, were 80, while the total number of available actuator positions $\mathbf{A_T}$, which were identified, were 52. The selected control unit for this aircraft, as well as weight considerations, limited the number of sensors **S** to 48 and the number of actuators **A** to 32.

Therefore, the definition of the present optimization problem is the following: given the total number of S_T available positions for sensors and $\mathbf{A_T}$ for actuators in the aircraft, the optimal positions for the positioning of **S** sensors and A actuators should be calculated simultaneously, where the number of chosen sensor positions S should be less than the total number of available sensor positions $\mathbf{S_T}$, and the number of chosen actuator positions **A** should be less than the total number of available actuator positions $\mathbf{A_T}$. Furthermore, the constraint of **S>A** which is required in order for the system to be overdetermined as a requisite from the ANC theory [2], obviously stands (48>32).

In the description of the optimization problem, above, the term 'optimal positions for the **S** sensors and **A** actuators', is stated. As optimal positions of **S** sensors and **A** actuators (optimal configuration) are defined the ones that produce the highest average noise reduction over the total number of $\mathbf{S_T}$ sensor positions in the control space, i.e., in the present application the passenger-cabin of the demonstrator aircraft. The average noise reduction of any given configuration of sensors and actuators in the aircraft can be calculated a-priori (before the ANCS installation) using a simulation model. This model, given the **S** sensor and **A** actuator positions, it calculates the optimum signal that should be fed to each of the actuators (excitation amplitude and phase) and then calculates the residual sound field at the $\mathbf{S_T}$ sensor positions as well as the average noise reduction. This model is based on the *Least Squares* method and is described in the next section.

It is obvious that in order to find the global optimal sensor-actuator configuration of the ANCS, the simplest way would be to calculate the effect of all the possible sensor-actuator configurations and select the best (exhaustive search in the sensor-actuator configuration space). Unfortunately, this is not possible due to practical reasons, as it will be evident.

The number of possible configurations of **S** sensors placed in $\mathbf{S_T}$ positions is given by the following formula:

$$[\mathbf{S_T}!/\{(\mathbf{S_T}\text{-}\mathbf{S})!\cdot\mathbf{S}!\}]. \tag{1}$$

Similarly, the number of possible actuator configurations is

$$[\mathbf{A_T}!/\{(\mathbf{A_T}\text{-}\mathbf{A})!\cdot\mathbf{A}!\}]. \tag{2}$$

From the above, the number of possible sensor-actuator configurations is:

$$[\mathbf{S_T}!/\{(\mathbf{S_T}\text{-}\mathbf{S})!\cdot\mathbf{S}!\}]\cdot[\mathbf{A_T}!/\{(\mathbf{A_T}\text{-}\mathbf{A})!\cdot\mathbf{A}!\}]. \tag{3}$$

For the test case at hand, $\mathbf{S_T}$=80, **S**=48, $\mathbf{A_T}$=52 and **A**=32, and the number of possible configurations estimated from (3) are:

$$[80!/(80\text{-}48)!\cdot 48!]\cdot[52!/(52\text{-}32)!\cdot 32!] \approx \mathbf{2.758\cdot 10^{36}}.$$

If we assume that for the calculation of the average noise reduction to-be-achieved by each of the configurations, the simulation model needs 0.001 seconds calculation time (an assumption which is far better than reality), then the evaluation of all possible configurations needs $2.758\cdot 10^{33}$ seconds, or $8.745\cdot 10^{25}$ *years*! From the above it becomes obvious that an optimization technique capable of solving large combinatorial problems is required and should be implemented in this case. Such an optimization method are Genetic Algorithms (GAs). A small introduction to the GAs, together with a description of the GA developed for this specific optimization problem, is given in the next section.

18.3 GENETIC ALGORITHMS

18.3.1 General description of GAs

Genetic Algorithms are part of a more general class of optimization techniques that are called *Evolutionary Algorithms*. These are algorithms that imitate the evolution of species in natural environments, according to the laws of evolution that C. Darwin had expressed. Most of the evolutionary algorithms were initially developed as artificial-life simulators. Soon it was realized that since the evolution process advances towards the optimal, in any natural environment, the evolutionary algorithms were also optimizers. Furthermore, the robustness of the evolution process (the fact that performs in any environment) was inherited also to the evolutionary algorithms since they perform regardless the optimization problem.

The main types of evolutionary algorithms that exist are: a) Genetic Algorithms, b) Evolution Strategies and c) Genetic Programming. Each of these techniques has its own particularities in its application but the basic rules of natural evolution exist in all three of them.

The characteristics that distinguish the GAs from the classic optimization techniques, are the following:

1. The GAs do not need any type of information other than a measure of how good each point in the optimization area is. Consequently, they may be applied in cases where only a black-box model of the system-to-be-optimized exists.

2. The GAs are robust, i.e., they perform regardless of the optimization problem.

3. The GAs provide more than one equivalently good near-global-optimum solutions. This allows the selection of the most appropriate solution for the problem at hand.

From the above it seems that for the application of a GA to a problem, a) nor a good knowledge of the problem, b) neither an experience using GAs is necessary. Even though this may be generally true, both (a) and (b), above, are necessary for the successful application of the GA. For an overview of GAs the reader is referred to Goldberg [3].

The GAs perform very well in the optimization of combinatorial problems [4]-[5], as well as multi-parameter ones [6]-[7], areas where the classic optimization techniques give poor results, or fail. Since the problem at hand was a combinatorial one, it was decided to develop a GA for its optimization.

18.3.2 The developed GA

A general overview of how a GA works is given in Fig. 18.2. In a few words, initially the parameter space of the problem is coded into a string (array of integers). Then the initial population is created randomly. After the initial population has been created, the generation phase begins, which consists of 3 stages. In the first stage of the generation phase, the fitness of each string in the random population is calculated (fitness is a number that shows how 'good' a string is in the population compared to the others, later on we will see how it is calculated for the problem at hand). In the second stage, certain operators, which are mathematical models of the simple laws of nature such as Reproduction, Crossover and Mutation, are applied to the population and result in the creation of a new population and in the third stage the new population replaces the older. This concludes the generation phase. Once the generation process has been completed, the number of generations that have been done up to now are compared with a predefined number of generations. If the number of generations has exceeded the predefined value, then the algorithm stops; in any other circumstance it continues until the necessary number of generations has been reached.

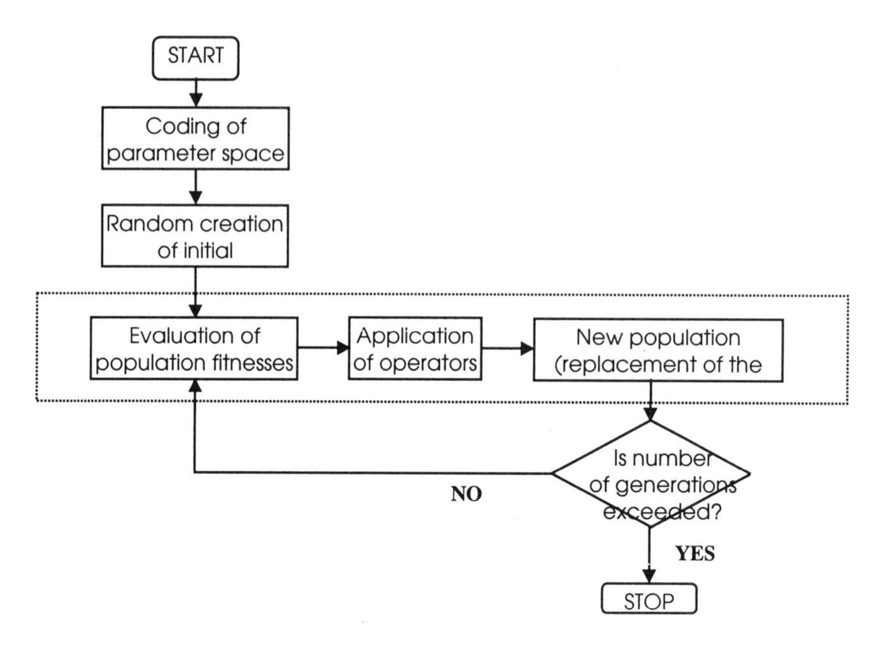

Fig. 18.2 General overview of how a GA works.

Therefore, as mentioned above, in order for the GA to be applied for the problem optimization, the parameter space was coded, i.e., the independent variables of the problem were transformed into a string. This string was used by the GA as optimization parameter.

Since the problem at hand was a combinatorial one, the following coding was selected: the available sensor positions were numbered from 1 to 80 and the actuator positions from 1 to 52. Any configuration of **S** sensors and **A** actuators was represented by a (**S**+**A**)-element string, consisting of two sub-strings; one containing the integers corresponding to the selected **S** sensor positions, and one containing the integers corresponding to the selected **A** actuator positions. Using this coding, each sensor-actuator configuration was represented by a (**S**+**A**)-element string.

Based on this coding the initial population of strings was created. Each of the strings comprising this population was created in random, as follows: Using a random number generator, S random integers belonging in [1, 80] were generated to take the positions of the sensor sub-string. Special care was taken that the S selected integers were different among them. Similarly, **A** different random integers were generated for the actuator sub-string. By linking together the two sub-strings, a complete string was formed. This procedure was repeated as many times as the number of the strings in the initial population.

The fitness of each string, as mentioned above, was calculated. This calculation is described analytically in the subsection 'fitness function'. The average noise reduction that was calculated for each string, served as its objective function value. In case that the string contained the same sensor, or actuator, location more than once it corresponded to a meaningless ANCS, since only one sensor, or actuator, may be positioned in each location. In this case the fitness value was set to be 0 (penalty). Since precautions were taken, such a string cannot exist in the initial population of strings. Nevertheless, such a string may arise from the application of the operators later.

The operators that were applied to the initial population, which are mathematical models of the laws of nature, were the following:

- **Reproduction**. The classic "roulette" reproduction operator was used. Practically, this operator is applied in the following steps:

  I. An order number, j, was assigned to the population strings.

  II. The sum of the fitnesses (F_{SUM}) of all the strings in the population was calculated.

  III. A random real number R_0 between 0 and F_{SUM} was selected.

  IV. From this number, the fitnesses of the strings F_j ($j = 1,2,\ldots$, population of strings) were subtracted, one after another, in the order they appeared in the population as follows:

  A. if $R_{j-1} - F_j = R_j > 0$, then the j string was discarded and $R_j - F_{j+1} = R_{j+1}$ was calculated.

  B. if $R_j \leq 0$ then the j string was kept for the next generation and the procedure started all over again for the next string from step III with the selection of a new R_0.

  For a more analytical description of this operator the reader is referred to Goldberg [3].

- **Crossover**. The crossover operator used was similar to the classic one, with the following variations (Fig. 18.3). It was not performed directly to the two strings selected for the crossover (parents), but between two temporal strings. The two temporal strings were formed by removing the sensors and the actuators that appeared in both parent strings. After the crossover between the temporal strings was performed, two temporal offsprings were created. Then the common sensors and actuators to the parent strings, that were removed temporarily, were placed back in their original positions in the strings giving rise to two new strings (offspings). This special crossover procedure was developed in order to avoid having as a result of the crossover a meaningless string. Finally, among the two parents and the two offsprings, the two best strings were kept for the next population (elitism). This crossover operator was applied successfully to combinatorial optimization problems by the LFME team [4], [5].

- **Mutation**. The classic mutation operator was used. This operator was applied to each string cell and it altered its content, with a small probability, as follows:

I. The value k of the cell, was read.

II. A random number R belonging in $[1,k)\cup(k,l]$ was selected, where l was 80 if the cell belonged to the sensor sub-string, or l was 52 if it belonged to the actuator sub-string.

III. The number R substituted the number k.

For a more analytical description of this operator the reader is referred to Goldberg [3].

a) mating

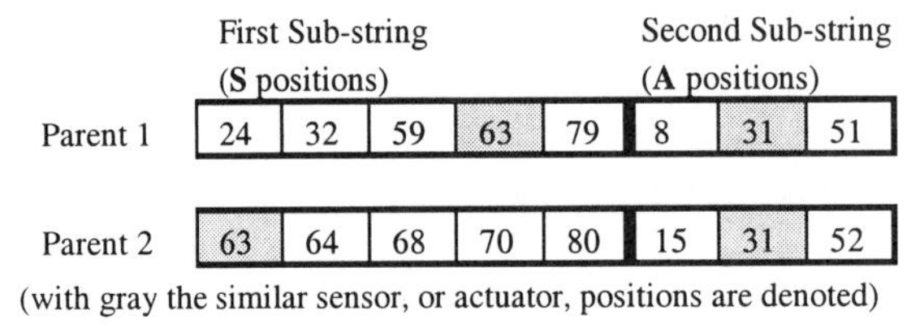

b) creation of temporal strings, crossing site selection and cutting

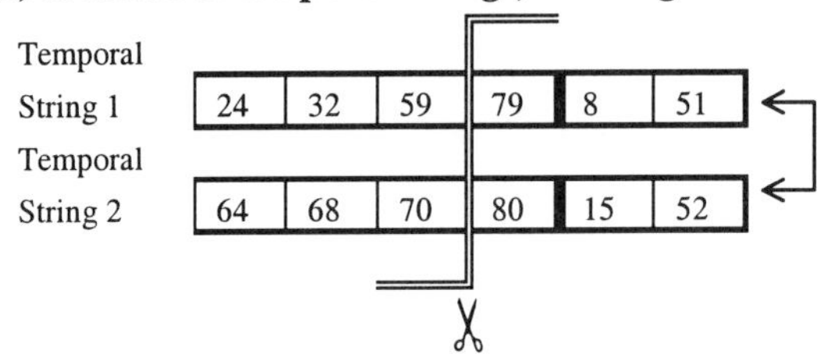

c) creation of the temporal offsprings

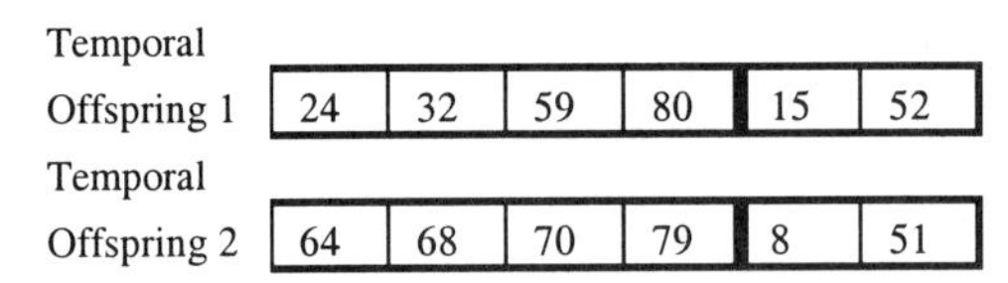

d) creation of the offsprings

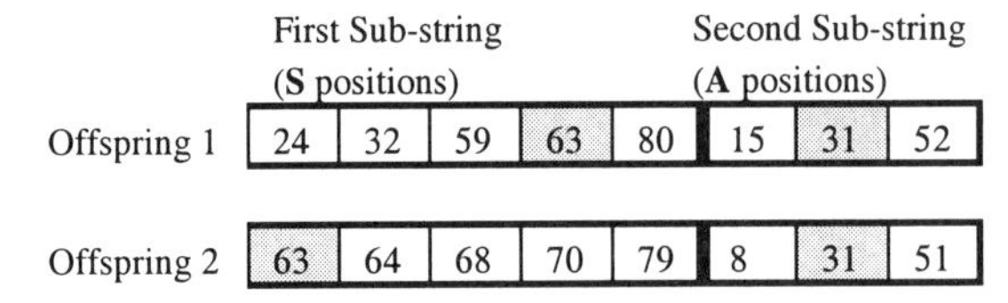

Fig. 18.3 The crossover operator.

From the application of these operators to the population, a new population was created, as it was mention earlier. The generation of the new populations, from the old ones, continued in a similar manner, until a maximum predefined number of generations was reached.

18.3.3 The fitness function

As it was already mentioned, in the simulation model a calculation of the optimal forces to the actuators of a given ANCS is performed (where given means that the configuration of sensors and actuators is a-priori selected). Then the resulting noise attenuation is calculated. An ANCS configuration is defined by the number and positions of the sensors, as well as the number and positions of the actuators.

For a given primary acoustic field and an ANCS of a given configuration, the optimal array of signals to the actuators can be calculated using the *Least Square* technique. Driving the actuators of the system with this optimal array of signals the ANCS will achieve its optimal performance, i.e., the maximum noise reduction will be achieved. In an installed ANCS this array of signals is calculated iteratively in order for the system to be able to adapt to any changes of the primary field. However, the performance achieved by an ANCS, using the optimal signals to the sources, can be calculated off-line. For this calculation, the Frequency Response Functions (FRFs) from every possible actuator location to every possible sensor location should be measured, along with the primary field in every possible sensor location.

Assuming that the number of possible sensor positions is $\mathbf{S_T}$ and the number of possible actuator positions is $\mathbf{A_T}$, then the FRFs between all the actuator and sensor positions compose a matrix $\mathbf{FRF(S_T,A_T)}$. Further assuming that the ANCS consists of **S** sensors and **A** actuators, then the FRFs between the actuators and sensors of the ANCS compose a matrix $\mathbf{FRF_{ANCS}(S,A)}$. Finally, the primary field is measured at the $\mathbf{S_T}$ possible sensor positions and comprises an array $\mathbf{PF(S_T)}$ [the array of the primary field at the **m** selected sensor positions will be noted as **PF(S)**].

Then, if a signal **SIG(A)** is applied at the actuators, the residual field at the sensors **RF(S)** will be given by the following formula:

$$\mathbf{RF(S)} = \mathbf{PF(S)} + \mathbf{FRF_{ANCS}(S,A) \cdot SIG(A)} \tag{4}$$

It should be kept in mind that all the above matrices/arrays consist of complex numbers. Therefore, in order for the residual field to be minimum, according to the Least Squares technique, the $\mathbf{RF(S)^H \cdot RF(S)}$ should be minimum [where $\mathbf{RF(S)^H}$ is the hermitian transpose of **RF(S)**]. Therefore, the $\mathbf{[RF(S)^H \cdot RF(S)]' = 0}$ is calculated and then solved for **SIG(A)**, as follows:

$$[\mathbf{RF(S)^H \cdot RF(S)}]' = 0 \Rightarrow \tag{5}$$

$$\{[\mathbf{PF(S)+FRF_{ANCS}(S,A) \cdot SIG(A)}]^H \cdot [\mathbf{PF(S)+FRF_{ANCS}(S,A) \cdot SIG(A)}]\}' = 0 \Rightarrow \tag{6}$$

$$\mathbf{SIG(A)} = -\{\mathbf{FRF_{ANCS}(S,A)^H \cdot FRF_{ANCS}(S,A)}\}^{-1} \cdot \mathbf{FRF_{ANCS}(S,A)^H \cdot PF(S)} \tag{7}$$

SIG(A) is now the array consisting of the optimal signal to be send to the actuators. The average noise reduction that will result from this signal is given by:

$$\text{average noise reduction} = 10 \cdot \log_{10} \frac{\left[\sum_{i=1}^{S_T} (\mathrm{RF}(i))^2\right]}{\left[\sum_{i=1}^{S_T} (\mathrm{PF}(i))^2\right]} \quad (8)$$

For a further theoretical background of this model and a more analytical description the reader is referred to [1].

Since the objective of the optimization problem was the calculation of the optimal sensor-actuator configuration of an ANCS, in such a way that this ANCS achieves the maximum average noise reduction, as fitness function of the GA was set the average noise reduction calculated from the above formula.

18.3.4 Necessary data

The necessary data for the calculations of the above model were: a) the FRFs between the 52 actuator positions and the 80 sensor positions, and b) the primary field measured in the 80 sensor positions. Both the FRFs and the primary field should be measured in the frequency of interest.

The frequency of interest for the case at hand was the Blade Pass Frequency (BPF) of the propeller of the aircraft. The FRFs and the primary field were measured and were provided in a matrix format.

18.4 RESULTS AND CONCLUSIONS

Several runs of the developed GA were performed with the following population sizes: 200, 400, 600, 800 and 1000 strings. The number of generations performed for each run was set to 1000.

From the description of the previous section it is obvious that the GA starts searching from many points in the optimization space, and continuously converges to the global optimum. In each generation there is a string that is the best of the population. The fitness of this best string versus the time it was achieved is given in Fig. 18.4, for each GA run. The maximum fitness achieved was 6.46 dBs for the case of 600 strings. Further rise of the population size does not lead to any better solution, thus the optimal population size is about 600 strings. Moreover, it is evident that almost all the GA runs converged to a fitness value between 6.3dBs and 6.5dBs, showing in this way that the global optimum should be near that value.

The following conclusions can be derived from the results:

1. GAs offer a promising, efficient and easy-to-apply optimization strategy for the problem of optimally positioning simultaneously sensors and actuators in an ANCS.

2. Even though it is not evident from the graphs, the GA developed provides more than one string, and therefore more than one sensor-actuator configurations, with similar near-global optimal average noise reduction. This is very important since it enables the engineer-designer to choose the configuration that fits better to the possible engineering constraints on the positioning of the actuators.

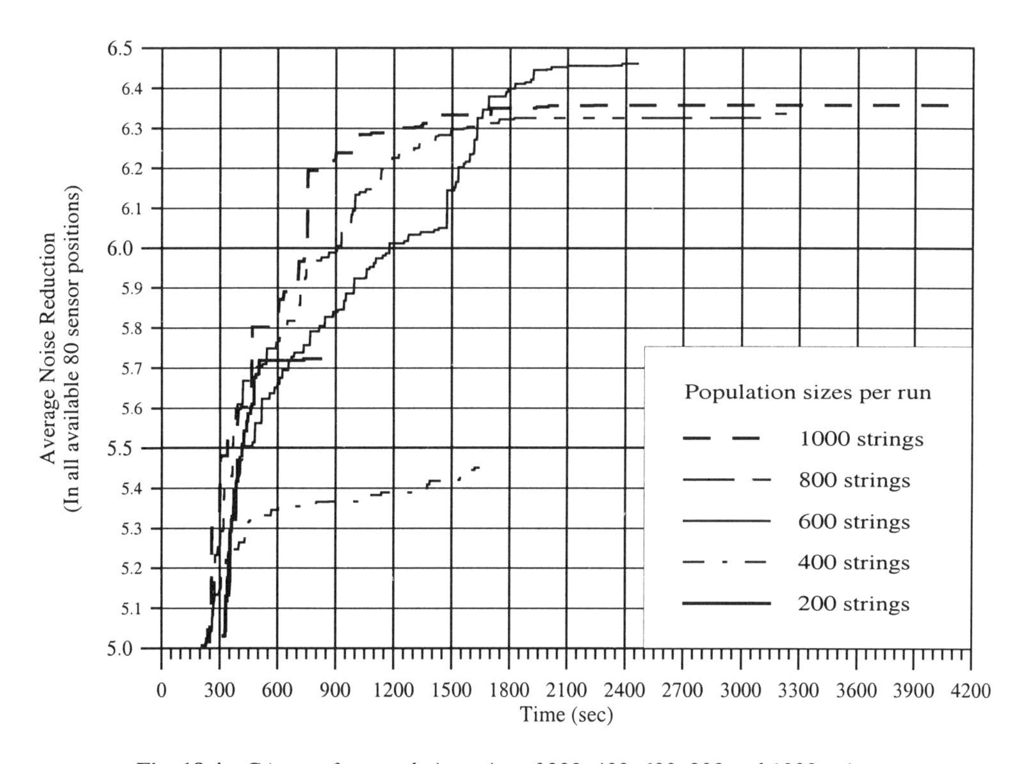

Fig. 18.4 GA runs for populations size of 200, 400, 600, 800 and 1000 strings.

Acknowledgments

The work reported in this paper was partially funded by the European Commission under the ASANCA II project, Contract Number AER2-CT94-0062 of the BRITE/EURAM programme. Also the presentation and preparation of the above paper is funded partially by the INGENET Thematic Network, Contract Number BRRT-CT97-5034, which is also funded by the European Commission.

REFERENCES

1. D. A. Manolas, I. Borchers, D. T. Tsahalis, "Simultaneous Optimization of the Sensor and Actuator Positions for an Active Noise and/or Vibration Control System Using Genetic Algorithms, Applied in a Dornier Aircraft", Proceedings of ECOMASS '98 Conference, Athens, Greece, pp. 204-208 (1998)
2. Nelson P. A., Elliott S.J., "Active Control of Sound" (Academic Press, London, 1993).
3. Goldberg D. E., "Genetic Algorithms in Search, Optimization, and Machine Learning", (Addison-Wesley Publishing Company Inc., 1989)

4. Katsikas S.K., Tsahalis D.T., Manolas D.A., Xanthakis S., "A Genetic Algorithm for Active Noise Control Actuator Positioning", Mechanical Systems and Signal Processing, November (1995)

5. Tsahalis D.T., Katsikas S.K., Manolas D.A., "A Genetic Algorithm for Optimal Positioning of Actuators in Active Noise Control: Results from the ASANCA Project", Proceedings of Inter-Noise 93, Leuven, Belgium, pp. 83-88 (1993).

6. T. Tsahalis, D. A. Manolas, T. P. Gialamas and C. A. Frangopoulos, "A Genetic Algorithm for Operation Optimization of an Industrial Cogeneration System", Journal of Computers & Chemical Engineering, Vol. 20, Suppl. B, pp. S1107-S1112, 1996.

7. T. Tsahalis, C. A. Frangopoulos, D. A. Manolas, T. P. Gialamas, "Operation Optimization of an Industrial Cogeneration System by a Genetic Algorithm", Energy Conversion and Management Journal, Vol. 38, No. 15-17, pp. 1625-1636, 1997.

19 Generator Scheduling in Power Systems by Genetic Algorithm and Expert System

B. GALVAN[1], G. WINTER[1], D. GREINER[1], M. CRUZ[2] and S. CABRERA[2]

[1]CEANI, University of Las Palmas de Gran Canaria

[2]UNELCO, Spain

19.1 INTRODUCTION

The unit commitment (UC) is the problem of determining the optimal set of generating units of a power system considering when to start up and/or shut down thermal generators and how to dispatch the committed units in order to serve the load demand and reserve requirements over a scheduling period at minimum operating cost [14]. This problem belongs to the class of complex combinatorial optimisation problem involving a large number of variables and a large and complex set of constraints (NP-hard combinatorial) thus, the problem is difficult to solve by traditional optimisation methods because of its nonconvex and combinatorial nature. Many methods have been proposed for solving the problem, but each method has involved one or more difficulties, such as suboptimal solutions, difficulty getting feasible solutions, etc. In the last decades intensive research efforts have been made because of potential cost savings for electrical utilities.

Exhaustive enumeration is the only technique that can find the optimal solution but is prohibitive the computational time for medium and large scale systems. The partial enumeration methods based on dynamic programming, branch-and-bound methods, etc., [8], [17], [18], [19], [20], [21]. Have large computational cost (computation time and memory requirements) according the problem size increases [2].

Some techniques include priority list methods, which are too much heuristic [15], [16]. More recently methods such as Benders decomposition [22] and Lagrangian relaxation [9], [10], [23], [24], [25] have been considered, but unfortunately these techniques require some approximations reducing the capacity to ensure a good solution of the problem. The above mentioned methods can only guarantee a local optimal solution and they have inability to obtain feasible solutions for some running problems.

The expert systems and artificial neural networks obtain results depending on the optimality of the rules of knowledge base [2]. Simulated Annealing (SA) was proposed by Zhang and Galiana to the UC problem [26]. The SA algorithm in its traditional form is computationally intensive and difficult to parallelise, and SA can be seen as a particular selection operator in Genetic Algorithms (GAs). Goldberg has shown how GAs can be used to give an exact copy of the behavior of SA in GAs [27]. Genetic algorithms have robustness, efficiency and flexibility to explore large space differing from others search algorithms (e.g., gradient methods, controlled random search, hill-climbing, simulated annealing, etc.) in that the search is conducted using the information of a population of structures instead of a direction or a single structure [5], [6]. GAs have some additional advantages over other search algorithms: are easy to implement, are easily parallelised, they can be used not only as a replacement of traditional optimisation methods, can extend and improve them. For example we can start the optimisation process with a GA and subsequently refine the best solutions found switching to other method (e.g., SA). In this paper the method proposed uses a very simple genetic algorithm. We provide three efficient strategies for solving unit commitment problem, two of them inspired on variance reduction techniques known as Importance Sampling and Expected Value, and the last one using a simple expert system rule based studying the history convergence to decide executing or not additional generations as stopping criteria in the GAs evolution process. Experiences from Scheble et al. [4] show that good results are obtained in the UC problem using hybrid technique of a GA and an expert system. Also see reference [12].

19.2 TEST SYSTEM. FITNESS FUNCTION AND CODING SCHEMES

We have considered as test system the example proposed and tested by Dasgupta [1], which consist of 10 thermal units as short term scheduling where the time horizon period is 24-hours. For each committed unit the involved cost are start-up cost (Si), Shut-down cost (Di) and Average Full Load cost (AFLC) per Mwh, with:

- Start-up cost is modelled for each units, as,

$$S_i = a_i \left(1 - e^{-c_i x_i^t}\right) + b_i$$

  being an exponential function that depends on the number of hours that the unit has been off prior to start up x_i^t. The factor $c_i = 1/\tau_i$ being τ_i time constant of cooling of the unit, b_i is the fuel cost requirement to start, a_i is the fuel cost to start cold unit boiler.

- Shut down cost is considered as a fixed amount for each unit per shut-down.

These state transition costs are applied in the period when the unit is committed or taken off-line respectively. The UC problem considered can be started as follows:

$$\min_{\substack{u_i(t) \\ P_i^t}} J \qquad \text{with} \quad J = F_1 + F_2.$$

The fitness function J is composed for cost function F_1 and penalty function F_2

$$F_1 = \sum_{t=1}^{T_{\max}} \sum_{i=1}^{u_{\max}} u_i^t (AFLC)_i + u_i^t \left(1 - u_i^{t-1}\right) S_i \left(x_i^t\right) + u_i^{t-1} \left(1 - u_i^t\right) D_i.$$

The constraints considered are:

a) The total power generated according to the unit's maximum capacity, such that for each period t,

$$\sum_{i=1}^{u_{\max}} u_i^t P_i^{\max} \geq R^t + L^t$$

where $u_i^t = 0, 1$ (off,on) being $P_i^{\max}$ the maximum output capacity of unit i, L^t is the energy demand and R^t is the spinning reserve, in the period t.

The capacity constraint is referred for each time period. The spinning reserve requirement in this test problem is assumed to be 10% of the expected hourly peak load.

The maximum capacity loading cost assumes that all the units are loaded to their maximum capacity, which is actually an over-production, but this approximate loading condition is acceptable to consider for a test problem, where the target is to compare solutions from different strategies and to evaluate performances by comparing different strategies.

b) Minimum up and minimum down times, implying that unit i must be kept on if it is up for less than the minimum up time, or be kept off if down for less than the minimum down time:

$$\begin{aligned} u_i^t &= 1 \quad \text{if } 1 \leq x_i(t) \leq \tau_{up}, \\ u_i^t &= 0 \quad \text{if } \tau_{down} \leq x_i(t) \leq 0. \end{aligned}$$

The best solution will be the solution doing minimum the cost function F_1 or objective function. Violations of the constraints are penalised through the penalty function F_2. The fitness function can be formulated using a weighted sum of the objective function F_1 and values of the penalty function F_2 based on the number of constraints violated and the extent of these violations.

19.3 THE CODING SCHEME

The generator scheduling problem suggests a convenient binary mapping in which zero denotes the off state and a one represents the on state. A candidate solution is a string whose length is the product of the number of generators and the scheduling period. In the test system the length of a chromosome will be of 240 bits. Two following coding schemes were tested in a similar unit commitment problem by Ma et al. [3]:

a) Coding scheme 1

$$\langle u_1 u_2 u_3 ... u_{\max} \rangle \langle u_1 u_2 u_3 ... u_{\max} \rangle \langle u_1 u_2 u_3 ... u_{\max} \rangle ....$$

status of $u_{\max}$ units at hour 1,..., status of $u_{\max}$ units at hour 24.

b) Coding scheme 2

$$\langle u_{1,1} u_{1,2} u_{1,3} ... u_{1,24} \rangle \langle u_{2,1} u_{2,2} u_{2,3} ... u_{2,24} \rangle \langle u_{3,1} u_{3,2} u_{3,3} ... u_{3,24} \rangle ......$$

status of unit 1 over 24 hours, ..., status of unit $u_{\max}$ over 24 hours.

The results obtained by Ma et al. [3] suggest that coding scheme 1 is much better suited for this problem than coding scheme 2, thus we have considered the coding scheme 1 for solving the test application.

On the other hand we have two major possibilities: Complete Planning Horizon Encode (CPHE) and Incomplete Planning Horizon Encode (IPHE) suggested by Dasgupta [1]. In the Complete Planning Horizon Encode (CPHE) the whole planning horizon time (24-hour) is encoded in a chromosome by concatenation of commitments spaces of all time periods, and performing a global search using GAs. This approach have three well know problems:

- A high number of generating units increase the size of encoding and the amount of space the algorithm has to explore to find good schemata.
- Since the optimal population size is a function of the string length for better schema processing a bigger size of population is needed.
- Such encoding makes appears the epistasis (amount of interdependency among genes encoding chromosome) problem.

If these drawbacks can be manageable due to the size of the problem or to efficient search algorithms, this approach can be the most adequate to solve the problem.

In the Incomplete Planning Horizon Encode (IPHE) the total horizon time (24 hours) is divided into periods (normally 24 periods, 1 hour each) and each chromosome is encoded in the form of a position-dependent genes (bit string) representing the number of power units available in the system, and allele value at loci give the state (on/off) of the units as a commitment decision at each time period.

$$u_1, u_2, ..., u_{\max} \qquad u_i = 0, 1 \ \text{(off,on)} \qquad i = 1, 2, ..., u_{\max}$$

The procedure go forward in time (period by period) retaining the best decision. In each period a fixed number of generations run till the best individual remains unchanged for along time.

This approach have the following drawbacks:

- Since the unit commitment problem is time-dependent can not be guaranteed to find the optimal commitment schedule.
- Lack of diversity due to the procedure itself.

The above drawbacks means that the approach is not robust. In order to make robust this approach it is necessary to create a mechanism capable to save a number of feasible commitment decisions with smaller associated costs at each time period and other mechanism to save better strategies (sequences of commitment decisions from the starting period to the current period with its accumulated cost) to be employed in forward periods.

Considering advantages and disadvantages we decide initially to explore the Complete Planning Horizon Encode (CPHE), the principal reasons were:

1. The optimisation problems with big search space ($> 2^{30}$) need more develop.
2. The Epistasis problem seems to be important is different applications and this test case can be employed to study it in deep.

After a new strategy have been considered based on Variance Reduction Techniques (VRT) in order to help the convergence of the optimisation process. In order to implement the fitness function proposed for the test case, we use the following expression:

$$J = F_1 + F_2,$$

$$F_1 = \sum_{t=1}^{T_{\max}} \sum_{i=1}^{N} \left[\left(u_i^t\right) AFLC_i + \left(u_i^t\right)\left(1 - u_i^{t-1}\right) S_i\left(x_i^t\right) + \left(u_i^{t-1}\right)\left(1 - u_i^t\right) D_i \right],$$

$$F_2 = \left\{ \sum_{j=1}^{r_{\max}} \left[A_j\left(NV_j\right) + B_j\left(SV_j\right) \right] \right\} \left[\frac{\alpha}{\beta} \right]^{\frac{\alpha-\beta}{Abs(\alpha-\beta)}},$$

$$\alpha = \sum_{t=1}^{t_{\max}} \left(D^t + R^t\right) \qquad \beta = \sum_{t=1}^{t_{\max}} \sum_{i=1}^{N} P_i^t$$

where

N = Number of generation units = 10

$T_{\max}$ = Number of hours considered (1 - 24 hrs.)

u_i^t = state (0/1) of generation unit "i" at hour "t"

u_i^{t-1}= state (0/1) of generation unit "i" at hour "t-1"

$AFLC_i$ = Average Full Load Cost per MWh of unit "i"

S_i = Start Up Cost of unit "i"

D_i = Shut-down cost of unit "i"
r_{max} = Number of restrictions considered = 3
A_j = Penalty factor for the Number of violations of the jth restriction
NV_j = Number of Violations of the jth restriction
B_j = Penalty factor for the Sum of violations of the jth restriction
SV_j = Sum of violations of the jth restriction
D^t = Power demand at hour "t"
R^t = Spinning reserve at time "t"
P_i^t= Power produced by unit "i" at time "t"
Abs = Absolute value
$\frac{\alpha}{\beta}$ = Utility Factor (proposed by Dasgupta [1])

19.4 METHOD PROPOSED

Two options were studied in order to increase the convergence of the optimisation process and to obtain better results: - Split the search space - New populations obtained from best individuals

Only the combination of last two options had positive influence on results. In fact they provide the best results we obtain as shown in Fig. 19.1.

19.4.1 Split the search space

The search space was divided in spaces of dimension:

$$2^{20}, 2^{30}, 2^{40}, 2^{50}, ..., 2^{120}, ..., 2^{240}$$

and the optimisation process was conducted from smaller to bigger. The Genetic Algorithm search first the optimum in the 2^{20} space, after search the optimum in the 2^{30} space and so on to the 2^{240}. Since smaller spaces are included into bigger spaces, the successive populations contain individuals more and more similar to the optimum. When the optimisation start the final search in the 2^{240} space the population have many individuals good candidates to became optimum. This technique can be derived form the Variance Reduction Technique (VRT) known as Importance Sampling, because in some sense we can say that the optimisation in the actual space is conducted principally over the most important individuals of the past spaces.

19.4.2 New populations obtained from best individuals

After a number of optimisation steps the GAs process had extremely low convergence. In order to increase this convergence we obtain a complete new population using single a and double mutation (one bit, two bits) of the best individual known and continuing the search. This new population explore the neighbourhood of the best known individual. This technique can be derived from the Variance Reduction Technique (VRT) known as Expected Value, because in some sense we can say the actual best individual have in his bit string the knowledge acquired by the GA process in the past generations. This approach was really efficient only in combination with the splitting of the search space previously described here.

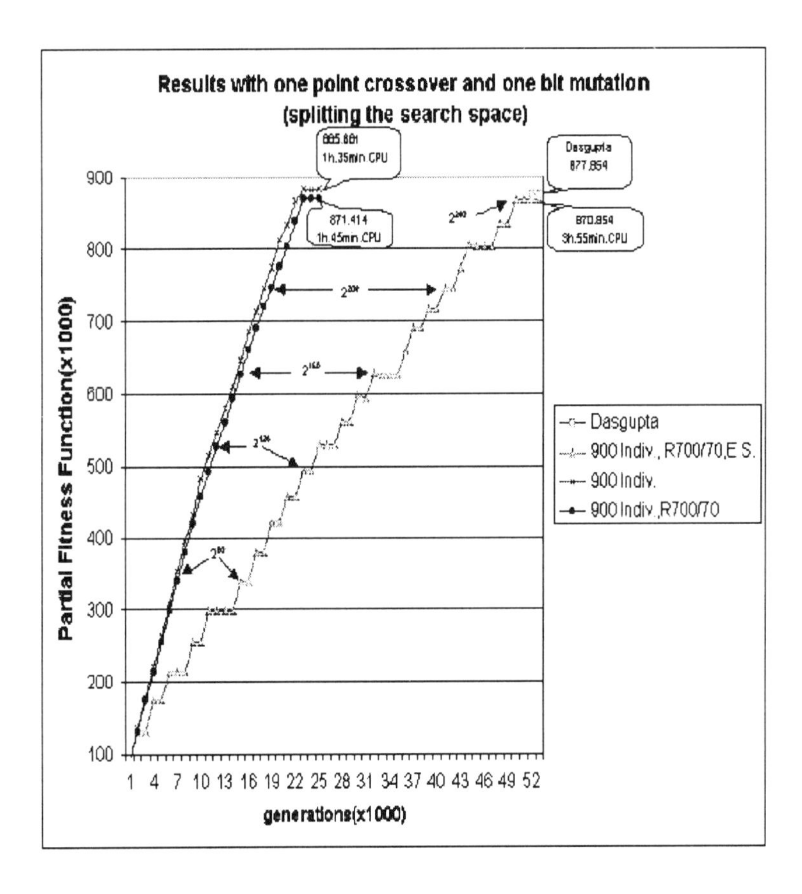

Fig. 19.1 Comparisons of the proposed method.

19.5 FINAL RESULTS

At the end of the process the fitness function is evaluated on total number of variables. The best solution (870,854) was founded using the first routine of one expert system (rule based), now new develop, designed to help the optimisation process (see Fig. 19.2). This routine study the history of convergence in the actual search space and decide continue some few additional generations on it or change to the next step toward following subespace.

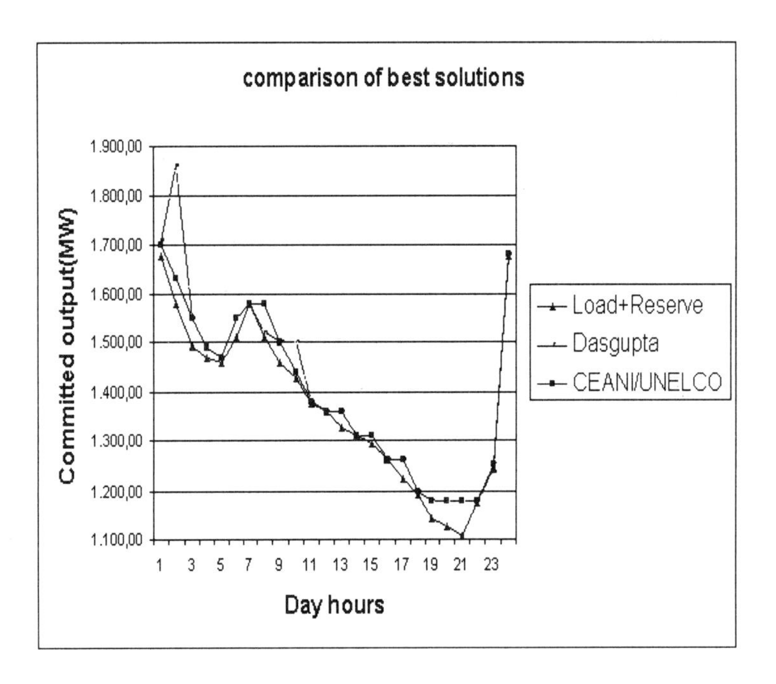

Fig. 19.2 Best results.

19.6 BEST SOLUTION

The best solution founded using the above methods is:

U	S	LR	CO	OE
1	1011111110	1677.850	1700.000	22.150
2	1111111010	1577.800	1630.000	52.200
3	1111101011	1493.850	1550.000	56.150
4	0111101011	1472.000	1490.000	18.000
5	1011101011	1461.650	1470.000	8.350
6	1111101011	1511.100	1550.000	38.900
7	1110101111	1577.800	1580.000	2.200
8	1110101111	1511.100	1580.000	68.900
9	1111101110	1461.650	1500.000	38.350
10	0111101110	1428.300	1440.000	11.700
11	1011110111	1376.550	1380.000	3.450
12	1101110111	1359.300	1360.000	0.700
13	1101110111	1327.100	1360.000	32.900
14	1111110011	1308.700	1310.000	1.300
15	1111110011	1292.600	1310.000	17.400
16	1111110110	1259.250	1260.000	0.750
17	1111110110	1225.900	1260.000	34.100
18	0111110110	1192.550	1200.000	7.450
19	1111100111	1141.950	1180.000	38.050
20	1111100111	1124.700	1180.000	55.300
21	1111100111	1107.450	1180.000	72.550
22	1111100111	1175.300	1180.000	4.700
23	0111010111	1243.150	1250.000	6.850
24	1111011011	1677.850	1680.000	2.150

U = Unit number

S = Unit Status

LR = Load + Reserve

CO = Commited Output

OE = Output Excess

Total Cost = 870,854.00

Parameter Values:

$A_1 = A_2 = A_3 = 100000$

$B_1 = 1000$ $B_2 = 20000$ $B_3 = 10000$

19.7 CONCLUSIONS

The summarize more relevant conclusions are:

- Our Incomplete Planning Horizon Encode proposed gives high efficiency to solve the optimal scheduling.
- The search space high dimension problem can be avoided using splitting-new population techniques, the strategy proposed produces improvement of convergence rate and we have got final results better than others reported.
- One point crossover and one bit mutation operators had good results (Pc=0.7, Pm=0.5)
- Populations of 900 individuals were sufficient to accomplish the optimisation
- Epistasis problem impact reported by Dasgupta have been minimised with the above method
- Best results were obtained without the Utility Factor

Acknowledgments

This work was partially financed by EC under the BRITE BRRT-CT-97-5034.

Appendix

The following data were reported at Dasgupta [1] (Time horizon period = 24 hours):

Un.	MCM	MUt	MDt	IS	SCa_i	SCb_i	SCc_i	SDC	AFLC
1	60	3	1	-1	85	20.588	0.2	15	15.3
2	80	3	1	-1	101	20.594	0.2	25	16
3	100	4	2	1	114	22.57	0.2	40	20.2
4	120	4	2	5	94	10.65	0.18	32	20.2
5	150	5	3	-7	113	18.639	0.18	29	25.6
6	280	5	2	3	176	27.568	0.15	42	30.5
7	520	8	4	-5	267	34.749	0.09	75	32.5
8	150	4	2	3	282	45.749	0.09	49	26.0
9	320	5	2	-6	187	38.617	0.130	70	25.8
10	200	5	2	-3	227	26.641	0.11	62	27

Un. = Unit number

MCM = Maximum Capacity (Mw)

MUt = Minimum Up time

MDt = Minimum Down time

IS = Initial Status

SCa_i =Startup cost constant a_i
SCb_i =Startup cost constant b_i
SCc_i =Startup cost constant c_i
SDC = Shut-Down cost
AFLC = Average Full Load Cost
(-) indicates unit is down for hours and positive otherwise

REFERENCES

1. D. Dasgupta, Optimal Scheduling of thermal Power Generation using Evolutionary Algorithms. Evolutionary Algorithms in Engineering Applications. Editors: D. Dasgupta.
2. S.O. Otero and M.R. Irving, A genetic algorithm for generator scheduling in power systems. Electric Power.
3. X. Ma, A.A. El-Keib, R.E. Smith and H. Ma, A genetic algorithm based approach to thermal unit commitment of electric power systems. Electric Power Systems Research, Vol. 34, pp. 29–36, 1995, Elsevier.
4. G.B. Scheble and T.T. Mai Feld, Unit commitment by genetic algorithm and expert system. Electric Power Systems Research vol. 30, pp. 115–121, 1994, Elsevier.
5. D.E. Goldberg, Genetic Algorithms in Search, Optimisation and Machine Learning. Addisson-Wesley, Reading MA, 1989.
6. J.H. Holland, Adaptation in natural and artificial systems. University of Michigan Press, Ann Arbor, 1975.
7. E. Khodaverdian, A. Bramellar and R.M. Dunnet, Semirigurous thermal unit commitment for large scale electrical power systems. Proc. IEEC Vol. 133 N°4, 1986, pp. 157–164.
8. C.L. Chen and S.C. Wang, Branch and bound scheduling for thermal units. IEEE Trans. Energy Conv. Vol 8 N°2, 1993, pp. 184–189.
9. J.F. Bard, Short term scheduling of thermal electric generators using Lagrangian relaxation. Operations Research Vol 36 N°5, 1988, pp. 756–766.
10. F. Zhuang and F.D. Galiana, Towards a more rigurous and practical unit commitment by Lagrangian relaxation. IEEE Trans. PWRS Vol 3, 1988, pp. 763–770.
11. A. Merlin and P. Sandrin, A new method for unit commitment at electricité de France. IEEE Trans Vol 102, 1983, pp. 1218–1225.
12. S. Li and S.H. Shahidehpour, Promoting the application of expert systems in short term unit commitment. IEEE Trans. PWRS, 1993, pp. 287–292.

13. H.S. Musoke, S.K. Biswas, E. Ahmed, P. Cliff and W. Kazibwe, Simultaneous solution of unit commitment and dispatch problems using artificial neural networks. Int. J. Elect. Power Energy Syst. Vol. 15 N°3, 1993, pp. 193–199.

14. G. Fahd and G.B. Sheblé, Unit commitment literature synopsis. IEEE Trans. Power Syst., 1994, 128–135.

15. C.J. Baldwin, K.M. Dale and R.F. Dittrich, A study of the economic shutdown of generating units in daily dispatch. Trans. AIEE Part 3, 78, 1959, 1272–1284.

16. H.H. Happ, R.C. Johson and W.J. Wright, Large scale hydrothermal unit commitment method and results. IEEE Trans. Power Appar Syst. PAS-90, 1971, pp. 1373–1383.

17. C.K. Pang, G.B. Sheblé and F. Albuyeh, Evaluation of dynamic programming based methods and multiple area representation for thermal unit commitment. IEEE Trans. Power Appar. Syst., PAS-100, 1981, pp. 1212–1218.

18. P.P.J. Van den Bosch and G. Hondred, A solution of the unit commitment problem via decomposition and dynamic programming. IEEE Trans. Power Appar. Syst., PAS-104, 1985, 16841690.

19. W.J. Hobs, G. Hermon, S. Warner and G.B. Sheblé, An enhanced dynamic programming approach for unit commitment, IEEE Trans. Power Syst., 3, 1988, 1201- 1205.

20. T.S. Dillon, Integer programming approach to the problem of optimal unit commitment with probabilistic reserve determination. IEEE Trans. Power Appar. Syst., PAS-97, 1978, 2154–2164.

21. A.I. Cohen and M. Yoshimura, Abranch-and-bound algorithm for unit commitment. IEEE Trans. Power Syst., PAS-102, 1983, pp. 444–451.

22. S.K. Tong and S.M. Shahidehpour, Combination of Lagrangian relaxation and linear-programming approaches for fuel constrained unit commitment. IEE Proc. C, 136, 1989, pp 162–174.

23. A.I. Cohen and S.H. Wan, A method for solving the fuel constrained unit commitment problem. IEEE Trans. Power Syst., PWRS-2 (3), 1987, 608–614.

24. S. Virmani, E.C. Anderian, K. Imhof and S. Mukherjee, Implementation of a Lagrangian relaxation based unit commitment problem, IEEE Trans. Power Syst., 4 (4), 1989, pp. 1373–1379.

25. S. Kuloor, G.S. Hope and O.P. Malik, Environmentally constrained unit commitment. IEEE Proc. C. 139, 1992, pp. 122–128.

26. F. Zhang and F.D. Galiana, Unit commitment by simulated annealing. IEEE Trans. Power Syst., 5 (1), 1990, pp. 311–317.

27. D.E. Goldberg, A note on Boltzmann tournament selection for genetic algorithms and population-oriented simulated annealing. TCGA Rep. N°90003, Clearinghouse for Genetic Algorithms, University of Alabama, Tuscalosa, AL, 1990.

28. A.I. Cohen and M. Yoshimura, A branch-and-bound algorithm for unit commitment. IEEE Trans. on Power Apparatus and Systems. PAS-102(2)444-449. Febraury, 1983.

20 Efficient Partitioning Methods for 3-D Unstructured Grids Using Genetic Algorithms

A.P. GIOTIS[1], K.C. GIANNAKOGLOU[1], B. MANTEL[2] and J. PÉRIAUX[1]

[1] National Technical University of Athens, Greece

[2] Dassault Aviation, France

Abstract. New partitioning methods for 3-D unstructured grids based on Genetic Algorithms will be presented. They are based on two partitioning kernels enhanced by acceleration and heuristic techniques to effectively partition grids, with reduced CPU cost. Five variants are presented and assessed using three 3-D unstructured grids.

20.1 INTRODUCTION

"Real" world problems which are modeled through partial differential equations require huge meshes and CPU-demanding computations to reach a converged numerical solution. Multiprocessing is a way to overcome the high computing cost. Using distributed memory computing platforms, data need to be partitioned to the processing units, so that evenly work-loaded processors with minimum exchange of data handle them concurrently.

In this paper, we are dealing with 3-D unstructured grids with tetrahedral elements, widely used in Computational Fluid Dynamics (CFD). Since there is always a mapping of any grid to a graph (see Fig. 20.1 (a)), the proposed partitioners are general. Working with the equivalent graph, grids with other types of elements (quadrilaterals, hexaedra, etc) or even hybrid grids can be readily partitioned. Apart from CFD, the grid or graph partitioning problem is recurrent in various disciplines, structural analysis, linear programming, etc. In the literature, the UGP problem is handled using greedy methods, coordinate-based bisections, inertial methods or recursive spectral bisection techniques, as discussed in [5]. Effective unstructured grid partitioners based on Genetic Algorithms (*GAs* [1]) have been presented in [4] and [3], where two partitioning kernels (in what follows they will be refered to as *BITMAP* and *FIELD*) have been introduced.

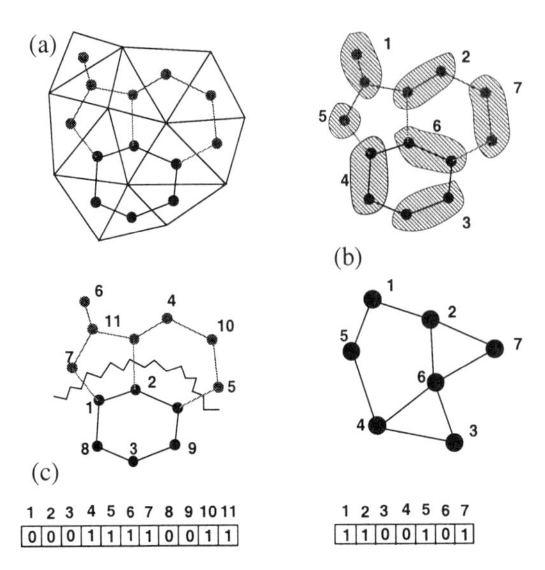

Fig. 20.1 (a) A simple 2-D unstructured grid and its equivalent graph, (b) a one-step coarsening procedure (and binary encoding at the coarse level), (c) Bisection encoding through the *BITMAP* method.

Based on the aforementioned kernels and heuristic methods, new partitioning algorithms have been devised. All of them use the recursive approach and the single-pass multilevel scheme. They employ new coarsening algorithms and, by appropriately scheduling the UGP kernels, they proved to be very effective partitioners. A noticeable advantage of the new partitioners is that they are faster, by at least one order of magnitude, than the methods presented in [4] and [3]. The new UGP methods will be demonstrated in the partitioning of 3-D unstructured meshes, where tetrahedron-based partitions will be sought for.

20.2 BASIC PARTITIONING TOOLS

In what follows, we will be dealing only with the equivalent graph of the grid. So, the terms "nodes" and "edges" will refer to graph nodes (i.e. tetrahedral elements) and graph edges (i.e. triangular faces separating adjacent elements), respectively.

20.2.1 Recursive bisections and the multilevel scheme

Starting from the graph to be partitioned G^0 (Fig. 20.1 (a)), a multilevel partitioner creates a sequence of graphs of smaller size, through the repetitive application of a coarsening procedure. During each coarsening step, graph nodes are clustered into groups, as in Fig. 20.1 (b). The total number M of coarsening steps is determined by the G^0 size and the desirable final graph size. With the same initial and (approximate) final graph sizes, different coarsening algorithms lead to different values of M. It is recommended that G^M should have about 50–150 nodes. During the coarsening procedure, weights are assigned to graph nodes and edges. At G^0, the weights for all graph nodes or edges are set to 1. Proceeding from level G^m to G^{m+1}, the weight of each newly formed graph node equals the sum of the weights of the constituent nodes. For the edges, new weights need to be computed due to the coarsening by taking into account the number of graph edges linking adjacent groups.

Edge-based and node-based coarsening procedures have been employed and tested in various test cases. Some of them are described, in detail, in [2]. Some other have been programmed using similar considerations, but they will be omitted in the interest of space. In general, edge-based schemes outperform the node-based ones. The most important features of the best coarsening procedure tested so far will be described below.

All G^m nodes are visited one-by-one, in random order. The edges emanating from a node and pointing to another G^m node are sorted by their weights. The edge having the maximum weight is selected and the two supporting nodes collapse and form a new G^{m+1} node with weight equal to the sum of the weights of the two G^m nodes. At the end, the remaining G^m nodes are directly upgraded to G^{m+1} ones, with the same weights. This method tends to maximize the weights of the collapsed graph edges and consequently to minimize the weights of the active G^{m+1} edges. Since the sought for separator will cut active G^{m+1} edges, separators of reduced length will be produced.

After series of tests, a major conclusion was that the coarsening procedure considerably affects the quality of the separator as well as the partitioning cost. Inappropriate coarsening is a serious threat to the partitioner's effectiveness, as it leads to "poor" coarse graphs. The optimum separator of a "poor" G^M is usually far from the optimum separator of G^0 and CPU-time consuming refinements at the finer graph levels should be employed. Unfortunately, the fast partitions of the coarser graphs cannot outweight the slow refinements at the finer ones.

20.2.2 Cost function

The cost function used for a single bisection should combine the requirements of equally loaded subdomains and of minimum interface. Assume that the bisection of a graph with N nodes gives rise to two subdomains with N_1 and N_2 nodes ($N_1 + N_2 = N$), respectively. Let N_{ie} be the number of graph edges cut by the separator. A cost function that scalarizes two objectives into a single one, reads

$$C_f(N_1, N_2, N_{ie}) = \frac{|N_1 - N_2|}{\sqrt{N_1 N_2}} + w\frac{N_{ie}}{N_{ge}} \tag{20.1}$$

where N_{ge} is the total number of graph edges and w is a weighting parameter. The optimum partition is, of course, the one that minimizes $C_f(N_1, N_2, N_{ie})$.

20.2.3 The *BITMAP* bisection method

The *BITMAP* method [4] is fully compatible with the binary encoding often used in *GAs*. It is based on the direct mapping of each graph node to a single gene or bit in the chromosome, as depicted in Fig. 20.1 (c). Genes marked with 0 or 1 denote graph nodes that belong to the one or the other partition. Concerning the genetic operations, a standard two-point crossover and a modified mutation operator [4] are employed. In each generation, the modified mutation operator takes into account the graph connectivity and the actual shape of the separator by assigning increased mutation probabilities to the graph nodes which are closer to the separator.

Tests have shown that a one-way mutation (only 0's turning to 1's, or vice-versa, are allowed during each generation) with variable probability enhance the method characteristics. Elitism is also used along with the correction scheme that applies whenever small groups of graph nodes fully surrounded by nodes belonging to another subdomain are found. The parent selection operations are carried out in part by linear fitness ranking and in part by probabilistic tournament. The cost function used in *BITMAP* is the one described in eq. (20.1). For the computation of the cost function at any level $m \neq 0$, the graph nodes' and edges' weights are considered.

20.2.4 The *FIELD* bisection method

The so-called *FIELD* method requires the mapping of the graph onto a parametric space, which could be carried out using different techniques. Two of them are described below.

Method P1: In the parametric space, each G^0 graph node is given the Cartesian coordinates of the barycenter of the corresponding grid element. At the coarser levels $m, (m \neq 0)$ the average Cartesian coordinates of the "constituent" elements are assigned to any graph node.

Method P2: Random $0 \leq \Phi, \Psi, \Omega \leq 1$ coordinates are firstly assigned to each graph node. Then, a Laplacian filter, is applied to smooth down the random mapping, without enforcing fixed (Φ, Ψ, Ω) values at the "boundary" nodes. The filter is iterative and each node is given the average of the adjacent nodes' (Φ, Ψ, Ω) values. Only a few iterations should be carried out, since the full convergence leads to a uniform (Φ, Ψ, Ω) field for the entire graph. The number of iterations required should be carefully chosen. Insufficient smoothing often results to multiply connected partitions.

In both parameterizations, the (Φ, Ψ, Ω) coordinates of any node are such that $0 \leq \Phi, \Psi, \Omega \leq 1$, defining thus a cubic space which encloses the graph.

In the parametric space (Φ, Ψ, Ω), two point-charges A and B are allowed to float within predefined limits, creating potential fields $F(\Phi, \Psi, \Omega)$ around them. At any point P in the (Φ, Ψ, Ω) space, the local potential value F_P is computed by superimposing scalar contributions from the charges. Different laws can be used. Among them, two laws with good physical reasoning are

$$F_P = F(\Phi_P, \Psi_P, \Omega_P) = e^{k_A r_{P_A}} - e^{k_B r_{P_B}} \tag{20.2}$$

or

$$F_P = F(\Phi_P, \Psi_P, \Omega_P) = \frac{k_A}{r_{P_A}^2} - \frac{k_B}{r_{P_B}^2} \tag{20.3}$$

where

$$r_{P_M} = \sqrt{(\Phi_P - \Phi_M)^2 + (\Psi_P - \Psi_M)^2 + (\Omega_P - \Omega_M)^2}, \qquad M = A, B. \tag{20.4}$$

Either k_A or k_B can be defined arbitrarily. So, $k_A = -1$ and a negative k_B value is sought for. Summarizing, the free-parameters are $(\Phi_A, \Psi_A, \Omega_A, \Phi_B, \Psi_B, \Omega_B, k_B)$ and these are controlled by the *GAs*. Standard parent selection, crossover and dynamic mutation operators are employed. Elitism is also used.

For any pair of point charges, a single potential value is computed at any graph node. Potential values are sorted and the graph nodes above and below the median are assigned to the first and second subdomain, respectively. At G^0, this is straightforward and satisfies the load balance requirement. In this case, the first term in the r.h.s. member of eq. (20.1) is automatically zeroed. At any other level, graph nodes' weights are involved in the computations and eq. (20.1) should be considered in its full expression.

20.3 ACCELERATION TECHNIQUES

Acceleration-refinement techniques have been devised in order to diminish the partitioning cost. Profound economy in CPU time is achieved when huge meshes are to be partitioned. In the context of the single-pass multilevel scheme, the acceleration tools are necessary during the bisection of the G^0 graph and, depending on the case, during some of the $G^m(m \ll)$ bisections.

The random initialization inherent in the *BITMAP* method inevitably creates badly organized distributions of 0's and 1's over the graph nodes. The modified genetic operators used in *BITMAP* help reduce the computing cost but the gain is not enough. For further acceleration, a technique which is well suited to *BITMAP* is the zonal refinement. This is activated whenever provisory, though distinct, subdomains have been formed and only local changes to the separator shape are expected. The concept is to encode and handle only a narrow zone of the closer to the interface graph nodes. Only these nodes are binary encoded, by forming chromosomes of reduced length. The economy in CPU time reflects the much lower CPU cost per generation (chromosomes are much shorter) and the small-scale alterations this method tolerates. Changes close to the separator are only allowed, without affecting the remaining parts of the graph which have already been assigned to subdomains.

In the *FIELD* method, each bisection is based on a potential field created by point charges. The potential field is always continuous and defines more or less (depending on the parameterization) distinct subdomains, even during the very first generations. Consequently, compared to *BITMAP* fewer generations are required to reach the converged solution but the cost of a single generation increases by the cost for computing the potential field and sorting nodal values. The number of generations required for convergence in the *FIELD*-based methods should be reduced through the regular shrinkage of the search space for all free-parameters. The new search space is always centered upon the location of the best current solution and extends symmetrically in each direction. It is smaller than the previous one by a user-defined factor.

An alternative, less effective way to accelerate a *FIELD*-based partition is through zonal refinement, similar to that used in *BITMAP*. The concept underlying this technique is based on the identification of the "negotiable" graph nodes. These are nodes which in the sorted list of nodal potential values, lie around its median. Only the negotiable nodes are allowed to change owner during the forthcoming generations. Their number should be either constant or a fixed percentage of the total number of nodes at this level.

20.3.1 Heuristic partitioning techniques

Apart from the acceleration-refinement techniques, a heuristic refinement method is additionally used. It can be used at any graph level, either as a supplement to the genetic optimization or as a substitute for it. It cannot create a partition from scratch, but effectively refines existing partitions. The proposed heuristic method in the form of a pseudocode is given below.

```
* Create a list with the graph nodes in contact with the separator
* According to the current imbalance, identify the preferable direction
   of node movement
iter = 0
do { changes = 0
       for (each list entry) {
              Extract a node from the tail
              if (node movement:
                     is towards desired direction
                     and the imbalance excess does not increase
                     and the interface length does not increase
              ) then
                     * Proceed with the node movement
                     * Update desired direction
                     * Update balance and interface
                     * Add new interfacial nodes to the head
                     changes = changes + 1
               else put node back to the head of the list
               endif
       }
       iter = iter+1
} while (list not empty and iter<maxiter and changes>0)
```

20.4 THE PROPOSED *UGP* ALGORITHMS

Fig. 20.2 shows a general flowchart for any UGP algorithm based on a single-pass multilevel scheme. This flowchart is representative of any partitioner described below, regardless the specific tools used to effect partitioning at each graph level. The first two proposed methods, already discussed in [3], are based on the standard kernels (*BITMAP* and *FIELD*) indiscriminately applied at any level. They will be referred to as *S–BITMAP* and *S–FIELD*. Unfortunately, both methods become too slow at the finest graph levels, and especially at G^0. The CPU cost per generation in *S–FIELD* is higher than in *S–BITMAP* though the latter requires less generations to reach the same fitness score.

A hybrid partitioner (*HYBRID*) was also presented in [3], where the *FIELD* kernel was applied for the partitioning of the coarser graphs, followed by the *BITMAP* one at the finer graph levels. Since G^M is handled by the *FIELD* method, good initial guesses are obtained that facilitate a lot the use of *BITMAP* in the subsequent finer graphs. The *HYBRID* method exploits the capabilities of *FIELD* to instantly create almost distinct subdomains and the excellent refinement properties of *BITMAP*.

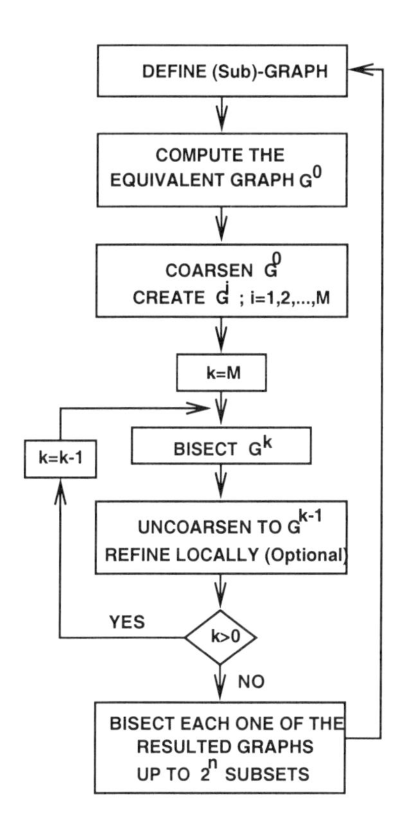

Fig. 20.2 Flowchart: Recursive bisections and the multilevel scheme.

The new two variants proposed in this paper will be referred to as *M–BITMAP* and *M–FIELD*, where M stands for Modified. In *M–BITMAP* the *BITMAP* kernel undertakes the partitioning of G^M. On completion of the G^M bisection, the best partition is directly uncoarsened to G^{M-1} where it is further improved using the aforementioned heuristic refinement. Its role is to smooth down the "wavy" parts of the interface. The same procedure is repeated up to G^0. A remark concering the *M–BITMAP* is that it may sometimes create multiple-connected subdomains. This occurs during the partitioning of G^M through the *BITMAP* method and cannot be recovered afterwards if heuristics are only applied.

Likely *M–BITMAP*, the *M–FIELD* was devised and tested. *M–FIELD* uses the *FIELD* method to partition G^M and then uncoarsens to the subsequent levels using heuristic refinement. In general, *M–FIELD* overcomes the problem of multiple-connected subdomains. Tests have shown that the use of the P2 parameterization in *M–FIELD* is favored.

20.5 APPLICATIONS - ASSESSMENT IN 3-D PROBLEMS

The five UGP variants discussed are evaluated in the partition of 3-D unstructured grids into 2^n subdomains. In all cases, the population size was 50 and the algorithm was considered as converged after 8 non-improving generations. The probabilities of mutation and crossover were equal to 0.15–0.5% and 90% respectively. The correction task applied on the 10% of the population and the upper bound for this correction was 20% of the G^0 nodes.

In the *FIELD*-based partitioners, eight bits per variable are used. The generations done per bisection are in the range of 150–500 for the *BITMAP*-based tools and 10–200 for the *FIELD*-based ones. The weighting factor w depends on the size of the graph to be partitioned. A recommended expression for w, which is the outcome of many *UGP* tests using graphs of various sizes, is given below

$$w \approx 9\ln(N_1 + N_2) - 55. \tag{20.5}$$

The first case deals with the unstructured grid modeling a 3-D compressor cascade. It consists of 111440 tetrahedra and was partitioned into $2^5 = 32$ subdomains. Tabulated results rather than views of the partitioned grids will be shown. For 2^n partitions, the parameter BAL is defined as $BAL = 2^n \cdot (larger\_partition\_size)/(graph\_size)$ while DD measures the number of Distict Domains that have been created. Here BAL, DD and the CPU cost in sec (on an Intel Pentium II 400MHz processor) are given. By examining the results shown in Table 20.1 (a), the superiority of the Modified versions of *BITMAP* and *FIELD* is obvious. They all provide simple-connected partitions (the fact that *M–BITMAP* and *M–FIELD*/P2 give 33 rather than 32 subdomains can be neglected) with a considerably smaller interface length than any other partitioner. Comparing *M–BITMAP* with *M–FIELD*, *M–FIELD* gives an interface length that is about 15-20% shorter than that of *M–BITMAP*. In this case, the graph space parameterizations P1 and P2 lead to separators of about the same size. Both *M–FIELD* variants are extremely fast. The standard versions result to much longer interfaces; besides *S–FIELD* provides a lengthy interface that defines 83 instead of 32 distinct domains. It is interesting also to note that the interface provided by *FIELD* is quite lengthy even during the first bisection. *HYBRID* outperforms both standard versions, giving a better interface with reduced CPU cost but its performance is still far from that of the Modified variants.

The second grid examined is formed around a 3-D isolated airfoil. This grid occupies a much larger physical space than the previous grid, the major part of which is filled in a very coarse manner. This is of interest as far as *FIELD*-based methods using the P1 parameterization are of concern. The grid consists of 257408 tetrahedra and has also been partitioned into 32 subdomains. Results are given in Table 20.1 (b). In general, the conclusions drawn from this study are similar to previous ones. The Modified variants perform better and faster than the standard or the *HYBRID* version. In this case, the difference between *M–FIELD* and *M–BITMAP* is more pronounced, not only due to the lengthy interface *M–BITMAP* produces but also due to the fact that *M–BITMAP* creates 43 instead of 32 distinct subdomains. It is also interesting to note that *M–BITMAP* provides more unbalanced partitions than any other method used. Finally, *HYBRID* is still better than the standard variants but, in this case, the qualities of the separators obtained by both methods are not far apart.

The third 3-D grid examined is formed around a complete aircraft. It consists of 255944 tetrahedra and a close view is shown in Fig. 20.3. Compared to the grid used in the previous case which was of about the same size, this grid is a pure unstructured grid, while the other was formed by stacking a real 2-D unstructured one. Results obtained by partitioning this grid through various methods are shown in Table 20.1 (c).

Table 20.1 Partitioning of (a) the 3-D compressor cascade grid, (b) the 3-D isolated airfoil grid and (c) the grid around an aircraft.

	Interface for 2^n subdomains					For n=5		
Partitioner	n=1	n=2	n=3	n=4	n=5	BAL	sec	DD
A–BITMAP	248	1146	1883	3082	6605	1.00	100.6	35
A–FIELD/P1	545	1764	3205	4870	7199	1.00	246.8	83
HYBRID	300	1139	2005	3295	5728	1.00	80.5	34
M–BITMAP	194	818	1533	2734	4733	1.11	13.4	33
M–FIELD/P1	180	719	1425	2361	4056	1.00	10.2	32
M–FIELD/P2	194	752	1484	2428	4064	1.03	8.7	33 (a)
A–BITMAP	2499	4420	7688	10759	14672	1.00	494	71
HYBRID/P1	1502	4081	7305	11073	14802	1.00	360	72
HYBRID/P2	1492	3190	6708	9951	13596	1.00	322	71
M–BITMAP	928	2233	3751	5767	8135	1.19	29.8	43
M–FIELD/P1	884	1875	2994	4544	6858	1.00	22.8	32
M–FIELD/P2	750	1642	2749	4497	6774	1.00	19.4	32 (b)
HYBRID/P2	3045	9594	15291	21215	29056	1.00	1688	71
M–BITMAP	1975	5720	92673	13408	18809	1.12	33.9	37
M–FIELD/P1	2021	4707	7365	11345	16278	1.00	22.4	32
M–FIELD/P2	1790	4519	7446	11639	16597	1.00	21.7	32 (c)

From Table 20.1 (c) we conclude that *M–FIELD* is better than any other separator, gives exactly 32 distinct domains with low CPU cost. *M–BITMAP* gives a slightly larger interface, slightly unbalanced partitions and slightly increased number of distinct subdomains. In general, both *M–BITMAP* and *M–FIELD* perform much better than *HYBRID*. The CPU cost of *HYBRID* is about 50 times that of the Modified versions.

20.6 CONCLUSIONS

Partitioning methods for 3-D unstructured grids, based on two standard kernels, have been presented. They all operate on the equivalent graph of the grid, so they are in fact graph partitioners appropriate for any grid type. They use Genetic Algorithms to minimize a properly defined cost function, either through direct (*BITMAP*) or indirect (*FIELD*) modeling.

The multi-level scheme is necessary for the proposed methods to efficiently undertake the partitioning of heavy industrial meshes. It uses a coarsening procedure which should be carefully chosen as it affects the quality of the partitions. It is recommended that the coarser graph G^M size be of about 50–150 nodes.

Some more remarks on the standard kernels follow. In general, *BITMAP* tends to create disjoint or slightly unbalanced partitions, requiring more generations than *FIELD* to converge. Even if *FIELD* converges faster, a *FIELD* generation requires more CPU time than a *BITMAP*

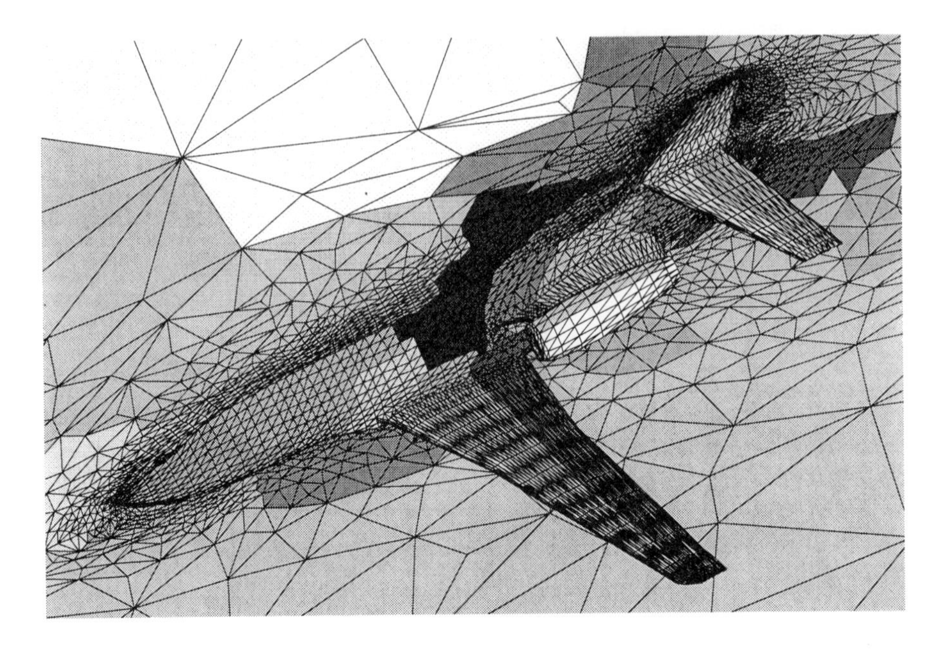

Fig. 20.3 A close view of the grid around an aircraft (Courtesy of Dassault Aviation).

one. The local refinement of the interpartition boundary created by *FIELD* is really necessary in order to improve its shape; this should be carried out at the post-processing level through heuristics. *M–FIELD* and *M–BITMAP* outperform any other partitioner. They produce short interface length, balanced partitions and avoid the formation of disjoint subsets. Among *M–FIELD* and *M–BITMAP*, tests have shown that *M–FIELD* is slightly better. Typical grids of the order of 260.000 tetrahedra can be partitioned in 32 subdomains within 20–25 secs on an Intel Pentium II 400MHz processor.

REFERENCES

1. Goldberg D.E. *Genetic Algorithms in search, optimization & machine learning*. Addison-Wesley, 1989.

2. Karypis G. and Kumar V. A fast and high quality multilevel scheme for partitioning irregular graphs. *SIAM Journal on Scientific Computing*, 1998.

3. K.C. Giannakoglou and A.P. Giotis. Unstructured 3-d grid partitioning methods based on genetic algorithms. ECCOMAS 98, John Wiley & Sons, 1998.

4. A.P. Giotis and K.C. Giannakoglou. An unstructured grid partitioning method based on genetic algorithms. *Advances in Engineering Software*, 1998.

5. Van Driessche R. and Roose D. Load balancing computational fluid dynamics calculations on unstructured grids. *AGARD*, R-807, 1995.

21 Genetic Algorithms in Shape Optimization of a Paper Machine Headbox

J.P. HÄMÄLÄINEN[1], T. MALKAMÄKI[2] and J. TOIVANEN[2]

[1] Valmet Corporation, Paper and Board Machines
P.O. Box 587, FIN-40101 Jyväskylä, Finland
E-mail: Jari.P.Hamalainen@valmet.com

[2] University of Jyväskylä
Department of Mathematical Information Technology
P.O. Box 35 (MaE), FIN-40351 Jyväskylä, Finland
E-mail: Timo.Malkamaki@mit.jyu.fi, Jari.Toivanen@mit.jyu.fi

21.1 INTRODUCTION

The headbox is located at the wet end of a paper machine. Its function is to distribute fibre suspension (wood fibres, filler clays and chemicals mixed in water) in an even layer, across the width of a paper machine. The fluid (fibre suspension) coming from a pump enters the first flow passage in the headbox, a header, which distributes the fluid equally across the machine width via a bank of manifold tubes. A controlled amount of the fluid is recirculated at the far end of the header to provide flow control along the header length.

A traditional designing problem in a paper machine headbox is tapering of the header such that the flow distribution is even across the full width of a paper machine. The problem has already been studied since 50's [1], [13], [16]. At that time CFD tools were not available and the models for flows in the header were simple and one-dimensional. Clearly, when the header design is based on a fluid flow model, the resulting design can only be as good as the model is. Thus, to answer today's headbox functioning demands, more sophisticated designing based on more accurate fluid flow modeling is needed.

Our approach is to formulate designing of the tapered header as a shape optimization problem based on today's CFD modeling and simulation capability. The cost function to be minimized describes how even the flow distribution from the header is. We have chosen it to be the square of the standard deviation of the outlet flow rate profile. Fluid flows in the header are solved by Valmet's own CFD software. A header model [8], [11], [12] is based on well-known fluid dynamics equations, i.e., the Reynolds averaged Navier-Stokes equations together with a turbulence model.

Tapering of the header is described with the help of a few design parameters to be optimized. Due to manufacturing reasons a part of design variables can have only discrete values. Without discrete variables similar shape optimization problems have been solved using gradient based methods; see, for example, [9], [10], [15]. Here, we perform the optimization using genetic algorithms (GAs) with a steady-state reproduction [3], real coded genes [14] and the roulette wheel selection [5].

The rest of this paper is organized as follows: In Section 21.2, we introduce the flow problem and the geometry of a header. The shape optimization problem is considered in Section 21.3. The used genetic algorithms are described in detail in Section 21.4. Numerical experiments have been performed with the described approach in Section 21.5. In the last Section 21.6, we give some concluding remarks and suggestions.

21.2 FLOW PROBLEM IN A HEADER

The header is the first component in the paper machine headbox. A two-dimensional illustration of a header is show in Fig. 21.1. The depth of the header is chosen to be 0.5 metres. The fibre suspension containing mixture of water (99 %) and wood fibres (1 %) flows into the header through the boundary Γ_{in}. The inflow velocity profile is assumed to be fully-developed parabolic profile. A major part of suspension goes to a tube bundle which begin from Γ_{out}. A small part of flow, say 10 %, is recirculated through the boundary Γ_{r}. Also, the recirculation profile is assumed to be parabolic.

The flow is incompressible and turbulent and, thus, the incompressible Reynolds-averaged Navier-Stokes equations with k-ε turbulence model is used to model the flow. The three-dimensional problem is reduced to two-dimension problem by using depth averaging. At the main outflow boundary, normal stresses are proportional to velocity: $\sigma \cdot n = Cu_n^2$, where C is a constant and u_n is a normal velocity on Γ_{out}. The outflow boundary condition results from the homogenization of the tube bundle (see [11] and [15]) and Cu_n^2 is approximately head losses in the tube bundle.

We have chosen the header's dimensions in metres to be as follows: $H_1 = 1.0$, $L_1 = 1.0$, $L_2 = 10.0$, $L_3 = 0.5$ and $H_2 = 0.2$. The average inlet and recirculation velocities are 4.44 and 2.2 m/s, respectively.

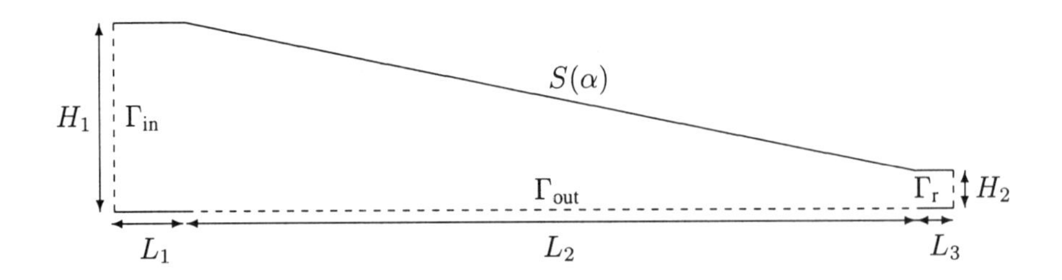

Fig. 21.1 The geometry of header and the related notations.

For the solution of the Navier-Stokes equations in a computer, they are discretized by using stabilized FEM (SFEM) [4] and bilinear elements for the velocity components and pressure. This leads to a large system of nonlinear equations. The solution is obtained by forming successive Newton linearizations of equations and solving them by using a direct solver based on the LU decomposition.

21.3 SHAPE OPTIMIZATION OF A HEADER

In order the produced paper by a paper machine to be of equal thickness, the outflow should be as equal as possible across the width of the paper machine. By changing the shape of the header the outflow profile can controlled. The problem is to find the shape which gives an even outflow. This shape optimization problem is solved iteratively by starting from the initial design [6]. For each design the outflow profile is obtained as the solution of the flow model in the header.

In our shape optimization problem, the shape of the back wall $S(\alpha)$ and the lengths of the inflow and recirculation boundaries (H_1 and H_2) determined by the design variable vector $\alpha = (\alpha_1\ \alpha_2\ \cdots\ \alpha_m)^T$ are allowed to change. This means that only the lengths L_1, L_2 and L_3 are fixed. It is beneficial in the manufacture if standard sized tubes can be used for the inflow and recirculation pipes of the header. Therefore, the lengths H_1 and H_2 may have only discrete values. To be more precise, we have chosen the steps of H_1 and H_2 to be 0.02 and 0.01 metres, respectively. They are not the actual sizes of existing standard tubes, but they are assumed to be such in this test example. This makes the optimization problem more complicated, since typically the gradient based optimization methods cannot handle discrete variables.

The cost function measuring the quality of the outflow flow profile is

$$J(\alpha) = \int_{\Gamma_{\text{out}}} (u_{\text{out}}(x) - u_{\text{c}})^2\, dx, \tag{21.1}$$

where $u_{\text{out}}(x)$ is the simulated velocity profile on Γ_{out} and u_{c} is the desired constant velocity. The shape optimization problem is equivalent to the minimization problem

$$\min_{\alpha \in U_{\text{ad}}} J(\alpha), \tag{21.2}$$

where U_{ad} is the set of admissible designs. In our case, U_{ad} is not a convex set, since some of the design variables may only have discrete values. Thus, we must use an optimization method which can cope with non convex sets.

21.4 GENETIC ALGORITHMS APPLIED TO SHAPE OPTIMIZATION

Genetic algorithms (GAs) are stochastic processes designed to mimic the natural selection based on the Darwin's principle of survival of the fittest. J.H. Holland [7] introduced a robust way of representing solutions to a problem in terms of a population of digital chromosomes that can be modified by the random operations of crossover and mutation on a bit string of information. In order to measure how good the solution is, a fitness function is needed. In shape optimization problems, the fitness function is naturally defined by the cost function given by (21.1).

GAs work with a population of individuals which are in our case designs. A traditional GA replaces the entire parent population by its offsprings. Here, we have chosen to use a steady-state reproduction [3]. With this approach only few individuals of population are replaced by offsprings. In our particular implementation, the worst individuals are replaced. This way we do not lose any good designs in reproduction.

In classical GAs, a binary coding is used for genes [5]. We have used real coding for the genes [14]. In our case, this is more natural, since the genes are real valued design variables. An offspring is obtained by either performing crossover of two parents or mutation to one individual. In our case, these both possibilities have the same probability $\frac{1}{2}$. After several numerical tests, we have decided to use a one point crossover and a uniform mutation. When applying a mutation to a discrete variable it must be taken care of that the resulting value also belongs to the set of possible discrete values. The discrete design variables of an offspring bred using one point crossover will have proper discrete values as long as the parents are proper with this respect. We use a roulette wheel selection with a linear normalization. The parameters for the GA are given in Table 21.1.

To form a new generation from the current one, the following steps are used:

1. Create n children through reproduction by performing n times with equal probability one of the following:
   (a) Select two parents using a roulette wheel selection and breed a child by using one point crossover.
   (b) Select one individual using a roulette wheel selection and make a child by cloning this and mutating each gene with a given probability.
2. Delete the worst n members of the population to make space for the children.
3. Evaluate and insert the children into the population.

Table 21.1 The parameters in the GA.

Parameter	Value
Number of genes	13
Number of individuals	30
Generations	2000
Replaced individuals	3
Mutation probability	0.2
Maximum mutation	0.03

21.5 NUMERICAL EXAMPLES

The shape of the back wall $S(\alpha)$ is defined using a Bézier curve [2], which has 13 control points. The control points are distributed uniformly in the x-direction and the y-coordinates of the control points are the 13 design variables. Thus, the first design variable α_1 and the last design variable α_{13} also define the lengths of the inflow and recirculation boundaries denoted by H_1 and H_2. These variables may have only discrete values and the steps of H_1 and H_2 are 0.02 and 0.01 metres, respectively. The values of design variables are constrained to be between 0.05 and 1.50.

In the optimization, the first one of the initial shapes of the header is the same one which was used to define the flow problem. Thus, the initial values of H_1 and H_2 are 1.0 and 0.2 metres and the back wall is linear. The another 29 designs in the initial population are obtained by making some predetermined changes to the first shape.

According to Table 21.1, one optimization run forms 2000 generations and, in each generation, the three worst individuals are replaced by new ones. Thus, the total number of flow problem solutions and cost function evaluations is around 6000. Due to the large amount of cost function evaluations, the GA was made parallel using MPI message passing. We were able to solved three flow problem in parallel, since three new individuals are bred and evaluated in each generation. One optimization run requires around 20 hours of wall clock time in a Sun Ultra Enterprise 4000 using three 250 MHz CPUs. The cost function value of the best individual in each generation is shown in Fig. 21.2.

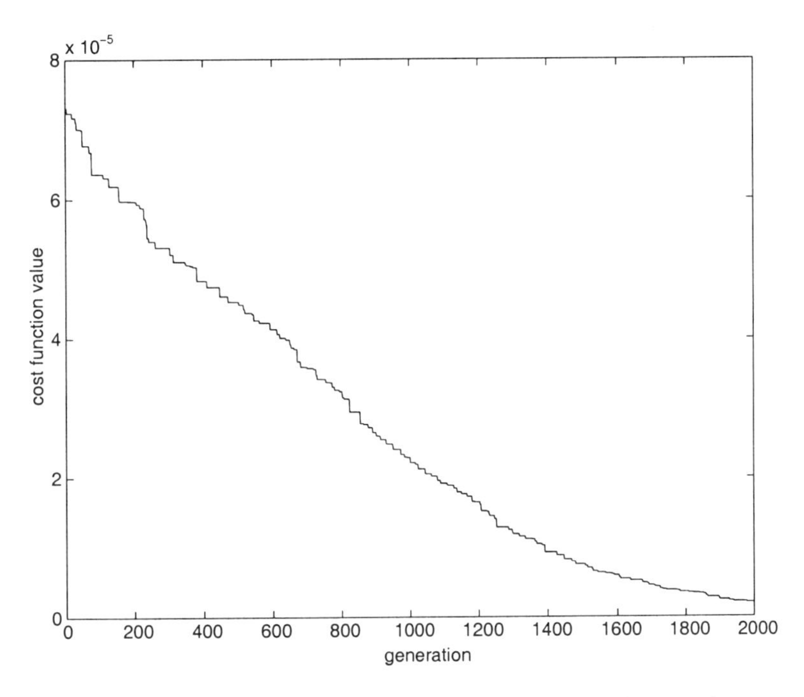

Fig. 21.2 The cost function value of the best individual in each generation.

The initial and optimized shape of the header with pressure profiles are shown in Fig. 21.4 and Fig. 21.5. The mesh for for the initial header illustrated in Fig. 21.3. The initial and optimized shape of the back wall is given in Table 21.2.

Fig. 21.3 The 171×23 mesh for the initial header.

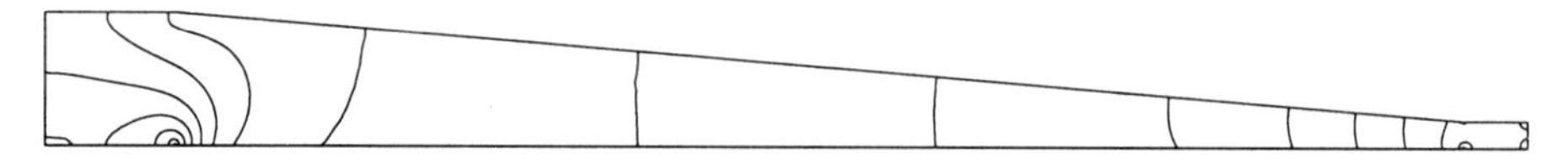

Fig. 21.4 The pressure profiles for the initial header.

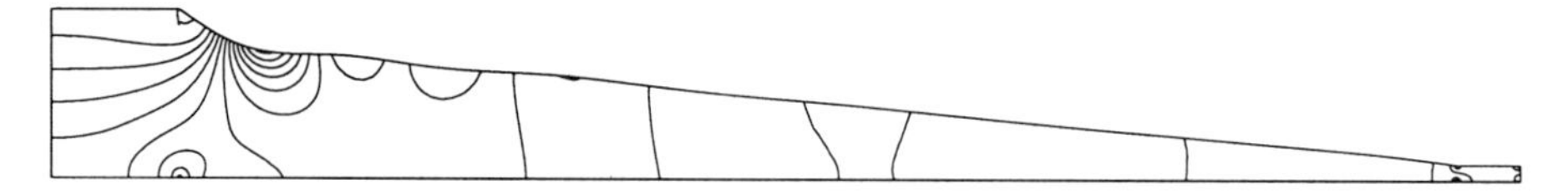

Fig. 21.5 The pressure profiles for the optimized header.

Table 21.2 The initial and optimized shape of the back wall.

x	Initial y	Optimized y
-1.000	1.000	1.280
0.000	1.000	1.280
0.500	0.960	1.006
1.250	0.900	0.932
2.000	0.840	0.845
3.000	0.760	0.781
4.000	0.680	0.671
5.000	0.600	0.584
6.000	0.520	0.492
7.000	0.440	0.399
8.000	0.360	0.314
8.750	0.300	0.250
9.500	0.240	0.181
10.000	0.200	0.120
10.500	0.200	0.120

The average inflow and recirculation velocities of the initial and optimized header are given in Table 21.3. The inflow velocity in the optimized header is lower, since the length of the inflow boundary H_1 is longer and the total inflow is kept fixed. The case for the recirculation is just the opposite. The velocity profile of the outflow for the initial and optimized header are shown in Fig. 21.6.

Table 21.3 The average inflow and recirculation velocities (m/s) of the initial and optimized header.

	Inflow	Recirculation
Initial	4.4	2.2
Optimized	3.5	3.7

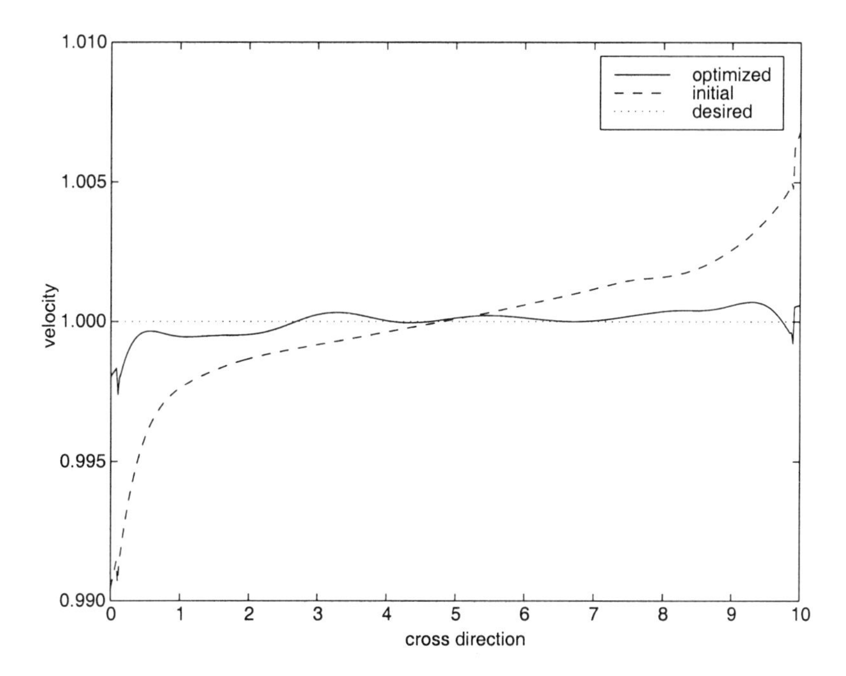

Fig. 21.6 The outlet velocity profile of the initial and optimized header.

The optimized shape of the header depends naturally on a cost function used in optimization. The very simple cost function (21.1) gives the shape shown in Fig. 21.5. Choosing a cost function is a common problem in optimization, not only for GAs. When shape optimization is utilized in designing real products in Valmet, an attention is paid on selecting a cost function depending on the velocity and pressure fields and, also, on the shape in order to find really the optimal solution with respect to the fluid dynamical performance and manufacturing costs.

21.6 CONCLUSIONS

Previous studies have proved the shape optimization with a gradient based optimization method to be an effective design tool for the header design problems. Here we have demonstrated GAs capability to treat a more complicated case with good results. Although the considered shape optimization problem was not a real industrial design problem, this simplistic example shows clearly the potential of GAs in shape optimization problems in designing the paper machine headbox.

Optimization together with CFD is still industrially and scientifically very challenging problem. In the future, flow models will be refined to describe the fluid motion more accurately and this makes the solution of CFD problems more expensive. Thus, the effectiveness of the solution procedures for the nonlinear CFD and optimization problems must be increased in order to this approach to be costeffective also in the future for industrial design problems. Particularly in the case of GAs, the convergence properties and robustness should be improved.

REFERENCES

1. W.D. Baines, C.I.H. Nicholl, R.C. Cook and J. Mardon, "The Taper-Flow Headboxes (A New Design Concept)," *Pulp and Paper Magazine of Canada,* pp. 139–148 (1956).

2. V. Braibant and C. Fleury, "Shape Optimal Design Using B–splines," *Comput. Meths Appl. Mech. Engrg.,* **44**, pp. 247–267 (1984).

3. L. Davis, *Handbook of Genetic Algorithms,* Van Nostrand Reinhold, 1991.

4. L. Franca and S. Frey, "Stabilized Finite Element Methods: II. The Incompressible Navier-Stokes Equations," *Comput. Meths Appl. Mech. Engrg.,* **99**, pp. 209–233 (1992).

5. D.E. Goldberg, *Genetic Algorithms in Search, Optimization and Machine Learning,* Addison-Wesley, 1989.

6. J. Haslinger and P. Neittaanmäki, *Finite Element Approximation for Optimal Shape, Material and Topology Design, 2nd edition,* Wiley, 1996.

7. J.H. Holland, *Adaptation in neural and artificial systems,* University of Michigan Press, 1975.

8. J. Hämäläinen, *Mathematical Modeling and Simulation of Fluid Flows in the Headbox of a Paper Machine,* Report 57, Doctoral thesis, University of Jyväskylä, Department of Mathematics, Jyväskylä, 1993.

9. J. Hämäläinen, J. Järvinen, Y. Leino and J. Martikainen, "High Performance Computing in Shape Optimization of a Paper Machine Headbox," *CSC News,* **10**, pp. 6–9 (1998).

10. J. Hämäläinen and P. Tarvainen, "CFD Coupled with Shape Optimization – A New Promising Tool in Paper Machine R&D," *Proceedings of 10th Conference of the European Consortium for Mathematics in Industry (ECMI),* Göteborg, 1998, to appear.

11. J. Hämäläinen and T. Tiihonen, "Flow Simulation with Homogenized Outflow Boundary Conditions," *Finite Elements in Fluids,* K. Morgan, E. Onate, J. Periaux, J. Peraire and O. Zienkiewicz (eds.), Pineridge Press, pp. 537–545 (1993).

12. J. Hämäläinen and T. Tiihonen, "Modeling and Simulation of Fluid Flows in a Paper Machine Headbox," *ICIAM – Third International Congress in Industrial and Applied Mathematics,* Issue 2: Applied sciences, especially mechanics (minisymposia), E. Kreuzer and O. Mahrenholtz (eds.), Akademie Verlag, pp. 62–66 (1996).

13. J. Mardon, D.W. Manson, J.E. Wilder, R.E. Monahan, A. Trufitt and E.S. Brown, "The Design of Manifold Systems for Paper Machine Headboxes, Part II – Taper Flow Manifolds," *TAPPI,* **46** (1963).

14. Z. Michalewicz, *Genetic algorithms + data structures = evolution programs,* Springer-Verlag, 1992.

15. P. Tarvainen, R. Mäkinen and J. Hämäläinen, "Shape Optimization for Laminar and Turbulent Flows with Applications to Geometry Design of Paper Machine Headboxes," *Proceedings of 10th Int. Conference on Finite Elements in Fluids,* M. Hafez and J.C. Heinrich (eds.), pp. 536–549 (1998).

16. A. Trufitt, "Design Aspect of Manifold Type Flowpreaders," *Pulp and Paper Technology Series, Joint Textbook Committee of the Paper Industry, C.P.P.A. and TAPPI* (1975).

22 A Parallel Genetic Algorithm for Multi-Objective Optimization in Computational Fluid Dynamics

N. MARCO[1], S. LANTERI[1], J.-A. DESIDERI[1] and J. PÉRIAUX[2]

[1]INRIA, 2004 Route des Lucioles
BP. 93, 06902 Sophia Antipolis Cédex, France

[2]Dassault-Aviation, AMD-BA
78, Quai Marcel Dassault, B.P. 300, 92214 Saint-Cloud Cédex, France

Abstract. We are developing and analyzing methodologies involving Genetic Algorithms for multi-objective optimum aerodynamic shape design. When several criteria are considered, the optimal solution is generally not unique since no point achieves optimality, w.r.t. all criteria simultaneously. Instead, a *set* of non-dominated solutions called *the optimal solutions of the Pareto set* can be identified.

22.1 INTRODUCTION

In everyday life, one is confronted with a variety of multi-objective optimization problems such as buying a new television set not too expensive and of good quality or chosing the optimal route to minimize time and maximize safety. In engineering, scientists attempt to build very performant spacecraft structures at low cost. In economics, one has to conciliate the utilities and the consumption of commodities to maximize the social welfare of each individual.

The common part of these optimization problems is that they have several criteria to be taken into consideration simultaneously. In these cases, methods of single objective optimization are no more suitable. Multi-objective optimization need be considered. A large number of application areas of multi-objective optimization has been presented in the literature.

In a simple optimization problem, the notion of optimality is straightforward. The best element is the one that realizes the minimum or the maximum of the objective function.

In a multi-objective optimization problem, the notion of optimality is not so obvious. If we agree in advance that we cannot link the values of the different objectives (i.e. if we refuse to compare apples and oranges), we must find a different definition of optimality, a definition which respects the entirety of each criterion. Then, the concept of Pareto optimality arises. There is no solution that is the best for all the criteria, but there exists a set of solutions that are better than other solutions in all the search space, when considering all the objectives. This set of solutions is known as *the optimal solutions of the Pareto set* or *non-dominated solutions*.

There are two different ways to solve a mutli-objective optimization problem. The first consists in forming a linear combination of the different criteria with different weights and to minimize the resulting function. This approach has two main drawbacks: (1) not all solutions are found, (2) the objective function is meaningless unless the weigths are selected very intellegently.
In contrast, Genetic Algorithms (GAs) operate on a population of potential solutions (potential, because they belong to the feasible space). Then, during the optimization, GAs explore all the feasible space and build a database of solutions determining a cloud. At convergence, the cloud no more evolves and its convex hull determines the optimal Pareto set.

This article is organized as follows: in Section 22.2, a brief presentation of a multi-objective optimization problem is given; in Section 22.3, a three-criteria optimization problem dealing with analytical functions is solved. Finally, in the last section, we consider a multi-objective optimization problem in Computational Fluid Dynamics, in which the goal is to exhibit a family of profiles that have a smooth transition in shape from a first profile to a second one. The flows on these two profiles have been evaluated at two different Mach numbers and incidences.

22.2 MULTI-OBJECTIVE OPTIMIZATION PROBLEM

22.2.1 A general multi-objective problem

A general multi-objective optimization problem consists of a number of objectives to be optimized simultaneously and is associated with a number of inequality and equality constraints (for a theoretical background, see for example the book of Cohon [1]). It can be formulated as follows:

$$\text{Minimize / Maximize} \quad f_i(x) \quad i = 1, \cdots, N$$

$$\text{subject to:} \begin{cases} g_j(x) = 0 & j = 1, \cdots, M \\ h_k(x) \leq 0 & k = 1, \cdots, K \end{cases} \quad \text{constraints.}$$

The f_i are the objective functions, N is the number of objectives, x is a p-dimensional vector whose p arguments are known as decision variables. In a minimization problem, a vector x^1 is partially less than another vector x^2 when:

$\forall i \; f_i(x^1) \leq f_i(x^2) \quad (i = 1, \cdots, N)$ and there exists at least one i such that $f_i(x^1) < f_i(x^2)$.

We say that *the solution x^1 dominates the solution x^2*.
Definition of non-dominance for 2 criteria in the case of a minimization:

$$\begin{cases} \text{Minimize} \quad f(x) = (f_1(x), f_2(x)) \\ \text{such that} \quad x \in X, \text{ the feasible region.} \end{cases}$$

x^1 and x^2 are two solutions to be compared. Then, x^1 dominates x^2 iff:

$$f_1(x^1) < f_1(x^2) \quad \text{and} \quad f_2(x^1) \leq f_2(x^2) \quad \text{or} \quad f_1(x^1) \leq f_1(x^2) \quad \text{and} \quad f_2(x^1) < f_2(x^2).$$

22.2.2 Ranking by fronts

GAs operate primarily by selection of the individuals according to their fitness values. But, in a multi-objective optimization problem, several criteria are assigned and *not* a single fitness value. Therefore, to evaluate the individuals, we need to build a fitness value, called *dummy fitness*.

After the application of the definition of non-dominance, the chromosomes are classified by fronts. The non-dominated individuals belong to front 1 (or rank 1), the following non-dominated individuals belong to front 2 (or rank 2), ... the worst individuals belong to front f, where f is the number of fronts (or ranks).

As an example, we consider a situation where the population of individuals is sorted according to two criteria (J_1, J_2). On Fig. 22.1 (left), the case of a population of 30 individuals corresponding to values of the criteria J_1 and J_2 picked randomly is presented. As a result, the figure indicates the 8 different fronts. On this example, the points belonging to fronts number *I, II, III, IV* have more chances to be reproduced than points belonging to fronts number *VII, VIII*.

Once the individuals have been ranked, they can be assigned the following values:

$$f_i = \frac{k}{\text{number of the rank to which } i \text{ belongs}} \qquad k > 0 \quad \text{in case of minimization,}$$

$$f_i = k * (\text{number of the rank to which } i \text{ belongs}) \qquad k > 0 \quad \text{in case of maximization.}$$

(i is the number of the individual.)

Because GAs operate on a whole population of points, they allow to find a number of optimal solutions of the Pareto set. A first application of GAs on multi-objective optimization problems has been studied by Schaffer in 1984 (see his experiments, called Vector Evaluated GA (VEGA), in his dissertation [15]). Schaffer's objective was to minimize a cost and maximize a reliability. He distributed the population in two half populations and optimized the two objectives on each half. After a lot of generations, the population converged towards the optima of each sub-region. This method gave a new avenue in the research of a method to solve a multi-objective problem. The results were encouraging but the obtained solutions only converged to the extreme points of the Pareto set (the first point is the first optimized criterion and the second point is the second optimized criterion), where one objective is maximal, since it never selects according to trade-offs among objectives. VEGA's results implied that a speciation through the population was developped. In order to get diversity (i.e to maintain individuals all along the non-dominated frontier), a non-dominated sorting procedure in conjunction with a sharing technique was subsequently implemented, first by Goldberg in [5] and more recently by Horn in [8], and Srinivas in [16]. Then, the goal has been to find a representative sampling of solutions all along the Pareto front.

Our results in multi-objective optimization are based on the algorithm of Srinivas and Deb [16], called the Non-dominated Sorting Genetic Algorithm (NSGA). In the next paragraph we expose this method.

22.2.3 Non-dominated sorting genetic algorithms

The non-dominated sorting procedure requires a ranking selection method which emphasizes the optimal points. The sharing technique, or niche method, is used to stablilize the subpopulations of the "good" points.

We use this method, because one of the main defects of GAs in a multi-objective optimization problem is the premature convergence. In some cases, GAs may converge very quickly to a point of the optimal Pareto set and as the associated solution is better than the others (it is called a "super individual"), it breeds in the population and, after a number of generations the population is composed of copies of this individual only! The non-dominated sorting technique with the niche method avoids this situation because as soon as a solution is found in multiple copies, its fitness value is decreased and in the next generation, new solutions that are different can be found, even if they are not so good. The fitness values decrease because the different niches to which belong the different optimal solutions are exhibited.

Technical aspects of the method

- From an initial population of solutions, one first identifies the non-dominated individuals (according to their criteria) as explained in paragraph 22.2.1. These individuals belong to the rank number 1. Recall that all the points belonging to front 1 are better than those not belonging to the front and are not comparable among themselves.

  These individuals have large probabilities to be reproduced.

- Then assign the same dummy fitness f_i to the non-dominated individuals i of front 1 (generally, the dummy fitness f_i is equal to 1).

- To maintain the diversity, the dummy fitness of the individuals is then shared : it is divided by a quantity proportional to the number of individuals around it, according to a given radius. If the individual has a lot of neighbours, this means that there is a lot of similar solutions. Then, the fitness is going to be shared in order to create some diversity in the next generation.

  We call this quantity a **niche** and it is evaluated as follows:

$$m_i = \sum_{j=1}^{M} Sh(d(i,j))$$

  where i is the number of the individual, M is the number of the individuals belonging to the current front (or rank).

$Sh(d)$ is the **Sharing function**, linear decreasing function of d defined by:

$$\begin{cases} Sh(0) = 1 \\ Sh(d(i,j)) = \begin{cases} 1 - \dfrac{d(i,j)}{\sigma_{share}} & \text{if } d(i,j) < \sigma_{share} \\ 0 & \text{if } d(i,j) \geq \sigma_{share} \end{cases} \end{cases}$$

σ_{share} is the maximum phenotypic distance allowed between any two individuals to become members of a niche. It has no universal value ; it is problem-dependent. In [7], Goldberg proposes to implement an adaptive niching scheme to overcome the limitations of fixed sharing schemes.

$d(i, j)$ is the distance between two individuals i and j. In some multi-objective optimization problems, the *Hamming distance* is used. It is a genotypic (at the string level) distance and it consists in counting the number of different bits in the two strings, divided by the string length. Such a distance has been used in electromagnetics system design optimization (see [12]) and in problems of finding an optimal distribution of active control elements in order to minimize the backscattering of a reflector in computationnal electromagnetics (see [4]). In this case, the Hamming distance has a true sense because the bits 1 and 0 do not correspond to a real codage but to the fact that the control elements are active (1) or not (0). When using real-value binary coding, the Hamming distance may be biased because two neighbouring solutions may have a very important Hamming distance (e.g., 0111 and 1000 correspond to 7 and 8, which are neighbours but their Hamming distance is maximal!). Then, in these cases, a phenotypic (at the real parameter level) distance is used, the *Euclidian distance.* is the number of individuals belonging to the current front. Then after having created a niche for each individual of the current front, the new dummy fitness $\frac{f_i}{m_i}$ is assigned.

- After this sharing applied, the non-dominated individuals of Front 1 are temporarily ignored from the population. The following non-dominated individuals belong to Front 2. One evaluates the minimum of the quantities $\frac{f_i}{m_i}$ taken over all chromosomes of Front 1. Then this value is assigned as a dummy fitness to the non-dominated individuals of Front 2. Then, a niche is evaluated for chromosomes of Front 2, and new dummy fitness values.

- This process is continued until the whole population has been visited.

- As all the individuals in the population have now a dummy fitness value, selection, crossover and mutation can be applied. The flowchart of the method is presented in Fig. 22.1 (right) (inspired from [16]).

Remark In the case of a minimization problem, the fact that we sort the population by assigning a greater dummy fitness to the best individuals (individuals belonging to Front 1 have a greater fitness than individuals belonging to Front 2 which have a greater fitness than the ones belonging to Front 3 ...) implies that the Pareto optimization becomes a *maximization problem*.

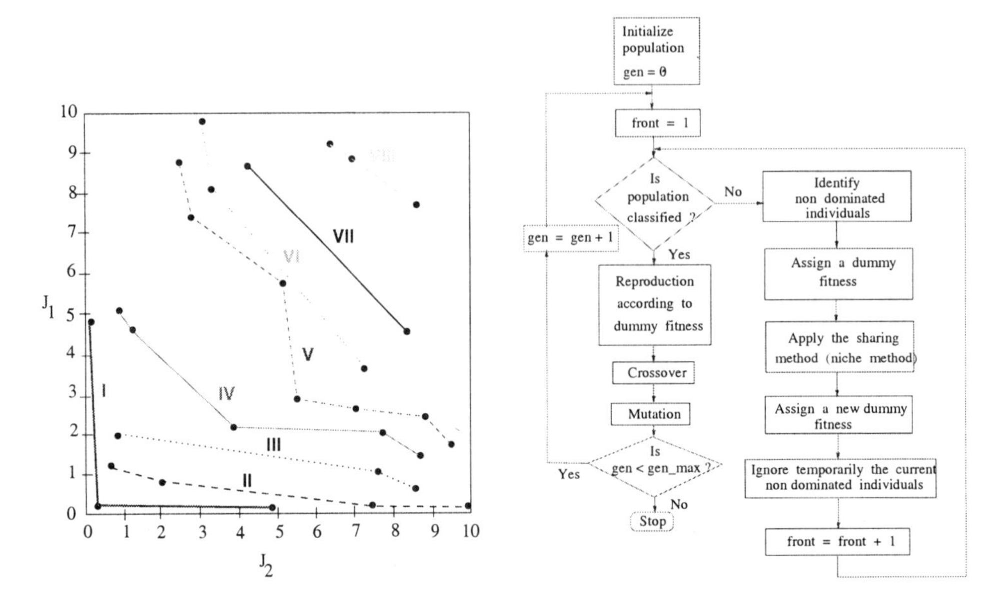

Fig. 22.1 Population ranking by fronts (left). Flowchart of the Non-dominated Sorting GA (right).

22.3 THREE-OBJECTIVE OPTIMIZATION PROBLEM

Three analytical functions depending on two parameters are considered below. The goal is to minimize these three functions simultaneously:

$$\begin{aligned} f_1(x,y) &= x^2 + (y-1)^2 \\ f_2(x,y) &= x^2 + (y+1)^2 + 1 \qquad \text{with } (x,y) \in [-2,2] \times [-2,2] \\ f_3(x,y) &= (x-1)^2 + y^2 + 2 \end{aligned}$$

Analytical solution The goal is to minimize $f(x,y) = \alpha f_1(x,y) + \beta f_2(x,y) + \gamma f_3(x,y)$ with $\alpha, \beta, \gamma \in [0,1]$ and $\alpha + \beta + \gamma = 1$.

The optimal Pareto front is obtained by solving:

$$\left\{ \begin{array}{l} \dfrac{\partial f(x,y)}{\partial x} = 0 \\ \dfrac{\partial f(x,y)}{\partial y} = 0 \end{array} \right. \Rightarrow \left\{ \begin{array}{l} 2\alpha x + 2\beta x + 2\gamma(x-1) = 0 \\ 2\alpha(y-1) + 2\beta(y+1) + 2\gamma y = 0 \end{array} \right. \Rightarrow \left\{ \begin{array}{l} x = \gamma \\ y = \alpha - \beta \end{array} \right.$$

Then, $x^{opt} \in [0,1]$ and $y^{opt} \in [-1,1]$, that means, $f_1^{opt} \in [0,5]$, $f_2^{opt} \in [1,6]$, $f_3^{opt} \in [2,4]$.

Solution by GAs The decision variables x and y are both coded by binary strings of length equal to 26 (corresponding to an accuracy of 10^{-7}) and a population of 40 chromosomes. The probability of crossover is $p_c = 0.9$, the probability of mutation is $p_m = 0.05$ and

$\sigma_{share} = 12$. An elitist strategy has been considered and a selection by 2-point tournament has been used.

Fig. 22.2 (left) depicts the evolution of the population in the objectives space from the first generation to the last one. We note that the optimal Pareto set is located between $f_1 \in [0, 5]$, $f_2 \in [1, 6]$ and $f_3 \in [2, 4]$.

On Fig. 22.2 (right), the optimal Pareto set obtained after 100 generations is also plotted; there are 1692 individuals on the set.

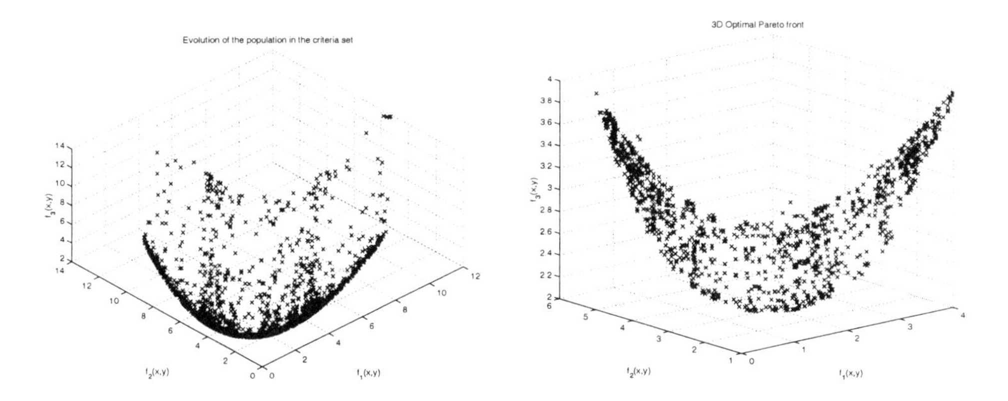

Fig. 22.2 Evolution of the population from the first generation to the 100th (left) and the Optimal 3D Pareto front (right).

22.4 MULTI-OBJECTIVE OPTIMIZATION IN SHAPE OPTIMUM DESIGN

22.4.1 The optimization problem

Two reference Euler-flow solutions are first obtained: a subsonic flow at a free-stream Mach number of 0.2 and an incidence of 10.8° on a "high-lift" profile (see Fig. 22.3); the corresponding calculated pressure is denoted P_1. Secondly, a transonic flow at a free-stream Mach number of 0.77 and an incidence of 1° is computed on a "low-drag" profile (see Fig. 22.3) ; the corresponding calculated pressure is denoted P_2.

Then, we consider the two-objective optimization problem consisting in minimizing the following two cost functionals:

$$\begin{cases} J_1(W(\gamma)) = \int_\gamma (P(W) - P_1)^2 \, d\sigma & \text{at Mach} = 0.2 \text{ and } \alpha = 10.8°, \\ J_2(W(\gamma)) = \int_\gamma (P(W) - P_2)^2 \, d\sigma & \text{at Mach} = 0.77 \text{ and } \alpha = 1°. \end{cases}$$

Following Poloni in [14], we want to determine a whole set of profiles existing between the low-drag profile and the high-lift profile. For each chromosome determining the profile of an airfoil γ, two Euler flows are computed at the two previously mentioned regimes.

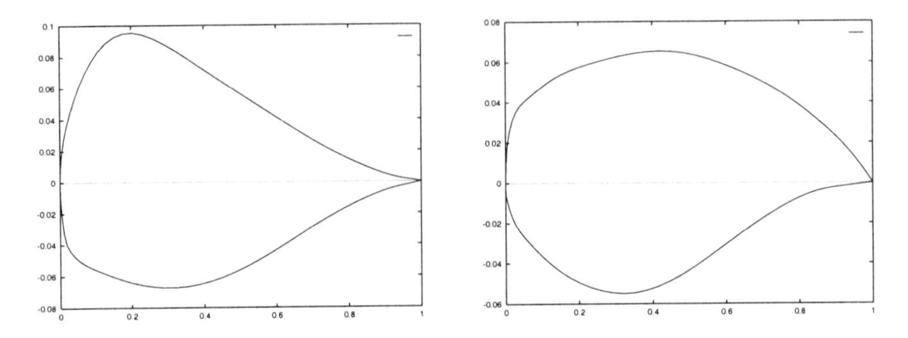

Fig. 22.3 High-lift and low-drag profiles.

The reference profile is the profile of a naca0012 airfoil.

22.4.2 The 2D Euler flow solver

The underlying flow solver solves the 2D Euler equations:

$$\frac{\partial W}{\partial t} + \vec{\nabla} \cdot \vec{\mathbf{F}}(W) = 0, \qquad W = \left(\rho, \rho\vec{U}, E\right)^T, \qquad \vec{\nabla} = \left(\frac{\partial}{\partial x}, \frac{\partial}{\partial y}\right)^T$$

where $\vec{\mathbf{F}}(W) = (F_1(W), F_2(W))^T$ is the vector of convective fluxes whose components are given by:

$$F_1(W) = \begin{pmatrix} \rho u \\ \rho u^2 + p \\ \rho uv \\ u(E+p) \end{pmatrix}, \quad F_2(W) = \begin{pmatrix} \rho v \\ \rho uv \\ \rho v^2 + p \\ v(E+p) \end{pmatrix}$$

In the above expressions, ρ is the density, $\vec{U} = (u, v)^T$ is the velocity vector, E is the total energy per unit volume and p denotes the pressure which is obtained using the perfect gas state equation $p = (\gamma_p - 1)(E - \frac{1}{2}\rho \parallel \vec{U} \parallel^2)$ where $\gamma_p = 1.4$ is the ratio of specific heats. The flow domain Ω is a polygonal bounded region of $I\!R^2$. Let $\mathcal{T}_h$ be a standard triangulation of Ω. At each mesh vertex S_i, a control volume C_i is constructed as the union of local contributions obtained from the set of triangles attached to S_i. The union of all these control volumes constitutes a dual discretization of the domain Ω. The spatial approximation method adopted here makes use of a mixed finite element/finite volume formulation. Briefly speaking, for each control volume C_i, one has to solve for n *one-dimensional Riemann problems* at the control volume boundary (finite volume part), n being the number of neighbors of S_i. The spatial accuracy of the Riemann solver depends on the accuracy of the interpolation of the physical quantities at the control volume boundary. For first order accuracy, the values of the physical quantities at the control volume boundary are taken equal to the values in the control volume itself. Extension to second order accuracy can be performed via a "Monotonic Upwind Scheme for Conservative Laws" (MUSCL) technique of Van Leer [17]. It consists in providing the Riemann solver with a linearly interpolated value, taking into account some gradient of the physical quantities (finite element part). Time integration relies on a linearised implicit formulation which is best suited to steady flow simulations [3]:

$$\left(\frac{\text{area}(C_i)}{\Delta t^n} + J(W^n)\right)(W^{n+1} - W^n) = -\Psi(W^n)$$

where $\Psi(W)$ is the flux depending on a numerical flux function and on the vector $F(W)$ of the convective fluxes. $J(W^n)$ denotes the associated Jacobian matrix. At each time step, the above linear system is solved using Jacobi relaxations. We refer to [9] and [11] for a more detailed description of the flow solver.

22.4.3 Parallelization strategy

The main concern related to the use of GAs in optimum shape design is the computational effort needed. In difficult problems, GAs need bigger populations and this translates directly into higher computational effort. In the case of the multi-objective optimization problem, two objectives in our case, we have to solve two Euler flows for each individual, which means that the computational effort is very high. One of the main advantages of GAs is that they can easily be parallelized. At each generation, the fitness values associated to each individual can be evaluated in parallel. We will not go into the details for the parallelization strategy we have adopted, it is not the topic of our paper. A complete description can be found in [11] or [10]. In summary, we have employed a two-level parallelization strategy: the first level is the parallelization of the flow solver, which combines domain partitioning techniques with a message-passing programming model (where calls to the MPI library are used). The second level is the parallelization of GAs. We have chosen to exploit the notion of process groups which is one of the main features of the MPI environment. Two groups are defined, each group is responsible for the evaluation of the criteria for one individual, and each group contains the same number of processes. Then, two individuals are taken into account in parallel. For more details, one can refer to [11].

22.4.4 Results

22.4.4.1 Data and parameters The "high-lift" mesh is made of 2130 nodes, the "low-drag" of 3847 nodes and the reference Naca0012 mesh of 1351 nodes.

Shape parametrization The individuals represent airfoil shapes. Following a strategy often adopted, the shape parametrization procedure is based on Bézier curves. Eigth order Bézier representation has been used (in extrados and intrados): 2 fixed points, P_0 and P_8, and 7 control points. The values $x_i \in [0, 1]$ are fixed and the only parameters that vary are the ordinates.

The GA Results have been obtained with a population of 40 individuals, after 100 generations. The probabilities of crossover and mutation have been set to $p_c = 0.8$ and $p_m = 0.007$, in addition, $\sigma_{share} = 0.7$. The selection operator is the binary tournament with an elitist strategy.

22.4.4.2 Numerical results Results have been obtained on a SGI Origin 2000 system equipped with Mips R10000/195 Mhz processors. By working with 2 groups and 4 processors by group, the CPU time was about 2 hours.

Fig. 22.4 (left) visualizes the evolution of the population in the course of the generations until a steady state is reached. We note the cloud of points and its convex hull. The optimal Pareto set is shown on Fig. 22.4 (right), it describes the convex hull. After 100 generations, there are 131 individuals on the Pareto front. All these individuals correspond to optimal solutions of the mutli-objective minimization problem. On Fig. 22.5, a sample of the different profiles corresponding to the optimal solutions of the two-objectives problem is depicted.

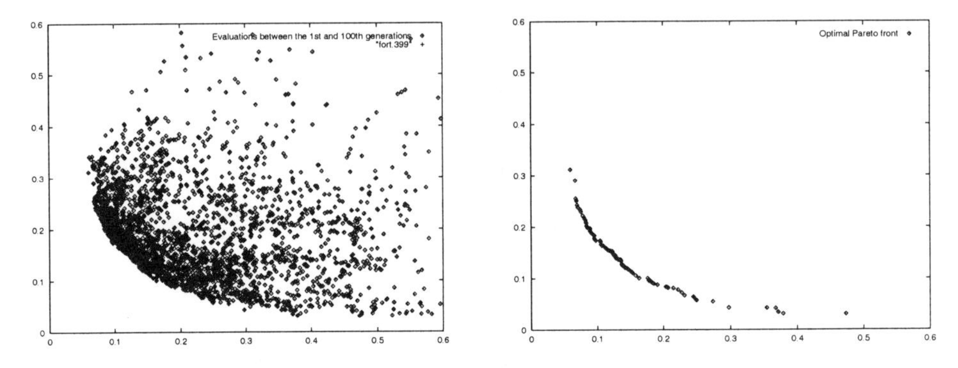

Fig. 22.4 Evolution of the individuals in the objectives plane, from the first generation to the last one (100 generations) (left) and optimal Pareto set with 131 individuals uniformly distributed (right).

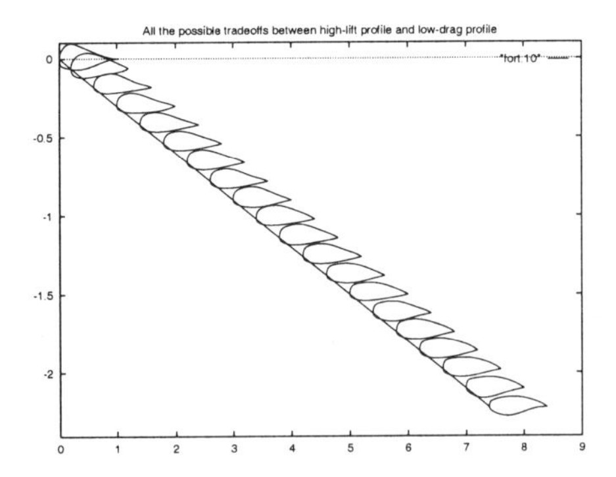

Fig. 22.5 A sample of profiles between high-lift and low-drag.

22.5 CONCLUDING REMARKS

The multi-objective optimization process by a genetic algorithm is carried independently of the decision maker's eventual options: a database is constructed as the population evolves and the pareto front is achieved at convergence of this database. In this way, the approach offers an alternative to optimization with penalty functions.

In some cases, it is difficult to identify numerically all feasible optimal solutions, particularly those corresponding to trade-offs between the objectives. A sharing procedure with a classification by fronts can then be introduced to increase diversity between the optimal solutions all along the Pareto front ([6], [2], [13], [7] and we have tested the method of [16]).

In this contribution, we have demonstrated the application of a GA for a practical 2D airfoil design optimization according to two criteria related to subsonic and supercritical conditions.

REFERENCES

1. J.L. Cohon, "*Multiobjective Programming and Planning*", Academic Press, New-York, R. Bellman Editor, 1978.

2. K. Deb and D.E. Goldberg, "An investigation of niches and species formation in genetic function optimization", In J.D. Schaffer, editor, *Proceedings of the Third International Conference on Genetic Algorithms*, pages 42–50, San Mateo (CA), Morgan Kaufmann, 1991.

3. L. Fezoui and B. Stoufflet, "A class of implicit upwind schemes for Euler simulations with unstructured meshes", *J. of Comp. Phys.*, 84:174–206, 1989.

4. R. Glowinski, J. Périaux, M. Sefrioui and B. Mantel, "Optimal backscattering of a reflector by means of Genetic Algorithms", In *ECCOMAS 96*, New-York, John Wiley and Sons, 1996.

5. D.E. Goldberg, "*Genetic Algorithms in Search, Optimization and Machine Learning*", Addison-Wesley Company INC., 1989.

6. D.E. Goldberg and J. Richardson, "Genetic Algorithms with sharing for multimodal function optimization", In *Genetic Algorithms and Their Applications: Proceedings of the Second International Conference on Genetic Algorithms*, pages 41–49, San Mateo (CA), Morgan Kaufmann, 1987.

7. D.E. Goldberg and L. Wang, "Adaptive niching via coevalutionary sharing", In C. Poloni D. Quagliarella, J. Périaux and G. Winter, editors, *Genetic Algorithms and Evolution Strategies in Engineering and Computer Science*, pages 21–38, Trieste (Italy), John Wiley and Sons, 1997.

8. J. Horn and N. Nafpliotis, "Multiobjective Optimization Using the Niched Pareto Genetic Algorithm", In *Proceedings of the First IEEE Conference on Evolutionary Computation, IEEE World Congress on Computational Intelligence*, Piscataway (NJ), Volume 1 (ICEC '94), 1994.

9. S. Lanteri, "Parallel solutions of compressible flows using overlapping and non-overlapping mesh partitioning strategies", *Parallel Comput.*, 22(7):943–968, 1996.

10. N. Marco and S. Lanteri, "Parallel Genetic Algorithms applied to Optimum Shape Design in Aeronautics", In M. Griebl, C. Langauer and S. Gorlatch, editors, *Euro-Par'97 Parallel Processing*, pages 856–863, Passau (Germany), Springer Verlag, 1997.

11. N. Marco and S. Lanteri, "A two-level parallelization strategy for Genetic Algorithms applied to shape optimum design", INRIA Report No 3463 - Submitted to *Parallel Comput.*, 1998.

12. E. Michielssen and D.S. Weile, "Electromagnetic System Design Using Genetic Algorithms", In M. Galàn G. Winter, J. Périaux and P. Cuesta, editors, *Genetic Algorithms in Engineering and Computer Science*, pages 345–368. John Wiley and Sons, 1995.

13. C.K. Oei, D.E. Goldberg and S.J. Chang, "Tournament selection, niching and the preservation of diversity", IlliGAL Report No 91011, 1992.

14. C. Poloni, "Hybrid GA for Multi-Objective Aerodynamic Shape Optimization", In M. Galàn G. Winter, J. Périaux and P. Cuesta, editors, *Genetic Algorithms in Engineering and Computer Science*, pages 397–415. John Wiley and Sons, 1995.

15. J.D. Schaffer, "Some experiments in machine learning using Vector Evaluated Genetic Algorithms", Unpublished doctoral dissertation, Vanderbilt University, Nashville (TN), 1984.

16. N. Srinivas and K. Deb, "Multiobjective Optimization Using Nondominated Sorting in Genetic Algorithm", *Evolutionary Computation*, 2(3):221–248, 1995.

17. B. Van Leer, "Towards the ultimate conservative difference scheme V: a second-order sequel to Godunov's method", *J. of Comp. Phys.*, 32:361–370, 1979.

23 Application of a Multi Objective Genetic Algorithm and a Neural Network to the Optimisation of Foundry Processes

G. MENEGHETTI[1], V. PEDIRODA[2] and C. POLONI[2]

[1] Engin Soft Trading Srl, Italy

[2] Dipartiento di Energetica, Università di Trieste, Italy

Abstract. Aim of the work was the analysis and the optimisation of a ductile iron casting using the Frontier software. Five geometrical and technological variables were chosen in order to maximise three design objectives. The calculations were performed using the software MAGMASOFT, devoted to the simulation of foundry processes based on fluid-dynamics, thermal and metallurgical theoretical approaches. Results are critically discussed by comparing the traditional and the optimised solution.

23.1 INTRODUCTION

A very promising field for computer simulation techniques is certainly given by the foundry industry. The possibility of reliably estimating both the fluid-dynamics, thermal and microstructural evolution of castings (from the pouring of the molten alloy into the mould till the complete solidification) and the final properties are very interesting. In fact if the final microstructure and then the mechanical properties of a casting can be predicted by numerical simulation, the a-priori optimisation of the process parameters (whose number is usually high) can be carried out by exploring different technological solutions with significant improvements in the quality of the product, managing of human and economical resources and time-savings.

This approach is extremely new in foundry and in this work an exploratory project aimed at the process optimisation of an industrial ductile iron casting will be presented.

23.2 THE SIMULATION OF FOUNDRY PROCESSES AND FOUNDAMENTAL EQUATIONS

From a theoretical point of view, a foundry process can be considered as the sequence of various events [1]–[4]:

- the filling of a cavity by means of a molten alloy, as described by fluid-dynamics laws (Navier-Stokes equation),
- the solidification and cooling of the alloy, according to the heat transfer laws (Fourier equation),
- the solid state transformations, related to the thermodynamics and the kinetics.

A full understanding of the whole foundry process requires an investigation throughout all these three phenomena. However under some hypotheses (regular filling of the mould cavity, homogeneous temperature distribution at the end of filling, etc.) the analyses of the solidification and the solid state transformation only can lead to reliable estimation of the final microstructure and of the properties of the casting.

The accuracy in simulating the solidification process depends mainly on:

- the use of proper thermophysical properties of the materials involved in the process, taking into account their change with temperature,
- the correct definition of the starting and boundary conditions, with particular regard to the heat transfer coefficients.

From a numerical point of view, the investigation of the solidification process could be carried out by means of a pure heat flow calculation described by Fourier's law of unsteady heat conduction :

$$\frac{\partial}{\partial t}(\rho C_p T) = \frac{\partial}{\partial x_j}\left[\lambda \frac{\partial T}{\partial x_j}\right].$$

However a more correct evaluation requires to incorporate the additional heat transport by convective movement of mass due to temperature dependent shrinkage of the solidifying mush. Doing that temperature-dependent density functions are needed, so that the shrinkage can be calculated basing on the actual temperature loss. The total metal shrinkage within one time interval will lead to a corresponding metal volume flowing from the feeder into the casting passing through the feeder-neck. The actual temperature distribution in the feeder neck can be calculated on the basis of the following equation :

$$\frac{\partial}{\partial t}(\rho C_p T) + \frac{\partial}{\partial x_j}(\rho C_p T u_j) = \frac{\partial}{\partial x_j}\left[\lambda \frac{\partial T}{\partial x_j}\right] + S$$

where the second term on the left hand side of the equation is the convective term while the first one on the right hand side is the conductive term. S denotes the additional internal heat source. The additional heat transport by convective movement of mass means that feeding may last much longer than being calculated by heat flow based uniquely on conduction.

Anyway, when the feeder-neck freezes to a certain temperature, the feeding mechanism locks. Therefore the solidification of any other portion of the cast, now insulated, will take place independently from one another and the feed metal required during solidification will come from the remaining liquid. The final volume shrinkage will result in a certain porosity, which typically will be located at the hot spots.

From the point of view of the real industrial interest, the above phenomena and the related equations can be approached only numerically: in fact complex 3D geometries have to be taken into account, as well as manufacturing parameters ensuring compliance with temperature-dependent thermophysical properties of the materials, production and process parameters. Finite elements, finite differences, control volumes or a combination of these are typical methods implemented in the software packages [2-3].

The final result of the simulation is the knowledge of the actual feeding conditions, which is the basis for correctly design the size of feeders. It must be recalled that this knowledge-based approach is often by-passed by the use of empirical rules and in most cases the optimisation of the feeder size is not really performed (so that the feeders are simply oversized) or it is carried out by means of expensive in-field trial-and-correction procedures.

The analyses were performed by using the MAGMASOFT® software, specifically devoted to the simulation of foundry processes, based on fluid-dynamics, thermal and metallurgical theoretical approaches. In particular MAGMASOFT has a module, named MAGMAIron, devoted to the simulation of mould filling, casting solidification, solid state transformation, with the related mechanical properties (such as hardness, tensile strength and Young Modulus) of cast irons [8].

23.3 OPTIMISATION TOOL

Formally, the optimisation problem addressed can be stated as follows.

Minimise: $F_j(\underline{X})$, $j = 1,\ldots,n$

$$F_1(\underline{X}),\ F_2(\underline{X}),\ldots,\quad F_n(\underline{X})$$

with respect to: $\underline{X}$

subject to $c_i(\underline{X}) \geq 0, \quad i = 1,\ldots,m$

where $\underline{X}$ is the design variables vector, $F_i(\underline{X})$ are the objectives, and $c_i(\underline{X})$ are the constraints.

FRONTIER's optimisation methods are based on Genetic Algorithms (GA) and hill climbing methods. These allow the user to combine the efficient local convergence of traditional hill climbers, with the strengths of GA's, which are their robustness to noise, discontinuities and multimodal characteristic, their ability to address multiple objectives cases directly, and their suitability for parallelisation.

23.3.1 GA general structure

A GA has the following stages:

1. initialise a population of candidate designs and evaluate them, then

2. create a new population by selecting some individuals which reproduce or mutate, and evaluate this new population

Stage 2 is repeated until termination.

23.3.2 GA mechanisms

Design variables are encoded into chromosomes by means of integer number lists. Though there is an inherent accuracy limitation in using integer values, this is not significant since accuracy can easily be refined using classical optimisation techniques. The initial selection of candidates is important especially when evaluations are so expensive that not many can be afforded in the total optimisation. Initialisation can be done in FRONTIER either by reading a user-defined set, or by random choice, or by using a Sobol algorithm [9] to generate a uniformly distributed quasirandom sequence. The optimisation can also be restarted from a previous population.

The critical operators governing GA performance are selection, crossover reproduction, and reproduction by mutation.

Four selection operators are provided, all based on the concept of Pareto dominance. They are; (1) Local Geographic Selection; (2) Pareto Tournament Selection; (3) Pareto Tournament Directional Selection; and (4) Local Pareto Selection. The user can choose from these though (4) is recommended for use with either type of crossover, and (2) to generate the proportion of the population which is sent to the next generation unmodified.

Most emphasis in FRONTIER is on use of directional crossover, which makes use of detected direction of improvement, and has some parallels with the Nelder & Mead Simplex algorithm. Classical two-point crossover algorithm are also provided.

Mutation is carried out when chosen, by randomly selecting a design variable to mutate, then randomly assigning a value from the set of all possibilities.

In all cases, GA probabilities can be selected by the user, in place of recommended defaults, if desired. All the algorithm are described in more detail in [10].

23.3.3 Operational user choice

Traditional GA's generate a complete new population of designs from an existing set, at each generation. This can be done in FRONTIER using its MOGA algorithm. An alternative strategy is to use steady state reproduction via a MOGASTD algorithm. In this case, only a few individuals are replaced at each generation. This strategy is more likely to retain best individuals. The FRONTIER algorithm removes any duplicates generated. Population size are under the user's control. FRONTIER case study work has usually used population from 16 to 64, due to the computational expense of the design evaluations.

Classical hill climbers can be chosen by the user not only the refine GA solution. They can be adopted from the start of the optimisation, if the user can formulate his problem suitably, and if he is confident that the condition are appropriate.

Returning to the problem of expansive design evaluation, many research have made use of response surface. These interpolate a set of computed design evaluation. The surface can then be used to provide objective functions which are much faster to evaluate. Interpolation of nonlinear functions in many variables, using polynomial or spline functions, becomes rapidly intractable however. FRONTIER provides a response surface option based on use of a neural net, with two nodal planes. Tests have shown this to be an extremely effective strategy when closely combined with the GA to provide a continuous update to the neural net.

23.3.4 Fitness and constraints handling

The objective values themselves are used as fitness values. Optionally, the user can supply weights to combine these into a single quantity. Constraints are normally used to compute a penalty decrementing the fitness. Alternatively, the combined constraint penalty can be nominated as an extra objective to be minimised.

23.3.5 Parallelisation of GA

The multithreading features of Java have been used to parallelise FRONTIER's GA's. The same code is usable in a parallel or sequential environment, thus enhancing portability. Multithreading is used to facilitate concurrent design evaluations, with analyses executed in parallel as far as possible, on the user's available computational resources.

23.3.6 Decision support

Even where there are a number of conflicting objectives to consider, we are likely to went to choose only one design. The Pareto boundary set of designs provides candidates for the final choice. In order to proceed further, the designer needs to focus on the comparative importance of the individual objectives. The role of decision support in FRONTIER is to help him to do this, by moving to a single composite objective which combines the original objectives in a way which accurately reflect his preferences.

23.3.7 Local utility approach

A wide range of methods has been tried for multiple criteria decision making . The main FRONTIER technique used is the Local Utility Approach (LUTA) [11]. This avoids asking the designer to directly weight the objectives relative to each other (though he can if he wishes), but rather asks him to consider some of the designs which have already been evaluated, and state which he prefers, without needing to give reasons. The algorithm then proceeds in two stages. First it decides if the preferences give are consistent in themselves, and guides the designer to change them if they are not. Then, it proposes a 'common currency' objectives measure, termed a utility, this being the sum of a set of piecewise linear utility functions, one for each individual objective. The preference information which the designer has provided can then be stated as a set of inequality relations between the utilities of designs. The algorithm uses the feasible region formed by these constraints to calculate the most typical composite utility function which is consistent with the designer's preferences.

This LUTA technique can be invoked after accumulating a comprehensive set of Pareto boundary designs as a result of a number of optimisation iterations. The advantage of the latter approach is that the focusing of attention on the part of the Pareto boundary which is of most interest can result in considerable computational saving, by avoiding computing information on the whole boundary.

In practice so far in FRONTIER, we have generally used the LUTA technique after a set number of design evaluations, after which the utility function for a local hill climber to rapidly refine a solution.

23.4 OBJECT OF THE STUDY AND ADOPTED OPTIMISATION PROCEDURE

The component investigated is a textile machine guide, for which both mechanical and integrity requirements are prescribed. Such requirements are satisfied, respectively, by reaching proper hardness values and by minimising the porosity content. Furthermore, from the industrial point of view, it is fundamental to maximise the process yield, lowering the feeder size.

The chemical composition of the ductile iron is the following:

C	Si	Mn	P	S	Cu	Sn	Ni	Cr	Mg
3.55	2.77	0.13	0.038	0.0037	0.048	0.045	0.017	0.030	0.035

The liquidus and solidus temperatures are 1155°C and 1120°C respectively. The thermo-physical properties of the material (thermal conductivity, density, specific heat, viscosity) are already implemented into the MAGMASOFT Materials Database.

The GA optimisation process was performed starting from a configuration of the casting system which is already the result of the foundry practise optimisation.

Only the solidification process was taken into account for the simulation, since it was considered to be more affected by the technological variables selected. Therefore the temperature of the cast at the beginning of the solidification process was set as a constant. Moreover the gating system was neglected in the simulation since its influence on the heat flow involved in the solidification process was thought to be negligible. As a consequence the numerical model considers only the cast, the feeder and the feeder-neck (see Fig. 23.1, referred to the starting casting system). The adopted mesh was chosen in such a way to balance the accuracy and the calculation time. As a consequence a number of metal cells ranging from 9000 to 12000 (resulting in a total number of cells approximately equal to 200000) was obtained in any analysed model.

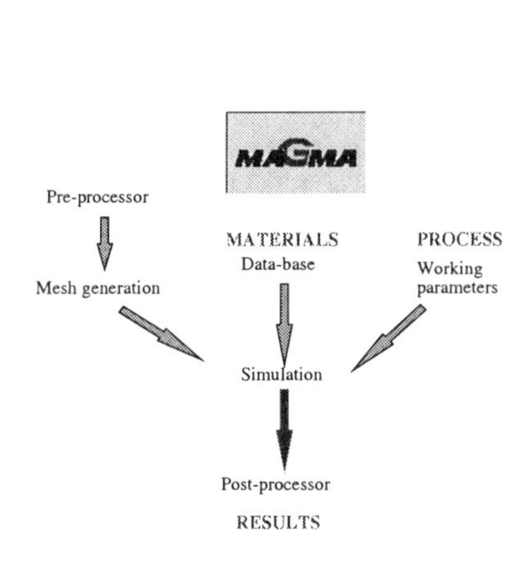

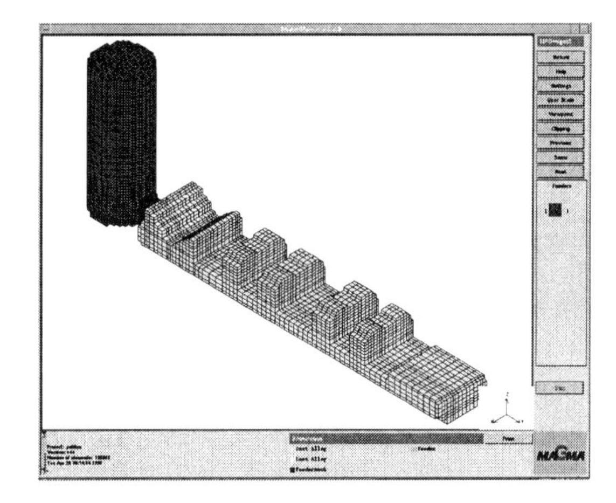

Fig. 23.1 Flow diagram of **MAGMASOFT** simulation procedure to study casting processes and adopted mesh for the cast and the feeder.

Five technological variables governing the solidification process have been taken into account and the respective ranges of possible variation were defined:

1. temperature of the cast at the beginning of the solidification process, $1300\ °C < T_{init} < 1380\ °C$;

2. heat transfer coefficients (HTC) between cast and sand mould, $400\ W/m^2K < HTC < 1200\ W/m^2K$;

3. feeder height, $80\ mm < H_f < 180\ mm$

4. feeder diameter, $30\ mm < D_f < 80\ mm$

5. section area of the feeder-neck, $175\ mm^2 < A_n < 490\ mm^2$.

These variables were considered to be representative of the foundry technology and significant in order to optimise the following design objectives:

1. hardness of the material in a particular portion of the cast,

2. casting weight (i.e. raw cast + feeder + feeder-neck),

3. porosity.

Aim of the analysis was to maximise the hardness and to minimise the total casting weight and the porosity. No constraints were defined for this analysis.

Generally speaking, the optimisation procedure should be performed running one MAGMA simulation for each generated individual. That implies the possibility to assign all the input parameters and start the analysis via command files. Similarly the output files should be available in the form of ascii files from which the output parameters can be extracted. However at this stage a complete open interface between MAGMASOFT and Frontier is not still available. As a consequence another solution was adopted. First of all 64 analyses were performed in order to get sufficient information in all the variable domain. After that a interpolation algorithm was used to build a response surface model basing on a Neural Network, "trained" on the available results. It has been verified that the approximation reached is lower than 1% for all the available set of solutions with the exception of one point only where the approximation is slightly lower than 5%. After that the response surface model was used in the next optimisation procedure to calculate the design objectives. In such a way further time-expensive work needed to run one MAGMASOFT interactive session for each simulation was avoided.

Concerning the Genetic Algorithm a mix between a classical and directional cross-over was used. The first population was created in a deterministic way.

23.5 RESULTS AND DISCUSSION

The first optimisation task was done for 4 generations with 16 individuals for each generation. Since a complete simulation required about 20 minutes of CPU time on a workstation HP C200, the total CPU time resulted in about 21 hours and 20 minutes. Figs 23.2 (a)–(c) report the obtained solutions. In particular, from the tables it can be noted that the hardness values increase as we move from the first to the fourth generation, while the weights decrease. Not the same for the porosity, whose values seems to be less stable to converge towards an optimum solution: in fact the same range between the minimum and the maximum value is maintained both in the first and in the last generation. Moreover, Fig. 23.2 (a) illustrates the strong correlation between the casting weight and the hardness: such correlation is due to the particular geometry of the casting under examination and to the position where the hardness value was determined. Anyway the dependency between these two variables is favourable, the hardness increasing as the casting weight decreases, due to the changed cooling conditions. Figs 23.2 (b) and (c) show that the other variables are not correlated to each other. From all these three figures it can be noted that the optimisation algorithm tends to calculate a greater number of solutions in a specific area of the design objectives plane, where the optimum solution can be expected to be located.

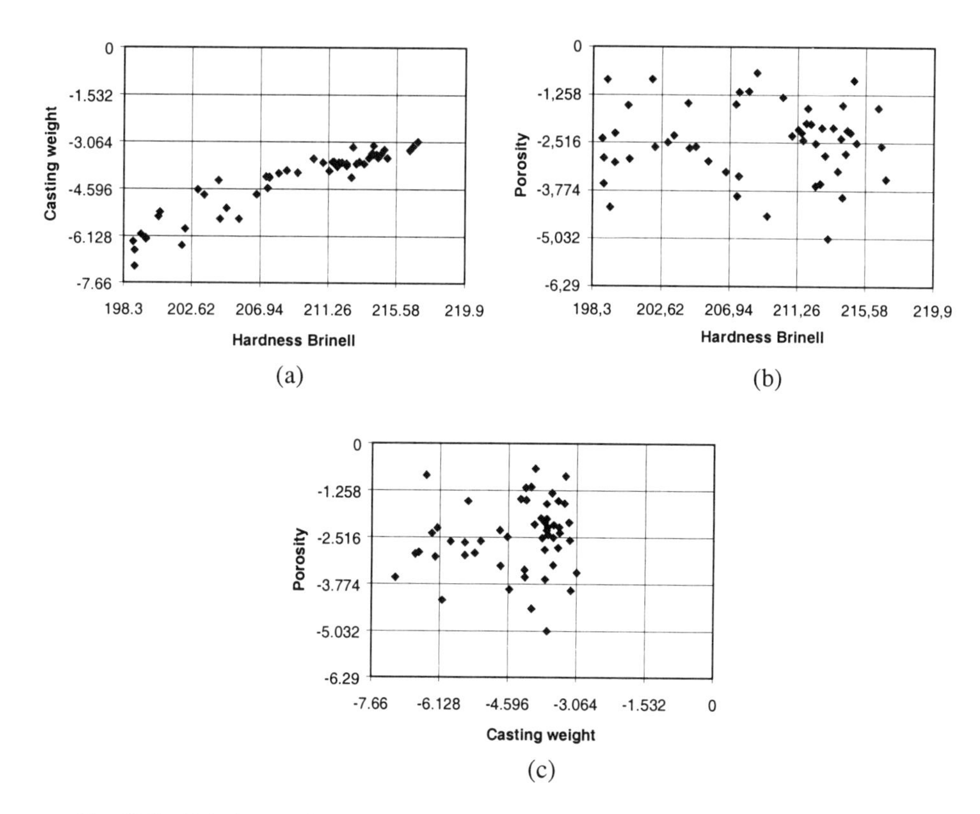

Fig. 23.2 Solutions in the design objectives space obtained using MAGMASOFT software.

As mentioned before the second optimisation step was performed using an approximation function consisting of three independent Neural Networks (one for each design objective) to fit the results obtained from the first optimisation procedure. Then to explore more extensively the variables domain, an optimisation task was done for 8 generation with 16 individuals for each generation.

Figs 23.3 (a)–(c) report the obtained solutions. By comparing this set of figures with the corresponding previous one (Figs. 23.2 (a)–(c)), it can be noted that the GA could reach better solutions, located at the top-right side area of each diagram. Since the raw casting weight was equal to about 2.5 kg and not influenced by any of the chosen variables, the casting weight resulted to be never lower than about 3 kg.

All these design objectives were further processed to obtain the results in the form of Pareto Frontier. The Pareto set is reported in Table 23.1, consisting of 11 non-dominated solutions. A direct comparison among them allowed for identifying three solutions (indicated with number 4, 7 and 8 in Table 23.1) which seemed to reach the best compromise among the three objectives.

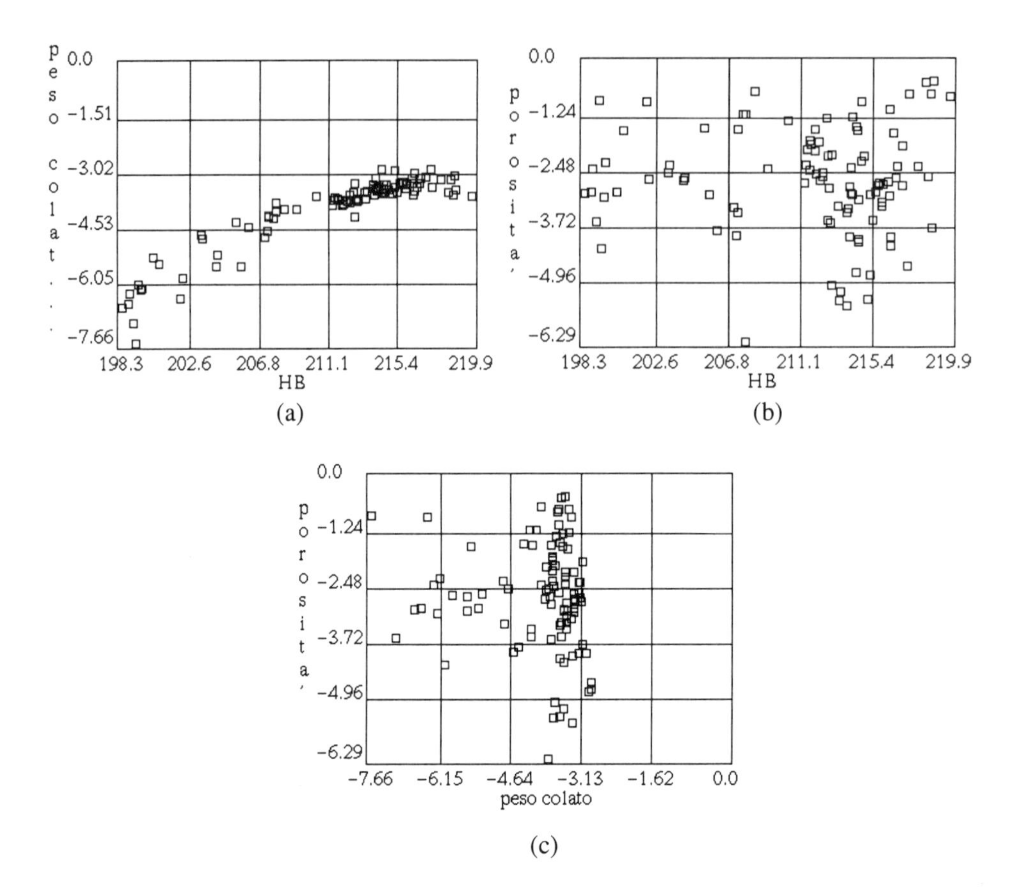

Fig. 23.3 solutions in the design objectives space using the Neural Network model.

These solutions were checked by running three MAGMAIron simulations. The comparison between the design objectives as predicted by the response surface model and as calculated by MAGMAIron is reported in Table 23.2. It can be noted that the hardness values are predicted with good approximation by the Neural Network, while the porosity values do not match satisfactorily those calculated by MAGMAIron. Anyway the optimum set of variables (4, 7 and 8) reported in Table 23.1 together with the objectives calculated by MAGMAIron were compared with the set of variables corresponding to the present foundry practise. The results, reported in Table 23.3, suggest to decrease the heat transfer coefficient and the feeder size and to increase the feeder-neck section, in order to reach the objectives. The initial temperature instead is already very near to the optimised value.

Finally Fig. 23.4 compares the sizes of the feeders and highlights the bigger feeder now adopted with respect to that of the optimised solution.

Table 23.1 Pareto Set extracted from the 128 available solutions obtained with the Neural Network.

	VARIABLES					DESIGN OBJECTIVES		
N°	T_{init} (°C)	HTC (W/m²□K)	H_{feeder} (mm)	D_{feeder} (mm)	A_{neck} (mm²)	Hardness Brinell	casting weight (kg)	porosity (%)
1	1300	1200	86	30	194	217	2.90	4.60
2	1380	811	97	36	289	215	3.34	0.87
3	1341	1037	87	32	276	218	3.17	2.35
4	1352	460	105	33	400	218	3.38	0.70
5	1371	940	80	30	176	219	3.09	3.75
6	1335	1200	85	31	225	216	3.01	3.93
7	1365	400	89	32	341	220	3.64	0.77
8	1336	400	112	31	400	219	3.47	0.43
9	1362	814	84	31	315	217	3.11	2.78
10	1346	1009	89	32	278	219	3.18	2.60
11	1335	1059	85	31	225	217	3.10	1.90

Table 23.2 Comparison between the design objectives as calculated by the Neural Network and by MAGMASOFT.

	INTERPOLATED			MAGMASOFT		
N°	Hardness Brinell	casting weight (kg)	porosity (%)	Hardness Brinell	casting weight (kg)	porosity (%)
4	218	3.38	0.70	217	3.28	2.34
7	220	3.64	0.77	218	3.11	1.80
8	219	3.47	0.43	218	3.24	3.30

Table 23.3 Optimum solutions compared with the present one as adopted in the foundry practice. All design objectives have been calculated by MAGMASOFT.

	VARIABLES					DESIGN OBJECTIVES		
	$T_{init.}$ (°C)	HTC (W/m²□K)	H_{feeder} (mm)	D_{feeder} (mm)	A_{neck} (mm²)	Hardness Brinell	casting weight (kg)	porosity (%)
Optimum Solutions	1352	460	105	33	400	217	3.28	2.34
	1365	400	89	32	341	218	3.11	1.80
	1336	400	112	31	400	218	3.24	3.30
Present Solution	1360	700	115	55	248	207	4.53	1.27

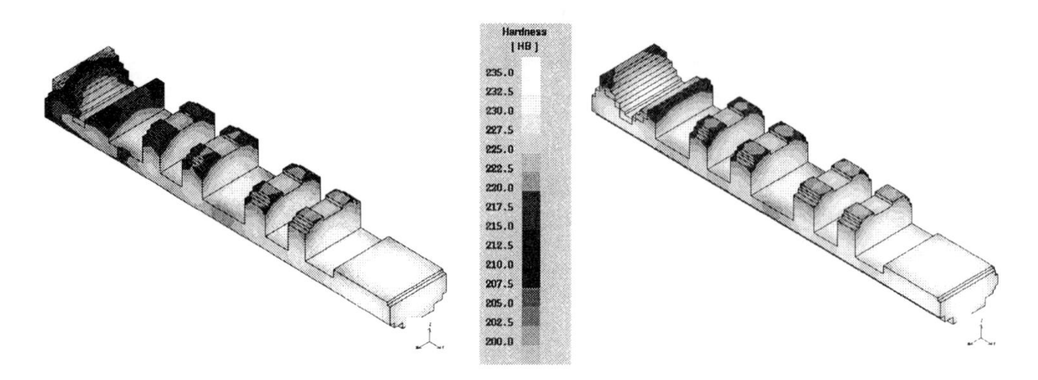

Fig. 23.4 Comparison between the present solution (left) and the optimised one (right).

23.6 CONCLUSIONS

Frontier was applied to MAGMASOFT code enabling the numerical simulation of mould filling and solidification of castings. On the other hand till now it was not possible to interface Frontier with MAGMASOFT since this software does not accept command files to input design parameters. As a consequence an initial optimisation procedure running MOGA for 4 generations with 16 individuals for each generation was performed and a Neural Network was built through the available design objectives. A second optimisation task running Frontier for 8 generations with 16 individuals for each generation was performed. Some design objectives belonging to the Pareto set were then checked running MAGMASOFT simulations. The following conclusions could be drawn:

- In this application the hardness could be increased from 207 HB up to 220 HB and the casting weight reduced from 4.53 kg to 3.11 kg with a slight increase in porosity from 1.27% to 1.80%.

- The approximation that could be reached with the Neural Network is probably limited by the small number of available "training points" considering that five design variables were treated. In fact, one of the three design objectives was not predicted satisfactorily, as compared with the solution obtained by MAGMASOFT.

REFERENCES

1. M.C. Flemings, "Solidification Processing", Mc Graw Hill, New York (1974).
2. ASM Metals Handbook, 9^{th}ed., vol. 15: Casting (1988), ASM - Metals Park, Ohio.
3. P.R. Sahm, P.N. Hansen, "Numerical simulation and modelling of casting and solidification processes for foundry and cast-house", CIATF (1984).

4. D.M. Stefanescu, "Critical review of the second generation of solidification models for castings: macro transport - transformation kinetics codes", Proc. Conf. "Modeling of Casting, Welding and Advanced Solidification Processes VI", TMS (1993), pp. 3-20.

5. T. Overfelt, "The manufacturing significance of solidification modeling", Journal of Metals, 6 (1992), pp. 17-20.

6. T. Overfelt, "Sensitivity of a steel plate solidification model to uncertainties in thermophysical properties", Proc. Conf. "Modelling of Casting, Welding and Advanced Solidification Processes - VI", 663-670.

7. F. Bonollo, N. Gramegna, "L'applicazione delle proprietà termofisiche dei materiali nei codici di simulazione numerica dei processi di fonderia", Proc. Conf. "La misura delle grandezze fisiche" (1997), Faenza, pp. 285-299.

8. MAGMAIron User Manual.

9. C.Poloni, V.Pediroda, "GA Coupled with Computationally Expensive Simulations: Tool to Improve Efficiency" in "Genetic Algorithms and Evolution Strategies in Engineering and Computer Science", J.Wiley and Sons 1998.

10. Paul Bratley and Bennett L. Fox, Algorithm 659, "Implementing Sobol's Quasirandom Sequence Generator", 88-100, ACM Transactions on Mathematical Software, vol.14,No. 1, March 1988.

11. Pratyush Sen, Jian Bo Yang, "Multiple-criteria Decision-making in Design Selection and Synthesis", 207-230, Journal of Engineering Design,vol.6 No. 3, 1995.

12. I.L. Svensson, E. Lumback, "Computer simulation of the solidification of castings", Proc. Conf. "State of the art of computer simulation of casting and solidification processes", Strasbourg (1986), pp. 57-64.

13. I.L. Svensson, M. Wessen, A. Gonzales, "Modelling of structure and hardness in nodular cast iron castings at different silicon contents", Proc. Conf. "Modeling of Casting, Welding and Advanced Solidification Processes VI", TMS (1993), pp. 29-36.

14. E. Fras, W. Kapturkiewicz, A.A. Burbielko, "Computer modeling of fine graphitc eutectic grain formation in the casting central part", Proc. Conf. "Modeling of Casting, Welding and Advanced Solidification Processes VI", TMS (1993), pp. 261-268.

15. D.M. Stefanescu, G. Uphadhya, D. Bandyopadhyay, "Heat transfer-solidification kinetics modeling of solidification of castings", Metallurgical Transactions, 21A (1990), pp. 997-1005.

16. H. Tian, D.M. Stefanescu, "Experimental evaluation of some solidification kinetics-related material parameters required in modeling of solidification of Fe-C-Si alloys", Proc. Conf. "Modeling of Casting, Welding and Advanced Solidification Processes VI", TMS (1993), pp. 639-646.

17. S. Viswanathan, V.K. Sikka, H.D. Brody, "The application of quality criteria for the prediction of porosity in the design of casting processes", Proc. Conf. "Modeling of Casting, Welding and Advanced Solidification Processes VI", TMS (1993), pp. 285-292.

18. S. Viswanathan, "Industrial applications of solidification technology", Journal of Metals, 3 (1996), p. 19.

19. F. Bonollo, S. Odorizzi, "Casting on the screen - Simulation as a casting tool", Benchmark, 2 (1998), pp. 26-29.

20. F. Bonollo, N. Gramegna, L. Kallien, D. Lees, J. Young, "Simulazione dei processi di fonderia e ottimizzazione dei getti: due casi applicativi", Proc. XIV Assofond Conf. (1996), Baveno.

21. F. Bonollo, N. Gramegna, S. Odorizzi, "Modellizzazione di processi di fonderia", Fonderia, 11/12 (1997), pp. 50-54.

22. F.J. Bradley, T.M. Adams, R. Gadh, A.K. Mirle, "On the development of a model-based knowledge system for casting design", Proc. Conf. "Modeling of Casting, Welding and Advanced Solidification Processes VI", TMS (1993), pp. 161-168.

23. G. Upadhya, A.J. Paul, J.L. Hill, "Optimal design of gating & risering for casting: an integrated approach using empirical heuristics and geometrical analysis", Proc. Conf. "Modeling of Casting, Welding and Advanced Solidification Processes VI", TMS (1993), pp. 135-142.

24. T.E. Morthland, P.E. Byrne, D.A. Tortorelli, J.A. Dantzig, "Optimal riser design for metal castings", Metallurgical Transactions, 26B (1995), pp. 871-885.

25. N. Gramegna, "Colata a gravità in ghisa sferoidale", Engin Soft Trading Internal Report (1996).

24 Circuit Partitioning Using Evolution Algorithms

J.A. MONTIEL-NELSON, G. WINTER, L.M. MARTINEZ and D. GREINER

Centre for Applied Microelectronics (CMA) &
Centre for Numeric Applications in Engineering (CEANI)
Universidad de Las Palmas de Gran Canaria
Campus Univ. de Tafira
E-35017 Las Palmas de Gran Canaria, Spain

Abstract. Complex electronic systems are implemented in more than one physical unit. Throughout the design of a system, engineers are faced with making decisions about how to divide a system's functionality. Partitioning is the design phase in which a designer divides the system's functionality into parts which fit into available physical units. Partitioning is necessary to accommodate varied design constraints, to take advantage of existing components and subsystems, to facilitate parallel development activities, to provide access, to environment protect, for testability, for maintainability, and to achieve reconfigurability. Heuristic procedures based on evolution programming have been applied to efficiently solve the problem. Here, we compare different evolution strategies in terms of convergence speed on a basis of a set of benchmark circuits. In particular, both experimental results using simulated annealing and genetic algorithm are introduced.

24.1 INTRODUCTION

A hierarchical design approach becomes essential to shorten the design period, as the complexity of VLSI (Very Large Scale of Integration) circuits increases. Partitioning plays an important role in finding the hierarchy of a circuit or system and reduce the complexity of a design problem improving both the performance and the reliability of the system.

Given a circuit or system consisting of interconnected components, there are three main types of partitioning problems. The first type of partitioning problem does not have a fixed underlying partition topology and therefore uses a **Ratio-Cut** cost function [1] as the objective. This is useful when we wish to determine the structure of the circuit and discover the so-called **Natural Clusters** of the circuit. The second type has a fixed, existing partition topology, which includes capacity for each partition, interconnection costs and delay models between partitions. This is the case for FPGA-type of partitioning and some MCM (Multi-Chip Module) types of partitioning problem. The third type of partition problem comes from the **Divide and Conquer** principle applied to a complex problem. In our system of interconnected components a previous partition at the floorplanning phase simplified subsequent synthesis procedures [2].

The organization for the rest of this paper is as follows. Previous related works are introduced in Section 24.2. We formulate the partitioning problem in Section 24.3 under timing and capacity constraints. Simulated annealing approach is presented in Section 24.4, and Section 24.5 presents the codification of the partitioning problem using genetic algorithm. Section 24.6 gives a description about the set of proposed benchmark circuits and systems, and introduces simulated annealing results. Experimental results using genetic algorithms are introduced in Section 24.7, and conclusions are presented in Section 24.8.

24.2 RELATED WORK

The graph partitioning problem belongs to the class of NP-hard problems [3]. The most well-known heuristic partitioning algorithm is [4], and a more efficient version of it [5] is widely used. However, many approaches have been proposed for attacking circuit partitioning problems, and they include clustering [6], [7], eigenvector decomposition [8], [9], [10], [11], [12], and ratio cut [1], [13].

Combinatorial optimization problems seek a global minimum of a real valued cost function defined over a discrete set. The set is called the state space and the elements of the set are referred to as states. A state space and an underlying neighborhood structure together form a solution space. Several optimization problems of interest are computationally intractable, i.e., their decision versions are NP-complete [3]. There have been several deterministic heuristics [14], [15], [16] suggested in the past for solving specific NP-complete problems. However, most of these heuristics are essentially descent algorithms with respect to the cost function; consequently, they are unable to escape local minima with respect to the underlying neighborhood structure. The simulated annealing (SA) algorithm introduced by Kirkpatrick et al. [17] is an iterative stochastic procedure for solving combinatorial optimization problems. Proofs of convergence to a global optimum [18], [19], [20], [21] and successful experimental implementations [22], [23] have led to the widespread use of SA and its acceptance as a viable general method for combinatorial optimization. However, in general, SA still suffers from two major drawbacks: an actual implementation of SA requires a careful tuning of some of its control parameters to achieve good results [24], [25], and SA uses excessive computation time and is often less effective when compared with some well-designed deterministic heuristics for specific problem being solved [26].

24.3 PROBLEM FORMULATION AND COST FUNCTION DEFINITION

Each component in the circuit and system may have variable size, reflecting the silicon area demand for realizing the component. On the other hand each partition provides a fixed amount of silicon area called the **Capacity** of the partition. The **Capacity Constraints** state that the total size of all components assigned to the same partition must be no greater than the capacity of that partition. The **Timing Constraints** are stated as a set of maximum routing delay allowed between a pair of components. These constraints are driven by system cycle time and can be derived from the delay equations and intrinsic delay in combinational circuit components.

Input to the problem includes the following:

- Descriptions of a System Consisting of Interconnected Components:
    1. J is a set of N circuit components. Let $j \in J$ be the index to the component.
    2. s_j is the size of component j, representing the silicon area component j requires.
    3. A is an $N \times N$ matrix, where $a_{i,j}$ is the number of interconnections from component i to j.
    4. D_C is an $N \times N$ matrix, where $D_C(i,j)$ is the maximum signal routing delay allowed from component i to j.
- Descriptions of Partitions:
    1. I is a set of M partitions. Let $i \in I$ be the index to a partition.
    2. c_i is the capacity of partition i.
    3. B is an $M \times M$ matrix, where $b_{i,j}$ is the cost of wire routing from partition i to j.
    4. D is an $M \times M$ matrix, where $D(i,j)$ is the routing delay from partition i to j. Notice that we don't assume any relationship between B and D in our formulation.
- Others:
    - P is an $M \times N$ matrix, where p_{ij} is the cost of assigning component j to partition i, M is the number of partitions, and N is the number of circuit components.

A solution to the problem is an assignment $A : J \longrightarrow I$ satisfying the following two sets of constraints:

1. C1: (Capacity Constraints)

$$\sum_{\forall j, A(j) \in i} s_j \leq c_i \qquad \forall i \in I,$$

2. C2: (Timing Constraints)

$$D(A(i), A(j)) \leq D_C(i,j) \qquad \forall i, j \in J.$$

The objective is to

$$\text{minimize} \quad \alpha \sum_{\forall i,j,A(j)=i} p_{ij} + \beta \sum_{\forall i,j} a_{ij} b_{A(i)A(j)} \qquad \text{s.t.} \qquad C1, C2$$

where p_{ij} means the cost of assigning component j to partition i, a_{ij} is the cost of wire routing from partition i to j, and a_{ij} is the number of interconnections from component i to j.

If we introduce a matrix $[x_{ij}]_{M\times N}$ of binary variables and let $x_{ij} = 1$ if $A(j) = i$, and $x_{ij} = 0$ otherwise, then each A corresponds to an unique $[x_{ij}]_{M\times N}$ such that $\sum_{i=1}^{M} x_{ij} = 1$, $\forall j \in J$, and vice versa. We say that $[x_{ij}]$ satisfies a certain constraint if its corresponding A does. Now we can rewrite the problem into:

1. C1: (Capacity Constraints)

$$\sum_{j=1}^{N} s_j x_{ij} \qquad \forall i \in I,$$

2. C2: (Timing Constraints)

$$D(i,j) \le D_C(k,l) \qquad \forall (i,k) \text{ and } (j,l) \qquad \text{s.t.} \qquad x_{ik} = 1 = x_{jl},$$

3. C3: (Generalized Upper Bound Constraints)

$$\sum_{i=1}^{M} x_{ij} = 1 \qquad \forall j \in J.$$

The transformation of $C2$ is obvious when we recognize the fact that $x_{ij} = 1$ means component j is assigned to partition i in the corresponding assignment A.

The objective becomes

$$\text{minimize} \quad \alpha \sum_{i=1}^{M} \sum_{j=1}^{N} p_{ij} x_{ij} + \beta \sum_{i=1}^{M} \sum_{k=1}^{N} \sum_{j=1}^{M} \sum_{l=1}^{N} a_{kl} b_{ij} x_{ik} x_{jl} \quad \text{s.t.} \quad C1, C2, C3.$$

The first term in the cost function is associated with the (constant) cost of assigning a particular component to a particular partition and is called the linear term. The leading α is a scaling factor for this term.

The second term in the objective function is associated with the interconnection cost between components. It is called the Quadratic term and can be used to model any type of interconnection cost metrics. For example, when B is a matrix of all 1's except all 0's on the main diagonal, this term equals the total number of wire crossings for the given assignment A. When b_{ij} is the Manhattan distance from partition i to j, this term equals the total Manhattan wire length. Similar arguments apply for quadratic wire length or other forms of cost metrics. The leading β is a scaling factor for this term.

24.4 SIMULATED ANNEALING ALGORITHM FOR CIRCUIT PARTITIONING

A double loop version of the simulated annealing algorithm [17] has been implemented as shown in Algorithm 24.4.1, as a circuit partitioning to minimize the capacity and timing constraint presented in Section 24.3.

24.4.1 Algorithm

Double loop version of the simulated annealing algorithm.

simulated annealing double loop version algorithm

```
S := S_0;                                         /* Initial state. */
T := T_0;                                         /* Initial temperature. */
repeat
   repeat
      S_New := PERTURB_SA(S);
      S := ACCEPT(S_New, S, T);
   until (inner-loop stopping criterion satisfied);
   T := UPDATE(T);
until (T < T_f);
End pseudo-code.
```

The function ***PERTURB***$_{SA}$(**S**) simply returns a random neighbor of ***S*** state after the neighborhood structure has been suitable defined. The ***ACCEPT***($\mathbf{S}_{New}$, **S**, **T**) function makes a stochastic decision on whether or not accept the generated neighbor as the next stage. The function ***UPDATE***(**T**) gradually lowers the temperature parameter **T** to zero according to a logarithm cooling schedule.

24.5 PARTITIONING CODIFICATION FOR GENETIC ALGORITHM SOLUTION

Evolution algorithms [27] arising from the classical genetic algorithms (GAs) [28], are search algorithms based on the mechanism of natural selection which use operations of reproduction, crossover and mutation on a population of chromosomes to create a new population. Each chromosome represents a possible solution to the problem. New individuals of the population are produced every generation by the repetition of a two-step cycle.

First each individual is decoded and its ability to solve the problem is assessed through the assignment of a fitness value. In the second stage the fittest individuals are preferentially chosen through a selection mechanism to undergo recombination and mutation to form the next generation population.

In the circuit partitioning problem each individual is a linear array x of n integer numbers. Each element of the array represents the index i of the component to be assigned to a partition. The contents of the element i of the array x is an integer number which represents the assigned partition to the component of index i.

Fig. 24.1 shows a full adder schematic which includes fourteen components numbered from 1 to 14. Then, an array of fourteen ordered elements represent one individual where each index matches the component number. Each element i of the array contents an integer number representing the assigned partition to the component number i.

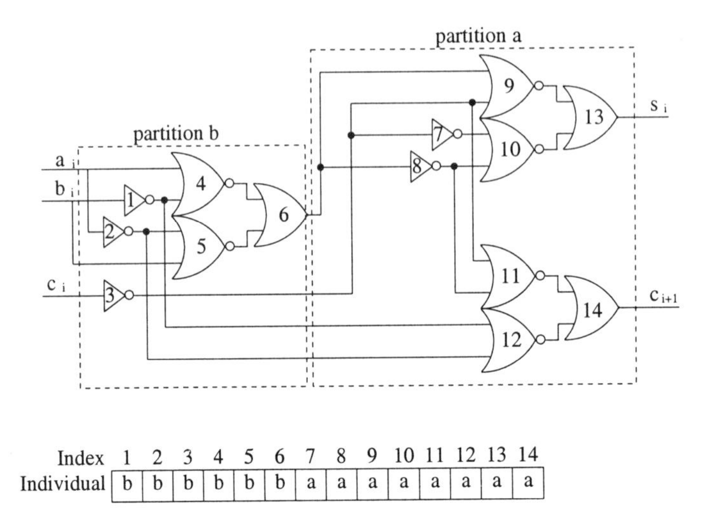

Index	1	2	3	4	5	6	7	8	9	10	11	12	13	14
Individual	b	b	b	b	b	b	a	a	a	a	a	a	a	a

Fig. 24.1 An example of problem codification for GA solution.

24.6 BENCHMARKS

A set of benchmark circuits are proposed in order to compare different optimizations based on evolution programming techniques applied to the partition problem. The set of benchmark circuits consists of several medium scale of complexity (MSI) circuits including: an edge triggered D-flipflop, a D-latch, a Toggle gate, a Muller C-element, a half adder, a full adder, a 4-2 compressor, and very large scale of complexity (VLSI) circuits including 4-bit Brent and Kung adder and 16-bit Brent and Kung adder, among others.

The MSI circuits were partitioned in two rows and implemented using a OLYMPO cell compiler [29], and the VLSI circuits were partitioned in four and eight partitions, respectively, and implemented using OLYMPO macro-cell generator [2], [30].

Table 24.1 gives previous results for the whole set of benchmark circuits when they were partitioned using SA technique presented at Section 24.4, and implemented by OLYMPO cell and macro-cell generators. Column 1 gives a brief description of the proposed circuit, and a short name is given in column 2.

Column 3 in the table is the time propagation delays (in picosecond) from the post-layout simulations. The given description of the circuit is partitioned and then implemented using the OLYMPO cell compiler. The obtained circuit was simulated using HSPICE electrical simulator, and the critical path time propagation delay was measured. Column 3 is the layout area of the obtained circuits using **H–GaAsII** and **H–GaAsIII** Vitesse processes, 0.8-μm and 0.6-μm design rule sets, respectively. Column 4 gives the density of the layout in transistors per millimeter square. Column 5 indicates the cost function number of evaluations that SA takes to do the partition of each circuit. Column 6 is the cost of the last configuration before the circuit layout generation.

Fig. 24.3 shows a layout of a 16-b Brent & Kung adder. It consist of 1133 transistors and is decomposed into 8 partitions. Hence, the cell compiler was invoked 8 times.

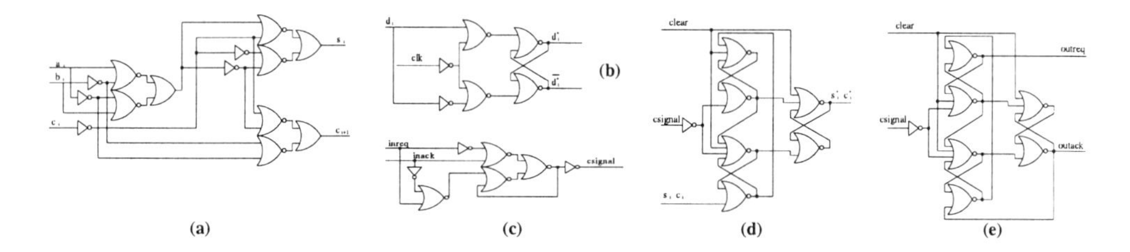

Fig. 24.2 Schematic circuits for: (a) Full adder, (b) D-latch, (c) Muller C-element with a bubble on its inAck input, (d) Edged triggered D-flipflop, and (e) Toggle element.

Table 24.1 Comparisons of experimental results. The technology is H-GaAsII† or H-GaAsIII‡ from Vitesse.

Circuit		Performance			Fitness	
Description	Name	Delay ps	Area $mm^2 \times$ 1E-3	Density $\frac{devices}{mm^2}$	Evaluations	Cost
Half adder‡	HA	179	3.2	4070	614	17
D-latch‡	DL	181	4.6	3478	1207	12.5
Muller C-element‡	MC	273	4.7	2909	2197	12.5
Toggle gate‡	TG	252	6.4	3714	2133	18.5
D-flipflop‡	DFF	243	6.5	3200	943	23
Full adder†	FA	425	10.7	4123	4679	12.5
1-b tap $\frac{x}{\sin(x)}$ FIR‡	TAP 1	592	20.2	3614	-	-
4-2 compressor†	COMP	790	25.0	3640	-	-
4-b Brent & Kung adder‡	BK 4	284	121.6	5468	-	-
16-b Brent & Kung adder†	BK 16	740	1331.9	5537	379024	445
11-tap FIR Filter†	TAP 11	2100	2962.0	6135	-	-

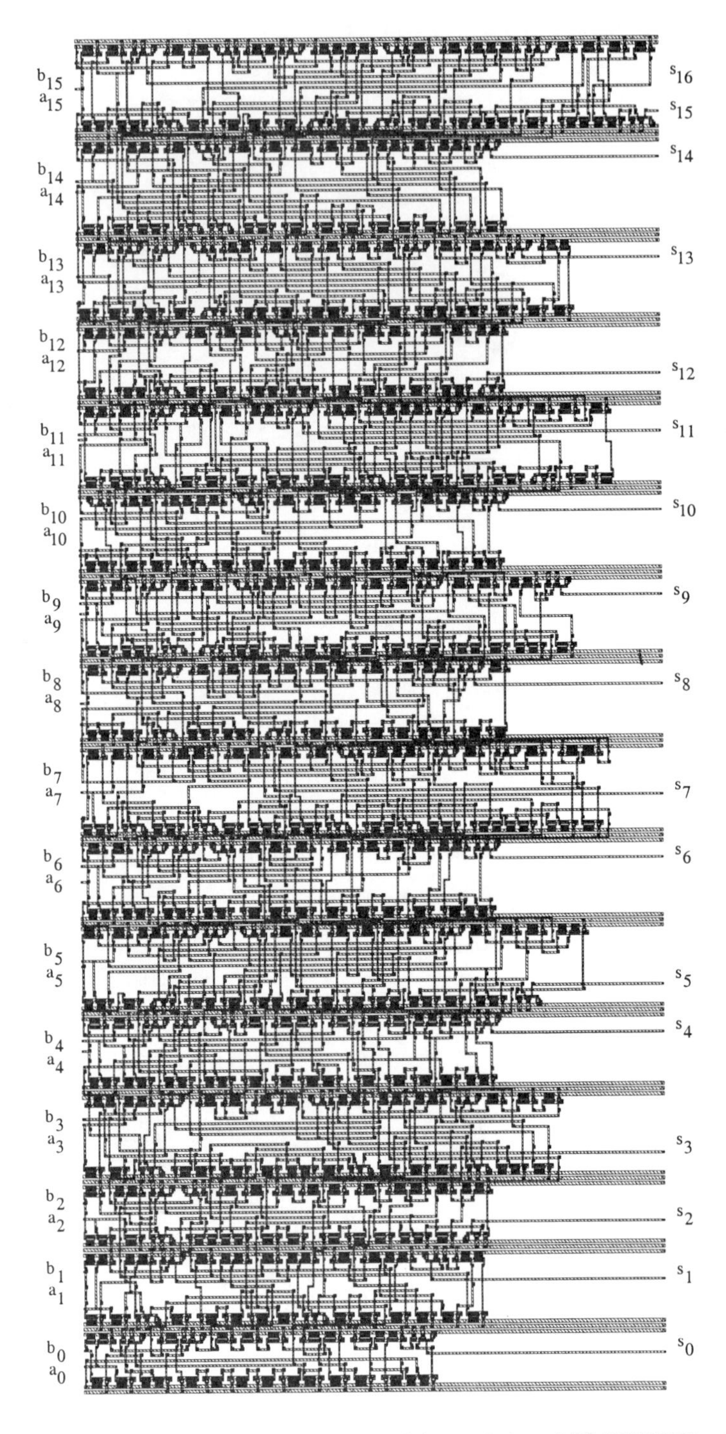

Fig. 24.3 A 16-b Brent & Kung adder module consisting of 16 OLYMPO cells.

24.7 EXPERIMENTS

Table 24.2 shows the results of the benchmark circuits in attempting to correctly partitioning the system's functionality into parts which fit into available physical units. The complexity of the benchmark circuits are measured in term of number of components, including medium scale of complexity (half adder, D-latch, Muller C-element, Toggle gate, D-flipflop, and full adder), and very large scale of complexity (16-bit Brent and Kung). In all of the experiments the total number of individual fitness evaluations was enough high to obtain the minimum fitness and not any progress in further generations.

In all the cases the strategy for old generation replacement after generating the subpopulation (offspring) was steady state. Column 2 in the table is the selection scheme. Population and generation number are listed in Column 3 and 4, respectively. The recombination (crossover) and mutation operators used in the generation of the offspring were multi-point and standard, respectively. The size of the proposed codification on the test case solution was $NE \times \log_2 NP$ (*NE: number of elements, NP: number of partitions*). Finally, Columns 5, 6 and 7 are the best, average and worst fitness individual, respectively.

Table 24.2 Comparisons of experimental results using genetic algorithms for partition of the benchmark circuits.

Circuit	Selection	Population	Generation	Best	Average	Worst
HA	Proportional	100	1×10^2	15.0	15.0	17.0
DL	Proportional	100	1×10^2	12.5	12.5	12.5
MC	Proportional	100	1×10^2	12.5	12.5	12.5
TG	Proportional	100	1×10^2	18.5	18.5	24.5
DFF	Proportional	100	1×10^2	14.0	14.0	18.0
FA	Proportional	100	1×10^2	12.5	12.5	12.5
BK 16	Proportional	100	4×10^3	1061	1128	1579
BK 16	Ranking	600	1×10^3	2043	-	-
BK 16	Ranking	600	3×10^3	1813	-	-
BK 16	Ranking	900	3×10^3	1512	-	-

Fitness for the partition of a 16-bit Brent and Kung adder using GA and SA is compared in Fig. 24.4. Both average and worst fitness value convergence of the benchmark MSI circuits are shown compared in Fig. 24.5 and 24.6, respectively. The fast convergence for the best fitness value in all the cases is an indication of the ability of the proposed algorithm to solve the proposed partition problems. Also, both average and worst fitness results converge in a few number of generations to the best value.

24.8 CONCLUSIONS

Partitioning is a combinatorial optimization problem, it seeks a global minimum of a real valued cost function defined over a discrete set. The set is called the state space and the elements of the set are referred to as states. A state space and an underlying neighborhood structure

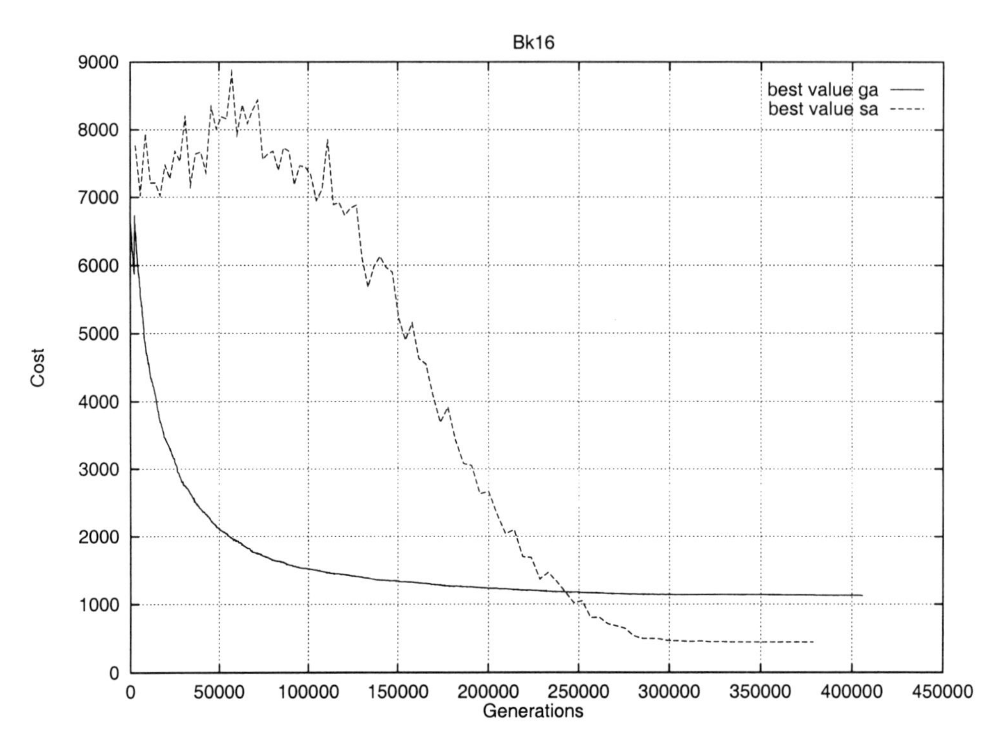

Fig. 24.4 Cost function evaluation results using GA and SA for a 16-bit Brent & Kung adder.

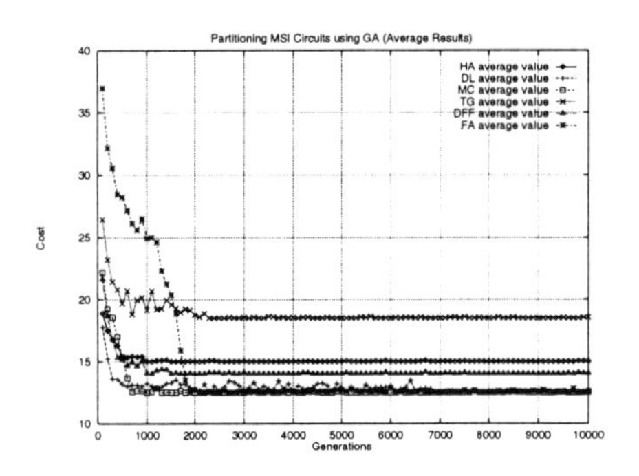

Fig. 24.5 Average fitness.

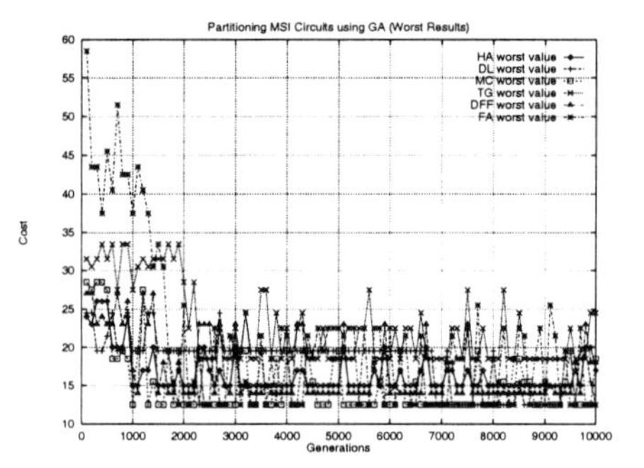

Fig. 24.6 Worst fitness.

together form a solution space. Several optimization problems of interest are computationally intractable, i.e., their decision versions are NP-complete [3].

There have been several deterministic heuristics [14], [15], [16] suggested in the past for solving specific NP-complete problems. However, most of these heuristics are essentially descent algorithms with respect to the cost function; consequently, they are unable to escape local minima with respect to the underlying neighborhood structure.

The simulated annealing (SA) algorithm introduced by Kirkpatrick et al. [17] is an iterative stochastic procedure for solving combinatorial optimization problems. Proofs of convergence to a global optimum [18], [19], [20], [21] and successful experimental implementations [22], [23] have led to the widespread use of SA and its acceptance as a viable general method for combinatorial optimization.

However, in general, SA still suffers from two major drawbacks: an actual implementation of SA requires a careful tuning of some of its control parameters to achieve good results [24], [25], and SA uses excessive computation time and is often less effective when compared with some well-designed deterministic heuristics for specific problem being solved [26].

It is seen in Table 24.2 that the genetic algorithm (GA) clearly outperforms the SA methods in terms of CPU time or Number of Generations in all the benchmarks for the circuit partitioning problem. GA was very quick to converge, making it a most suitable algorithm to apply to this type of problem, when it is compared with results presented in Table 24.1.

Acknowledgments

This work was supported by the Directorate-General for Science, Research and Development Industrial and Materials Technologies, DG XII/C/4, European Commission, under project INGENET (contract number BRRT-CT-97-5034).

REFERENCES

1. Y. Wei and C. Cheng. Toward Efficient Hierarchical Designs by Ratio Cut Partitioning. In *Conf. on Computer Aided Design*, pages 298–301, 1989.

2. J.A. Montiel-Nelson, V. Armas, R. Sarmiento, and A. Nunez. A Cell and Macrocell Compiler for GaAs VLSI Full-Custom Design. In *Proc. of Design, Automation and Test in Europe Conference and Exhibition*, pages 284–290, Feb. 1998.

3. M.R. Garey and D.S. Johnson. *Computers and Intractability, A Guide to the Theory of NP–Completeness*. W. H. Freeman, 1979.

4. B.W. Kernighan and S. Lin. An Efficient Heuristic Procedure for Partitioning Graphs. *Bell Syst. Tech. J.*, 49(2):291–307, February 1970.

5. C.M. Fiduccia and R.M. Mattheyses. A Linear Time Heuristic for Improving Network Partitions. In *Proc. 19th Design Automation Conference*, pages 175–181, 1982.

6. H.R. Charney and D.L. Plato. Efficient Partitioning of Components. In *Proc. 1st Design Automation*, pages 16.0–16.21, July 1968.

7. D.M. Schuler and E.G. Ulrich. Clustering and Linear Placement. In *Proc. 9th Design Automation Workshop*, pages 50–56, 1972.

8. J. Frankle and R.M. Karp. Circuit Placement and Cost Bounds by Eigenvector Decomposition. In *Proc. Int. Conf. on Computer Aided Design*, pages 414–417, 1986.

9. B. Krishnamurthy. An Improved Min-Cut Algorithm for Partitioning VLSI Networks. *IEEE Trans. on Computer*, C-33:438–446, May 1984.

10. L.A. Sanchis. Multi-way Network partitioning. *IEEE Trans. on Computer*, 38:62–81, January 1989.

11. D.G. Schweikert and B.W. Kernighan. A Proper Model for the Partitioning of Electrical Circuits. In *Proc. 9th Design Automation Workshop*, pages 57–62, 1972.

12. C. Sechen and D. Chen. An Improved Objective Function for Mincut Circuit Partitioning. In *Proc. Int. Conf. on Computer-Aided Deign*, pages 502–505, 1988.

13. Y. Wei and C. Cheng. Ratio Cut Partitioning for Hierarchical. In *IEEE Trans. on Computer Aided Design*, volume 10, pages 298–301, July 1991.

14. A.E. Dunlop and B.W. Kernighan. A Procedure for Placement of Standard-Cell VLSI Circuits. *IEEE Transactions on Computer-Aided Design of Integrated Circuits and Systems*, pages 92–98, Jan. 1985.

15. K.S. Arun and V.B. Rao. A New Principle-Components Algortihm for Graph Partitioning with Applications in VLSI Placement Problems. In *20th Asilomar Conf. on Signals, Systems, and Computers*, 1986.

16. P. Suaris and G. Kedem. An Algorithm for Quadrisection and its Application to Standard Cell Placement. *IEEE Trans. Circuit and Systems*, (35), March 1988.

17. S. Kirkpatrick, C.D.Jr. Gelatt, and M.P. Vecchi. Optimization by Simulated Annealing. *Science*, 220(4598):671–680, May. 1983.

18. B. Gidas. Non-Stationery Markov Chains and Convergence of the Annealing Algorithm. *Journal Statistical Phys.*, (39):73–131, 1985.

19. D. Mitra, F. Romeo, and A. Sangiovanni-Vicentelli. Convergence and Finite-Time Behaviour of Simulated Annealing. *Advances Appl. Probability*, (18):747–771, 1986.

20. D. Connors and P. Kumar. Simulated Annealing and Balance of Recurrence Order in Time-Inhomogeneus Markov Chains. In *26th Conf. on Decision and Control*, pages 2261–2263, December 1987.

21. B. Hajek. A tutorial survey of theory and applications of simulated annealing. In *Proc. IEEE Conf. on Decision and control*, Dec. 1988.

22. M. Vecchi and S. Kirpatrick. Global Wiring by Simulated Annealing. *IEEE Trans. on Computer Aided Design*, CAD(2):215–222, October 1985.

23. C. Sechen and A. Sangiovanni-Vicentelli. The TimberWolf Placement and Routing Package. *IEEE Journal of Solid-State Circuits*, SC–20:510–522, 1985.

24. S. Nahar, S. Sahni, and E. Shragowitz. Experiments with Simulated Annealing. In *22nd Design Automation Conference*, pages 748–752, 1985.

25. S. Nahar, S. Sahni, and E. Shragowitz. Simulated Annealing and Combinatorial Optimization. In *23rd Design Automation Conference*, pages 293–299, 1986.

26. Y. Saab and V. Rao. Linear Ordering by Stochastic Evolution. In *4th CSI IEEE Intl. Symp. on VLSI Design*, pages 130–135, January 1991.

27. Z. Michaelwicz. *Genetic Algorithms + Data Structures = Evolution Programs*. Springer Verlag, 3rd edition, 1996.

28. D. Goldberg. *Genetic Algorithms in Search, Optimisation and Machine Learning*. Addison Wesley, 1989.

29. J.A. Montiel-Nelson. *Cell Synthesis and Compilation in GaAs Technology*. PhD thesis, ETSI Industriales, Univ. Las Palmas de G.C., Jul. 1994.

30. J. A. Montiel-Nelson, V. de Armas, R. Sarmiento, and A. Nunez. OLYMPO: A GaAs Compiler for VLSI Design. In *Conf. on Electronics, Circuits and Systems*, volume 3, pages 385–388, September 1998.